HEPARIN AND RELATED POLYSACCHARIDES

ADVANCES IN EXPERIMENTAL MEDICINE AND BIOLOGY

Recent Volumes in this Series

Volume 309A
PURINE AND PYRIMIDINE METABOLISM IN MAN VII, Part A:
Chemotherapy, ATP Depletion, and Gout
Edited by R. Angus Harkness, Gertrude B. Elion, and Nepomuk Zöllner

Volume 309B
PURINE AND PYRIMIDINE METABOLISM IN MAN VII, Part B:
Structural Biochemistry, Pathogenesis, and Metabolism
Edited by R. Angus Harkness, Gertrude B. Elion, and Nepomuk Zöllner

Volume 310
IMMUNOLOGY OF MILK AND THE NEONATE
Edited by Jiri Mestecky, Claudia Blair, and Pearay L. Ogra

Volume 311
EXCITATION-CONTRACTION COUPLING IN SKELETAL,
CARDIAC, AND SMOOTH MUSCLE
Edited by George B. Frank, C. Paul Bianchi, and Henk E. D. J. ter Keurs

Volume 312
INNOVATIONS IN ANTIVIRAL DEVELOPMENT AND THE DETECTION OF VIRUS INFECTIONS
Edited by Timothy M. Block, Donald Jungkind, Richard L. Crowell,
Mark Denison, and Lori R. Walsh

Volume 313
HEPARIN AND RELATED POLYSACCHARIDES
Edited by David A. Lane, Ingemar Björk, and Ulf Lindahl

Volume 314
CELL-CELL INTERACTIONS IN THE RELEASE OF
INFLAMMATORY MEDIATORS: Eicosanoids, Cytokines, and Adhesion
Edited by Patrick Y-K Wong and Charles N. Serhan

Volume 315
TAURINE: Nutritional Value and Mechanisms of Action
Edited by John B. Lombardini, Stephen W. Schaffer, and Junichi Azuma

---

HEPARIN AND RELATED POLYSACCHARIDES

Edited by

David A. Lane

Charing Cross and Westminster Medical School
London, United Kingdom

Ingemar Björk

The Swedish University of Agricultural Sciences
Uppsala, Sweden

and

Ulf Lindahl

Uppsala University
Uppsala, Sweden

PLENUM PRESS • NEW YORK AND LONDON

Proceedings of an international symposium on Heparin and Related Polysaccharides, held September 1–6, 1991, in Uppsala, Sweden

ISBN 0-306-44212-4

A Division of Plenum Publishing Corporation
233 Spring Street, New York, N.Y. 10013

Printed in the United States of America

PREFACE

This volume is a record of a meeting entitled "Heparin and Related Polysaccharides" that was held at the Biomedical Center, Uppsala, Sweden between 1-6 September 1991. The meeting was hosted by U. Lindahl, L. Kjellén and I. Björk, who were helped in their preparations by a scientific advisory panel that included U. Abildgaard, B. Casu, E. Holmer and D. Lane. Altogether, 230 participants from 18 countries attended the meeting, and most were present to be included in a photograph, which is to be found at the end of this volume.

The selection of presentations for inclusion in the Symposia of the meeting was made on the basis of the known high quality of the work of the individuals or groups involved. This, we believe, is reflected in the contents of the enclosed articles, which collectively give a comprehensive overview of the present state of knowledge of heparin and related compounds.

Some of the areas covered are evolving or controversial and the views expressed in each article should be regarded as those of the author(s). The authors have taken various amounts of care to define all their abbreviations and some familiarity with the different forms of nomenclature used in the fields of polysaccharide chemistry and of the coagulation proteinases and their inhibitors will assist the reader.

The meeting in Uppsala was made possible by financial contributions from the following sponsors: Alfa Wassermann S.p.a., Carmeda AB, The City Council of Uppsala, Crinos Industria Farmacobiologica S.p.A., Diosynth bv, AB Draco, Erik Jorpes Memorial Fund of the Swedish Society of Medicine, Italfarmaco S.p.A., Kabi Pharmacia Cardiovascular, Lilly Research Laboratories, Mediolanum Farmaceutici S.r.l., The Nobel Committee for Chemistry of the Royal Swedish Academy of Sciences, Novo Nordisk A/S, Opocrin S.p.A., Rhône-Poulenc Rorer S.A., Sanofi Recherche, The Swedish Medical

Research Council, The Swedish Ministry of Education, The Swedish University of Agricultural Sciences and Wyeth-Ayerst Research. These contributors are all thanked for their help.

November 1991

U. Lindahl
I. Björk
D.A. Lane

CONTENTS

HISTORICAL ASPECTS

STRUCTURAL PROPERTIES OF HEPARIN AND HEPARAN SULPHATE

HEPARIN AND HEPARAN SULPHATE PROTEOGLYCANS-CELLULAR ASPECTS

NEW PERSPECTIVES
(E. Jorpes Symposium)

HEPARIN - AN INTRODUCTION

Lennart Rodén, Sandya Ananth, Patrick Campbell, Tracy Curenton, Göran Ekborg, Stephen Manzella, Dennis Pillion, and Elias Meezan

Metabolic Diseases Research Laboratory, The University of Alabama at Birmingham, Birmingham, Alabama 35294, U.S.A.

INTRODUCTION

Seventy-five years after its discovery (1,2), heparin remains an important tool in medicine. Among its established uses are the prevention of postoperative thrombosis, the treatment of acute venous thrombosis, and the prevention of clot formation in the heart-lung machine (3). Little was known about the structure of heparin when the first clinical trials began in the mid-1930s, and it is only during the 1980s that the detailed structural basis of heparin action has been elucidated through the characterization of a specific, antithrombin-binding pentasaccharide segment in the polysaccharide molecule. In the following, we shall retrace some of the steps that have led to this goal.

THE YEAR 1935

The year 1935 was a remarkable time in the history of heparin. By then, it was apparent that heparin was not a phospholipid, as had initially been assumed on the basis of the original mode of isolation. Analysis by the naphthoresorcinol reaction in Howell's laboratory at Johns Hopkins University had shown that heparin contained carbohydrate, more specifically a uronic acid, and this conclusion was confirmed by Jorpes at the Karolinska Institute, who published a landmark paper on the chemistry of heparin in 1935 (4). In addition to the colorimetric test, which was not always reliable, Jorpes used decarboxylation in strong acid for the qualitative and quantitative analysis of uronic acid in heparin. (It is of interest to note that, in 1936, two other pioneers in heparin research, Charles and Scott, in Toronto, still did not believe that heparin contained uronic acid.) Other basic features of the composition of heparin were also established by Jorpes in 1935. The newly developed Elson-Morgan procedure helped provide evidence for the presence of hexosamine in heparin, and, in accordance with the belief at the time that all hexosamine residues in complex carbohydrate molecules were N-acetylated, the presence of acetyl groups was also verified experimentally. A major finding in Jorpes' work was the high content of sulfate groups in heparin, determined by analysis of the ash, and this observation led Jorpes to conclude that heparin was the most highly charged polyanion in Nature.

Heparin and Related Polysaccharides
Edited by D.A. Lane *et al.*, Plenum Press, New York, 1992

In 1935, there were two major centers in heparin research in the world, Stockholm and Toronto. Under the direction of Charles Best, a team in Toronto had developed methods for the purification of heparin, and large-scale preparations were being undertaken at the Connaught Laboratories, affiliated with the University of Toronto. In Stockholm, Jorpes was building on the experience gained by Charles and Scott of the Toronto team, and large-scale preparations were being made possible through the participation of the Vitrum Company in this endeavor. (The difficulties in obtaining useful amounts of heparin on a laboratory scale were made abundantly clear to the senior author when, early in his career, he was given 80 kg of pig intestinal slime by Jorpes and eventually succeeded in isolating 112 mg of heparin with an anticoagulant activity of 56 units per mg (L. Rodén, unpublished results). Despite the considerable effort, this accomplishment did not a thesis make.)

The obvious goal of the researchers in Stockholm and Toronto was to develop heparin into a therapeutically useful drug. As a young surgeon in Stockholm, Clarence Crafoord had observed only too many times how the best efforts at the operating table had subsequently been thwarted by the development of postoperative venous thromboses resulting in fatal lung emboli. His zeal in wanting to eliminate this seemingly unnecessary complication helped Jorpes focus his efforts on the rapid development of heparin preparations fit for use in humans. After such preparations had been tested on medical students working in the Department of Medical Chemistry at the Karolinska Institute, Crafoord began clinical trials in August of 1935. A few months earlier, the team in Toronto had embarked on similar studies, and the efforts of the two groups soon proved the value of heparin as a clinically useful drug (5-8). Crafoord was the first to publish the results of this work (5) and provided convincing evidence that heparin was effective in the prevention of postoperative thrombosis. Early ambulation has subsequently lessened the incidence of postoperative thrombosis in general, but certain groups of patients are still at high risk, and the development of appropriate antithrombotic prophylaxis remains an area of active investigation even today.

Yet another line of research, relevant to heparin, came to fruition in 1935. Charles A. Lindbergh - the pilot of Spirit of St. Louis - had been working for several years in the laboratory of Nobel Laureate Alexis Carrel at the Rockefeller Institute on the construction of a mechanical heart pump, and in June of 1935 it was announced that this endeavor had succeeded. The Lindbergh apparatus allowed investigators to keep organs from experimental animals alive for more than one month by perfusing them with nutrient fluids of appropriate composition under strictly sterile conditions. Although perfusion of organs had been carried out earlier by many investigators, this was the first time that such experiments could be extended beyond a period of only a few hours. Whole blood was not used in Lindbergh's and Carrel's experiments in 1935, but in their book, "The culture of organs," published in 1938 (9), they reported that they had obtained heparin from the Connaught Laboratories in Toronto and had thus been able to compare the performance of whole blood and other nutrient fluids in their systems.

In the cardiovascular literature, the Lindbergh apparatus is described as one of the forerunners of the heart-lung machines, and its successful use in Carrel's laboratory and by other investigators undoubtedly stimulated the development of the technology necessary for application in humans. When heart-lung machines were first used for open-heart surgery in the early 1950s, Crafoord was again one of the leaders and reported the second successful operation of this kind in 1954.

THE PIECES OF THE PUZZLE

Monosaccharide Composition of Heparin

Five different monosaccharides have been identified as components of the heparin molecule: D-glucosamine, D-glucuronic acid, L-iduronic acid, D-galactose, and D-xylose. The definitive identification of the monosaccharide components of heparin began in 1936 and was not completed until 1964. In 1936, Jorpes and Bergström (10) (who was then a medical student like McLean) isolated the hexosamine component of heparin and showed that it was glucosamine. Jorpes' initial belief that heparin was an oversulfated chondroitin sulfate thus proved incorrect, and it was concluded, instead, that heparin was akin to mucoitin sulfuric acid, a glucosamine-containing polysaccharide isolated from gastric mucosa. (Mucoitin sulfuric acid has subsequently receded into oblivion.)

The basis of the successful identification of glucosamine by Jorpes and Bergström was the resistance of this sugar to treatment with strong acid at high temperature, which made it possible to subject heparin to acid hydrolysis and to isolate the sugar in good yield in crystalline form from the hydrolysate. In contrast, uronic acids are not as stable to acid treatment, and the conditions required for their release from a polymer are such that free uronic acids are rapidly destroyed. Today, it would have been relatively easy to show the presence of free glucuronic acid in a heparin hydrolysate, since we now have access to more sensitive analytical methods. At the time, however, it was not possible to detect the small quantities present, and it was only through indirect methodology that Wolfrom and Rice (11) succeeded in obtaining information about the identity of the uronic acid component of heparin. In 1946, these authors reported results of experiments in which bromine had been included in the reaction mixture during acid hydrolysis (bromine-sulfuric acid at about 3 °C for one week), thus effecting immediate oxidation of the released uronic acid to the more stable dicarboxylic acid. Having isolated and identified the latter as glucaric acid (saccharic acid), Wolfrom and Rice concluded that the uronic acid of heparin was D-glucuronic acid. It should be noted, however, that glucaric acid may also be derived from L-guluronic acid, but this possibility was not mentioned in the paper by Wolfrom and Rice and was perhaps considered too remote. Subsequently, more stringent proof for the D-gluco configuration was obtained, as has been described by Brimacombe and Webber (ref. 12 and refs. cited therein).

The third monosaccharide component of heparin was discovered by Cifonelli and Dorfman in 1962 (13). Facing the same difficulty as Wolfrom and Rice, they chose another approach and first removed the N-sulfate groups (see below) by mild acid hydrolysis; the free amino groups were then acetylated to yield a structure in which the glucosaminidic linkages were more susceptible to hydrolytic cleavage than in the starting material. After acid hydrolysis of the modified polysaccharide and repeated N-acetylation and hydrolysis, paper chromatography showed the presence of iduronic acid and iduronolactone in the hydrolysate. Further characterization of the iduronolactone was carried out by measurement of its optical rotation (indicative of the L-configuration) and reduction to idonolactone, which was clearly separated from gluconolactone, gulonolactone, and mannonolactone by paper chromatography. Initially, the work of Cifonelli and Dorfman was met with skepticism, but their important discovery was eventually confirmed by several investigators (14-17), including Perlin and his collaborators (15,17), whose

NMR studies showed that L-iduronic acid was the major uronic acid component of heparin.

The discovery of the two remaining monosaccharides, galactose and xylose, was incidental and occurred in 1964 in the course of a project aimed in a different direction (18,19). Although it had been known since Mörner's investigations in 1889 (20) that chondroitin sulfate was associated with protein, the prevalent notion during the first several decades of this century was that this association was effected by strong ionic interactions. It was only in the 1950s that convincing evidence for a covalent linkage between the protein and the polysaccharide emerged, mainly through the investigations of Schubert and Mathews and their collaborators. Based on amino acid analyses of chondroitin sulfate that had been subjected to extensive proteolysis, Helen Muir (21) concluded in 1958 that the polysaccharide was attached to serine residues in the core protein of the proteoglycan. The same conclusion was subsequently reached by Meyer and his collaborators on the basis of characterization of the products formed on alkaline cleavage of the protein-polysaccharide bond. As a consequence of the investigations of the chondroitin sulfate proteoglycan and its carbohydrate-protein linkage, the possibility naturally had to be explored that other polysaccharides, including heparin, were also covalently bound to protein. To test this hypothesis, Lindahl et al. (22) subjected partially processed heparin preparations to amino acid analysis and found that some such preparations contained essentially only serine. Mild acid hydrolysis yielded two tell-tale fragments, xylosylserine and galactosylxylosylserine, and it was therefore concluded that not only was native heparin covalently bound to peptide or protein, but the carbohydrate-protein linkage region contained two sugars that had not previously been recognized among the monosaccharide components of heparin (18,19).

Other Components

Substituents of three kinds are found on the sugar residues of heparin: acetyl, sulfate, and phosphate groups. The first analyses of the purified polysaccharide in Jorpes' laboratory indicated that, on a molar basis, the acetyl content approximately equalled that of hexosamine. At the time, this finding was almost a foregone conclusion, since in all hexosamine-containing complex carbohydrates examined previously, the amino groups of the hexosamine components had always been found to be N-acetylated. When Jorpes' landmark paper of 1935 (4) was published, however, it had become clear that the N-acetyl content of some heparin preparations was much lower than expected. The explanation of these results was not immediately obvious, but with his dogged perseverance, Jorpes did not let go of the problem and eventually arrived at the solution in 1950 (23) in the course of studies of the second class of substituents, the sulfate groups.

Although the presence of sulfate in heparin preparations was known to Howell, it was only through Jorpes' meticulous analytical work that it became clear that the sulfate was part of the polysaccharide molecule and that the sulfate content was much higher than in any other sulfated macromolecule known at the time. We now know that the sulfate groups may be found in five different locations in mammalian heparin, i.e., linked to the amino groups of the glucosamine moiety (N-sulfate) and as ester sulfate (O-sulfate) on C-6 and C-3 of glucosamine and on C-2 of iduronic and glucuronic acid. The assignment to these five positions was not easy and spanned over half a century, following Jorpes' discovery of the sulfate groups in 1935 (on average one position every 10 years!). The N-sulfate groups, which are unique to heparin and the closely related heparan sulfate, were discovered by Jorpes, Boström, and Mutt in 1950 (23) and the last assignment (to C-2 of some of the glucuronic acid residues) was made in 1985 by Bienkowski and Conrad (24). (See ref. 2 and refs. cited therein for details on other structural studies concerning the sulfate groups.)

Since heparin was originally thought to be a phospholipid and the earliest preparations were undoubtedly contaminated with such substances, it was not surprising that Howell's first analyses showed the presence of phosphorus. As the isolation procedures improved and heparin became a complex carbohydrate, the phosphorus disappeared. Recently, however, following the discovery by Oegema and his collaborators (25) that a phosphate group may be present at C-2 of the xylose residue in chondroitin sulfate, similar analyses by Fransson et al. (26) have shown that the same is true for heparan sulfate from bovine lung, and Rosenfeld and Danishefsky (27) have shown the presence of phosphate in the same location in heparin. The function of the phosphate group is not known, nor has the reaction by which it is introduced into the polysaccharide molecules been demonstrated.

<u>Bigger Pieces of the Puzzle</u>

The heteropolysaccharides found in Nature are often (but not always) composed of small repeating subunits, containing two or more monosaccharides. The determination of their detailed structure, including the positions and anomeric configurations of the glycosidic linkages, therefore may be facilitated by depolymerization and isolation of small fragments that are more amenable to analysis than the intact macromolecules. Acid hydrolysis has for many years been a mainstay in the degradation of polysaccharides by chemical methodology and is, of course, generally applicable to the study of all polysaccharides. For example, acid hydrolysis of hyaluronan produces the disaccharide, hyalobiuronic acid, which is composed of glucuronic acid and glucosamine, linked by a ß1,3 linkage. Our knowledge of heparin structure is, likewise, based in part on the characterization of hydrolytic fragments generated from native or chemically modified heparin, and the results of this work have indicated that heparin is composed of alternating glucosamine and uronic acid residues. Interestingly, the hydrolysis pattern of heparin is different from that of hyaluronan, partly because most of the glucosamine amino groups in heparin carry an N-sulfate group, while those in hyaluronan are N-acetylated. The rapid removal of the N-sulfate groups upon acid hydrolysis and the ensuing protonation of the amino groups render the glucosaminidic linkage highly resistant to hydrolysis, and the heparin molecule, therefore, breaks preferentially at the uronidic linkages. As a consequence, the predominant disaccharide fragments generated from heparin are glucosaminyl-iduronic acid and glucosaminyl-glucuronic acid. This is in contrast to the hydrolysis of hyaluronan and other N-acetylated glycosaminoglycans, which proceeds by preferential cleavage of the N-acetylhexosaminidic linkages. The extensive studies of the disaccharides from native and chemically modified heparin are described in detail elsewhere (2,12,28,29).

By virtue of the presence of the unique N-sulfate groups in heparin, the polysaccharide may also be degraded by a chemical method which specifically targets these groups, i.e., deaminative cleavage by nitrous acid. Well known in organic chemistry as a reagent that converts a primary amine to the corresponding alcohol and free nitrogen, nitrous acid has also been found to react with unsubstituted as well as N-sulfated glucosamine amino groups. If the glucosamine is polymer-bound, as in heparin, the treatment with nitrous acid also results in cleavage of the adjacent glucosaminidic linkage and in the conversion of the glucosamine to anhydromannose. Deaminative cleavage by nitrous acid, as used by the Birmingham group in England in the 1950s and subsequently in the laboratories of Cifonelli, Lindahl, and Conrad, has been the single most useful method for the generation of heparin disaccharides and larger fragments and has facilitated greatly the structural studies as well as metabolically oriented investigations (30-32). Recently, its application to the production of low-molecular-weight heparin has also opened up a new field in the therapeutic use of heparin (33). It should be noted that an obvious but important difference between acid hydrolysis and deaminative cleavage is that the latter procedure does not affect the O-sulfate groups and that the fragments obtained therefore represent the original structure of the polymer, albeit with anhydromannose at the reducing terminus instead of glucosamine.

In addition to chemical methods of degradation, enzymatic depolymerization has also played an important role in the elucidation of the structures of mammalian polysaccharides. Testicular hyaluronidase cleaves the endo-hexosaminidic linkages of hyaluronan and the chondroitin sulfates to yield tetrasaccharides as the major products, and leech hyaluronidase specifically cleaves the endoglucuronidic linkages of hyaluronan, yielding "reverse" oligosaccharides with glucuronic acid at the reducing terminus. The study of heparin structure has, likewise, benefited from the use of degradative enzymes of mammalian as well as of bacterial origin. A host of exoenzymes are present in mammalian tissues, which, in concerted action, are capable of degrading the polymer to its constituents (34). In addition, an endoglycosidase has also been found that specifically attacks endoglucuronidic linkages in the heparin molecule and whose action results in the formation of relatively large heparin fragments with a reducing terminal glucuronic acid residue (34).

Bacterial endoglycosidases of three types are known, which all cleave glucosaminidic linkages in heparin by an eliminase reaction that yields a Δ4,5-unsaturated uronic acid residue at the nonreducing terminus of the pertinent fragment (34). The bacterial enzymes may, in a sense, be regarded as tools that are complementary to deaminative cleavage, since the glucosamine moiety participating in the reaction remains intact, while the uronic acid residue at the cleavage point undergoes a structural change. When needed, the unsaturated uronic acid may be removed by a short treatment with mercuric acetate at room temperature (35).

In summary, the several procedures described above - acid hydrolysis, deaminative cleavage, and degradation with mammalian and bacterial enzymes - have allowed heparin researchers to degrade the polysaccharide to fragments of various sizes - disaccharides and larger - which are amenable to structural analysis. Aided by a wide range of additional methods - e.g., methylation analysis, periodate oxidation, and NMR spectroscopy - they have succeeded in establishing the basic structural features of heparin, including its monosaccharide composition, the positions and configurations of the glycosidic linkages, and the locations of the sulfate groups (Fig. 1). This work has eventually led to a deeper insight into the molecular mechanisms of action of heparin as an anticoagulant, and this aspect of heparin biology will be discussed briefly in the following.

THE STRUCTURAL BASIS OF THE ANTICOAGULANT ACTION OF HEPARIN

In 1935, Jorpes (4) pointed out that heparin is the most highly charged polyanion in Nature, and this is probably still true. It was then reasonable

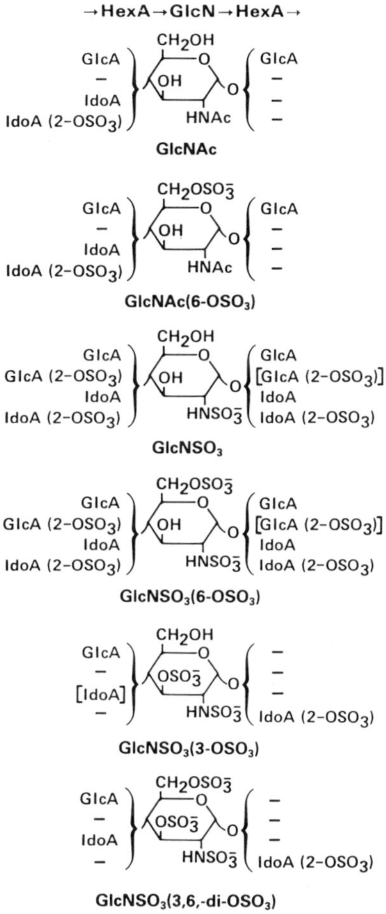

Structure of hexuronic acid residues

Abbreviations

Hexuronic acid residues

HexA	unspecified hexuronic acid
GlcA	D-glucaronic acid
GlcA(2-OSO_3)	D-glucaronic acid 2-0-sulphate
IdoA	L-iduronic acid
IdoA(2-OSO_3)	L-iduronic acid 2-0-sulphate

Glucosamine residues

GlcN	2-deoxy-2-amino-D-glucose (D-glucosamine)
GlcNAc	N-acetyl-D-glucosamine
GlcNAc(6-OSO_3)	N-acetyl-D-glucosamine 6-O-sulphate
GlcNSO_3	D-glucosamine-N-sulphate
GlcNSO_3(6-OSO_3)	D-glucosamine-N-,6-O-sulphate
GlcNSO_3(3-OSO_3)	D-glucosamine-N-,3-O-sulphate
GlcNSO_3(3,6,-di-OSO_3)	D-glucosamine-N-,3-O,6-O-sulphate

Fig. 1. Scheme of disaccharide sequences identified in heparin and heparan sulphate. The six variously substituted glucosamine residues in the middle (GlcN), are combined with the hexuronic acid units (HexA) at C4 and Cl to give 17 (possibly 18) different →HexA→GlcN→ sequences, and 10 (possibly 12) →GlcN→HexA→ sequences, respectively. The structures of the hexuronic acid residues are indicated in the upper right hand corner. Gaps indicated by (-) denote combinations that have not been found, either because they do not exist (e.g.→ GlcNAc→IdoA→), or because they occur very infrequently. Probable sequences, still to be uneqivocally demonstrated, are indicated by [] around the hexuronic acid concerned. Reproduced with permission of the author (U. Lindahl) and the publisher from "Heparin", D.A. Lane and U. Lindahl, eds., Edward Arnold, London (1989).

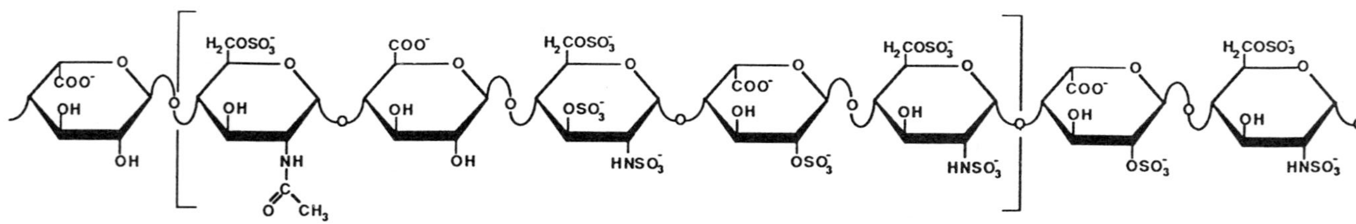

Fig. 2. Structure of a heparin octasaccharide sequence that displays most of the variously substituted monosaccharide components identified to date. The pentasaccharide sequence within brackets represents the antithrombin-binding region.

to assume that the anticoagulant activity of heparin was due, in part at least, to its high negative charge density, and this hypothesis was supported by the finding that chemical sulfation of neutral polysaccharides endows them with a measure of anticoagulant activity. A high negative charge density, however, is not the sole requirement for a high anticoagulant activity, and we now know that this property resides in a segment of specific structure in the heparin molecule. The first indication to this effect came in 1976, when Lam, Silbert, and Rosenberg (37) found that only a portion of the molecules in a standard heparin preparation interacted with antithrombin, as shown by centrifugation in a sucrose gradient. The same conclusion was reached by Höök, Björk, Hopwood, and Lindahl (38), who demonstrated that only about one third of a similar heparin preparation was bound strongly to an antithrombin-agarose column and that the anticoagulant activity was associated with this fraction of the material. Since the chemical composition, including the sulfate content, of the antithrombin-binding fraction did not differ substantially from that of the inactive material, it was apparent that charge density was not the only determining factor in the interaction between antithrombin and heparin. A search for the specific structural features responsible for the biological activity was then launched, which culminated in the discovery by Lindahl and collaborators (39,40) that a sulfate group is present at C-3 of a glucosamine unit in the antithrombin-binding segment of the molecule (Fig. 2).

Through careful dissection of the heparin molecule in Lindahl's laboratory, it is now well established that the antithrombin-binding segment of heparin is a sulfated pentasaccharide of unique structure, in which not only the 3-O-sulfate group but also other sulfate residues are essential for the biological activity (40,41). The analytical studies have been complemented, in an impressive manner, with chemical synthesis of the antithrombin-binding pentasaccharide and other relevant oligosaccharides (42), and the chemical manipulation of the pertinent structures has set the stage for detailed conformational studies of the interaction between antithrombin and heparin. A description of the results to date is beyond the scope of this presentation, and reference is made to other articles (40,43) and works cited therein.

BIOSYNTHESIS OF HEPARIN

In the earliest studies of heparin biosynthesis - where credit goes particularly to Jeremiah Silbert - the formation of a nonsulfated polysaccharide was shown to occur when the microsomal fraction of a mouse mastocytoma was incubated with UDP-glucuronic acid and UDP-N-acetylglucosamine (44-47). Although this polysaccharide had the same charge density and, presumably, the same composition as hyaluronan, it was not degraded by testicular hyaluronidase. Instead, its susceptibility to digestion with heparinase from *Flavobacterium heparinum* indicated that it was related to heparin. In the presence of 3'-phosphoadenylylsulfate (PAPS), a sulfated polysaccharide was formed, which was similar to heparin in its charge density and contained N-sulfate groups. The uronic acid composition of the polysaccharide products was not determined, and at the time there seemed to be no particular reason to do so, since it was well known that the uronic acid component of heparin was D-glucuronic acid. (Even in 1967 (45-47) it was not yet known that the predominant uronic acid is L-iduronic acid.)

Following Silbert's pioneering investigations, Lindahl and his collaborators adopted a similar experimental system, using the Furth mastocytoma rather than the Dunn-Potter tumor, and conducted a series of elegant studies that have led to our current picture of the biosynthetic process (36,48-51).

In brief, it was demonstrated that the assembly of a nonsulfated polysaccharide chain, composed of alternating glucuronic acid and N-acetylglucosamine residues, is followed by several polymer modifications that occur in an ordered sequence, regulated in large part by the substrate specificities of the enzymes involved. The first of these reactions, N-deacetylation, is catalyzed by an interesting bivalent enzyme which, when PAPS is present in the medium, also carries out the next step in the sequence, i.e., N-sulfation (52). A second protein, apparently of nonenzymatic nature, is required for the N-deacetylase activity to be expressed, but this protein is not necessary for N-sulfation to occur. (It should be noted that the N-sulfotransferase has previously been purified to homogeneity from rat liver by Brandan and Hirschberg (53).) Whereas the N-deacetylation of the heparin precursor polysaccharide never goes to completion, the N-sulfation of the free amino groups generated by the N-deacetylase is essentially quantitative, and the few free amino groups that may occasionally be detected in heparin preparations are probably the result of some loss of the labile N-sulfate groups during the manufacturing process.

The replacement of the N-acetyl groups with N-sulfate groups generates a structure that is recognized as a substrate by the enzyme catalyzing the next reaction in the sequence of modifications, i.e., 5-epimerization of the D-glucuronosyl residues. This reaction is unique among the metabolic sugar interconversions, inasmuch as it takes place after rather than before the assembly of the complex carbohydrate molecule. The new structure formed in this reaction is the substrate for sulfation of the iduronic acid residues at C-2, and O-sulfation then occurs at C-6 of glucosamine residues, preferentially those that are N-sulfated. At some point, an O-sulfate group is introduced at C-2 of some glucuronic acid residues, but the exact place of this reaction in the modification sequence is not yet known. It is clear, however, that the limited O-sulfation at C-3 of glucosamine residues, which completes the formation of the antithrombin-binding structure, takes place after the sulfation of iduronic acid and the C-6 position of glucosamine.

When a particular process is examined in vitro, it must ultimately be determined whether the behavior of the in vitro system truly reflects the physiological modus operandi of the tissue under study. This problem has not yet been addressed in investigations of heparin biosynthesis, but in expanded studies of the microsomal system Lidholt et al. (49-51) have examined the relationship between polymerization and sulfation to answer the question whether formation of complete polysaccharide chains precedes sulfation or whether polymerization occurs concomitantly with sulfation. Although a nonsulfated GlcUA-GlcNAc polymer and its partially N-deacetylated derivative are synthesized in reaction mixtures without PAPS and rapid conversion to sulfated products occurs when PAPS is added (36,48), it has not been known whether the nonsulfated species are also formed when PAPS is present from the beginning of the incubation. Lidholt et al. (49-51) have now shown that this is indeed the case. Furthermore, it was observed that the nonsulfated chains were considerably smaller than the sulfated polysaccharides formed in the same reaction mixture. It was therefore concluded that polymerization and sulfation may also occur concomitantly and that sulfation actually promotes the polymerization process. The conclusion that the two processes can occur simultaneously was supported by pulse-chase experiments, which showed that the sulfated chains were capable of being elongated further.

HEPARIN(S), HEPARAN SULFATE(S), AND REGULATION OF BIOSYNTHESIS

The investigations of the biosynthesis of heparin have given us not only knowledge about this process per se but also a deeper understanding of the structure of heparin and an insight into the relationship between heparin and its close relative, heparan sulfate (55,56). A basic lesson learned is that the polymer modifications do not go to completion, with the possible exception of N-sulfation. As a consequence, the final product of the biosynthetic process is not a single molecular species but a mixture of closely related molecules. We should perhaps sometimes speak of "heparins" to remind ourselves of this fact.

In all heparins, the polymer modifications are extensive, and, according to a recent definition (54), heparin is "a polysaccharide in which more than 80 per cent of the glucosamine residues are N-sulphated and the concentration of O-sulphate groups exceeds that of the N-sulphate groups." Substances of this nature are produced only by connective tissue mast cells.

The heparan sulfates, produced by most other cell types, are assembled by a synthetic apparatus that is presumably identical, qualitatively at least, to that involved in heparin biosynthesis. Thus, the fundamental compositions and structures of the two polysaccharide types are the same, but substantial quantitative differences exist: there is more glucuronic acid and, consequently, less iduronic acid in heparan sulfate than in heparin, the N-acetyl content is higher and the N-sulfate content lower, and the degree of O-sulfation is lower (55,56). In many heparan sulfates, the N-acetyl and N-sulfate contents are approximately equal. All of these differences may be viewed as consequences of a difference in the degree of N-deacetylation during the polymer modifications, and it should be recalled that in the absence of N-deacetylation, none of the subsequent modifications can take place.

We do not yet know much about the organization of the biosynthetic enzymes in the Golgi membranes, where both chain growth and modifications take place, nor do we have much information about the physiological mechanisms by which these processes are regulated. It is reasonable to assume, however, that the N-deacetylation step is under metabolic control, as are many other reactions that initiate a multistep sequence. Indeed, evidence to this effect has recently been obtained by Unger et al. (57,58), who showed that the N-deacetylase activity in hepatocytes from streptozotocin-diabetic rats was 40% lower than in control cells. In contrast, the activity of another modifying enzyme, the uronosyl 5-epimerase, was unchanged. Although a lowered level of N-deacetylase activity cannot, a priori, be presumed to affect the extent of N-deacetylation during the polymer modifications, analysis of the substitution pattern of the heparan sulfate produced by the diabetic cells showed that its N-acetyl content was higher and the N-sulfate content lower than in the polysaccharide synthesized by the normal control cells (57). It thus appears that the level of N-deacetylase activity, at least under certain conditions, determines the degree of N-deacetylation. Continued pursuit of this line of investigation should prove rewarding.

Although the disparity in the extent of polymer modification is an important difference between heparins and heparan sulfates, it is not the only one. On closer scrutiny, a difference in the organization of the N-acetylated glucosamine residues has also been observed, inasmuch as these residues are, to a large extent, arranged in blocks in the heparan sulfates, while the N-acetylated residues in heparin are mostly located singly between N-sulfated disaccharide units.

An important aspect of the structure of heparins and heparan sulfates has been ignored up to this point. All heparins and heparan sulfates are initially synthesized as part of proteoglycan molecules, and, as has been mentioned earlier, the heparin chains are linked to serine residues in the core protein of the proteoglycan. The native proteoglycan has been an elusive entity, but dramatic progress has been made in several laboratories during the past few years, and the primary structures of the core proteins in the heparin proteoglycan from the Furth mastocytoma and the heparan sulfate proteoglycans from several sources have now been determined (60-65). These investigations have established that the core protein of the heparin proteoglycan differs from those found in heparan sulfate proteoglycans and that the latter are also different from one heparan sulfate proteoglycan to another. Of particular interest is the finding that the heparin proteoglycan (60) contains 10 ser-gly groups in uninterrupted sequence, while the heparan sulfate proteoglycans contain single ser-gly groups separated by other amino acids (exception: glypican (65) that contains three ser-gly repeats).

Not so long ago, we believed that anticoagulant activity was an exclusive property of heparin, displayed by all molecules of the polysaccharide, and that heparan sulfates were never endowed with this characteristic. Anticoagulant activity was therefore yet another criterion by which we could distinguish heparin from heparan sulfate. But no more. Since 1976, we know that not all heparins have anticoagulant activity (37,38), and in 1986 we learned that some heparan sulfates do, as shown by Marcum et al. (66), who isolated a heparan sulfate proteoglycan with an antithrombin-binding site from bovine aortic endothelial cells.

POLYSACCHARIDE CHAIN INITIATION

Aspects of heparin and heparan sulfate biosynthesis discussed above have been focused on those processes that are unique to the formation of these polysaccharides. Nine enzymatic steps have been mentioned so far, i.e., formation of a GlcUA-GlcNAc polymer by the combined action of a glucuronosyltransferase and an N-acetylglucosaminyltransferase, N-deacetylation/N-sulfation, catalyzed by a single enzyme with two active sites, uronosyl 5-epimerization, and O-sulfation in four different positions. Before these reactions can take place, however, the polysaccharide chains are initiated by the transfer of xylose from UDP-xylose to specific serine residues in the core proteins of the heparin and heparan sulfate proteoglycans (67). We now believe that, in addition to the xylosyltransferase catalyzing this reaction, five more enzymes are required for completion of the region near the carbohydrate-protein linkage, which has the following structure: -GlcNAc-GlcUA-Gal-Gal-Xyl(2-P)-Ser. The five enzymes are: 1) an enzyme that introduces the phosphate group at C-2 of the xylose; 2) galactosyltransferase I; 3) galactosyltransferase II; 4) glucuronosyltransferase I; and 5) N-acetylglucosaminyltransferase I. The existence of the two galactosyltransferases and glucuronosyltransferase I is firmly established, and it has been shown that the latter is different from the glucuronosyltransferase catalyzing formation of the repeating disaccharide units that constitute the bulk of the polysaccharide chain (67). Phosphorylation of the xylose residue, however, has not yet been demonstrated directly, and the existence of N-acetylglucosaminyltransferase I is postulated on the basis of indirect evidence. Thus, in the formation of the analogous segment of chondroitin sulfate proteoglycans, a separate enzyme catalyzes transfer of N-acetylgalactosamine to the glucuronic acid residue of the linkage region, and this transferase is distinct from that involved in the synthesis of the remainder of the chain (68). It should be mentioned in this context that the glycosyltransferases catalyzing complex carbohydrate formation rarely recognize a monosaccharide as a substrate (galactosyltransferase I is an exception and transfers galactose to xylose), and a disaccharide unit is more often required for the reaction to occur. The linkage region disaccharide presented as a substrate in the first N-acetylglucosaminyltransfer reaction is GlcUA-Gal-, while the corresponding structure participating in the polymerization reactions is GlcUA-GlcNAc-; in view of the anticipated substrate specificities of the enzymes involved, it is therefore reasonable to assume that two different N-acetylglucosaminyltransferases are required for the transfer reactions in question.

TYING UP SOME LOOSE ENDS

Endogenous Xylose Acceptors

The first studies of the chain-initiating xylosyltransferase reaction were carried out by Grebner, Hall and Neufeld (69,70) and by Robinson, Telser and Dorfman (71). Since an exogenous substrate for the enzyme was not available at the time, this work was carried out in the hope that a sufficient amount of core protein would be present in the crude tissue extracts so as to allow detection of the reaction product. In the studies of Robinson et al. (71), it was shown that an extract of embryonic chick cartilage catalyzed transfer of xylose from UDP-xylose to endogenous acceptors that had the properties expected of a proteoglycan core protein. Thus, more than 90% of the xylose was released from the product by treatment with alkali under conditions that were known to cleave the xylose-serine linkage in the chondroitin sulfate proteoglycan. In contrast, Grebner et al. (69,70), using hen's oviduct and the Dunn-Potter mastocytoma as enzyme sources, found that at times more than half of the incorporated xylose could not be released by alkali treatment. They therefore postulated that two separate xylosyltransferase activities were present in their crude extracts, one that catalyzed the reaction that they had set out to demonstrate, and one that yielded an alkali-stable product. Despite careful examination of the properties of the alkali-stable material, it was not possible to establish its identity at the time.

In 1978, Kimura and Caplan (72) likewise observed formation of an alkali-stable xylose-labeled product in their studies of proteoglycan synthesis in limb buds from embryonic chicks, and extensive characterization of this material led to the conclusion that it was glycogen. The interpretation offered for this result was that glycogen synthase may use not only UDP-glucose but also UDP-xylose as a donor substrate. This hypothesis, however, was never tested directly by Kimura and Caplan, and, in our experience, glycogen synthase cannot substitute UDP-xylose for UDP-glucose as a glycosyl donor.

Some time ago, we began a study of proteoglycan biosynthesis in the rat kidney, with the goal of determining the regulatory basis of differences seen between normal and diabetic kidneys. When xylosyltransferase was assayed, with silk as an exogenous substrate (73), activity was readily detected, but it was also apparent that the crude kidney extract contained substantial amounts of endogenous acceptor (74-76), since xylose incorporation into trichloroacetic acid-precipitable material in the absence of silk was about one third of that observed in the complete system. All or most of the product of transfer to the endogenous acceptor was alkali-stable, and it was therefore apparent that it was not the core protein of a xylose-serine linked proteoglycan. Further characterization showed that the product was susceptible to digestion with trypsin and amylase, and SDS-PAGE followed by autoradiography showed the presence of a single labeled product that had migrated to a position corresponding to a M_r of about 33,000. Upon purification of the xylosyltransferase catalyzing the formation of the alkali-stable product, the enzyme could never be separated from the acceptor substrate, and a highly purified preparation yielded essentially a single protein band on SDS-PAGE, which migrated to the same position as the radioactive product. From these data, together with the finding that the purified enzyme could also use UDP-glucose as a glycosyl donor, it was concluded that the endogenous xylose acceptor was identical to glycogenin, a protein that is at the same time the core protein of glycogen proteoglycan and an enzyme that glucosylates itself (77-81).

On the basis of the work described above, we consider it highly likely that the endogenous xylose acceptors observed by Grebner et al. (69,70) and Kimura and Caplan (72) were identical to glycogenin.

A comparison between UDP-xylose and UDP-glucose, using the purified renal xylosyltransferase/acceptor, showed that UDP-xylose was approximately one third as efficient as UDP-glucose as a donor substrate, and, in view of these results, the question may be asked whether xylose residues are present in glycogen under physiological conditions. Whereas this seems theoretically possible, the number of xylose residues in a full-grown glycogen molecule is, however, likely to be extremely low. First, the physiological concentration of UDP-xylose is probably in the order of 10% of that of UDP-glucose, and in view of the relative donor efficiencies of the two nucleotide sugars, the incorporation of xylose into glycogenin would therefore be expected to be only about 3% of that observed with UDP-glucose as the donor. Second, the fraction of the glycogen molecule synthesized by glycogenin is small (8-9 glucose residues), and since glycogen synthase, which catalyzes the continued growth of the molecule, cannot utilize UDP-xylose, no xylose can be incorporated into the bulk of the glycogen chains. Nevertheless, glycogenin has an affinity for UDP-xylose which is comparable to that of the xylosyltransferase involved in glycosaminoglycan chain initiation, and glycogen is a more abundant product in many cells than proteoglycans. Therefore, it is possible that, under certain metabolic conditions, glycogenin could compete appreciably with xylosyltransferase for the available pool of UDP-xylose, thus possibly influencing the extent of proteoglycan synthesis.

Glucuronosyl Transfer to Galactose Residues

As has already been indicated, an early step in the assembly of heparin and heparan sulfate is the transfer of glucuronic acid from UDP-glucuronic acid to a galactose residue in the carbohydrate-protein linkage region (67,82). The same reaction occurs in the course of chondroitin sulfate biosynthesis, and the enzyme involved - glucuronosyltransferase I - was first detected in embryonic chick cartilage (83). Brandt et al. (84) reported the presence of the same activity in a partially purified enzyme preparation from embryonic chick brain and observed, in addition, that N-acetyllactosamine was also a substrate for glucuronosyl transfer. In 1969, when these studies were carried out, a physiologically occurring structure with glucuronic acid attached to an N-acetyllactosamine unit was not known, and it seemed most reasonable to assume that the transfer to N-acetyllactosamine was the result of a less than absolute substrate specificity on the part of glucuronosyltransferase I. Recently, however, two glycolipids have been discovered in nerve tissue, which display reactivity towards the HNK-1 antibody and contain a terminal glucuronic acid residue that is sulfated at C-3 and is linked by a ß1,3 linkage to the galactose moiety of an N-acetyllactosamine group (85,86). Transfer of glucuronic acid to two glycolipids, which may be regarded as the naturally occurring precursors of the two HNK-1 reactive compounds, was recently reported by Das et al. (87). In view of the similarity of this reaction to that catalyzed by glucuronosyltransferase I, we examined the question whether a single enzyme or two separate glucuronosyltransferases are involved in these processes (88). Assay of the two activities, with Gal-

Gal-Xyl-Ser and Gal-GlcNAc (N-acetyllactosamine) as substrates, in fractions of a homogenate of embryonic chick brain showed that the proteoglycan-related activity was firmly membrane-associated, while the majority of the activity towards N-acetyllactosamine was present in the soluble fraction. No activity towards N-acetyllactosamine was found in embryonic chick cartilage, which is a rich source of the proteoglycan-related enzyme. Furthermore, Gal-Gal-Xyl-Ser did not significantly inhibit transfer to N-acetyllactosamine in mixed substrate experiments. On the basis of these results, we conclude that two separate enzymes are involved in the glucuronosyl transfer to the galactose residues of the two substrates, and the results also support the notion that a disaccharide is needed for substrate recognition by the enzymes.

CONCLUDING REMARKS

In conclusion, it is interesting to recall some of the concepts of heparin structure expressed by Jorpes and his contemporaries. When Jorpes found that highly purified preparations of heparin (meaning preparations with high anticoagulant activity) contained approximately 2.5 sulfate groups per glucosamine residue, he then assumed that he was dealing with a mixture of two molecular species, heparin di- and trisulfuric acid, and he strived to fractionate the material further. The underlying thought was obviously that heparin consisted of disaccharide subunits and that each unit had the same number of sulfate groups throughout the entire molecule, two for one species and three for the other. In response to this thought, Sir Alexander Todd (89) pointed out that Jorpes had overlooked the possibility that the subunit might be a tetrasaccharide with five sulfate groups rather than a disaccharide. Generalization and simplification are powerful tools that help scientists move forward, but when Nature refuses to cooperate, the scientists' insistence on simplification may become an impediment to progress. The question of the N-acetyl groups is another case in point. When Jorpes et al. (23) had eventually shown that heparin contained N-sulfate groups, Jorpes' conviction that the structure of a pure polysaccharide was uniform led him to conclude that heparin did not contain any acetyl groups, even though he had never been able to isolate any such material and the actual analyses did not support his conclusion. It was difficult to break with the traditional views of what a polysaccharide molecule ought to look like, and it does not behoove us to gloat over our deeper understanding of the intricacies and the innate heterogeneity of the heparin molecules. We should, instead, remember the fundamental contributions of the pioneers that have led us to this insight through many trials and some errors and continue to build on the edifice that they have given us.

REFERENCES

1. J. McLean, The thromboplastic action of cephalin, Am. J. Physiol. 41:250 (1916).
2. L. Rodén, Highlights in the history of heparin, in: "Heparin," D. A. Lane and U. Lindahl, eds., Edward Arnold, London (1989).
3. J. Hirsch, Heparin, New Engl. J. Med. 324:1565 (1991).
4. E. Jorpes, The chemistry of heparin, Biochem. J. 29:1817 (1935) .
5. C. Crafoord, Heparin and post-operative thrombosis, Acta Chir. Scand. 82:319 (1939).
6. C. Crafoord and E. Jorpes, Heparin as a prophylactic against thrombosis, J. Am. Med. Assoc. 116:2831 (1941).
7. C. H. Best, Preparation of heparin and its use in the first clinical cases, Circulation 19:79 (1959).
8. E. Jorpes, Heparin: A mucopolysaccharide and an active antithrombotic drug, Circulation 19:87 (1959).

9. A. Carrel and C. A. Lindbergh, "The Culture of Organs," Paul B. Hoeber, New York (1938).
10. E. Jorpes and S. Bergström, Der Aminozucker des Heparins, Hoppe-Seyler's Z. Physiol. Chem. 244:253 (1936).
11. M. L. Wolfrom and F. A. H. Rice, The uronic acid component of heparin, J. Am. Chem. Soc. 68:532 (1946).
12. J. S. Brimacombe and J. M. Webber, "Mucopolysaccharides," Elsevier Publishing Company, Amsterdam (1964).
13. J. A. Cifonelli and A. Dorfman, The uronic acid of heparin, Biochem. Biophys. Res. Commun. 7:41 (1962).
14. U. Lindahl, Further characterization of the heparin-protein linkage region, Biochim. Biophys. Acta 130:368 (1966).
15. A. S. Perlin, M. Mazurek, L. B. Jaques, and L. W. Kavanagh, A proton magnetic resonance spectral study of heparin. L-Iduronic acid residues in commercial heparins, Carbohyd. Res. 7:369 (1968).
16. M.L. Wolfrom, S. Honda, and P. Y. Wang, The isolation of L-iduronic acid from the crystalline barium acid salt of heparin, Carbohyd. Res. 10:259 (1969).
17. A. S. Perlin and G. R. Sanderson, L-Iduronic acid, a major constituent of heparin. Carbohyd. Res. 12:183 (1970).
18. U. Lindahl and L. Rodén, The linkage of heparin to protein, Biochem. Biophys. Res. Commun. 17:254 (1964).
19. U. Lindahl and L. Rodén, The role of galactose and xylose in the linkage of heparin to protein, J. Biol. Chem. 240:2821 (1965).
20. C. T. Mörner, Chemische Studien über den Trachealknorpel, Scand. Arch. Physiol. 1:210 (1889).
21. H. Muir, The nature of the link between protein and carbohydrate of a chondroitin sulphate complex from hyaline cartilage, Biochem. J. 69:195 (1958).
22. U. Lindahl, J. A. Cifonelli, B. Lindahl, and L. Rodén, The role of serine in the linkage of heparin to protein, J. Biol. Chem. 240:2817 (1965).
23. J. E. Jorpes, H. Boström, and V. Mutt, The linkage of the amino group in heparin. Alleged acetyl content of heparin, J. Biol. Chem. 183:607 (1950).
24. M. J. Bienkowski and H. E. Conrad, Structural characterization of the oligosaccharides formed by depolymerization of heparin with nitrous acid, J. Biol. Chem. 260:356 (1985).
25. T. R. Oegema, Jr., E. L. Kraft, G. W. Jourdian, and T. R. Van Valen, Phosphorylation of chondroitin sulfate in proteoglycans from the Swarm rat chondrosarcoma, J. Biol. Chem. 259:1720 (1984).
26. L.-Å. Fransson, I. Silverberg, and I. Carlstedt, Structure of the heparan sulfate-protein linkage region. Demonstration of the sequence galactosyl-galactosyl-xylose-2-phosphate, J. Biol. Chem. 260:14722 (1985).
27. L. Rosenfeld and I. Danishefsky, Location of specific oligosaccharides in heparin in terms of their distance from the protein linkage region in the native proteoglycan, J. Biol. Chem. 263:262 (1988).
28. W. D. Comper, "Heparin (and related polysaccharides. Structural and Functional Properties," Gordon and Breach, New York (1981).
29. B. Casu, Structure and biological activity of heparin, Adv. Carbohyd. Chem. Biochem. 43:51 (1985).
30. J. E. Shively and H. E. Conrad, Formation of anhydrosugars in the chemical depolymerization of heparin, Biochemistry 15:3932 (1976).
31. J. E. Shively and H. E. Conrad, Nearest neighbor analysis of heparin: identification and quantitation of the products formed by selective depolymerization procedures, Biochemistry 15:3943 (1976).

32. S. R. Delaney, M. Leger, and H. E. Conrad, Quantitation of the sulfated disaccharides of heparin by high performance liquid chromatography. Anal. Biochem. 106:253 (1980).
33. E. Holmer, Low molecular weight heparin, in: "Heparin," D. A. Lane and U. Lindahl, eds., Edward Arnold, London (1989).
34. J. J. Hopwood, Enzymes that degrade heparin and heparan sulphate, in: "Heparin," D. A. Lane and U. Lindahl, eds., Edward Arnold, London (1989).
35. U. Ludwigs, A. Elgavish, J. D. Esko, E. Meezan, and L. Rodén, Reaction of unsaturated uronic acid residues with mercuric salts, Biochem. J. 245:795 (1987).
36. U. Lindahl, Biosynthesis of heparin and related polysaccharides, in: "Heparin," D. A. Lane and U. Lindahl, eds., Edward Arnold, London (1989).
37. L. H. Lam, J. E. Silbert, and R. D. Rosenberg, The separation of active and inactive forms of heparin, Biochem. Biophys. Res. Commun. 69:570 (1976).
38. M. Höök, I. Björk, J. Hopwood, and U. Lindahl, Anticoagulant activity of heparin: Separation of high-activity and low-activity heparin species by affinity chromatography on immobilized antithrombin. FEBS Lett. 66:90 (1976).
39. U. Lindahl, G. Bäckström, L. Thunberg, and I. G. Leder, Evidence for a 3-O-sulfated D-glucosamine residue in the antithrombin-binding sequence of heparin, Proc. Natl. Acad. Sci. USA 77:6551 (1980).
40. I. Björk, S. T. Olson, and J. D. Shore, Molecular mechanisms of the accelerating effect of heparin on the reactions between antithrombin and clotting proteinases, in: "Heparin," D. A. Lane and U. Lindahl, eds., Edward Arnold, London (1989).
41. U. Lindahl, L. Thunberg, G. Bäckström, J. Riesenfeld, K. Nordling, and I. Björk, Extension and structural variability of the antithrombin-binding sequence in heparin, J. Biol. Chem. 259:12368 (1984).
42. M. Petitou, Chemical synthesis of heparin, in: "Heparin," D. A. Lane and U. Lindahl, eds., Edward Arnold, London (1989).
43. P. D. J. Grootenhuis and C. A. A. van Boeckel, Constructing a molecular model of the interaction between antithrombin III and a potent heparin analogue, J. Am. Chem. Soc. 113:2743 (1991).
44. J. E. Silbert, Incorporation of ^{14}C and ^{3}H from nucleotide sugars into a polysaccharide in the presence of a cell-free preparation from mouse mast cell tumors, J. Biol. Chem. 238:3542 (1963).
45. J. E. Silbert, Incorporation of $^{35}SO_4$ into endogenous heparin by a microsomal fraction from mast cell tumors, J. Biol. Chem. 242:2301 (1967).
46. J. E. Silbert, Biosynthesis of heparin. III. Formation of a sulfated glycosaminoglycan with a microsomal preparation from mast cell tumors, J. Biol. Chem. 242:5146 (1967).
47. J. E. Silbert, Biosynthesis of heparin. IV. N-Deacetylation of a precursor glycosaminoglycan, J. Biol. Chem. 242:5153 (1967).
48. U. Lindahl, D. S. Feingold, and L. Rodén, Biosynthesis of heparin, Trends Biochem. Sci. 11:221 (1986).
49. K. Lidholt, "A New Model for the Biosynthesis of Heparin," SLU Info/Repro, Uppsala (1991).
50. K. Lidholt, J. Riesenfeld, K.-G. Jacobsson, D. S. Feingold, and U. Lindahl, Biosynthesis of heparin. Modulation of polysaccharide chain length in a cell-free system, Biochem. J. 254:571 (1988).
51. K. Lidholt, L. Kjellén, and U. Lindahl, Biosynthesis of heparin. Relationship between the polymerization and sulfation processes, Biochem. J. 261:999 (1989).

52. I. Pettersson, M. Kusche, E. Unger, H. Wlad, L. Nylund, U. Lindahl, and L. Kjellén, Biosynthesis of heparin. Purification of a 110 kDa mouse mastocytoma protein required for both glucosaminyl N-deacetylation and N-sulfation, J. Biol. Chem. 266:8044 (1991).
53. E. Brandan and C. B. Hirschberg, Purification of rat liver N-heparan-sulfate sulfotransferase, J. Biol. Chem. 263:2417 (1988).
54. D. A. Lane and U. Lindahl, Preface, in: "Heparin," D. A. Lane and U. Lindahl, eds., Edward Arnold, London (1989).
55. L.-Å. Fransson, Heparan sulphate proteoglycans: structure and properties, in: "Heparin," D. A. Lane and U. Lindahl, eds., Edward Arnold, London (1989).
56. J. T. Gallagher and M. Lyon, Molecular organization and functions of heparan sulphate, in: "Heparin," D. A. Lane and U. Lindahl, eds., Edward Arnold, London (1989).
57. E. Unger, "Proteoglycans in Diabetes," SLU Info/Repro, Uppsala (1991).
58. E. Unger, I. Pettersson, U. J. Eriksson, U. Lindahl, and L. Kjellén, Decreased activity of the heparan sulfate modifying enzyme glucosaminyl N-deacetylase in hepatocytes from streptozotocin-diabetic rats, J. Biol. Chem. 266:8671 (1991).
59. L. Kjellén and U. Lindahl, Proteoglycans: structures and interactions, Annu. Rev. Biochem. 60:443 (1991).
60. L. Kjellén, I. Pettersson, P. Lillhager, M.-L. Steen, U. Pettersson, P. Lehtonen, T. Karlsson, E. Ruoslahti, and L. Hellman, Primary structure of a mouse mastocytoma proteoglycan core protein, Biochem. J. 263:105 (1989).
61. D. M. Noonan, E. A. Horigan, S. R. Ledbetter, G. Vogeli, M. Sasaki, Y. Yamada, and J. R. Hassell, Identification of cDNA clones encoding different domains of the basement membrane heparan sulfate proteoglycan, J. Biol. Chem. 263:16379 (1988).
62. S. Saunders, M. Jalkanen, S. O'Farrell, and M. Bernfield, Molecular cloning of Syndecan, an integral membrane proteoglycan, J. Cell. Biol. 108:1547 (1989).
63. M. Mali, P. Jaakola, A.-M. Arvilommi, and M. Jalkanen, Sequence of human Syndecan indicates a novel gene family of integral membrane proteoglycans, J. Biol. Chem. 265:6884 (1990).
64. P. Marynen, J. Zhang, J.-J. Cassiman, H. Van den Berghe, and G. David, Partial primary structure of the 48- and 90-kilodalton core proteins of cell surface-associated heparan sulfate proteoglycans of lung fibroblasts, J. Biol. Chem. 264:7017 (1989).
65. G. David, V. Lories, B. Decock, P. Marynen, J.-J. Cassiman, and H. Van den Berghe, Molecular cloning of a phosphatidylinositol-anchored membrane heparansulfate proteoglycan from human lung fibroblasts, J. Cell Biol. 111:3165 (1990).
66. J. A. Marcum, D. H. Atha, L. M. S. Fritze, P. Nawroth, D. Stern, and R. D. Rosenberg, Cloned bovine aortic endothelial cells synthesize anticoagulantly active heparan sulfate proteoglycan, J. Biol. Chem. 261:7507 (1986).
67. L. Rodén, Structure and metabolism of connective tissue proteoglycans, in: "The Biochemistry of Glycoproteins and Proteoglycans," W. J. Lennarz, ed., Plenum Press, New York (1980).
68. K. Rohrmann, R. Niemann, and E. Buddecke, Two N-acetylgalactosaminyltransferases are involved in the biosynthesis of chondroitin sulfate, Eur. J. Biochem. 148:463 (1985).
69. E. E. Grebner, C. W. Hall, and E. F. Neufeld, Glycosylation of serine residues by a uridine diphosphate-xylose: protein xylosyltransferase from mouse mastocytoma, Arch. Biochem. Biophys. 116:391 (1966).

70. E. E. Grebner, C. W. Hall, and E. F. Neufeld, Incorporation of D-xylose-^{14}C into glycoprotein by particles from hen oviduct, Biochem. Biophys. Res. Commun. 22:672 (1966).
71. H. C. Robinson, A. Telser, and A. Dorfman, Studies on biosynthesis of the linkage region of chondroitin sulfate-protein complex, Proc. Natl. Acad. Sci. USA 56:1859 (1966).
72. J. H. Kimura and A. I. Caplan, Identification of glycogen as the major xylose acceptor in polysomal preparations from chick embryo cartilage cultures, Arch. Biochem. Biophys. 191:687 (1978).
73. P. Campbell, I. Jacobsson, L. Benzing-Purdie, L. Rodén, and J. H. Fessler, Silk - A new substrate for UDP-xylose:proteoglycan core protein β-D-xylosyltransferase, Anal. Biochem. 137:505 (1984).
74. E. Meezan, S. Ananth, S. Siegal, S. Manzella, D. Pillion, and L. Rodén, Glucose transfer to a glycogen-like glycoprotein from rat kidney, J. Cell Biol. 107:191a (1988).
75. E. Meezan, S. Ananth, D. Pillion, S. Siegal, S. Manzella, P. Campbell, and L. Rodén, Effect of streptozotocin diabetes on a self-glycosylating protein from rat renal cortex, Diabetes 38 (Suppl. 2):214A (1989).
76. S. Manzella, S. Ananth, T. R. Oegema, Jr., L. C. Rosenberg, L. Rodén, and E. Meezan, Specific inhibition by cytidine 5′-diphosphate (CDP) of xylosyl transfer to the self-glycosylating protein glycogenin, Pharmacologist 33:39 (1991).
77. W. J. Whelan, The initiation of glycogen synthesis, Bioessays 5:136 (1986).
78. J. Lomako, W. M. Lomako, and W. J. Whelan, A self-glucosylating protein is the primer for rabbit muscle glycogen biosynthesis, FASEB J. 2:3097 (1988).
79. J. Pitcher, C. Smythe, D. G. Campbell, and P. Cohen, Identification of the 38-kDa subunit of rabbit skeletal muscle glycogen synthase as glycogenin, Eur. J. Biochem. 169:497 (1987).
80. J. Pitcher, C. Smythe, and P. Cohen, Glycogenin is the priming glycosyltransferase required for the initiation of glycogen biogenesis in rabbit skeletal muscle, Eur. J. Biochem. 176:391 (1988).
81. C. Smythe and P. Cohen, The discovery of glycogenin and the primary mechanism for glycogen biogenesis, Eur. J. Biochem. 200:625 (1991).
82. T. Helting, Biosynthesis of heparin. Solubilization and partial purification of uridine diphosphate glucuronic acid: acceptor glucuronosyltransferase from mouse mastocytoma, J. Biol. Chem. 247:4327 (1972).
83. T. Helting and L. Rodén, Biosynthesis of chondroitin sulfate. II. Glucuronosyl transfer in the formation of the carbohydrate-protein linkage region, J. Biol. Chem. 244:2799 (1969).
84. A. E. Brandt, J. Distler, and G. W. Jourdian, Biosynthesis of the chondroitin sulfate-protein linkage region: purification and properties of a glucuronosyltransferase from embryonic chick brain, Proc. Natl. Acad. Sci. USA 64:374 (1969).
85. K. H. Chou, A. A. Ilyas, J. E. Evans, R. H. Quarles, and F. B. Jungalwala, Structure of a glycolipid reacting with monoclonal IgM in neuropathy and with HNK-1, Biochem. Biophys. Res. Commun. 128:383 (1985).
86. T. Ariga, T. Kohriyama, L. Freddo, N. Latov, M. Saito, K. Kon, S. Ando, M. Suzuki, M. E. Hemling, K. L. Rinehart, Jr., S. Kusunoki, and R. K. Yu, Characterization of sulfated glucuronic acid containing glycolipids reacting with IgM M-proteins in patients with neuropathy, J. Biol. Chem. 262:848 (1987).

87. K. K. Das, M. Basu, S. Basu, D. K. H. Chou, and F. B. Jungalwala, Biosynthesis in vitro of GlcAβ1-3nLcOse4Cer by a novel glucuronyltransferase (GlcAT-1) from embryonic chicken brain, J. Biol. Chem. 266:5238 (1991).

88. T. Curenton, G. Ekborg, and L. Rodén, Glucuronosyl transfer to galactose residues in the biosynthesis of HNK-1 antigens and xylose-containing glycosaminoglycans: one or two transferases?, Biochem. Biophys. Res. Commun. 179:416 (1991).

89. A. F. Charles and A. R. Todd, Observations on the structure of the barium salt of heparin, Biochem. J. 34:112 (1940).

CHEMICAL SYNTHESIS AND HEMISYNTHESIS IN THE FIELD OF GLYCOSAMINOGLYCANS

Maurice Petitou
Sanofi Recherche - Centre Choay
9, rue du Président Allende
F-94256 Gentilly Cedex, France

INTRODUCTION

Glycosaminoglycans are complex polysaccharides having several types of functional groups: acetamido, sulfamido, sulfate ester, carboxylate, primary and secondary hydroxyls. The biological function of these glycosaminoglycans is at least in part related to precise distributions of these functional groups on unique oligosaccharide sequences as illustrated by the binding of heparin to antithrombin III which involves a unique pentasaccharide sequence[1].

Most glycosaminoglycans also exhibit pharmacological properties. Some of them are probably linked to their biological function inasmuchas they involve peculiar domains of the molecules (like the anticoagulant activity of heparin and dermatan sulphate). Others have not been assigned to precise structural features, like the antiproliferative activity of heparin (smooth muscle cell growth inhibition) or its antiviral activity (for reviews on the various pharmacological activities of glycosaminoglycans, see references 2 and 3).

All the functional groups mentioned above can either be modified or substituted to introduce structural changes on glycosaminoglycans. Several such changes have been already performed on heparin molecules (O- and/or N-desulfation, esterification, O- and N-acylation) in order to modulate the pharmacological properties. Thus the anticoagulant activity, that is considered as a drawback for the clinical use in other indications, can be practically abolished by chemical degradation of the antithrombin binding site using periodate oxidation, while the antiproliferative activity is preserved[4]. Here we describe specific O-acylation of heparin and dermatan sulphate and the influence of this chemical modification on their pharmacokinetic and antithrombotic properties.

In a different approach, chemical synthesis has been used to reproduce the pentasaccharide sequence of heparin which is responsible for binding to AT III[5]. It has been demonstrated that this compound activates AT III selectively against blood coagulation factor Xa[6] and is an effective antithrombotic agent in animal models[7-9]. Here we show that chemical modification of the functional groups borne by this sequence provides a way to modulate its biological properties and allows an easier preparation of this kind of molecules.

As a whole chemical synthesis in the field of glycosaminoglycans is a means to obtain new pharmacological agents with possible therapeutic use.

SELECTIVE O-ACYLATION OF GLYCOSAMINOGLYCANS

Chemical preparation

As shown in figure 1 several acyl radicals can be introduced on a heparin molecule in different positions, namely 2 and 3 positions of unsulphated uronic acids, 3 position of sulphated uronic acids and glucosamine and 6 position of 6-unsulphated glucosamine units.

Several procedures have been reported in the literature for acylation of glycosaminoglycans[10-13]. However in our hands we found that these procedures either led to partial and hardly reproducible substitution or to products that were at the same time N- as well as O-acylated.

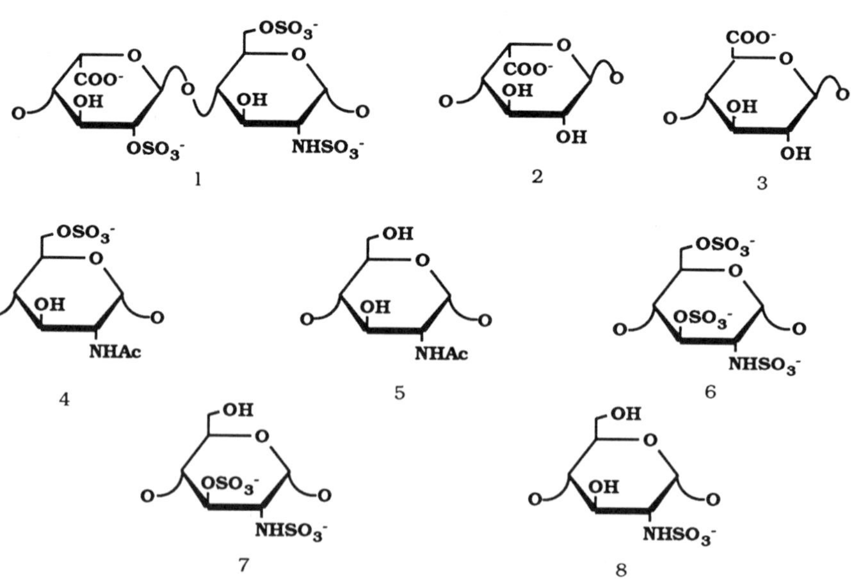

Figure 1. Heparin structure. The major part of the polysaccharide is accounted for by the repetition of disaccharide 1. The monosaccharide units 2-8 are present to a lesser extent. All hydroxyl groups are susceptible to be acylated.

For these reasons and in order to keep the starting polysaccharide intact, we decided to investigate first the behaviour, under acylation conditions of model, well defined monosaccharides bearing the various functional groups occurring in glycosaminoglycans (Petitou et al., submitted for publication). Thus we found that neither acylation by an acid anhydride nor by an acid chloride affected O-sulphate groups. On the contrary, the use of an acid chloride led to more or less N-desulphated N-acylated derivatives from initially N-sulphated glucosamine units. Similarly N-acetylated glucosamine units were partially converted into imides in the presence of acid chlorides. The carboxylate groups of glucuronic acids were prone to conversion into mixed anhydrides particularly when an acid chloride was used.

From these experiments we concluded that smooth O-acylation of glycosaminoglycans should be achievable using a tributyl or a tetrabutylammonium salt of the corresponding glycosaminoglycan, dimethylformamide as solvent, carboxylic acid anhydride as acylating agent, in the presence of a catalytic amount of 4-dimethylaminopyridine (scheme 1).

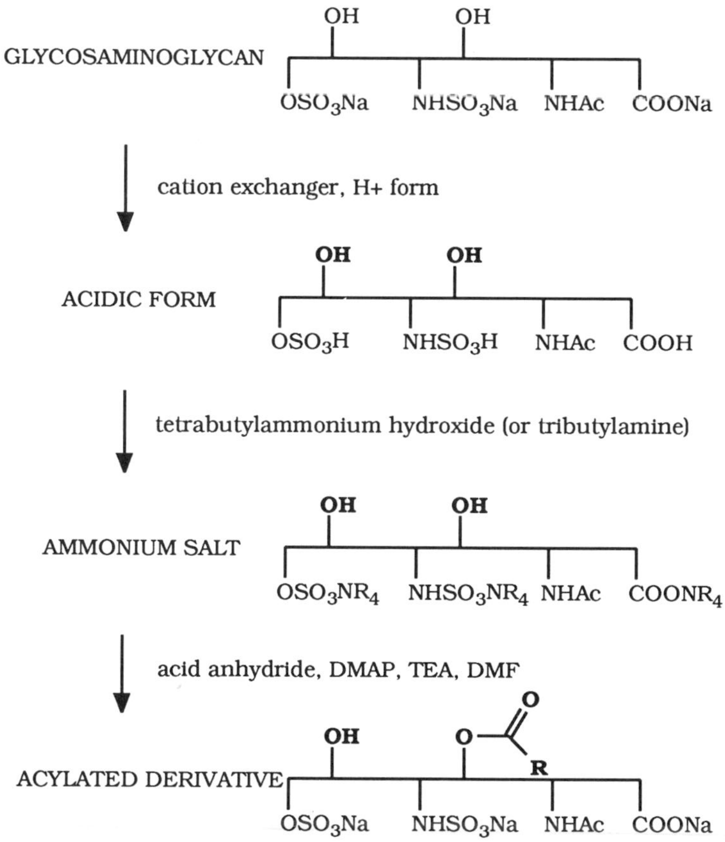

Scheme 1. Selective O-acylation of glycosaminoglycans. The carbohydrate backbone is represented as a straight line. DMAP dimethylaminopyridine. TEA: triethylamine. DMF: dimethylformamide. R depends on the nature of the acid anhydride used.

These experimental conditions were applied to different heparin preparations, low molecular weight heparin, periodate oxidized heparin fragments, dermatan sulphate etc.[14]. As shown in table 1 different reaction conditions led to different degrees of substitution determined by ^{1}H-NMR spectroscopy.

TABLE 1

Degree of acylation of heparin fragments

Time (h)	25°C/5 eq	25°C/10 eq	50°C/5eq
1	-	-	0.93
2	-	-	1.07
4	0.07	-	1.05
8	1.13	1.08	1.39
24	1.30	1.28	1.54
48	1.45	1.53	1.60
72	155	1.57	-
96	1.65	1.54	-

Degree of acylation (number of acyl chains per disaccharide unit) of heparin fragments under different experimental conditions. Reactions were carried out at 25°C with 5 and 10 molar equivalents of anhydride per hydroxyl function and at 50°C with 5 molar equivalents. The degree of acylation was obtained by ^{1}H-NMR by integration of the acyl signals on the one hand and of the ring protons signals taken altogether on the other hand.

The structure of the product was checked by ^{13}C-NMR spectroscopy in order to distinguish N- and O-acylation. This analysis clearly indicated that selective O-substitution was obtained under our experimental conditions (Petitou et al., submitted for publication).

Influence of O-butanoylation on the biological properties of a low molecular weight heparin preparation (Saivin et al., submitted for publication).

The low molecular weight heparin CY 216 was converted as described above into its butanoylated derivative. An average of 1.7 acyl chains were introduced per disaccharide unit. Investigation of the biological properties of the derivatized CY 216 compared to that of the parent compound showed that in a purified system their ability to catalyse thrombin and factor Xa was comparable. After bolus and continuous intravenous injection to rabbits, the clearances of the two activities of butanoylated CY 216 were on average half the corresponding values of CY 216. After subcutaneous injection the bioavailability of both compounds were comparable. The butanoylated derivative was as potent as CY 216 in preventing venous thrombosis in the thromboplastin-Wessler model and the duration of the antithrombotic effect was longer than that of the parent compound. Thus, after a subcutaneous injection of 31.3 mg/kg, 90 ± 6% thrombose prevention could be observed after 16 hours, compared to 9 ± 26% for the parent unsubstituted preparation. Therefore this kind of derivative constitutes a long lasting form of low molecular weight heparin.

Influence of O-succinylation on the biological properties of a dermatan sulphate derivative (Saivin et al., submitted for publication)

Several attempts to improve the specific anticoagulant activity of dermatan sulphate by oversulphation have failed, mainly because the products thus obtained had pronounced bleeding properties[15]. We have prepared a succinylated dermatan sulphate derivative in which the anionicity is also increased, but by carboxylate groups instead of sulphates. This compound contained 0.6 succinyl residues per basic disaccharide unit. In vitro it was two times more potent than the parent dermatan sulphate in catalysing the inhibition of thrombin in the presence of heparin cofactor II. After bolus injection it was also two times more potent to prevent experimental venous thrombosis in a Wessler model. In contrast with oversulphated dermatan sulphate preparations the bleeding effect as determined in the rat tail transsection model was not increased after succinylation. Thus this chemical modification leads to a new dermatan sulphate derivative with an improved benefit/haemorragic risk ratio.

Other pharmacological activities of glycosaminoglycans like their activity on cell growth and their antiviral activity can also be improved by O-acylation (Barzu et al., de Clercq et al., to be published).

O-ALKYLATED O-SULPHATED HEPARIN OLIGOSACCHARIDE FRAGMENTS*

The antithrombotic properties of heparin pentasaccharide fragments have been clearly demonstrated in animal models of venous thrombosis and these purely antifactor Xa compounds are new candidates for the development of antithrombotic drugs. The "natural" sequence (**1** fig 2) was the first to be investigated[7-9] but it is now challenged[16] by the more potent analog **2** which has an extra sulphate group at the 3 position of the reducing end glucosamine unit[17]. As shown in table 2 this extra sulphate confers a higher affinity for AT III which results in a longer duration of action in plasma[16].

*This part of our research is being carried out in collaboration with Organon International bv, Oss, The Netherlands; and is funded by the EEC Eurêka programme (project EU 237).

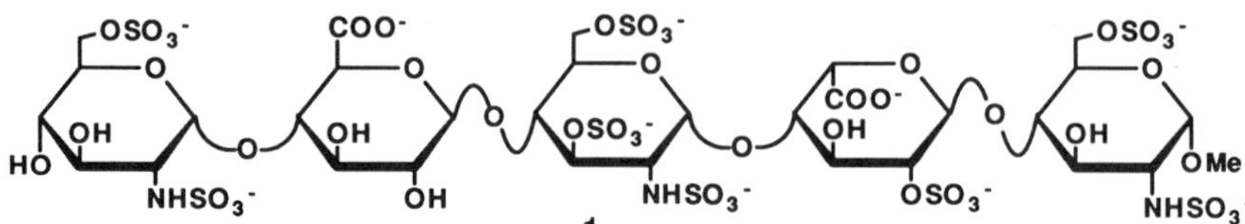

1

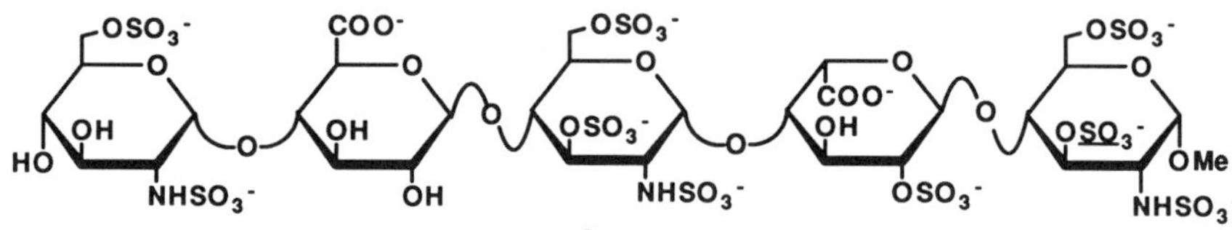

2

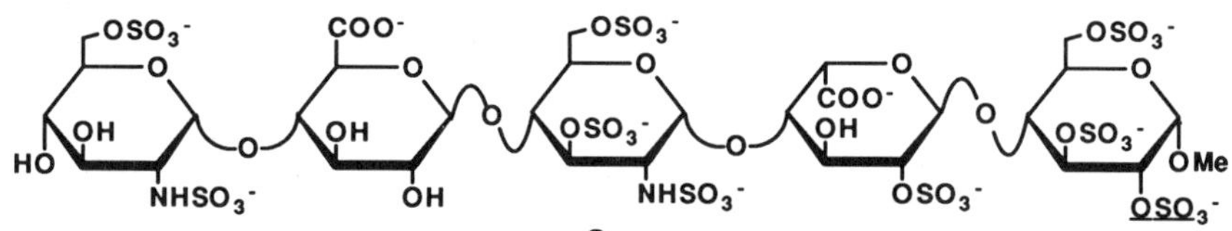

3

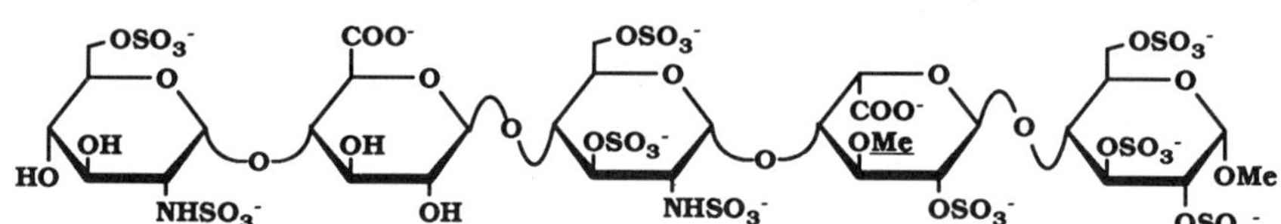

4

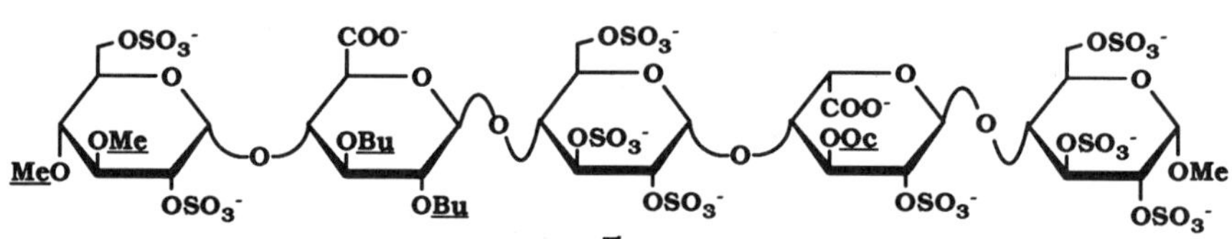

5

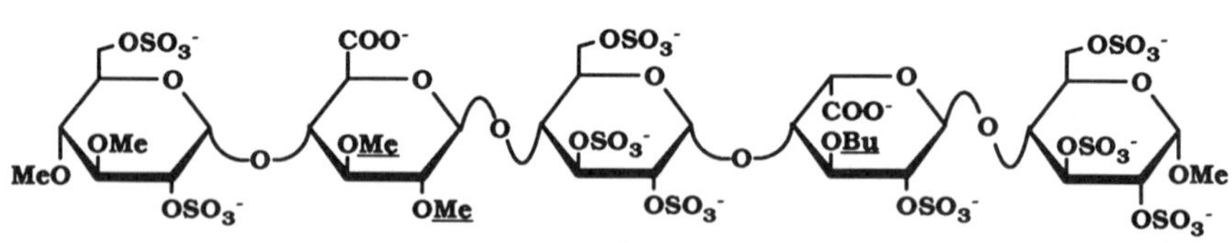

6

Figure 2 : The "basic" pentasaccharide sequence (1) and its analogs. Modifications from one molecule to the next are underlined on the formula. Me = methyl; Bu = butyl; Oc = octyl.

However, the preparation of this kind of products involves a large number of steps, which is a drawback for their development as drugs. In order to overcome this difficulty we first analysed the precise role of the different structural features in **1** since minor modifications of the structure could result in a dramatic simplification of the chemical synthesis. No real improvement regarding our goal resulted from this research [5].

In a second approach we investigated the role of N-sulphate groups with the view to replace them by O-sulphates. This would avoid the need to discriminate between both types of sulphates and at the same time this would eliminate the rather labile N-sulphates. In this respect we first synthesized compound **3** containing a glucose unit in place of the reducing end glucosamine[18]. The biological properties of **3** were identical with that of **2**.

In a third step we investigated the influence of alkyl substitution of hydroxyl groups on the biological properties, having in mind that the presence of alkyl groups in the final compound would obviate temporary protection of numerous alcohol functions and thereby simplify the chemical procedure. Several compounds like **4** (in which only the 3 position of L-iduronic acid unit is substituted by a methyl ether) were thus prepared and their biological properties were shown to be preserved.

TABLE 2

Properties of synthetic oligosaccharides

Compound (fig. 2)	K_D/ATIII nM	Axa (u/mg)	T1/2 (hours)	ED_{50} (µg/kg)
1	50.0	700	1.45	-
2	1.3	1260	16.06	14.8
3	3.0	1302	nd	nd
4	4.0	1110	nd	nd
5	1.6	700	nd	58.0
6	0.3	1080	26.86	19.8

Binding constants were determined by fluorescence spectroscopy [19]. Antifactor Xa was measured according to the method of Teien and Lie [20]. Half lives were determined in rabbit after i.v. bolus injection of 500 µg/kg and antithrombotic activity in the Wessler model in rat using thromboplastin as thrombogenic agent. nd : not determined.

Finally combining both types of modifications, several analogs of the potent pentasaccharide 2 were prepared in which all the N-sulphate groups have been replaced by O-sulphate and all hydroxyl functions by various alkyl groups. It is clear from the data shown in table 2 that such drastic changes are compatible with the activity and that the nature as well as the place of the substituents have an influence on the final properties of the products.

CONCLUSION

The examples presented in this publication further illustrate that the biological properties of glycosaminoglycans can be modulated by appropriate chemical modifications. It is shown also that chemical synthesis provides new efficient ways to obtain biologically active analogs of specific glycosaminoglycan sequences otherwise practically inaccessible, thus offering new active substances for drug development.

ACKNOWLEDGEMENTS

The data reported here were obtained in collaboration with my colleagues whose names appear in the references. Pharmacological data presented on O-acylated derivatives were obtained by Pr Boneu, Dr Sié and their coworkers. Products and data in the O-alkylated-O-sulphated Heparin Oligosaccharides series were obtained in collaboration with the Organon Medicinal Chemistry Group (C.van Boeckel and coworkers).

REFERENCES

1. For a review, see I. Björk and U. Lindahl, Mechanism of the anticoagulant action of heparin, Mol. Cell. Biochem. 48: 161 (1982)

2. D.A. Lane and U. Lindahl, "Heparin: Chemical and Biological Properties, Clinical Applications", E. Arnold, London (1989)

3. F.A. Ofosu, I. Danishefsky and J. Hirsh, "Heparin and Related Polysaccharides: Structure and Activities", Ann. NY Acad. Sci. vol 556 (1989)

4. J.C. Lormeau, M. Petitou and J. Choay, Low-molecular-weight heparins with a regular structure, their preparation and biological uses, Chem.Abstr. 111 (1989) abstr 17717x

5. For a review, see M. Petitou, Chemical synthesis of heparin, in: "Heparin: chemical and biological properties, clinical applications", D.A. Lane and U. Lindahl, Eds.,
E. Arnold, London (1989)

6. J. Choay, M. Petitou, J.C. Lormeau, P. Sinay, B. Casu, G. Gatti, Structure-activity relationship in heparin: a synthetic pentasaccharide with high affinity for antithrombin III and eliciting high anti-factor Xa activity, Biochem. Biophys. Res. Commun. 116: 492 (1983)

7. J. Walenga, J. Fareed, M. Petitou, M. Samama, J.C. Lormeau and J. Choay, Intravenous antithrombotic activity of a synthetic heparin pentasaccharide in a human serum induced stasis thrombosis model, Thromb. Res. 43: 243 (1986)

8. J.M. Walenga, M. Petitou, J.C. Lormeau, M. Samama, J. Fareed and J. Choay, Antithrombotic activity of a synthetic heparin pentasaccharide in a rabbit stasis thrombosis model using different thrombogenic challenges, Thromb. Res. 46: 187 (1987)

9. P.M.J. Hobbelen, T.G. van Dinther, G.M.T. Vogel, C.A.A. van Boeckel, H.C.T. Moelker and D.G. Meuleman, Pharmacological profile of the chemically synthesized antithrombin III binding fragment of heparin (pentasaccharide) in rats, Thromb. Haemost. 63: 265 (1990)

10. H.J. Bell and L.B. Jaques, An acetyl derivative of heparin, Can. J. Res. 25B: 472 (1947)

11. O. Akira, Y. Eiji, T. Katsuhiko, Heparin esters, Chem. Abstr. 88 (1978) abstr 41689s

12. K.M. Foley, G.C. Campbell, A. Eduardo, Heparin esters, Chem. Abstr. 112 (1990) abstr 191955d

13. J. Mardiguian and P. Fournier, Heparin esters with prolonged coagulation-inhibiting action, Chem. Abstr. 78 (1973) abstr 140393r

14. M. Petitou, J.C. Lormeau and J. Choay, Selective O-acylation of glycosaminoglycans for use in pharmaceuticals, Chem. Abstr. 113 (1990) abstr 174401y

15. J.van Ryn-McKenna, F.A. Ofosu, J. Hirsh and M.R. Buchanan, Antithrombotic and bleeding effects of glycosaminoglycans with different degrees of sulphation, Brit. J. Haematol.71: 265 (1989)

16. D.G. Meuleman, P.M.J. Hobbelen, T.G. van Dinther, G.M.T. Vogel, C.A.A. van Boeckel and H.C.T. Moelker, Antifactor Xa activity and antithrombotic activity in rats of structural analogues of the minimum antithrombin III binding sequence: discovery of compounds with a longer duration of action than of the natural pentasaccharide, Seminars Thromb. Hemost. 17: 112 (1991)

17. C.A.A. van Boeckel, T. Beetz and S.F. van Aelst, Synthesis of a potent antithrombin activating pentasaccharide: A new heparin-like fragment containing two 3-O-sulphated glucosamines, Tetrahedron Lett. 29: 803 (1988)

18. M. Petitou, G. Jaurand, M. Derrien, P. Duchaussoy and J. Choay, A new highly potent heparin-like pentasaccharide fragment containing a glucose residue instead of a glucosamine, BioMed. Chem. Lett. 1: 95 (1991)

19. T. Barzu, M. Petitou, G. Jaurand, J.C. Lormeau, J.P. Hérault and J. Choay, Binding to AT III of synthetic oligosaccharides derived from the high affinity pentasaccharide sequence of heparin, Thromb. Haemost. 65: 645 (1991)

20. A.N. Teien and M. Lie, Evaluation of an amidolytic heparin assay method: increased sensivity by adding purified AT III, Thromb. Res. 10: 399 (1977)

STRUCTURAL ANALYSIS OF PERIODATE-OXIDIZED HEPARIN

H. Edward Conrad and Yuchuan Guo

Glycomed, Inc.

Alameda, CA 94501, USA

ABSTRACT

Treatment of heparin with HONO at pH 1.5 cleaves the polymer at N-sulfated, but not at N-acetylated GlcN residues, and yields di- and tetrasaccharides. The $GlcNSO_3$ residues at the sites of cleavage are converted into anhydromannose (AMan) residues. Reduction of heparin cleavage products with NaB^3H_4 yields mixtures of di- and tetrasaccharides with reducing terminal [^{3}H]anhydromannitol residues. The identification and quantification of these oligosaccharides by HPLC procedures have been described[1,2] These procedures have been used to determine the rates of periodate oxidation of the susceptible unsulfated GlcA and IdoA residues in heparin by measuring the disappearance of the di- and tetrasaccharides that contain GlcA and IdoA. Complete oxidation with IO_4^- results in the total loss of the unsulfated uronic acid-containing oligosaccharides, but kinetic studies reported here show that IdoA is oxidized much more rapidly than the major fraction of the GlcA under all reaction conditions. As the pH is lowered from 7 down to 3, the overall rate of the oxidation slows markedly, but the relative rates of GlcA and IdoA oxidation do not change. The slow rate of oxidation of GlcA residues at all pH's yields oxidation products early in the reaction progress in which all of the unsulfated IdoA residues are oxidized while 70-80 % of the pH 1.5 nitrous acid-releasable GlcA→AMan(3,6-$(SO_4)_2$) are retained. The anticoagulant activity (APTT) of the partially oxidized product is reduced from 170 IU/mg to 38 IU/mg. Further studies show that the GlcA residue in the antithrombin III binding pentasaccharide is oxidized much more rapidly than the bulk of the GlcA residues in heparin. The results suggest that heparin contains GlcA→AMan(3,6-$(SO_4)_2$) sequences that lie outside of the antithrombin-binding pentasaccharide.

Heparin and Related Polysaccharides
Edited by D.A. Lane *et al.*, Plenum Press, New York, 1992

PERIODATE OXIDATION OF THE ATIII-BINDING SEQUENCE OF HEPARIN

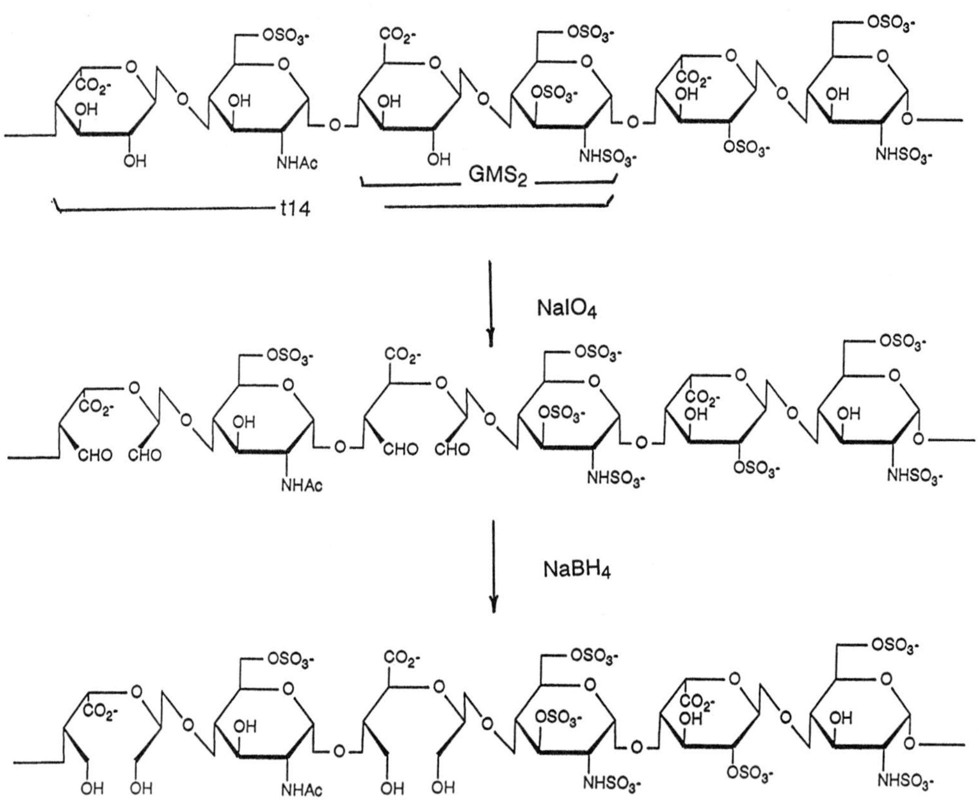

Figure 1

INTRODUCTION

Periodate oxidation of heparin was described in early literature[3], even before some of the important structural features of heparin were recognized. More recent work[4,5] showed that polyanionic polysaccharides, including heparin, were oxidized more slowly than uncharged polysaccharides and that the rates of oxidation of these charged substrates were increased by addition of salt to suppress the effect of the negative electrostatic field. Although these workers did not distinguish the rates of oxidation of the GlcA and the IdoA residues in heparin, they did show that dermatan SO_4, which contained largely IdoA residues, was oxidized more rapidly than chondroitin SO_4's, which contained only GlcA residues, an observation confirmed by Fransson[6] who showed in addition that the GlcA residues in these chondroitin SO_4/dermatan SO_4 polymers were oxidized at pH 7 and 37 ^oC, but not at pH 3 and 4 ^oC, whereas the IdoA-containing polymers were oxidized under both conditions. However, Fransson and co-workers[7,8] found that, when heparin was oxidized under these latter conditions, GlcA, but not IdoA was oxidized, whereas both uronic acids were oxidized at pH 7 and 37 ^oC. Furthermore, oxidation at pH 3 and 4 ^oC did not destroy the anticoagulant activity, an observation that was not consistent with the fact that the antithrombin III-binding sequence of heparin contains a critical GlcA residue[9]. The present study re-examines this apparent contradiction.

METHODS

<u>Periodate oxidation</u>. Hog mucosa heparin (anticoagulant activity = 170 USP units/mg) was oxidized with $NaIO_4$ at 4 ^oC in 50 mM Na citrate buffer (pH 3.0) or at 37 ^oC in 50 mM Na phosphate (pH 6.5), essentially as described by Fransson *et al*[7,8]. Aliquots were removed from the reaction mixtures at intervals and treated with ethylene glycol to destroy unreacted $NaIO_4$. The samples were dialyzed vs water, dried by lyophilization, and reduced with $NaBH_4$.

<u>Analysis of oxidized heparin samples</u>. The disaccharide and tetrasaccharide compositions of the original heparin and the oxidized samples were measured to follow the destruction of the uronic acid residues of the heparin during the oxidation. Each sample was treated with nitrous acid at pH 1.5[1,2] and the resulting di- and trisaccharides were quantified using the reversed phase ion pairing HPLC method described previously[2]. Anticoagulant activity was determined by APTT and anti-Xa assays. These assays were kindly performed for us by Dr. Betty Yan, Lilly Research Laboratories, Indianapolis IN, USA.

RESULTS AND DISCUSSION

Figure 1 shows the reaction sequence for the $NaIO_4$ oxidation of a segment of the heparin chain containing the antithrombin III binding pentasaccharide. This Figure shows the abbreviations used below for the trisulfated disaccharide that is critical for the anticoagulant activity, and a tetrasaccharide, t14, that contains this disaccharide[1]. The trisulfated disaccharide, when released from the N-deacetylated heparin by treatment with nitrous acid, yields GlcA→AMan(3,6-$(SO_4)_2$), abbreviated here as GMS_2. Note that direct nitrous acid cleavage of this segment of heparin yields the tetrasaccharide, t14, and not free GMS_2. On the other hand, when the GMS_2 in heparin is situated in a position with a $GlcNSO_3$ residue linked to the C4 position of its GlcA, it will be released by nitrous acid without prior N-deacetylation as the free disaccharide.

Figure 2 shows a comparison of the rates of disappearance of the major disaccharide units of heparin at pH's 3.0 (4 ^{o}C) and 6.5 (37 ^{o}C). (Other abbreviations in Figures 2-4 are:

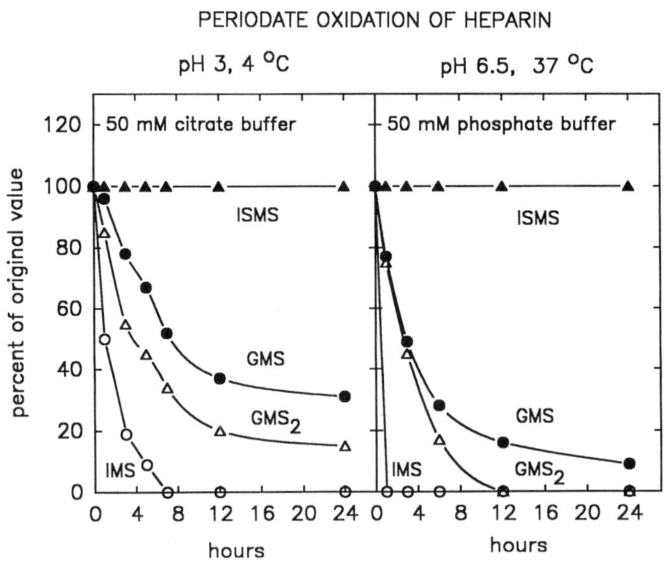

Figure 2

ISMS, IdoA(2-SO_4)→AMan(6-SO_4); ISM, IdoA(2-SO_4→AMan; IMS, IdoA→AMan(6-SO_4); and GMS, GlcA→AMan(6-SO_4)). For these measurements, the oxidized samples were N-deacetylated and then cleaved with nitrous acid at both pH's 1.5 and 4.0 to give total disaccharide release. Since all of the GlcN residues in heparin are resistant to $NaIO_4$ oxidation, the disappearance of each susceptible disaccharide is due to the oxidation of its uronic acid residue. Only those uronic acid residues that lack a SO_4 substituent at C2 are susceptible to IO_4^-. The results in Figure 2 show (a) that the overall oxidation of susceptible uronic acids proceeds more rapidly at pH 6.5 and 37 ^{o}C than at pH 3.0 and 4 ^{o}C, and (b) that under both oxidation conditions the unsulfated IdoA residues are oxidized much more rapidly than the unsulfated GlcA residues. A study of heparin oxidation at pH 5 and 4 ^{o}C (not shown) gave rates similar to those observed at pH 6.5. Since the ratios of the rates of IdoA and GlcA oxidation were similar at both pH's 3.0 and 6.5, the pH 3 conditions were chosen for further examination of the oxidation of heparin, since, under the latter conditions, the progression of the reaction could be observed over a more extended time interval, allowing better control of the reaction.

The slow and incomplete oxidation of GMS_2 at pH 3 suggests than the anticoagulant activity of heparin should also be lost slowly and incompletely, as reported previously[10]. To explore this further, the rates of loss of anticoagulant activity and GMS_2 were compared, as shown in Figure 3. The results shown in the upper left panel (total

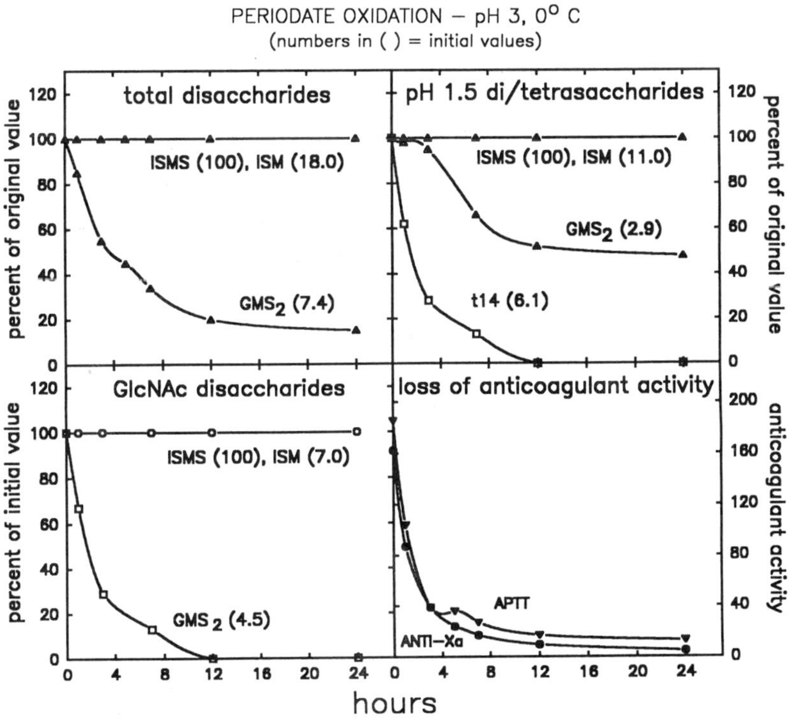

Figure 3

disaccharides) are the same as those shown in Figure 2, whereas the GMS_2 and t14 that were released by pH 1.5 nitrous acid treatment without prior N-deacetylation are shown in the upper right panel. The GMS_2 formed under the latter conditions represents "$GlcNSO_3$-linked GMS_2"; i.e., this GMS_2 is formed only when there is a $GlcNSO_3$ residue linked to the GlcA. Finally, the subtraction of the $GlcNSO_3$-linked GMS_2 from the total GMS_2 yields the GlcNAc-linked GMS_2, shown in the lower left panel, all or most of which is derived from the antithrombin III-binding pentasaccharide. The results in Figure 3 show that theGlcNAc-linked GMS_2 and the $GlcNSO_3$-linked GMS_2 are oxidized at markedly different rates, and that the rate of loss of anticoagulant activity parallels the rates of disappearance of only the GlcNAc-linked GMS_2 and the t14 tetrasaccharide (which contains the GlcNAc-linked GMS_2), and not the $GlcNSO_3$-linked GMS_2.

The work reported here was facilitated by the recently developed methods[1,2] for quantification of di- and tetrasaccharides formed when heparin is cleaved with nitrous acid before or after N-deacetylation. These approaches give a more precise measure of uronic acid residues as the oxidation proceeds than the previously used colorimetric[6-8] or spectrophotometric measurements[11]. Thus, it was possible to demonstrate that IdoA in heparin is indeed oxidized much more rapidly than most of the GlcA, just as reported previously for the comparison of dermatan SO_4 and chondroitin SO_4. Furthermore, the rate of IO_4^- oxidation of GlcA is markedly influenced by the substitution on the amino group of the GlcN residue that is linked to C4 of the GlcA, as previously suggested[8]. Finally, the oxidation of the $GlcNSO_3$-linked GlcA residues that yield GMS_2 on direct pH 1.5 nitrous acid treatment appears to be (at least) biphasic, suggesting that this type of disaccharide unit occurs in several different environments in heparin, and that one or more of these environments is not in an antithrombin III binding pentasaccharide. Interestingly, the final slow rate of GMS_2 oxidation parallels the slow loss of the residual anticoagulant activity.

REFERENCES

1. M. J. Bienkowski and H. E. Conrad (1985) J. Biol. Chem. 260,356-365.
2. Y. Guo and H. E.Conrad (1989) Analyt. Biochem.176, 96-104.
3. A. B. Foster, R. Harrison, T. D. Inch, M. Stacey, and J. M. Webber (1963) J. Chem. Soc. 2279-2287).
4. J. E. Scott and R. J. Harbinson (1963) Histochemie 14, 215-220.
5. J. E. Scott and R. J. Harbinson (1969) Histochemie 19, 155-161.
6. L.-A. Fransson (1974) Carbohyd. Res. 36, 339-348.
7. L.-A. Fransson (1978) Carbohyd. Res. 62, 235-244.
8. L.-A. Fransson and W. Lewis (1979) FEBS Lett. 97, 119-123.
9. M. Petitou, J.-P. Lormeau, J. Choay (1991) Suppl. to Nature 350, 30-33.
10. L.-A. Fransson, A. Malmstrom, and I. Sjoberg, (1980) Carbohyd. Res. 80, 131-145.
11. B. Casu, G. Diamantini, G. Fedeli, M. Mantovani, P. Oreste, R. Pescador, R. Porta, G. Prino, G. Torri, and G. Zoppetti (1986) Arzneim-Forsch./Drug Res. 36, 637-642.

NEW METHODOLOGIES IN HEPARIN STRUCTURE ANALYSIS AND THE GENERATION OF LMW HEPARINS

Robert J. Linhardt, Hui-ming Wang and Stephen A. Ampofo

Division of Medicinal and Natural Products Chemistry
College of Pharmacy, University of Iowa
Iowa City, IA 52242, USA

INTRODUCTION

This chapter examines the methods used to prepare heparin, low molecular weight (LMW) heparins, and immobilized heparins. These heparins and heparin derivatives are polydisperse heterogeneous mixtures that require structural characterization. Analyses rely on integrated enzymatic, chemical, chromatographic, electrophoretic and spectroscopic techniques. The goal is to establish the physical-chemical properties of a heparin preparation as well as the exact composition, distribution and sequence of the polysaccharide chains that are present. Ultrasensitive analytical techniques are also described for the analysis of heparin in tissues and biological fluids.

Heparin Structure

Although heparin has been used clinically as an anticoagulant for 56 years, its precise structure remains unknown.[1] The structural complexity of heparin can be considered at several levels. At the proteoglycan (PG) level the number, position and nature of the polysaccharide chains attached to the protein core can be examined. At the level of glycosaminoglycan (GAG) heparin some of the structural complexity results from its polydispersity. GAG heparin has a MW ranging from 5-40 kD (degree of polymerization (DP) 10-80) with an MW (average) of 13 kD. Even the heparin chain corresponding to the most prevalent DP represents a mere 5 mole% of a typical GAG heparin preparation.[2] GAG heparin has a second level of structural complexity associated with its primary structure or sequence.[3] The structural features of PG heparin have been primarily established by studying its biosynthesis.[4] The structural features of GAG heparin have relied on chemical, enzymatic, and spectroscopic techniques. Recent efforts in our laboratory have sought to use techniques originally developed to sequence nucleic acids and proteins to establish the sequence of heparin.

Heparin's Biological Activities

PG heparin is primarily found in the granules of mast cells[5] and there is no direct

Heparin and Related Polysaccharides
Edited by D.A. Lane *et al.*, Plenum Press, New York, 1992

evidence that endogenous heparin plays a role in maintaining blood flow through the vasculature.[6] A structurally related PG, heparan sulfate, from the endothelial surface has "heparin-like" sequences and may be responsible for a variety of heparin related activities[7,8] including the binding of growth factors,[9] hormones,[10] and various regulators.

Although the biological roles of endogenous heparin is not completely understood, this has not precluded its use for a variety of medical applications. Heparin, the most commonly used anticoagulant, is administered *iv.* during most extracorporeal therapy and a variety of other surgical procedures.[7,8] It is also given by the *sc.* route and, despite its reduced bioavailability, a low level of heparinization can often be maintained for prolonged periods of time.[11] Although it would be desirable to prepare an orally active heparin, which could be administered outside a hospital setting, no such oral formulation is currently available.[12,13] Heparin's major side-effect, hemorrhagic complications,[14] is closely linked to its anticoagulant activity.[8] Heparin also causes the release of lipoprotein lipase from the endothelium[15,16] and inhibits the proliferation of smooth muscle cells, important components of atherogenesis.[17] Heparan sulfate, present on the endothelium, probably plays a physiological role in the regulation of smooth muscle proliferation.[15] Heparin's principal location in man is the granules of tissue mast cells and basophils. Because heparin's primary location is so closely linked to the immune response, its ability to regulate complement has become an active area of interest.[18,19]

Heparin may play a variety of roles in angiogenesis. Immediately before capillary ingrowth, mast cells, containing heparin, congregate. Heparin from these mast cells can stimulate endothelial cell migration.[20] Heparin may increase the activity, stability or binding of growth factors such as fibroblast growth factor (FGF), and endothelial cell growth factor (ECGF).[21] Heparin and heparin oligosaccharides also inhibit angiogenesis in the presence of angiostatic steroids.[,22,23] To fully understand the role of heparin in the process of angiogenesis, specific saccharide sequences within the heparin polymer which bind growth factors need to be elucidated.

PREPARATION OF HEPARIN AND HEPARIN DERIVATIVES

Heparin's biological activities are primarily mediated through its binding and regulation of proteins. To substitute for heparin, LMW heparin, heparin oligosaccharides and heparinized surfaces must bind to these proteins at the same sites and regulate these same activities. Consensus peptides (small linear binding sequences) are found in many heparin-binding proteins.[24] Heparin's binding is primarily through electrostatic interactions and depends on its high charge density. Only the heparin ATIII binding site has been sufficiently studied to develop a well defined structure activity relationship and this is the only heparin-protein interaction that has been demonstrated to be restricted to a specific saccharide sequence. An increased understanding of heparin's interaction with other proteins to which it binds and regulates will be required to systematically develop heparin substitutes.

Extraction of Heparin from Tissues

Glycosaminoglycan heparin is prepared from animal tissues that are rich in mast cells.[25] A human heparin has also been recently prepared from a vascular tumor containing a high concentration of mast cells.[26] The basic approach for the commercial processing of heparin involves proteolytic treatment, extraction and complexing with ion pairing reagents followed by fractional precipitation. Treatment with base to remove residual protein and/or bleaching is commonly used to prepare the drug form of heparin. The major criteria for purity remains a high specific activity expressed as USP or BP units per milligram.[25] Heparin prepared using mild conditions is indistinguishable from the commercial one.[25]

Low Molecular Weight Heparins

The most important of the heparin derivatives, are the LMW heparins. These typically consist of GAG chains of molecular weights ranging from 2 to 8 kD (average of 5 kD).[27] Although the major application of LMW heparin has been as antithrombotic agents, other uses can be envisaged that exploit other biological activities. These alternative applications may also include smaller heparin oligosaccharides. While LMW heparin is a polydisperse mixture, small oligosaccharides can be prepared that are pure, discrete chemical entities of defined structure.[8]

GAG heparin is polydisperse and contains <15 wt% chains having MW$\leq$5 kD.[28,29] The content of low molecular weight chains can be enriched[28,30] but because of their small percentage in heparin, yields are low. Heparin can be depolymerized by hydrolysis using mineral acids but this results in desulfation and deacetylation. The loss of labile N-sulfate groups during acid hydrolysis can be circumvented by hydrolyzing glycosidic linkages in 94% sulfuric acid and 5% chlorosulfonic acid at -4°C.[31] This results in a supersulfated LMW heparin which retains substantial pharmacological activity.[32] On esterification and mild treatment with base heparin undergoes facile β-elimination with cleavage of this glycosidic linkage resulting in the formation of a Δ4,5 unsaturated uronic acid.[27,33] This method is used to prepare a commercial LMW heparin, PK10169 (Enoxaparin/Lovenox Pharmuca, France).[27,33]

Treatment of heparin with nitrous acid at pH 1.5 cleaves glycosidic linkages to glucosamine and N-sulfated glucosamine residues. The 2,5-anhydro-D-mannose residue at the reducing end of the oligosaccharide is usually stabilized by reduction with $NaBH_4$ to anhydromanitol.[34] Nitrous acid treatment for the partial depolymerization of heparin has been used to prepare several commercial LMW heparins including Fraxiparin and CY222 (Sanofi, France), Fragmin (KabiVitrum, Sweden), and standard LMW-heparin (NIBSC, England).[27]

The unsulfated uronic acid residues in heparin are susceptible to periodate oxidation.[35,36] The small number of these residues reduces heparin's susceptibility towards this method of depolymerization. Hydrogen peroxide at low pH oxidatively depolymerizes heparinic acid.[37] RD-heparin (Hepar, US) currently in clinical trials is prepared by this process. Heparin can be depolymerized with hydroxyl radicals prepared with ascorbate, Cu^{++}, and hydrogen peroxide at pH 6.[38] The commercial LMW heparin Floxum (Opocrin, Italy) is made using this process.[8,27] Oxidative reductive depolymerization, has been recently reported that uses oxygen at neutral pH in the presence of Fe^{++} to depolymerize heparin.[39] Hypochlorous acid can be used in the absence of metals to generate oxygen radicals to prepare LMW heparin.[40] Little or no loss of sulfation is observed if these processes are carefully controlled and significant anticoagulant activity remains.

Heparin can be depolymerized under mild conditions (pH 6-8, 20-40°C) using polysaccharide lyases from *Flavobacterium heparinum*. The glycosidic linkages between hexosamine and uronic acid are broken resulting in an unsaturated uronic acid at the non-reducing end.[41] Purified heparin lyase has been used to prepare LMW heparins with retention of anticoagulant activity.[41] A commercial LMW heparin, Logiparin (Novo, Denmark), is also prepared using this method. A major disadvantage of this method is the unusual unsaturated uronic acid residue found in the non-reducing end of the chains.[42] This residue may affect the rate of *in vivo* biotransformation of these chains. It is possible to remove this unusual residue by treatment of the product with ozone under mildly acidic conditions.[43]

Immobilized Heparins

Heparin has been covalently immobilized to polymeric matrices using a variety of chemical linkages[44,45] such as (i) an amide through its carboxyl groups by carbodiimide activation; (ii) an amide to carboxylated supports or to cyanogen bromide activated supports through free amino groups prepared by N-desulfation or N-deacetylation or by attaching an amino containing coupling arm; (iii) an ether through hydroxyl groups reacting with epoxidated supports; (iv) linking heparin at its reducing end by reductive amination or through an anhydromannose residue; and (v) through direct polymerization of a vinyl or acrylated heparin derivative to form an insoluble polymer. Although all of these methods are effective at immobilizing heparin, matrices prepared by each method have different physical, chemical and biological properties. There are a few important aspects with respect to an immobilized heparin that need to be considered, such as; (i) chemical and enzymatic stability (particularly important for biomedical device applications)[46]; (ii) the quantity of the heparin that can be coupled; (iii) the maintenance of biological activity; (iv) the orientation of the coupled chains; (v) the density and uniformity of polymer coverage; (vi) the physical, chemical and biological properties of the underlying polymer.

DETERMINING THE MOLECULAR WEIGHT OF HEPARINS

The molecular weight of LMW heparins affect their biological activity and bioavailability.[47] The molecular weight of heparin has been estimated using a number of different physical methods. The number of average molecular weight (M_n), weight average molecular weight (M_W), and polydispersity of heparin and LMW heparin are typically obtained using gel permeation HPLC.[48] The molecular weight of heparin has been estimated by viscosimetry and ultracentrifugation.[49] Light scattering methods used to determine heparin's molecular weight are difficult to apply to the study of LMW heparins.[48] A new approach towards analyzing the molecular weight distribution of heparin and LMW heparins uses gradient polyacrylamide gel electrophoresis (PAGE).[27,50]

CARBOHYDRATE COMPOSITION AND OLIGOSACCHARIDE MAPPING

The glucosamine residue in heparin is primarily N-sulfated (D-GlcNS) and the uronic acid is L-iduronic acid-2-sulfate (L-IdoA2S). The most frequently occurring disaccharide sequence in heparin is →4)-α-D-GlcNS6S(1→4)-α-L-IdoA2S(1→. Heparan sulfate contains primarily N-acetylated glucosamine (D-GlcNAc) and glucuronic acid (D-GlcA) and its most common disaccharide sequence is →4)-α-D-GlcNAc(1→4)-ß-D-GlcA(1→.

Heparinase (EC 4.2.2.7 or heparin lyase I), heparin lyase II (no EC number), and heparitinase (EC 4.2.2.8 or heparin lyase III) each acts eliminatively to cleave specific linkages[51,52,53] within heparin and heparan sulfate and afford a mixture of oligosaccharide products. These oligosaccharide products can be analyzed by strong anion exchange (SAX)-HPLC,[27,51,54] reversed phase ion pairing (RPIP) HPLC[55] and by PAGE.[27,50,56] The analysis of GAG-derived oligosaccharides prepared using the heparin lyases has been termed oligosaccharide mapping.[54,56]

Recently a rapid, high resolution and highly sensitive method was developed to determine the disaccharide composition of chondroitin and dermatan sulfates.[57,58] These GAGs were first depolymerized to a mixture of disaccharides using chondroitin lyases and then analyzed by capillary zone electrophoresis (CZE). This approach has also been extended to the analysis of the disaccharide products obtained on treating heparin and heparan sulfate with lyases.[59] The disaccharide products were identified using newly available commercial disaccharide standards and structure was confirmed by fast atom bombardment mass spectrometry (FAB-MS) and their purity ($>90\%$) assessed using nuclear magnetic resonance spectroscopy (NMR). An equimolar mixture of the eight heparin/heparan sulfate derived disaccharide standards could be separated in a 60 cm capillary with a detection sensitivity of 29.3 pg (52 fmol).[59]

Oligosaccharide mapping of heparins and LMW heparins can provide structural information necessary to begin to establish a structure-activity relationships. Oligosaccharide mapping of acidic polysaccharides is an approach comparable to the peptide mapping of proteins.[50,54] LMW heparin is first depolymerized enzymatically and then fractionated by either using PAGE or SAX-HPLC. By using standard oligosaccharides, specific sequences within the polymeric substrate can be identified and quantitated. Oligosaccharide maps for commercial porcine heparins looked remarkably similar, with each showing several major oligosaccharides.[54] Oligosaccharide maps can also be prepared with as little as 1 μg of heparin.

The picture of heparin that emerges from oligosaccharide maps is of a polymer primarily comprised of →4)IdoA2S(1→4)GlcN2S6S(1→, but containing a small number of additional oligosaccharides.[54] The quantitation of oligosaccharides with known structure typically result in an average mass balance recovery of >90 wt.% for porcine mucosal heparins. The presence of dermatan sulfate (DS), a contaminant of heparin preparations, brings the average mass balance to >95 wt. %. The remaining 5 wt%, uncharacterized minor oligosaccharides, may represent important parts of heparin's fine structure.

One unusual feature of the oligosaccharide maps of LMW heparins (not previously observed in commercial heparins) is the presence of major unidentified oligosaccharides.[27] These are the result of a chemical modifications introduced during the depolymerization reaction used to prepare LMW heparins.[27] A new oligosaccharide with an anhydromannose reducing end is formed on nitrous acid depolymerization in amounts consistent with it arising from the reducing end of each polymer chain. An oligosaccharide formed in the oxidative depolymerization of heparin was found in <50% of the chains. Several attempts to unambiguously establish the structure of this unusual oligosaccharide have failed.[27] LMW heparins prepared by either enzymatic or chemical ß-eliminative cleavage contain unsaturated uronic acid at the non-reducing end of each chain.

Oligosaccharide mapping[50,54,56] has also recently been used to study DS.[60] This method was able to pinpoint a disaccharide that occurred at elevated levels in DS having high heparin cofactor II activity. In an extension of this study, oligosaccharide mapping by gradient PAGE provided an added dimension to GAG analysis. The dominant feature observed in PAGE analysis is the larger oligosaccharides. PAGE analysis suggested that iduronic acid content may play a role in DS *in vivo* antithrombotic activity. The presence of iduronic acid may have importance in DS as it does in heparin by permitting greater conformational flexibility of the polymer and self-aggregation.[61]

SPECTROSCOPIC CHARACTERIZATION OF OLIGOSACCHARIDE STRUCTURE

FAB-MS

Carbohydrate structure determination is increasingly relying on spectroscopic methods such as NMR spectroscopy and FAB-MS.[1,8] In the absence of large amounts of sample, MS analysis represents the best method for obtaining structural information on complex oligosaccharides.[43,62,63] ^{252}Cf plasma desorption (PD)-MS[64] analysis on heparin oligosaccharides using a surfactant resulted in positively charged molecule ions.[64] However, the use of surfactant increases the molecular weight of a hexasaccharide, from 1603 to 7510 amu and resulted in poor resolution.[64] No significant fragment ions were observed suggesting PD-MS may be impractical for molecule ion analysis and only provide limited structural information.

FAB ionization has been used for the analysis of sulfated oligosaccharides.[43,62,65,66] Reinhold and coworkers reported on the negative ion FAB-MS analysis of heparin oligosaccharides prepared by chemical synthesis.[65] The dominant molecule ions observed were $[M+Na_x-H_{x+1}]^-$ (where M is the fully protonated, acid form and x=2-9), and the spectra contained no structurally significant fragment ions.[65] Dell and coworkers described a strategy for sequencing sulfated oligosaccharide mixtures using positive ion FAB-MS, which requires multiple derivatization steps.[66] Negative ion FAB-MS analysis in our laboratory[43,62] on enzymatically prepared, non-derivatized heparin oligosaccharides uses triethanolamine as the FAB matrix. In addition to the clear presence of monoanionic sodiated molecule ions, structurally significant (sequence) fragment ions are observed.

Thirteen GAG-derived disaccharides (20 nmoles) analyzed by FAB-MS gave appropriate molecule ion peaks assignable to $[M+Na_x-H_{x+1}]^-$, x=0-2. The most characteristic fragment ion observed is associated with the loss of sulfation[62,65] but many of the disaccharides also produced fragment ions resulting from cleavage at glycosidic linkages. Examination of a homologous series of oligosaccharides, based on heparin's major repeating trisulfated disaccharide, demonstrated that their fragmentation patterns were similar, easily interpretable and generally resulted in sufficient information to definitively assign the sequence. The heparin-derived octasaccharide, having twelve sulfate groups and sixteen negative charges, represents the largest oligosaccharide yet analyzed in our laboratory.[62] A molecule ion of m/z 2637 and its characteristic fragmentation pattern confirm the structure assigned to it using NMR spectroscopy. It may be possible to analyze even larger sulfated oligosaccharides by FAB-MS if sufficient sample is available and its purity (lack of both saccharidic and salt contaminants) can be assured. Successful MS/MS experiments are also improving our understanding of the effect of sulfate position on fragment ion formation.[67]

NMR Spectroscopy

High-field NMR spectroscopy is useful to assign the structure of heparin oligosaccharides. One-dimensional (1D) ^{1}H-NMR spectrum is first recorded in ^2H_2O to obtain single-proton resonances corresponding to anomeric protons. Comparison of the chemical shift and coupling constant data of these signals with those previously reported for heparin and heparin-derived oligosaccharides can be useful in assigning structure.[63] Dependence on published shift on coupling constant data obtained on related molecules, however can lead to misassignment of structure. Two-dimensional (2D) NMR offers a less ambiguous method. The 600 MHz ^{1}H-NMR 2D COSY-45 spectrum affords complete spin connectivity information for each sugar residue within an oligosaccharide. Having accomplished the complete proton chemical shift assignment of all residues present in an oligosaccharide the next step is to establish their sequence. We have demonstrated the usefulness of the 2D rotating frame nuclear Overhauser enhancement spectroscopy (ROESY) to obtain sequence information for a heparin-derived tetrasaccharides and hexasaccharides.[63,68] The ROESY spectrum of an oligosaccharide results in intense cross-peaks corresponding to the through space coupling H1 and H4 that are in close proximity in each glycosidic linkage permitting unambiguous assignment of sequence.

^{13}Carbon NMR spectroscopy often requires more sample than is routinely available. The information offered by this technique is also valuable in understanding physical-chemical properties. Oligosaccharide standards were recently analyzed by ^{13}C-NMR to determine the pKa of the different carboxyl groups present within their structures.[69] Definitive, unambiguous assignments of carbon resonances of carboxyl carbon resonances were made using 2D COSY-45 and heteronuclear correlation experiments. Heparin itself provided the most interesting application of this methodology. The ^{13}C NMR of heparin through a range of pH values was used to calculate the pKa of heparin's major uronic acid residues. The pKa of the carboxyl group in the IdoA2S and GlcA residues of heparin polymer were 3.13 and 2.79, respectively. Based on the oligosaccharide model compounds a pKa of 3.0 for the IdoA residue of heparin was predicted. With a 20 mM sample concentration it took one week to determine each pKa. Acquisition times might be substantially reduced using newer proton detected ^{13}C NMR spectroscopy methods such as HMBC. Recently, ^{1}H-detected ^{13}C-NMR was taken of an 8 mg sample of human heparin prepared from a large hemangioma and when compared to the ^{13}C-NMR spectrum of porcine heparin it was found to be nearly identical.[26]

POLYSACCHARIDE SEQUENCE ANALYSIS

The structural complexity of heparin is associated with both its polydispersity and microheterogeneity. The polydispersity of these molecules makes it difficult to approach their sequencing as one would a protein or nucleic acid. Two approaches are possible to sequence heparin. In the first, a single heparin chain would be purified to homogeneity and then sequenced. The pitfalls associated with this approach are the difficulty in preparing such a monodisperse heparin chain and the question of whether such a chain would be truly representative of polydisperse heparin mixture. The microheterogeneity of heparin suggests that it is important to examine saccharide sequence to understand structure and function. Heparin can have three different types of sequence: 1) random, like that seen in synthetic polymers prepared by free radical polymerization; 2) partially ordered, where blocks of ordered sequence are interspersed between regions of random sequence; and 3) ordered, like those of proteins and nucleic acids.

Three approaches have been developed to sequence heparin: they are computer or mathematical-based simulation studies, kinetic or multienzyme sequencing and reading-frame sequencing. The first of these approaches focuses on the search for ordered domains or sequence information within the chains of the polydisperse heparin. This approach uses information theory to search for ordered domains (or signals) in the presence of random domains (or noise). A polydisperse ensemble of number-chains, where one digit numbers correspond to each type of saccharide unit, are simulated.[3,70,71] The sequence of these chains can be randomly structured or can take into account our knowledge of heparin biosynthesis. Once the substrate's structure has been simulated enzyme action is simulated. The simulation collects data continuously about the number, size and sequence of both product and substrate number chains and compares these to experimental data. Although computer and mathematical simulations can not establish a sequence, they can rule out sequences that are incompatible with experimental observations.

A second method to sequence heparin is kinetic or multienzyme sequencing. Heparin is partially, enzymatically depolymerized.[19,50] Oligosaccharides still containing enzyme cleavable sites are then purified from the reaction mixture. Each "transient" oligosaccharides is tagged at one end and treated a second time with the same enzyme. Analysis of a banding pattern on a PAGE results in the sequences of the transient oligosaccharides.[43,50] Reading frame sequencing requires a common reading frame from which a sequence is determined. The most obvious reading frame is at the reducing-end of the heparin or the point of attachment between core protein and heparin chain.[72] PG is treated to release GAG chains from core protein exposing a specific sugar residue in the linkage region. This residue, common to all the released GAG chains, is then tagged. Treatment with enzyme and analysis on PAGE should result in a banding pattern giving sequence. The similarity to the Maxam and Gilbert[73] approach for sequencing nucleic acids makes its application appealing.

MICROANALYSIS IN TISSUES AND BIOLOGICAL FLUIDS

Most analysis of GAGs and PGs in tissues have been performed on tissue culture samples metabolically labeled with ^{35}S or ^{3}H. Although these studies are valuable they provide a limited view of the actual distribution and structure of GAG in a whole animal. Substantial quantities of GAG and PG are found in tissues but extraction and characterization have been limited by the presence of contaminating biopolymers (proteins and nucleic acids) and their structural complexity. It is surprising that monoclonal antibodies have played a limited role in the detection and quantitation of GAGs in tissue. This is primarily due to the low immunogenicity of these molecules.

GAGs are found in both blood and urine.[74] Because of the high concentrations of protein present in blood (~60 mg/mL) it has been difficult to analyze GAGs other than by bioassay (with100 ng/mL sensitivity).[75] However, these assays can be subject to interference and they tell little about the structure of the heparin being measured or about the presence of other inactive GAGs.

Before it is possible to analyze tissue or biological fluid for GAG, contaminating substances must be removed. One approach relies on exhaustive proteolysis to reduce tissue or biological fluid to a mixture of peptide fragments, nucleic acids, salts and GAG.[25,26] Ion exchange chromatography is then used to recover GAG. This approach recently resulted in the purification of milligram quantities of human heparin from a vascular tumor.[26] Chemical and chromatographic assays are being developed for the analysis of GAGs in the urine of patients with metabolic disorders[76] or with rheumatoid arthritis.[74] Recent developments in PAGE analysis have made it possible to obtain oligosaccharide maps of endogenous (ng/mL) GAGs.[74]

FUTURE PROSPECTS AND CONCLUSIONS

Heparin has been in clinical use for over 50 years. Despite complications associated with its use, no replacement for this drug is on the horizon. The discovery of many new biological activities of heparin have been reported in the past decade spurring additional research on heparin. Immobilized heparins are also becoming more important with the increased demand for biomedical devices that require heparin or heparinized biomaterials. All these factors suggest that heparin will be widely used well into the twenty-first century.

The chemical and enzymatic methods described in this chapter need refinement to prepare improved LMW heparins and modified heparins. The rapid progress in heparin analysis suggest that complete structural analysis of small amounts of heparin may soon be possible. With these advances, the attention of heparin researchers is now turning to endogenous heparin in an effort to understand its true biological role.

ACKNOWLEDGEMENTS

This work was supported by National Institutes of Health grants GM 38060, HL 29797 and AI 22350 and funding from Merck Sharp & Dohme, Italfarmaco S.p.A., Celsus and Wyeth-Ayerst Research Laboratories.

REFERENCES

1. B. Casu, Adv. Carbohydr. Chem. Biochem. 43:51 (1985).
2. R.J. Linhardt, Z.M. Merchant, K.G. Rice, Y.S. Kim, G.L. Fitzgerald, A.C. Grant and R. Langer, Biochemistry 24:7805 (1985).
3. R.J. Linhardt, D.M. Cohen and K.G. Rice, Biochemistry 28:2888 (1989).
4. U. Lindahl, D.S. Feingold and L. Roden, TIBS 221 (1986).
5. R.L. Stevens, C.C. Fox, L.M. Lichtenstein and K.F. Austen, Proc. Natl. Acad. Sci. USA 85:2284 (1988).
6. S. Vanucchi, M. Ruggiero and V. Chiarugi, Biochem. J. 227:57 (1985).
7. R.J. Linhardt, Chemistry and Industry 2:45-50 (1991).
8. R.J. Linhardt and D. Loganathan, "Biomimetic Polymers" (C.G. Gebelein, Ed.), Plenum Press, New York, pp 135-175 (1990).

9. J. Folkman and M. Klagsbrun, Science 235:442, (1987).
10. K.M. Krianciunas, F. Grigorescu and C.R. Kahn, Diabetes 36:163 (1987).
11. H.F. Schran, D.W. Fitz, F.J. DiSerio and J. Hirsh, Thromb. Res. 31:51 (1983).
12. S.E. Lasker and M.L. Chium, U.S. Patent No. 3,766,167 (1973).
13. T.K. Sue, L.B. Jaques and E. Yeun, Can. J. Physiol. Pharmacol.54:613 (1976).
14. A.S. Gervin, Surg. Gyn. Obstet. 140, 789 (1975).
15. H. Engelberg, Pharmacol. Rev. 36:91 (1984).
16. Z.M. Merchant, E.E. Erbe, W.P. Eddy, D. Patel and R.J. Linhardt, Atherosclerosis 62:151 (1986).
17. J.J. Castellot, Jr., J. Choay, J.-C. Lormeau, M. Petitou, E. Sache and M.J. Karnovsky, J. Cell. Biol. 102:1979 (1986).
18. M.D. Sharath, J.M. Weiler, Z.M. Merchant, Y.S. Kim, K.G. Rice and R.J. Linhardt, Immunopharmacol. 9:73 (1985).
19. R.J. Linhardt, K.G. Rice, Y.S. Kim, J.D. Engelken and J.M. Weiler, J. Biol. Chem. 263:13090 (1988).
20. R.G. Azizkhan, J.C. Azizkhan, B.R. Zetter and J. Folkman, J. Exp. Med. 152:931 (1980).
21. A.B. Schreiber, J. Kenney, W.J. Dowalski, R. Friesel, T. Mehlman and T. Maciag, Proc. Natl. Acad. Sci. 82: 6138 (1985).
22. J. Folkman, R. Langer, R.J. Linhardt, C. Haudenschild and S. Taylor, Science 221:719 (1983).
23. R. Orem, S. Szabo and J. Folkman, Science 230:1375 (1985).
24. A.D. Cardin and H.J.R. Weintraub, Atherosclerosis 9:21 (1989).
25. E. Coyne, Fed. Proc. 36:32 (1977).
26. R.J. Linhardt, S.A. Ampofo, J. Fareed and J. Folkman, J. Biol. Chem. 1991, submitted.
27. R.J. Linhardt, D. Loganathan, A. Al-Hakim, H.M. Wang, J.M. Walenga, D. Hoppensteadt and J. Fareed, J. Med. Chem. 33:1639 (1990).
28. Y.S. Kim and R.J. Linhardt, Thromb. Res. 53:55 (1989).
29. R.E. Edens, A. Al-Hakim, J.M. Weiler, J. Fareed and R.J. Linhardt, J. Pharm. Sci. submitted (1991).
30. A.C. Grant, R.J. Linhardt, G.L. Fitzgerald, J.J. Park and R. Langer, Anal. Biochem. 137:25 (1984).
31. A. Naggi, G. Torri, B. Casu, J. Pangrazzi, M. Abbadini, M. Zametta, M.B. Donati, J. Lansen and J.P. Maffrand, Biochem. Pharmacol. 36:1895 (1987).
32. A. Naggi and G. Torri, U.S. Pat. No. 4,948,881 (1990).
33. J.S. Mardiguan, U.S. Pat. No. 4,440,926 (1984).
34. M.J. Bienkowski and H.E. Conrad, J. Biol. Chem. 260:356 (1985).
35. J.A. Cifonelli and J.A. King, Biochemistry 16:2137 (1977).
36. Z.M. Merchant, Y.S. Kim, K.G. Rice and R.J. Linhardt, Biochem. J. 229:369 (1985).
37. F. Fussi and G. Fedeli, U.K. Pat. No. GB 2,002,406B (1982).
38. P. Bianchini and G. Mascellani, U.S. Pat. No. 4,791,195 (1988).
39. H. Uchiyama, Y. Dobashi and K. Nagasawa, XVth International Carbohydrate Symposium, Yokohama, Japan, August 1990.
40. V.B. Diaz, R.H. Domanico and F. Fussi, U.S. Pat. No. 4,977,250 (1990).
41. R.S. Langer, R.J. Linhardt, C.L. Cooney and G. Fitzgerald and A. Grant, U.S. Pat. No. 4,396,762 (1983).
42. A.K. Larsen, R.J. Linhardt, K.G. Rice and G. Wogan and R.Langer, J. Biol. Chem. 264:1570 (1989).

43. L.M. Mallis, H.M. Wang, D. Loganathan and R.J. Linhardt, Anal. Chem. 61:1453 (1989).
44. O. Larm, R. Larsson and P. Olsson, "Heparin: Chemical and Biological Properties, Clinical Applications," (D.A. Lane, U. Lindahl, Eds.), CRC Press, Inc., Boca Raton, Florida, pp 159-189 (1989).
45. R.J. Linhardt, A. Al-Hakim, J. Liu, "Polymers from Biotechnology" (C.G. Gebelein, Ed.), Plenum Press, New York, in press (1991).
46. L. Nilsson, K.E. Storm, S. Thelin, L. Bagge, J. Hultman, J. Thorelius and U. Nilsson, Artificial Organs 14:46, 1990.
47. R.M. Emanuele and J. Fareed, Fed. Proced. 46:868 (1987).
48. J. Fareed, J. Walenga, A. Racanelli, D. Hoppensteadt, X. Huan and H.L. Messmore, Haemostasis (1988).
49. I. Nieduszynski in Heparin, Chemical and Biological Properties, Clinical Applications, Lane, D.A. and U. Lindahl, CRC Press, Boca Raton, FL, 1989.
50. K.G. Rice, M.K. Rottink and R.J. Linhardt, Biochem. J. 244:515 (1987).
51. R.J. Linhardt, J.E. Turnbull, H.M. Wang, D. Loganathan and J.T. Gallagher, Biochem. J. 29:2611 (1990).
52. R.J. Linhardt, C.L. Cooney and P.M. Galliher, Appl. Biochem. Biotechnol. 12:135 (1986).
53. K.G. Rice and R.J. Linhardt, Carbohydr. Res. 190:219 (1989).
54. R.J. Linhardt, K.G. Rice, Y.S. Kim, D.L. Lohse, H.M. Wang and D. Loganathan, Biochem. J. 254:781 (1988).
55. R.J. Linhardt, K.N. Gu, D. Loganathan and S.R. Carter, Anal. Biochem. 181:288-296 (1989).
56. J.E. Turnbull and J.T. Gallagher, Biochem. J. 251:597 (1988).
57. A. Ali-Hakim and R.J. Linhardt, Anal. Biochem. 195:68 (1991).
58. S.L. Carney and D.J. Osborne, Anal. Biochem. 195:132 (1991).
59. S.A. Ampofo, H.M. Wang and R.J. Linhardt Anal. Biochem. 1991, in press.
60. R.J. Linhardt, A. Al-Hakim, J. Liu, D. Hoppensteadt, J. Fareed, G. Mascellani and P. Bianchini, Biochem. Pharmacol., 1991, in press.
61. B. Casu in: "New Trends in Hemostasis" J. Harenberg, D.L. Heene, G. Stehle and G. Schettler, Eds., Springer Verlag, Heidelberg, 1990.
62. R.J. Linhardt, H.M. Wang, D. Loganathan, D. Lamb and L.M. Mallis, Carbohydr. Res., 1991, in press.
63. D. Loganathan, H.M. Wang, L.M. Mallis and R.J. Linhardt Biochemistry 29:2611 (1990).
64. C.J. McNeal, R.D. Macfarlane and I. Jardine, Biochem. Biophys. Res. Commun. 139:18 (1986).
65. V.N. Reinhold, S.A. Carr, B.N. Green, M. Petitou, J. Choay and P. Sinay, Carbohydr. Res. 161:305 (1987).
66. A. Dell, M.E. Rogers, J.E. Thomas-Oates, T.N. Huckerby, P.N. Sanderson and I.A. Nieduszynski, Carbohydr. Res. 179:7 (1988).
67. D.J. Lamb, L.M. Mallis, R.J. Linhardt, H.M. Wang and D. Loganathan, Am. Soc. for Mass Spectrometry, Nashville, TN, May 1991.
68. R.J. Linhardt, H.M. Wang and D. Loganathan, J.H. Bae, J. Biol. Chem., 1991, submitted.
69. H.M. Wang, D. Loganathan and R.J. Linhardt, Biochem J., 1991, in press.
70. R.J. Linhardt, Z.M. Merchant, K.G. Rice, Y.S. Kim, G.L. Fitzgerald, A. Grant and R. Langer, Biochemistry 24:7805-7810 (1985).
71. D.M. Cohen and R.J. Linhardt, Biopolymers, 30:733 (1990).

72. R.J. Linhardt, D. Loganathan, A. Al-Hakim and S.A. Ampofo in :"New Trends in Hemostasis," J. Harenberg, D.L. Heene, G. Stehle and G. Schettler, Eds., Springer Verlag, Heidelberg, 1990.
73. A.M. Maxam and W. Gilbert, Proc. Nat. Acad. Sci. (USA) 74:560 (1977).
74. A. Al-Hakim and R.J. Linhardt, Appl. Theoret. Electrophor. 1:305 (1991).
75. J.M. Walenga, J. Fareed, D. Hoppensteadt and R.M. Emanuele, CRC Crit. Rev. Clin. Lab. Sci. 22:361, 1986.
76. R.S. Varma and R. Varma, Eds, "Glycosaminoglycans and proteoglycans in physiological and pathological processes of body systems." Karger AG, Basel. (1982).

HEPARAN SULPHATE PROTEOGLYCANS: MOLECULAR ORGANISATION OF MEMBRANE-ASSOCIATED SPECIES AND AN APPROACH TO POLYSACCHARIDE SEQUENCE ANALYSIS

John T. Gallagher, Jeremy E. Turnbull and Malcolm Lyon

Cancer Research Campaign Department of Medical Oncology
Christie Hospital NHS Trust
Manchester M20 9BX UK

INTRODUCTION

Heparan sulphate proteoglycans (HSPGs) are normally the most abundant PG components of cell surfaces and basement membranes (1, 2). At the molecular level five core proteins have been described that can be glycanated with HS chains (2,3) but evidence exists for several other probably distinct species including two cell surface HSPGs in which the core proteins have been shown to bind either TGF-ß (4) or fibroblast growth factor (FGF;5); the former has been named betaglycan. A recent surprising finding was that the secretory granule PG, named serglycin, which normally appears as a heparin or chondroitin sulphate PG, is synthesised as an HSPG in an erythroid cell line (6). The topographical distribution of HSPGs is therefore variable and extensive and probably reflects a broad functional spectrum to which both protein and polysaccharide components will make significant contributions (1,2,7).

MEMBRANE HSPGS

Four membrane HSPG core proteins have been cloned and they all have interesting structural features (Fig. 1). Two of these core proteins, syndecan (33kD) and fibroglycan (23kD) which were originally cloned from mouse mammary epithelial cells (8) and human lung fibroblasts (9) respectively, have highly homologous transmembrane and cytoplasmic domains but the ectodomain sequences are generally quite distinct apart from two similar features, one being a common putative protease-cleavage site close to the membrane surface and the other appearing in the form of a close resemblance in the arrangement of three potential ser-gly glycanation sites in the distal region of the ectodomain (Fig. 1). However, unlike fibroglycan, syndecan normally contains both HS and CS chains (10), and there are two additional ser-gly sequences in the syndecan ectodomain that may prime the synthesis of CS to produce this hybrid PG structure. Three membrane HSPG core proteins have been detected in material purified from extracts of rat liver (11) and the

Heparin and Related Polysaccharides
Edited by D.A. Lane *et al.*, Plenum Press, New York, 1992

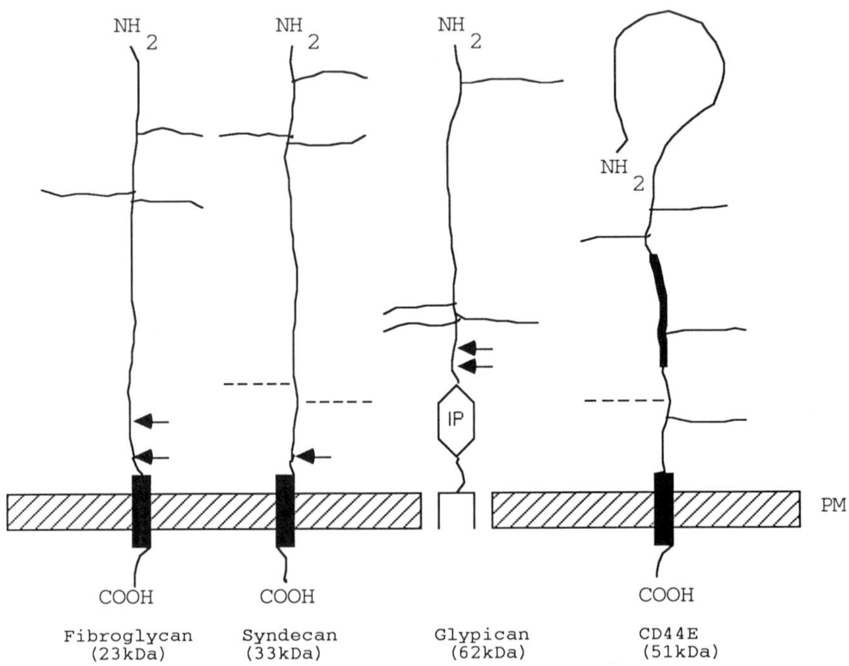

Fig.1. Diagrammatic representation of the four cloned cell surface HSPGs. Core protein molecular weights, as determined from their cDNA sequences, are shown in brackets. Syndecan possesses two more potential glycanation sites than fibroglycan, and these are shown substituted with CS (- - - -) rather than HS (______) chains, though the actual disposition of specific GAG chains between these sites is unknown. Glypican is unusual in that it is anchored in the membrane by an inositol phoshate (IP)-linked lipid rather than a hydrophobic protein sequence (shaded box). The epithelial variant of CD44 contains a 13kDa spliced insert (thick line) which is absent in the haemopoietic molecule. Of the five potential glycanation sites, one is contained within the insert while the other four flank it. Glycanation sites are absent from the N-terminal globular domain which may function as a hyaluronate-binding site. The actual distribution of HS and CS between the different sites is not known. All the core proteins, except CD44E, possess dibasic sequences (arrows) close to the plasma membrane (PM) which may be potential sites for proteolytic release of the glycanated ectodomain from the cell surface.

major species has recently been identified as fibroglycan (12). Northern blotting analysis has indicated that hepatocytes are an important source of this PG (12). Fibroglycan and syndecan are likely to be co-expressed on the hepatocyte surface membrane (13 and our data) and fibroglycan therefore appears not to be restricted solely to fibroblastic cells. Articles in this volume by Jalkanen and David will provide more extensive information on these particular PGs.

Considerable interest has been shown in HSPGs that may be intercalated in plasma membranes by an inositol phospholipid (IP) anchor. However, only one IP-anchored species has been cloned and sequenced and this HSPG has been named glypican (14). The predicted core protein sequence (62kD) contains four putative HS attachment sites, three of which are in a contiguous sequence. Glypican was identified in the same human lung fibroblast cell line that also synthesises fibroglycan. IP-HSPG components have also been detected in hepatoma, neural and ovarian granulosa cell cultures (15,16,17) and it will be important to establish the structure of the core proteins in these PG species. Glypican could thus be the prototype for a new family of hydrophobic membrane-associated PGs which may function differently from the alternative transmembrane protein family. The susceptibility of glypican to the action of phosphatidylinositol-specific phospholipase C also provides for a distinct mechanism of regulation of expression and release from the plasma membrane in addition to that provided by proteolytic cleavage adjacent to the membrane (Fig 1).

The most recent, and in some ways the most intriguing membrane protein to be identified as a potential HSPG is the CD44 or Hermes antigen, which was first described as a lymphocyte homing receptor and adhesion molecule (18). It is now known that CD44 is not confined to lymphocytes. It has been identified on mesenchymal and epithelial cells, and is strongly expressed on some carcinomas (19). Several protein isoforms can be produced through alternative mRNA splicing, and differential glycosylation and glycanation patterns may impose further variability on the expressed proteins. In lymphocytes, a minor fraction of CD44 is glycanated with CS chains (20), but in cultured human keratinocytes CD44 is mainly synthesised as an HSPG (18) and this may correlate with the production of a specific isoform containing a large 132 amino acid insert in the major splice site (Fig.1). The remarkable capacity of CD44 isoforms to be transformed into different PGs may provide some clues on the long-standing enigma of how core proteins direct the synthesis of specific GAG species.

The absolute number of different HSPG core proteins is yet to be ascertained but the possibility that they may fit into a relatively small number of distinct families suggests that cellular variations in HSPG properties may be modulated predominantly by differences in the fine structure of the HS chains superimposed on these core proteins. Plasma membrane-associated endo-heparanases may also have a role to play in regulating cell surface expression of HS (21).

STRUCTURE OF HS

HS is perhaps the most complex of mammalian polysaccharides and the cell-type related variations in molecular fine structure (23,24) are indicative of organ- and tissue-specific functions. The basic polymeric structure is composed of a repeating sequence of disaccharides consisting of glucosamine (GlcN) and hexuronate (HexA). GlcN is either N-acetylated or N-sulphated and although these substituents are present in

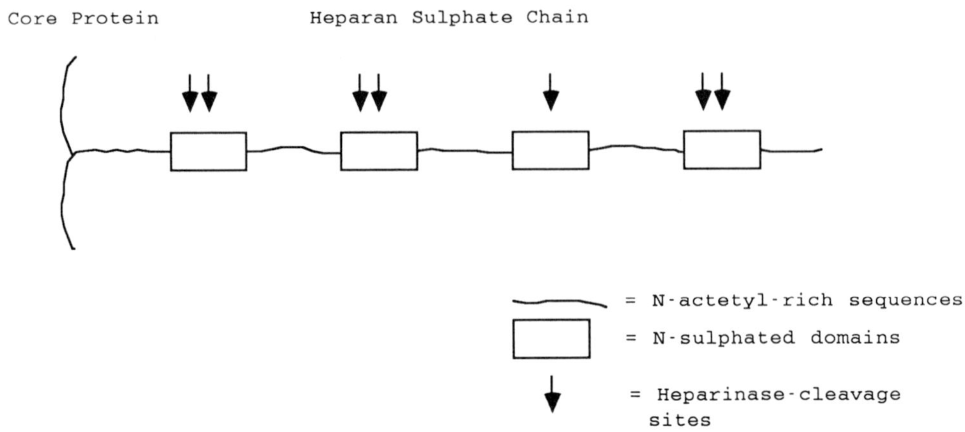

Fig.2. Model of the domain structure of heparan sulphate. The model depicts a speculative structure for skin fibroblast HS. It is proposed that the HS chain comprises a series of alternate sulphate-rich (mainly N-sulphated) and sulphate-poor (mainly N-acetylated) domains. The model reflects data indicating the wide spacing of N-sulphated domains containing clusters of heparinase-susceptible disaccharides (26). The structure represents a simplified picture since variations in the precise positions and spacing of the sulphated domains occurs, and N-sulphate groups are sparsely distributed in the N-acetylated domains. Adapted with permission of the Biochemical Society (see ref. 26).

approximately equal proportions in HS, they are commonly segregated into distinct block sequences from 2-9 disaccharides in length (1,2,22). Two isomers of hexuronate are present, glucuronate being associated with GlcNAc residues (GlcNAc α 1,4 GlcA) whereas a substantial proportion of hexuronates in the N-sulphated disaccharides are iduronate ($GlcNSO_3$ α 1,4 IdoA). Structural complexity is further amplified by varying levels of ester(O)-linked sulphates which are largely, though not entirely, restricted to the N-sulphated domains. Ester sulphates are most frequent at C-6 of $GlcNSO_3$ and C-2 of IdoA, though C-6 of GlcNAc, C-3 of $GlcNSO_3$ and C-2 of GlcA may also be O-sulphated with important consequences for the biological properties of the polymer (2,22).

The tendency for co-clustering of sulphate groups and iduronate residues creates regions of high fixed charged density and conformational flexibility (22,25) which are likely to play key roles in determining the disposition and reactivities of HS in the pericellular domain. Recent findings indicate that these highly sulphated regions are distributed in an ordered manner along the polysaccharide chain, separated by N-acetyl rich sequences 20-25 disaccharides in length (26; Fig 2).

Despite the predominant segregation of N-acetylated and N-sulphated disaccharides in HS, about 25% of the polymer chain comprises mixed sequences of type:

$GlcNSO_3$ ___ HexA ___ GlcNAc___ GlcA ___ $GlcNSO_3$ structure 1

in which solitary N-acetylated units are flanked by N-sulphates. The location of such mixed sequences in HS is unknown but we believe that they may lie at the interface between contiguous N-sulphated and N-acetylated regions. Mixed sequences are biologically significant. For example, they lead into the anti-thrombin III (AT-III) binding sequence (27; in such instances the $GlcNSO_3$ adjacent to GlcA in structure 1 would be sulphated at C-3) and in the mixed sequences in HS from tumour and transformed cells significant reductions in C-6 and C-3 sulphation of glucosamine residues have been detected (28,29). Loss of 3-O-sulphate was correlated with a reduced affinity for AT-III (29) but the significance of this finding is uncertain. It has also been demonstrated that HS from transformed rat fibroblasts, specifically undersulphated in the above type of sequence, has a lower affinity for surface membrane receptors than the normal polysaccharide (30). This reduced affinity could impair an important mechanism for HS-mediated cellular interactions. The nature of the membrane receptor is unknown. Is the receptor homologous to AT-III? The low binding affinity of tumour HS provides an important clue about the structural requirements for receptor recognition and receptor isolation would be a significant step in understanding some of the functions of HS.

SEQUENCE ANALYSIS OF HS

The presence of structurally-ordered domains within the HS chain (Fig.2) prompted the development of approaches for analysing the precise sequence of sugars along the polysaccharide. The sequence analysis of HS presents a difficult analytical problem. Peptide sequencing is achieved by stepwise removal and analysis of N-terminal amino acids but comparable methods are not available for sequential removal of the sugars in HS. However the arrangement of sugars relative to a defined reference point (usually the reducing terminal xylose residue) can be assessed by various end-labelling techniques (for reviews, see 31,32). We have developed an indirect approach to the problem of HS sequence analysis in which the PG is first treated with reagents that cleave the polysaccharide chain at known sugar residues (see below). Recovery of oligosaccharides originating at the point of attachment to the protein core is then achieved by coupling the treated PG to CNBr-Sepharose via the protein (33,34). After extensive washing to remove free oligosaccharides the end-referenced oligosaccharides are then released by base-borohydride. Analysis of their size distribution provides information on the position of the cleaved linkages downstream from the protein (Fig.3). However clear separation and accurate molecular weight determination of large HS oligosaccharides, which is necessary for complete sequence analysis of partially-depolymerised chains, is extremely difficult even using high resolution gradient PAGE (23,35). Therefore, we have initially restricted the analysis to determining the locations of specific sugar units most proximal to the core protein.

HSPGs used in our studies were prepared from cultures of human skin fibroblasts. The PGs were degraded with nitrous acid or heparinase for sufficient time to scission all susceptible linkages. Oligosaccharides recovered from the core protein following treatment with nitrous acid, which specifically cleaves at $GlcNSO_3$ residues, fell into a narrow molecular size range (approx. 5kDa) indicating that an extended N-acetylated sequence of about 10 disaccharide units separated the first downstream $GlcNSO_3$ from the core protein (33,34).

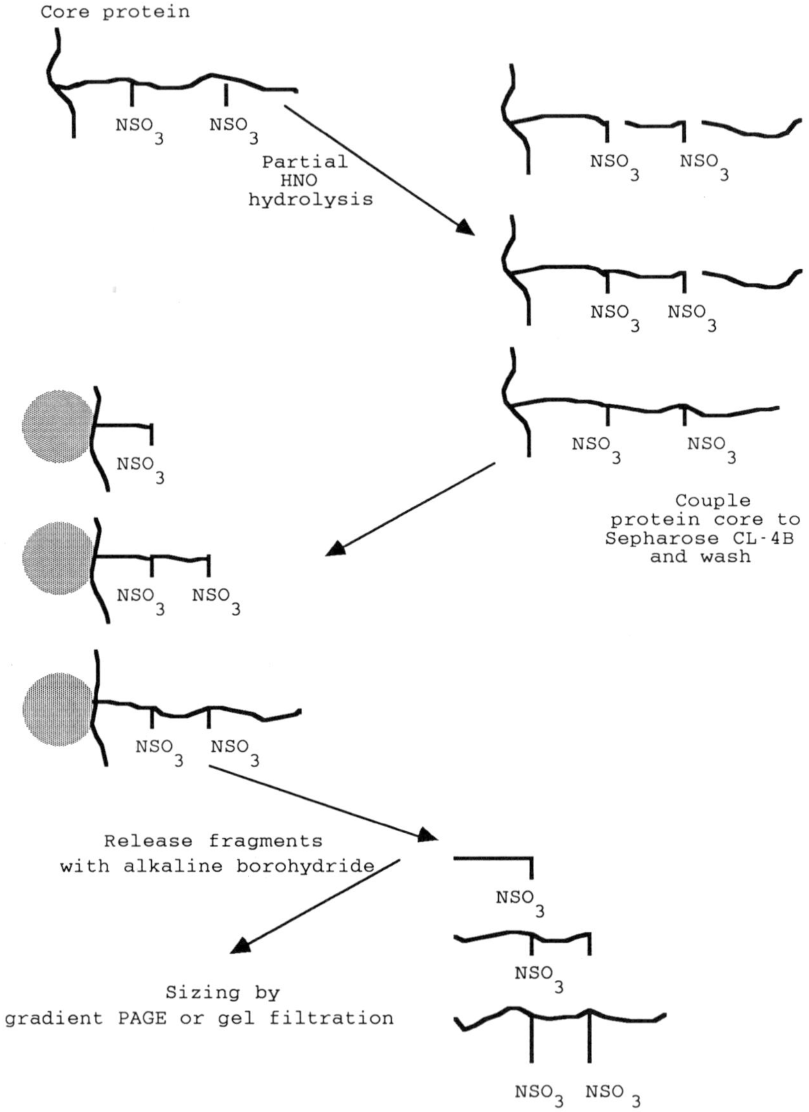

Fig.3. End Referencing - A sequencing strategy for HS. Using this strategy, oligosaccharide size defines the location of susceptible linkages "downstream" from the reducing end of the chains located at the point of attachment to the core protein. If depolyermisation is complete, the first downstream positions of linkages are located (for details see text and ref. 26).

Despite minor variations in length (range 8-12 disaccharides) the inner N-acetylated domain was well conserved in the HS population. Length variations do however reflect sequence microheterogeneity in relation to the position of the most proximal N-sulphate group.

When the HSPG was treated with heparinase to detect the most proximal IdoA,2S (36) an unexpected polymorphism in the structure of HS was detected. Complete heparinase scission indicated that the IdoA,2S residues were more remote from the core protein than the first $GlcNSO_3$ occupying two well-defined locations in different HS chains at about 15 and 35 disaccharides from the protein (34).

The results of sequence analysis with both heparinase and nitrous acid clearly indicate that sulphation of HS, at least in the regions near to the core protein, is not a random process - distinct areas appear to be targetted for N-sulphation and O-sulphation of iduronate, and in the latter case differential targetting occurs. The general structure of the heparinase-resistant oligosaccharides on the PG core proteins is given by the formula:

$$IdoA,2S-GlcNSO_3-(HexA-GlcNR)_n-HexA-GlcNSO_3-(GlcA-GlcNAc)_{8-12}-GlcA-Gal-Gal-Xyl$$

in which n = approx. 2 or 22. We should stress that the structures described here refer to HSPGs from human skin fibroblasts and it remains to be established whether the proximal regions of HS chains from other sources have similar sulphation patterns.

At the present time we have little information on the mechanism of biosynthesis of HS (27). Structural data demonstrate that the process is carefully regulated, operating within quite strict 'tolerance' limits. Although close similarities to the synthesis of heparin may be anticipated (27, eg the same enzymes may be used), the distinctions in composition and molecular organisation between HS and heparin suggest that some important differences exist in the activity and/or regulation of these enzymes in connective tissue mast cells, which synthesise heparin, and in the majority of other mammalian cells which synthesise HS.

ACKNOWLEDGEMENTS

We would like to thank the Cancer Research Campaign for their continued support and Roberta Ellis for her secretarial assistance.

REFERENCES

1. Gallagher, J.T., Lyon, M. and Steward, W.P. (1986). Structure and function of heparan sulphate proteoglycans. Biochem. J. 236: 313.
2. Kjellen, L. and Lindahl. U. (1991). Proteoglycans - Structures and Interactions. Annu. Rev. Biochem. 60: 443.
3. Gallagher, J.T. (1989). The extended family of proteoglycans: social members of the pericellular zone. Current Opinion in Cell Biology 1: 1201.
4. Cheifetz. S., Andres J.L., and Massague, J. (1988.) The transforming growth factor ß-receptor type III in a membrane proteoglycan J. Biol. Chem. 263: 16984.
5. Sakaguchi, K., Yanagishita, M., Takeuchi, Y. and Aurbach, G.D. (1991).Identification of heparan sulphate proteoglycan as a high affinity receptor for acidic fibroblast growth factor (aFGF) in a parathyroid cell line. J.Biol. Chem. 266: 7270.

6. Okayama, M., Oguri, K., Yoshida, K., and Ohkita, T. (1991). Purification and characterisation of novel heparan sulphate proteoglycans produced by murine erythroleukaemia cells in the growing phase. J. Biol. Chem. 266: 38080.
7. Ruoslahti, E. (1988). Structure and biology of proteoglycans. Annu. Rev. Cell Biology 4: 229.
8. Saunders, S., Jalkanen, M., O'Farrell, S., and Bernfield, M. (1989). Molecular cloning of Syndecan, an integral membrane proteoglycan. J. Cell Biol. 108: 1547.
9. Marynen, P., Zhang, J., Cassiman, J-J., Van den Berghe, H., and David, G. (1989). Partial primary structure of the 48 and 90 kilodalton core proteins of cell surface associated heparan sulphate proteoglycans of lung fibroblasts. J. Biol. Chem, 264: 7017.
10. Rapraeger, A., Jalkanen, M., Endo, E., Koda, J. and Bernfield, M. (1985). The cell surface proteoglycan from mouse mammary epithelial cells bears chondroitin sulphate and heparan sulphate glycosaminoglycans. J. Biol. Chem. 260: 11046.
11. Lyon, M., and Gallagher, J.T. (1991). Purification and partial characterisation of the major cell-associated heparan sulphate proteoglycan of rat liver. Biochem. J. 273: 415.
12. Pierce, A., Cowling, G., Lyon, M. and Gallagher, J.T.-unpublished observations.
13. Hayashi, K., Hayashi, M., Jalkanen, M., Firestone,J.H., Trelstad, R.L., and Bernfield, M. (1987). Immunocytochemistry of cell surface heparan sulphate proteoglycan in mouse tissues. A light and electron microscopic study. J. Histochem. Cytochem. 35, 1079.
14. David, G., Lories, V., Decock, B., Marynen, P., Cassiman, J-J., and Van den Berghe, H. (1999)Molecular cloning of a phosphatidylinositol-anchored membrane heparan sulphate proteoglycan from human lung fibroblasts. J. Cell. Biol. 111: 3165.
15. Ishihara, M., Fedarko, N. and Conrad, H.E. (1983). Involvement of phosphatidylinositol and insulin in the coordinate regulation of proteoheparan sulphate metabolism and hepatocyte growth. J. Biol. Chem. 262: 4708.
16. Carey, D.J and Todd, D.M. (1989). Membrane anchoring of heparan sulphate proteoglycan by phosphatidylinositol and kinetics of synthesis of peripheral and detergent solubilised proteoglycans in Schwann cells. J. Cell. Biol. 108: 1891.
17. Yanagishita, M. and McQuillan, D.J. (1989). Two forms of plasma membrane - intercalated heparan sulphate proteoglycan in rat ovarian granulosa cells. Labelling of proteoglycans with a photoactivatable hydrophobic probe and effect of the membrane anchor-specific phospholipase C. J. Biol. Chem. 264: 17551.
18. Brown, T.A., Bouchard, T., St John, T., Wayner, E. and Carter, W.G. (1991). Human keratinocytes express a new CD44 core protein (CD44E) as a heparan sulphate intrinsic membrane proteoglycan with additional exons. J. Cell. Biol. 113: 207.
19. Haynes, B.F., Telen, M.J., Hale, L.P., and Denning, S.M. (1989). CD44 - A molecule involved in leukocyte adherence and T-cell activation. Immunology Today 10: 423.
20. Jalkanen, S., Jalkanen, M., Bargatze, R., Tammi, M. and Butcher, E.C. (1988). Biochemical properties of glycoproteins involved in lymphocyte recognition of high endothelial venules in man. J. Immunol. 141: 1615.
21. Gallagher, J.T., Walker, A., Lyon, M. and Evans, W.H. (1988). Heparan sulphate-degrading endoglycosidase in liver plasma membranes. Biochem. J. 250: 719.

22. Gallagher, J.T. and Walker, A. (1985). Molecular distinctions between heparan sulphate and heparin. Analysis of sulphation patterns indicates that heparan sulphate and heparin are separate families of N-sulphated polysaccharides. Biochem. J. 230: 665.
23. Turnbull, J.E. and Gallagher, J.T. (1988). Oligosaccharide mapping of heparan sulphate by polyacrylamide-gradient gel electrophoresis and electrotransfer to nylon membrane. Biochem. J. 251: 597.
24. Gallagher, J.T., and Lyon, M. (1989). Molocular organisation and functions of heparan sulphate. In Heparin, ed by Lane,D.A. and Lindahl, U. pp 135 Edward Arnold, London.
25. Casu, B., Petitou, M., Provasoli, M., and Sinay, P. (1988). Conformational flexibility: a new concept for explaining binding and biological properties of iduronic acid - containing glycosaminoglycans. Trends Biochem. Sci. 13: 221.
26. Turnbull, J.E. and Gallagher, J.T. (1991). Distribution of iduronate-2-sulphate residues in heparan sulphate. Evidence of an ordered polymeric structure. Biochem. J. 273: 553.
27. Lindahl, U. (1989). Biosynthesis of heparin and related polysaccharides. In: Heparin, ed. by Lane D.A., and Lindahl, U. pp 159, Edward Arnold, London.
28. Winterbourne, D.J. and Mora, P.T. (1981). Cells selected for high tumourigenicity or transformed by SV40 synthesise heparan sulphate with a reduced degree of sulphation. J. Biol. Chem. 256: 4310.
29. Pejler, G., and David, G. (1987). Basement membrane heparan sulphate with high affinity for antithrombin synthesised by normal and transformed mouse mammary epithelial cells. Biochem. J. 248: 69.
30. Winterbourne, D.J. (1982). Binding of heparan sulphate and heparin to control and virus-transformed cells. Bioscience Reports 2: 1009.
31. Gallagher, J.T., Turnbull, J.E. and Lyon, M. (1991). Patterns of sulphation in heparan sulphate: polymorphism based on a common structural theme. Int. J. Biochem. - in press.
32. Turnbull, J.E. (1991). Oligosaccharide mapping and sequence analysis of glycosaminoglycans. In: Methods in Molecular Biology - Membrane Methods. Humana Press, New Jersey - in press.
33. Lyon, M., Steward, W.P., Hampson, I.N. and Gallagher, J.T. (1987). Identification of an extended N-acetylated sequence adjacent to the protein-linkage region of fibroblast heparan sulphate. Biochem. J. 242: 493.
34. Turnbull, J.E. and Gallagher, J.T. (1991). Sequence analysis of heparan sulphate indicates defined locations of N-sulphated glucosamine and iduronate 2-sulphate residues proximal to the protein-linkage region. Biochem. J. 277: 297.
35. Rice, K.G., Rottink, M.K. and Linhardt, R.J. (1987). Fractionation of heparin-derived oligosaccharides by gradient polyacrylamide gel electrophoresis. Biochem. J. 244: 515.
36. Linhardt, R.J., Turnbull, J.E., Wang, H., Loganathan, D. and Gallagher, J.T. (1990). Examination of the substrate specificities of heparin and heparan sulphate lyases. Biochemistry 29: 2611.

REGULATION OF THE GENE THAT ENCODES THE PEPTIDE CORE OF HEPARIN PROTEOGLYCAN AND OTHER PROTEOGLYCANS THAT ARE STORED IN THE SECRETORY GRANULES OF HEMATOPOIETIC CELLS

Donald E. Humphries and Richard L. Stevens

Department of Medicine, Harvard Medical School; the Department of Veterans Affairs OPC, Boston; and the Department of Rheumatology and Immunology, Brigham and Women's Hospital, Boston, Massachusetts

INTRODUCTION

Proteoglycans are a complex family of macromolecules that consist of a protein core to which at least one glycosaminoglycan is covalently attached through the neutral trisaccharide, Galß1—>3Galß1—>4Xylß—>Ser (for review, see Rodén, 1980). The heparin/heparan sulfate family of glycosaminoglycans is formed in the Golgi by the sequential addition of alternating GlcA and GlcNAc onto this trisaccharide, whereas the chondroitin sulfate family of glycosaminoglycans is formed by the sequential addition of alternating GlcA and GalNAc. After addition of the precursor saccharides, the polymer undergoes a number of enzymatic modifications including N- and O-sulfation and epimerization of GlcA to IdoA to yield mature glycosaminoglycans. The combination of different core proteins, varied numbers, types, and sizes of glycosaminoglycans, and possible N- and O-linked oligosaccharides results in a diverse family of proteoglycans. This diversity probably reflects the proteoglycans' numerous functions in different cells and tissues. Proteoglycans are major constituents of the extracellular matrix where they help to determine both the physical properties and function of connective tissues. Proteoglycans are also important components of plasma membranes where they bind growth factors and often participate in intercellular binding and cell/matrix interactions.

In addition to the extracellular matrix-localized and membrane localized proteoglycans, a distinct family of proteoglycans has been identified inside many hematopoietic cells, including mast cells. These proteoglycans are unique in that they have a protease resistant core containing a glycosaminoglycan-attachment region rich in serine and glycine. In mast cells, these novel proteoglycans contain either heparin or highly sulfated chondroitin glycosaminoglycans. They are major constituents of the mast cell's secretory granules where they are stored complexed to basically-charged proteases that are enzymatically active at neutral pH (Schwartz et al., 1981; Serafin et al., 1986 and 1987). Upon activation of the mast cell, these proteoglycan/protease macromolecular complexes are exocytosed intact.

Heparin and Related Polysaccharides
Edited by D.A. Lane *et al.*, Plenum Press, New York, 1992

MAST CELL HEPARIN AND CHONDROITIN SULFATE PROTEOGLYCANS

The finding that mucosal mast cells can be distinguished histochemically from connective tissue mast cells when stained with the cationic dye safranin (Enerbäck, 1966) led to the discovery that mast cells vary considerably in the type of proteoglycan found in their secretory granules. The predominant proteoglycan found in connective tissue mast cells contains heparin, the most highly sulfated glycosaminoglycan and the most negatively charged macromolecule synthesized by mammals. The most common disaccharide within heparin contains three sulfates and has the structure IdoA-2-SO_4 —>GlcNSO_4-6-SO_4. Heparin proteoglycan isolated from rat mast cells has a molecular weight of 700-900 kDa and contains several heparin chains of ~80 kDa (Yurt et al., 1977; Robinson et al., 1978; Metcalfe et al., 1980). Mucosal mast cells that increase in the intestines of helminth infected rats produce a proteoglycan that contains predominately chondroitin sulfate di-B glycosaminoglycans made up of the disulfated disaccharide, IdoA-2-SO_4 —>GalNAc-4-SO_4 (Stevens et al., 1986). Other types of highly sulfated chondroitins have been found in cultured mouse mast cells. For example, *in vitro*-differentiated mouse bone marrow-derived mast cells synthesize chondroitin sulfate E glycosaminoglycans which contain abundant amounts of the disulfated disaccharide, GlcA—>GalNAc-4,6-diSO_4 (Razin et al., 1982). On the other hand, *in vitro*-differentiated mouse lymph node-derived mast cells synthesize chondroitin sulfate D-like glycosaminoglycans containing GlcA-2-SO_4—>GalNAc-6-SO_4 (Davidson et al., 1990). The function of these different glycosaminoglycans is probably dependent on their unique sulfate positioning rather than simply net negative charge. The importance of defined sulfated structures in glycosaminoglycans may be best demonstrated by the specific oligosaccharide structure found in some forms of heparin which is required for high affinity binding to antithrombin III (Lindahl et al., 1979 and 1980).

Mast cell proteoglycans contain several glycosaminoglycans attached to a small peptide core that is distinguished by its high content of serine and glycine [termed SG-PG (also known as serglycin)]. Although not investigated in detail, it has been demonstrated that at least some of these proteoglycans contain low molecular weight O-linked oligosaccharides (Stevens et al., 1985). The peptide core of mature rat mast cell heparin proteoglycan was the first proteoglycan found to contain almost exclusively serine and glycine (Robinson et al., 1978; Metcalfe et al., 1980). Because it had been known for some time that heparin glycosaminoglycans were linked to peptide cores at serine-glycine sequences (Lindahl and Rodén, 1964), it was proposed that the peptide core of heparin proteoglycan contained a region that was essentially alternating serine and glycine (Robinson et al., 1978). Mast cell proteoglycans that contain either chondroitin sulfate di-B (Seldin et al., 1985) or chondroitin sulfate E (Stevens et al., 1985) glycosaminoglycans were later found to have similar serine/glycine-rich peptide cores. The glycosaminoglycan attachment region of SG-PG is resistant to proteolytic attack because no known protease will cleave an alternating serine and glycine sequence.

MOLECULAR BIOLOGY OF THE SECRETORY GRANULE PROTEOGLYCAN PEPTIDE CORE

Isolation of cDNAs that Encode Rat, Mouse, and Human SG-PG

cDNA clones were first isolated from a rat L2 yolk sac tumor cell cDNA library that encoded a novel proteoglycan peptide core containing a region consisting of 49 alternating serines and glycines (Bourdon et al., 1985 and 1986). As assessed by RNA blot analysis using a gene-specific fragment of this rat L2 cell cDNA, all rat and mouse mast cells examined to date express this ~1 kb transcript, irrespective of the type of glycosaminoglycan attached to the peptide core (Tantravahi et al., 1986). The presence of

SG-PG mRNA in human eosinophils (Rothenberg et al., 1988), T lymphocytes (Harrigan, 1989), and platelets (Alliel et al., 1988) has led to the proposal that most, if not all, granule proteoglycans of hematopoietic cells contain the same peptide core, and that the type of glycosaminoglycan found on this peptide core varies with cell type. ß-Xyloside treatment of rat serosal mast cells results in an inhibition of heparin proteoglycan biosynthesis and an induction of chondroitin sulfate E biosynthesis onto the exogenous acceptor (Stevens and Austen, 1982). Thus, these mast cells have the biosynthetic capability to synthesize both chondroitin sulfate and heparin families of glycosaminoglycans, but only add heparin onto SG-PG. The determination of the type of glycosaminoglycan attached to the SG-PG peptide core must therefore be controlled in some way by the peptide core or trisaccharide linkage region. Presumably SG-PG can be differentially modified post-translationally so that it is targeted to the proper polymerases and processing enzymes within the Golgi.

cDNAs that encode SG-PG have recently been cloned and sequenced from human promyelocytic leukemia HL-60 cells (Stevens et al., 1988; Stellrecht and Saunders, 1989), human platelets (Alliel et al., 1988), rat basophilic leukemia (RBL)-1 cells (Avraham et al., 1988), rat natural killer (NK) cells (Giorda et al., 1990), nontransformed mouse bone marrow-derived mast cells (Avraham et al., 1989a), and mouse mastocytoma P815 cells (Kjellén et al., 1989). Although the cDNAs isolated from rat L2 cells, rat NK cells, and RBL-1 cells encode the same protein, the L2 cell-derived transcript is larger due to this cell's usage of an alternative transcription-initiation site. Based on the deduced amino acid sequences of their cDNAs, SG-PGs of human, mouse, and rat are initially translated as 17.6, 16.7, and 18.6 kDa peptide cores, respectively. Although a serine/glycine repeat region is found in all three species, rat SG-PG has the longest repeat consisting of 49 amino acids. Mouse SG-PG has a serine/glycine repeat region of 21 amino acids, whereas this region in human SG-PG is only 18 amino acids long and one of the serines has been replaced by phenylalanine. Comparison of human and mouse SG-PG has revealed that only 51% of the amino acids are identical, and that the N-terminus is the most highly conserved region of the peptide cores. All three species contain two cysteines near their N-termini, suggesting the presence of a disulfide bond. It was somewhat of a surprise to find that a conserved sequence within the 5′ untranslated regions of human, mouse, and rat SG-PG is nearly identical to corresponding sequences in the 5′ untranslated regions of cDNAs that encode several serine proteases which also reside in the secretory granules of mast cells (Benfey et al., 1987; Serafin et al., 1990). It is possible that this untranslated nucleotide sequence is involved in the translational regulation of granule-destined proteins, and may be important for efficient binding of transcripts to ribosomes in hematopoietic cells.

The deduced amino acid sequence of the HL-60 cell-derived cDNA was used to select peptides for the production of antibodies to study the biosynthesis of human SG-PG (Nicodemus et al., 1990). A 16-mer peptide, corresponding to a region of the proteoglycan core preceding the serine/glycine repeat region, was synthesized and employed to elicit antibodies in rabbits. Using the resulting antibodies, an ~20 kDa protein was immunoprecipitated after a 2 min incubation of HL-60 cells with [^{35}S]methionine. After 10 min of incubation, a [^{35}S]methionine-labeled 150 kDa chondroitin sulfate proteoglycan was precipitated. Pulse-chase experiments indicated that within 1 hr the radiolabeled proteoglycan could no longer be precipitated. The inability to immunoprecipitate the mature proteoglycan from HL-60 cells presumably is the result of post-translational proteolysis of the antigenic region of the peptide core once it is in the secretory granule. Although the N-terminus of SG-PG remains intact in platelets (Alliel et al., 1988), this proteoglycan peptide core appears to be rapidly degraded at its N- and C-termini in most hematopoietic cells.

Isolation of the Mouse and Human SG-PG Genes

The genes that encode mouse (Avraham et al., 1989b, Angerth et al., 1990) and human (Nicodemus et al., 1990) SG-PG have been cloned and analyzed. The mouse gene is ~15 kb in size, while the human gene is ~17 kb. Both genes are located on chromosome 10 (Avraham et al., 1988; Stevens et al., 1988) and contain 3 exons (Avraham et al., 1989b; Nicodemus et al., 1990). Exon 1 encodes the 5′ untranslated region and the hydrophobic signal peptide. Exon 2 encodes the 49 amino acid sequence that is predicted to be the N-terminus of the protein after it leaves the endoplasmic reticulum. Because this is the most conserved exon, it may encode determinants which are critical for the processing and/or targeting of this proteoglycan to various intracellular compartments. Exon 3 is the largest exon and contains the nucleotide sequence encoding the serine/glycine glycosaminoglycan attachment region and the 3′ untranslated region of the mRNA. S1 nuclease mapping and primer extension analysis revealed that the primary transcription-initiation site of the gene in mouse mast cells (Avraham et al., 1989b) and human HL-60 cells (Nicodemus et al., 1990) resides ~40 and ~50 nucleotides, respectively, upstream of the translation-initiation sites. Although neither the mouse nor the human SG-PG genes contain a classical proximal promoter element such as a TATA box (Breathnach and Chambon, 1981) or GC-rich element (Sehgal et al., 1988), the ~300 bp sequence immediately preceding their transcription-initiation sites is nearly identical (Nicodemus et al., 1990).

Sequencing of the entire human SG-PG gene has recently been completed (Humphries et al., unpublished results). Intron 1 and intron 2 are 8.8 and 6.7 kb, respectively. Both introns begin with the nucleotide sequence GTAAG and end with the sequence CAG (Nicodemus et al., 1990), in accordance with the nucleotide consensus sequence for intron splicing (Mount, 1982). Each intron contains several *Alu* elements which are a family of repetitive DNAs present in at least 500,000 copies in the human genome (Schmid and Shen, 1985). They have a dimeric structure, are ~300 bp in length, and are thought to be derived from the 7SL RNA gene (Ullu et al., 1982). While *Alu* elements are actively transcribed *in vitro* by RNA polymerase III, no transcript that consists of just an *Alu* region has been detected in humans (Schmid and Shen, 1985). The *Alu* elements in the human SG-PG gene are spaced at relatively constant intervals and account for ~28% and ~42% of introns 1 and 2, respectively.

Transcriptional Regulation of the SG-PG Gene

Primary control of protein biosynthesis generally occurs at the level of transcription, and is regulated by DNA-binding proteins (termed *trans-acting* factors) which bind to specific nucleotide sequences (termed *cis-acting* elements) in the gene. Often the most important *cis-acting* elements are found in the 1 to 2 kb nucleotide sequence immediately upstream of the gene's transcription-initiation site. Because *trans-acting* factors can function as either enhancers or suppressors, at least one of those present in non-hematopoietic cells was predicted to suppress transcription of the SG-PG gene. In contrast, hematopoietic cells must possess *trans-acting* factors which stimulate transcription of the SG-PG gene at the appropriate time during differentiation of progenitor cells.

To study the expression of the SG-PG gene during differentiation of mouse bone marrow stem cells into mast cells, RNA was prepared from the starting cells, and from cells cultured up to 3 wk in the presence of either recombinant interleukin 3 (Ihle et al., 1983) or recombinant c-kit ligand (Tsai et al., 1991). These cytokines induce progenitor cells to become immature mast cells. Analysis of the resulting RNA blots revealed that transcription of the SG-PG gene occurred before transcription of any of the

7 protease genes that are eventually expressed by mast cells (Gurish et al., unpublished observation). Starting bone marrow cells contain abundant amounts of the SG-PG transcript, but do not express detectable levels of transcripts that encode the high affinity IgE receptor or any secretory granule protease. Only after 1 to 2 wk of culture are detectable levels of these latter transcripts obtained. The early expression of the SG-PG transcript suggests that these proteoglycans must be present before the proteases are synthesized and transported to the granules.

Stable transfection of rat fibroblasts with up to 11 copies of a 20 kb mouse genomic clone that contained the SG-PG gene resulted in low level expression of the mouse SG-PG transcript (Avraham et al., 1989b). This finding indicated that the clone contained the entire gene and at least some of its regulatory elements. Although a classical promoter was not found, the degree of conservation in the 5′ flanking region of the corresponding mouse and human SG-PG genes suggested that this region contained *cis-acting* elements that regulate its transcription.

To determine if this conserved region did in fact contain regulatory elements, a series of DNA constructs was prepared by ligating various regions of the 5′ flanking region of the mouse SG-PG gene to plasmid DNA which contained the human growth hormone gene (Avraham et al., 1991). Hematopoietic cells and mesenchymal cells which express and do not express SG-PG, respectively, were transiently transfected with the varied constructs to determine the precise location of any *cis-acting* elements that regulate transcription of the SG-PG gene. By comparing the amount of human growth hormone in the culture media of the transfected cells, three *cis-acting* elements that regulate constitutive transcription were discovered in the 5′ flanking region. No growth hormone was detected when cells were transfected with a construct that lacked a 20 bp fragment 20-40 nucleotides upstream from the transcription-initiation site, and thus it was concluded that this region contained the proximal promoter of the mouse SG-PG gene. Although no classical TATA box is present in this nucleotide sequence, it was speculated that the nucleotide sequence TCTAAAA might function as a TATA box-equivalent. Mutational analysis confirmed that this sequence is part of the proximal promoter and that the cytosine is critical for the transcription activity of this promoter.

Two other regulatory elements were found in the 5′ flanking region of the mouse SG-PG gene. An element ~100 bp upstream of the transcription-initiation site was found to enhance transcription of the human growth hormone gene in hematopoietic cells and fibroblasts. Like most other enhancers, its orientation in the plasmid and distance from a generic promoter did not alter its activity. An element 190-250 nucleotides upstream of the transcription-initiation site was found to suppress transcription of this gene. This suppressor element is much more active in rat-1 and mouse 3T3 fibroblasts (two mesenchymal cells that do not express the SG-PG transcript) than in RBL-1 cells and WEHI-3 myelomonocytic cells (two hematopoietic cells that express the transcript). Gel-mobility-shift analyses of nuclear extracts revealed that RBL-1 cells and rat-1 fibroblasts contain several DNA-binding proteins which specifically bind to these three regulatory elements. While some of the DNA-binding proteins appear to be common to both cell types, others are more limited in their distribution (Avraham et al., 1991).

Recently it was discovered that mast cells express various amounts of three transcription-regulatory proteins, designated GATA-1, GATA-2, and GATA-3 (Martin et al., 1990; Zon et al., 1991). Although transfection of GATA-1 into mouse mastocytoma cells resulted in increased mast cell carboxypeptidase A promoter activity, the amount of SG-PG mRNA remained the same. GATA-1 is therefore a *trans-acting* factor that regulates transcription of protease genes in mast cells but not the SG-PG gene.

Since most *trans-acting* DNA-binding proteins bind to *cis-acting* motifs in the 5′ flanking regions of genes, a computer search was performed to identify sequences within the conserved 5′ flanking region of the SG-PG gene that might interact with known transcription factors. Several potential regulatory sites were found. These include the element that binds the yeast *trans-acting* regulatory protein GCN4 (Arndt and Fink, 1986), the serum-regulated element for a heat shock gene (Wu et al., 1987), a cAMP response element (Montminy and Bilezikjian, 1987; Sassone-Corsi, 1988), an *ets*-binding domain (Klemsz et al., 1990), and a glucocorticoid binding motif (Scheidereit et al., 1983). Mouse T-lymphocytes (Harrigan et al., 1989) and mouse bone marrow-derived mast cells (Humphries et al., unpublished observations) have increased levels of SG-PG mRNA after treatment with dexamethasone or other glucocorticoids. In dexamethasone-treated mouse mast cells, the cellular levels of SG-PG mRNA increased within 3 hr of exposure to steroid and remained high for several days. Because these kinetics of mRNA induction are similar to those obtained with other steroid-inducible genes, the glucocorticoid/receptor complex probably interacts directly with the glucocorticoid binding motif in the 5′ flanking region of the SG-PG gene. Treatment of mouse T-lymphocytes with forskolin resulted in increased SG-PG mRNA levels (Harrigan et al., 1989), suggesting that the cAMP response element is also active in these hematopoietic cells.

FUTURE DIRECTIONS

While much progress has been made in the study of the SG-PG gene over the last decade, much work remains. Now that the two introns of the human gene have been sequenced, the function of these introns can be addressed. It is likely that other *cis-acting* elements will be discovered and that the role of cytosine methylation in transcriptional regulation of this gene will be examined. The identification of *cis-acting* elements that either enhance or suppress transcription of the SG-PG gene will, no doubt, result in the isolation, characterization, and cloning of the novel *trans-acting* factors that bind to these regulatory motifs.

REFERENCES

Alliel, P. M., Périn, J-P., Maillet, P., Bonnet, F., Rosa, J-P., and Jollès, P., 1988, Complete amino acid sequence of a human platelet proteoglycan., FEBS Lett., 236:123.

Angerth, T., Huang, R., Aveskogh, M., Pattersson, I., Kjellén, L., and Hellman, L., 1990, Cloning and structural analysis of a gene encoding a mouse mastocytoma proteoglycan core protein; analysis of its evolutionary relation to three cross hybridizing regions in the mouse genome., Gene, 93:235.

Arndt, K., and Fink, G. R., 1986, GCN4 protein, a positive transcription factor in yeast, binds general control promoters at all 5′ TGACTC 3′ sequences., Proc. Natl. Acad. Sci. USA, 83:8516.

Avraham, S., Stevens, R. L., Gartner, M. C., Austen, K. F., Lalley, P. A., and Weis, J. H., 1988, Isolation of a cDNA that encodes the peptide core of the secretory granule proteoglycan of rat basophilic leukemia-1 cells and assessment of its homology to the human analogue., J.Biol. Chem., 263:7292.

Avraham, S., Stevens, R. L., Nicodemus, C. F., Gartner, M. C., Austen, K. F., and Weis, J. H., 1989a, Molecular cloning of a cDNA that encodes the peptide core of a mouse mast cell secretory granule proteoglycan and comparison with the analogous rat and human cDNA., Proc. Natl. Acad. Sci., 86:3763.

Avraham, S., Austen, K. F., Nicodemus, C. F., Gartner, M. C., and Stevens, R. L., 1989b, Cloning and characterization of the mouse gene that encodes the peptide core of secretory granule proteoglycans and expression of this gene in transfected rat-1 fibroblasts., J. Biol. Chem., 264:16719.

Avraham, S., Avraham, H., Austen, K. F., and Stevens, R. L., 1991, Negative and positive *cis-acting* regulatory elements in the 5' flanking region of the mouse gene that encodes the serine/glycine-rich peptide core of proteoglycans found in the cytoplasmic granules of hematopoietic cells., J. Biol. Chem., in press.

Benfey, P. N., Yin, F. H., and Leder, P., 1987, Cloning of the mast cell protease, RMCP II. Evidence for cell-specific expression and a multi-gene family., J. Biol. Chem., 262:5377.

Bourdon, M. A., Oldberg, A., Pierschbacher, M., and Ruoslahti, E., 1985, Molecular cloning and sequence analysis of a chondroitin sulfate proteoglycan cDNA., Proc. Natl. Acad. Sci. USA, 82:1321.

Bourdon, M. A., Shiga, M., and Ruoslahti, E., 1986, Identification from cDNA of the precursor form of a chondroitin sulfate proteoglycan core protein., J. Biol. Chem., 261:12534.

Breathnach, R., and Chambon, P., 1981, Organization and expression of eukaryotic split genes coding for proteins., Ann. Rev. Biochem., 50:349.

Davidson, S., Gilead, L., Amira, M., Ginsburg, H., and Razin, E., 1990, Synthesis of chondroitin sulfate D and heparin proteoglycans in murine lymph node-derived mast cells., J. Biol. Chem., 265:12324.

Enerbäck, L., 1966, Mast cells in rat gastrointestinal mucosa. 2. Dye-binding and metachromatic properties., Acta Pathol. Microbiol. Scand., 66:303.

Giorda, R., Chambers, W. H., Dahl, C. A., and Trucco, M., 1990, Isolation and characterization of a cDNA that encodes the core protein of the cytolytic granule proteoglycan in rat natural killer cells., Nat. Immun. Cell. Growth Regul., 9:91.

Harrigan, M. T., Baughman, G., Campbell, N. F., and Bourgeois, S., 1989, Isolation and characterization of glucocorticoid- and cyclic AMP-induced genes in T lymphocytes., Mol. Cell. Biol., 9:3438.

Ihle, J. N., Keller, J., Oroszlan, S., Henderson, L. E., Copeland, T. D., Fitch, F., Prystowsky, M. B., Goldwasser, E., Schrader, J. W., Palaszynski, E., Dy, M., and Lebel, B., 1983, Biologic properties of homogeneous interleukin 3. I. Demonstration of WEHI-3 growth factor activity, mast cell growth factor activity, P cell-stimulating factor activity, colony-stimulating factor activity, and histamine-producing cell-stimulating factor activity., J. Immunol., 131:282.

Kjellén, L., Pettersson, I., Lillhager, P., Steen, M.-L., Pettersson, U., Lehtonen, P., Karlsson, T., Ruoslahti, E., and Hellman, L., 1989, Primary structure of a mouse mastocytoma proteoglycan core protein., Biochem. J., 263:105.

Klemsz, M. J., McKercher, S. R., Celada, A., Van Beveren, C., and Maki, R. A., 1990, The macrophage and B cell-specific transcription factor PU.1 is related to the *ets* oncogene., Cell, 61:113.

Lindahl, U., and Rodén, L., 1964, The linkage of heparin to protein., Biochem. Biophys. Res. Commun., 17:254.

Lindahl, U., Bäckström, G., Höök, M., Thunberg, L., Fransson, L-Å., and Linker A., 1979, Structure of the antithrombin-binding site in heparin., Proc. Natl. Acad. Sci. USA, 76:3198.

Lindahl, U., Bäckström, G., Thunberg, L., and Leder, I. G., 1980, Evidence for a 3-O-sulfated D-glucosamine residue in the antithrombin-binding sequence of heparin., Proc. Natl. Acad. Sci. USA, 77:6551.

Martin, D. I. K., Zon, L. I., Mutter, G., and Orkin, S. H., 1990, Expression of an erythroid transcription factor in megakaryocytic and mast cell lineages., Nature, 344:444.

Metcalfe, D. D., Smith, J. A., Austen, K. F., and Silbert, J. E., 1980, Polydispersity of rat mast cell heparin., J. Biol. Chem., 255:11753.
Montminy, M. R., and Bilezikjian, L. M., 1987, Binding of a nuclear protein to the cyclic-AMP response element of the somatostatin gene., Nature, 328:175.
Mount, S., 1982, A catalogue of splice junction sequences., Nucleic Acids Res., 10:459.
Nicodemus, C. F., Avraham, S., Austen, K. F., Purdy, S., Jablonski, J., and Stevens, R. L., 1990, Characterization of the human gene that encodes the peptide core of secretory granule proteoglycans in promyelocytic leukemia HL-60 cells and analysis of the translated product., J. Biol. Chem., 265:5889.
Razin, E., Stevens, R. L., Akiyama, F., Schmid, K., and Austen, K. F., 1982, Culture from mouse bone marrow of a subclass of mast cells possessing a distinct chondroitin sulfate proteoglycan with glycosaminoglycans rich in N-acetylgalactosamine-4,6-disulfate., J. Biol. Chem., 257:7229.
Robinson, H. C., Horner, A. A., Höök, M., Ögren, S., and Lindahl, U., 1978, A proteoglycan form of heparin and its degradation to single-chain molecules., J. Biol. Chem., 253:6687.
Rodén, L., 1980, Structure and metabolism of connective tissue proteoglycans., in: **The Biochemistry of Glycoproteins and Proteoglycans**, W. J. Lennarz, ed., Plenum, New York.
Rothenberg, M. E., Pomerantz, J. L., Owen, W. F., Avraham, S., Soberman, R. J., Austen, K. F., and Stevens, R. L., 1988, Characterization of a human eosinophil proteoglycan, and augmentation of its biosynthesis and size by interleukin 3, interleukin 5, and granulocyte/macrophage colony stimulating factor., J. Biol. Chem., 263:13901.
Sassone-Corsi, P., 1988, Cyclic AMP induction of early adenovirus promoters involves sequences required for E1A trans-activation., Proc. Natl. Acad. Sci. USA, 85:7192.
Scheidereit, C., Geisse, S., Westphal, H. M., and Beato, M., 1983, The glucocorticoid receptor binds to defined nucleotide sequences near the promoter of the mouse mammary tumour virus., Nature, 304:749.
Schmid, C. W., and Shen, C-K. J., 1985, The evolution of interspersed repetitive DNA sequences in mammals and other vertebrates, in: **Molecular Evolutionary Genetics**, R. J. MacIntyre, ed., Plenum, New York.
Schwartz, L. B., Riedel, C., Caulfield, J. P., Wasserman, S. I., and Austen, K. F., 1981, Cell association of complexes of chymase, heparin proteoglycan, and protein after degranulation by rat mast cells., J. Immunol., 126:2071.
Sehgal, A., Patil, N., and Chao, M., 1988, A constitutive promoter directs expression of the nerve growth factor receptor gene., Mol. Cell Biol., 8:3160.
Seldin, D. C., Austen, K. F., and Stevens, R. L., 1985, Purification and characterization of protease-resistant secretory granule proteoglycans containing chondroitin sulfate di-B and heparin-like glycosaminoglycans from rat basophilic leukemia cells., J. Biol. Chem., 260:11131.
Serafin, W. E., Katz, H. R., Austen, K. F., and Stevens, R. L., 1986, Complexes of heparin proteoglycans, chondroitin sulfate E proteoglycans, and [^{3}H]diisopropyl fluorophosphate-binding proteins are exocytosed from activated mouse bone marrow-derived mast cells., J. Biol. Chem., 261:15017.
Serafin, W. E., Dayton, E. T., Gravallese, P. M., Austen, K. F., and Stevens, R. L., 1987, Carboxypeptidase A in mouse mast cells: Identification, characterization, and use as a differentiation marker., J. Immunol., 139:3771.
Serafin, W. E., Reynolds, D. S., Rogelj, S., Lane, W. S., Conder, G. A., Johnson, S. S., Austen, K. F., and Stevens, R. L., 1990, Identification and molecular cloning of a novel mouse mucosal mast cell serine protease., J. Biol. Chem., 265:423.

Stellrecht, C. M., and Saunders, G. F., 1989, Nucleotide sequence of a cDNA encoding a hemopoietic proteoglycan core protein., Nucleic Acids Res., 17:7523.

Stevens, R. L., and Austen, K. F., 1982, Effect of p-nitrophenyl-β-D-xyloside on proteoglycan and glycosaminoglycan biosynthesis in rat serosal mast cell cultures., J. Biol. Chem., 257:253.

Stevens, R. L., Otsu, K., and Austen, K. F., 1985, Purification and analysis of the core protein of the protease-resistant intracellular chondroitin sulfate E proteoglycan from the interleukin 3-dependent mouse mast cell., J. Biol. Chem., 260:14194.

Stevens, R. L., Lee, T. D. G., Seldin, D. C., Austen, K. F., Befus, A. D., and Bienenstock, J., 1986, Intestinal mucosal mast cells from rats infected with *Nippostrongylus brasiliensis* contain protease-resistant chondroitin sulfate di-B proteoglycans., J. Immunol., 137:291.

Stevens, R. L., Avraham, S., Gartner, M. C., Bruns, G. A. P., Austen, K. F., and Weis, J. H., 1988, Isolation and characterization of a cDNA that encodes the peptide core of the secretory granule proteoglycan of human promyelocytic leukemia HL-60 cells., J. Biol. Chem., 263:7287.

Tantravahi, R. V., Stevens, R. L., Austen, K. F., and Weis, J. H., 1986, A single gene in mast cells encodes the core peptides of heparin and chondroitin sulfate proteoglycans., Proc. Natl. Acad. Sci. USA, 83:9207.

Tsai, M., Takeishi, T., Thompson, H., Langley, K. E., Zsebo, K. M., Metcalfe, D. D., Geissler, E. N., and Galli, S. J., 1991, Induction of mast cell proliferation, maturation, and heparin synthesis by the rat c-kit ligand, stem cell factor., Proc. Natl. Acad. Sci. USA, 88:6382.

Ullu, E., Murphy, S., and Melli, M., 1982, Human 7SL RNA consists of a 140 nucleotide middle-repetitive sequence inserted in an *Alu* sequence., Cell, 29:195.

Wu, B. J., Williams, G. T., and Morimoto, R. I., 1987, Detection of three protein binding sites in the serum-regulated promoter of the human gene encoding the 70-kDa heat shock protein., Proc. Natl. Acad. Sci. USA, 84:2203.

Yurt, R. W., Leid, R. W., Jr., Austen, K. F., and Silbert, J. E., 1977, Native heparin from rat peritoneal mast cells., J. Biol. Chem., 252:518.

Zon, L. I., Gurish, M. F., Stevens, R. L., Mather, C., Reynolds, D., Austen, K. F., and Orkin, S. H., 1991, GATA-binding transcription factors in mast cells regulate the promoter of the mast cell carboxypeptidase A gene., J. Biol. Chem., in press.

STRUCTURAL AND FUNCTIONAL DIVERSITY OF THE HEPARAN SULFATE PROTEOGLYCANS

Guido David

Center for Human Genetics
University of Leuven
Leuven, Belgium

Several enzymes, cell adhesion molecules, growth factors, proteinase inhibitors and extracellular matrix components possess heparin-binding domains, and are profoundly affected in their reactivities with third parties in the presence of this glycosaminoglycan. Heparin, e.g. markedly accelerates the reaction of antithrombin III with thrombin[1], and allows bFGF to interact with its receptor at the cell surface[2,3]. This implies that a whole series of biological processes may be modulated by the availability of heparin or heparin-like polysaccharides. Heparin is, however, not likely to be physiologically involved in most of these situations. Heparin is mainly a product of mast cells, which is stored intracellularly and is released upon degranulation of these cells at sites of inflammation. In contrast, the surfaces of most cells and the extracellular matrix are decorated by heparan sulfate, a glycosaminoglycan that shares several structural and functional features with heparin.

Support for the contention that heparan sulfate is a regulatory polysaccharide may be found in the studies that report dramatic and dynamic changes in the expression of heparan sulfate proteoglycans during embryonic development[4,5,6]. This process calls upon controlled cell proliferations and migrations, the establishment of selective cellular adhesions, and the stabilisation of the generated forms and associations through the deposition and remodelling of the extracellular matrix, engaging several of the known ligands for heparan sulfate. One of the acquisitions of the recent years is the realisation that several distinct proteins of the cell surface and of the extracellular matrix carry heparan sulfate chains, and that these are expressed in specific patterns, implying that each of these proteins may assume specific aspects of these heparan sulfate-'driven' systems.

CELL SURFACE-ASSOCIATED HEPARAN SULFATE PROTEOGLYCANS

Structural Heterogeneity

Most of the adherent cell types express heparan sulfate proteoglycans at their cell surfaces. Some of these proteoglycans have outspoken lipophilic properties and can easily be incorporated into

Heparin and Related Polysaccharides
Edited by D.A. Lane *et al.*, Plenum Press, New York, 1992

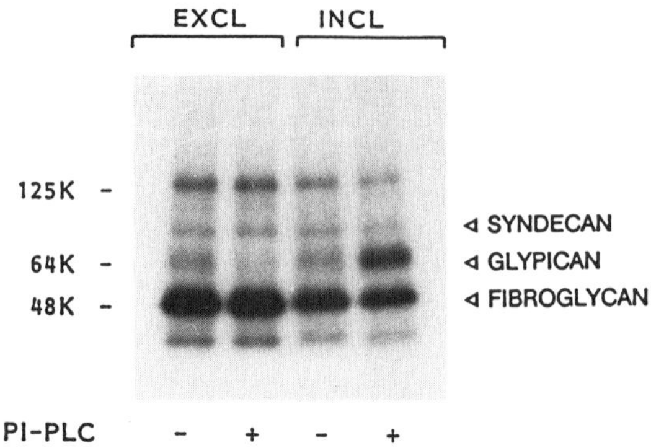

Fig. 1. Heterogeneity of the cell surface heparan sulfate proteoglycans from human lung fibroblasts. The cell surface proteoglycans were purified by ion exchange chromatography, gel filtration and incorporation into liposomes, and labelled with ^{125}I. The labelled proteoglycans were treated with PI-specific phospholipase C (+) or left untreated (–) and mixed with phosphatidylcholine in the presence of octylglucoside. After dialysis, to remove the detergent, the liposome-proteoglycan mixtures were fractionated by gel filtration over Sepharose CL4B. Equal amounts of labelled proteoglycan eluting in the EXCLuded and INCLuded volumes (70–80% and 20–30% of the applied label respectively) were treated with heparitinase, fractionated by SDS-PAGE and submitted to autoradiography. The phospholipase specifically eliminated the 64 kDa protein from the excluded fractions, causing it to elute as the core of a non hydrophobic proteoglycan in the included fractions. The different bands correspond to fibroglycan, glypican, syndecan and two other as yet unidentified proteoglycan cores with apparent Mr of 35 and 125 kDa (from David et al.; reproduced from the Journal of Cell Biology, 1990, 111, 3165 by copyright permission of the Rockefeller University Press).

liposomes, a property which has often been exploited to isolate these components from detergent extracts of the cells[7,8,9]. When this approach is followed for cultured human foetal lung fibroblasts and the hydrophobic proteoglycan fraction is treated with heparitinase, a core protein preparation is generated which can be resolved into a series of protein bands by electrophoresis in SDS polyacrylamide gradient gels[10] (see also fig. 1). Similar analyses with other cultured cell types or with tissue extracts have generally yielded similarly complex but not necessarily identical patterns, implying that this proteoglycan fraction is heterogeneous. Moreover, the mode of association of these proteins with the cell surface seems to vary, as some proteoglycans can be released from the cell surface or be depleted from liposome preparations by treatment with phosphatidylinositol-specific phospholipase C[11,12]. This suspicion of heterogeneity has now been substantiated by the molecular cloning of several representatives of this group.

```
                                    O                      ▽ ▽ ▽
MRRAWILLTLGLVACVSAESRAELTSDKDMYLDNSSIEEASGVYPIDDDDYASASGSGAD    60
EDVESPELTTTRPLPKILLTSAAPKVETTTLNIQNKIPAQTKSPEETDKEKVHLSDSERK   120
                                                 *        *
MDPAEEDTNVYTEKHSDSLFKRTEVLAAVIAGGVIGFLFAIFLILLLVYRMRKKDEGSYD   180
          *      *
LGERKPSSAAYQKAPTKEFYA                                          201
```

Fig. 2. Primary sequence of human fibroglycan.
The protein is 201 amino acids long and features a signal peptide (single underlining), an ectodomain with one site for N-glycosylation (O) and three sites for heparan sulfate chain attachment (▽), a transmembrane segment (double underlining), and a short cytoplasmic domain. The four conserved tyrosines that seem characteristic for the members of this family are indicated by an asterix.

Fibroglycan. One of the most prominent protein bands obtained by digesting the hydrophobic proteoglycans from cultured fibroblasts with heparitinase migrates with an apparent Mr of approximately 48 kDa (Fig. 1). Monoclonal antibodies that specifically react with this protein have been used to screen cDNA expression libraries and have led to the isolation of recombinants hat code for the peptide which forms the backbone of this proteoglycan[13]. The cDNA sequence and the analysis of the aminoterminal amino acid sequence of the immunopurified proteoglycan core predict that this 'fibroglycan' is a type I integral membrane protein of 201 amino acids, with an aminoterminal extracellular domain that may carry up to three heparan sulfate chains, a membrane-spanning segment, and a carboxylterminal cytoplasmic domain that is only 32 amino acids long (Fig. 2). Its only cysteine occurs in the aminoterminal signal peptide. The structure of this protein, in particular that of its cytoplasmic domain, seems to be highly conserved in man and mouse, as revealed by cDNA sequencing, and probably in other species as well, as suggested by the wide cross reactivities of the monoclonal antibodies raised against human fibroglycan. Fibroglycan is expressed at high levels in cultured lung or skin fibroblasts and in some cell lines of mesodermal and neurectodermal origin, but is virtually absent from most epithelial cell lines. Immunohistochemically, fibroglycan is barely detectable in adult quiescent tissues, but strong cell surface stainings are observed in the perichondrial tissues and in certain areas of the facial, lung and gut mesenchyme during early murine embryonic development, which seems to lend some justification to its name. Interestingly, fibroglycan appears to be a representative of a larger family of cell surface proteoglycans with transmembrane orientations whose cytoplasmic domains show extensive structural similarities.

Syndecan. One of the other members of this family is syndecan. Originally identified in mouse mammary epithelial cells as a hybrid proteoglycan which carries both heparan sulfate and chondroitin sulfate chains[14,15], this protein has now been identified in a number of cells, including fibroblasts. The ectodomain of syndecan, which can accommodate up to five glycosaminoglycan chains, is larger and distinct from that of fibroglycan, but the transmembrane and cytoplasmic domains of the two proteins are very similar, showing up to 65 percent of sequence identity, including the conservation of four tyrosine residues (fig. 3). Like fibroglycan, syndecan has been highly conserved in mouse [16] and man[17]. Contrary to fibroglycan, syndecan appears to be more abundant in epithelial cells than in mesenchymal cells[18], although the latter may become transiently positive during embryonic development at moments of intense morphogenesis when condensation of the mesenchyme occurs as a

```
                                    ▽    O  ▽ ▽
MRRAALWLWLCALALSLQPALPQIVATNLPPEDQDGSGDDSDNFSGSGAGALQDITLSQQ   60
TPSTWKDTQLLTAIPTSPEPTGLEATAASTSTLPAGEGPKEGEAVVLPEVEPGLTAREQE  120
ATPRPRETTQLPTTHQASTTTATTAQEPATSHPHRDMQPGHHETSTPAGPSQADLHTPHT  180
                         ▽         ▽
EDGGPSATERAAEDGASSQLPAAEGSGEQDFTFETSGENTAVVAVEPDRRNQSPVDQGAT  240
                                   *         *            *
GASQGLLDRKEVLGGVIAGGLVGLIFAVCLVGFMLYRMKKKDEGSYSLEEPKQANGGAYQ  300
        *
KPTKQEEFYA                                                    310
```

Fig. 3. Primary sequence of human syndecan. Copy DNA was obtained by reverse transcribing mRNA from human lung fibroblasts and amplification of the syndecan sequences by a polymerase chain reaction. The predicted protein sequence which is shown here is identical to that reported for Mali et al[17], except at one position. The difference is probably a structural polymorfism, rather than a polymerase error. The symbols are as in fig. 2.

result of tissue interactions[6]. However, the most differentiated epithelia of mature tissues also show little expression of syndecan. There is some evidence for the possible existence of syndecan structural variants or additional related proteoglycans as one of the monoclonal antibodies raised against the cell surface proteoglycans from human lung fibroblasts reacts with both syndecan and one other unique proteoglycan. The core protein of this proteoglycan has an apparent Mr of 125 kDa (see Fig. 1) and also harbours peptide epitopes that are not present on syndecan.

Glypican. Treating the proteoglycans from lung fibroblasts with phosphatidylinositol-specific phospholipase C specifically depletes the protein which migrates with an apparent Mr of ~ 64 kDa from the liposome-intercalable fractions, leaving the other cores, including that of fibroglycan and syndecan unaffected (Fig. 1). This glypiated proteoglycan, 'glypican' in shorthand notice, is a cysteine-rich protein of 558 amino acids which lacks a characteristic transmembrane and cytoplasmic domain (fig. 4). The carboxyl terminus of the protein consists of a hydrophobic sequence which is too short to span a membrane and which resembles the signal peptide-like sequences identified in the carboxyl termini of other proteins with known glycosyl phosphatidylinositol membrane anchors. These sequences seem to function as part of a recognition signal for a transpeptidase reaction which transfers the polypeptide to the glycolipid and eliminates the carboxyl terminal residues from the nascent protein. Glypican is also made by several epithelial cells, and histochemical analyses suggest that in certain polarized cell types this proteoglycan may accumulate at the apical cell surface.

Glypican seems to also differ from the transmembrane proteoglycans in its metabolic fate. In culture, glypican is rapidly and quantitatively shed to the growth medium of the cells, whereas the transmembrane forms appear to be cleared from the cell surface by endocytosis and degradation[19]. This shedding removes a ~ 3 kDa fragment from the carboxyl terminus of the protein, which suggests that this 'constitutive' release from the cell surface involves a proteolytic cleavage of the protein. In this context it is interesting to note that tyrosine recognition signals for internalisation through coated pits have been identified in the cytoplasmic domains of a number of transmembrane

```
                                                          ▽
MELRARGWWLLCAAAALVACARGDPASKSRSCGEVRQIYGAKGFSLSDVPQAEISGEHLR   60
                  O                                     O
ICPQGYTCCTSEMEENLANRSHAELETALRDSSRVLQAMLATQLRSFDDHFQHLLNDSER  120
TLQATFPGAFGELYTQNARAFRDLYSELRLYYRGANLHLEETLAEFWARLLERLFKQLHP  180
QLLLPDDYLDCLGKQAEALRPFGEAPRELRLRATRAFVAARSFVQGLGVASDVVRKVAQV  240
PLGPECSRAVMKLVYCAHCLGVPGARPCPDYCRNVLKGCLANQADLDAEWRNLLDSMVLI  300
TDKFWGTSGVESVIGSVHTWLAEAINALQDNRDTLTAKVIQGCGNPKVNPQGPGPEEKRR  360
RGKLAPRERPPSGTLEKLVSEAKAQLRDVQDFWISLPGTLCSEKMALSTASDDRCWNGMA  420
RGRYLPEVMGDGLANQINNPEVEVDITKPDMTIRQQIMQLKIMTNRLRSAYNGNDVDFQD  480
     ▽ ▽ ▽
ASDDGSGSGSGDGCLDDLCGRKVSRKSSSSRTPLTHALPGLSEQEGQKTSAASCPQPPTF  540
                                                          ↑
LLPLLLFLALTVARPRWR                                            558
```

Fig. 4. Primary sequence of human glypican
The carboxyl terminal hydrophobic sequence which probably serves as a recognition signal for the glypiation of the protein has been indicated by double underlining. The single arrow points to a possible attachment site for the glycosyl phosphatidylinositol anchor. Other symbols are as in fig. 2.

surface glycoproteins[20], whereas the glypiated folate receptor has been shown to enter non-coated pits, to recycle to the surface, and ultimately to be shed[21]. This suggests that the conserved tyrosines in the cytoplasmic domains of the transmembrane proteoglycans may direct molecules like syndecan and fibroglycan to the lysosomes, to be degraded, whereas the phospholipid tail may direct glypican to an endocytotic compartment which does not merge with the catabolic pathway but ultimately delivers the protein at the cell surface after a limited cleavage of the protein and release from its membrane anchor. Unlike fibroglycan, glypican seems to be synthesized and shed by a large variety of cultured fibroblastic and epithelial cell lines, but whether shedding occurs in vivo is not known.

Functional Similarities

The heparan sulfate chains of the cell surface proteoglycans potentially bind to a variety of components of the interstitial extracellular matrix, cell adhesion molecules and growth factors. This has been particularly well documented for syndecan, which has been shown to bind to several matrix proteins[6,22], including fibrillar collagens (types I, III and V), fibronectin, thrombospondin and tenascin, but also to basic fibroblast growth factor[23]. Each time the binding appears mediated through the heparan sulfate chains. However, the structure of the glycosaminoglycan moiety of syndecan varies according to the tissue of its origin[24,25], and this variability seems to affect the potential interactions of this proteoglycan[25]. Moreover, collagen type I, fibronectin and antithrombin III-binding studies of the cell surface proteoglycans of human lung fibroblasts and umbilical vein endothelial cells, have failed to identify significant functional differences amongst the different proteoglycan forms.

Fibroglycan, syndecan and glypican from human lung fibroblasts apparently bind with similar affinities (Kd = ~ 2.10^{-9}) to type I collagen fibers, and seem to be equally represented amongst the cell surface proteoglycans that bind with 'high' (Kd = 10^{-10}) or 'low' (Kd = $2x10^{-9}$) affinity to fibronectin, a binding which, for all these cell surface forms, is completely abolished by a heparitinase-treatment. The results obtained with antithrombin III-binding in endothelial cells, however, are less clear. The proteoglycan fraction which binds with high affinity to the proteinase inhibitor contains only a small fraction of each of the expressed proteoglycan forms, but compared to the starting material, appears to be enriched in glypican, suggesting that glypican may have a somewhat higher incidence of antithrombin III-binding chains.

These functional studies, so far, suggest that, in vitro, in reconstitution experiments, the binding interactions of these proteoglycans are characterized by promiscuity and lack of specificity. The advantages and rationale for expressing different and apparently highly conserved proteoglycan forms at the cell surface may therefore not reside in the intrinsic differences in the (heparan sulfate-mediated) functional properties of these components, but in their specific and regulated expression patterns, possibly allowing the cells to respond differentially to environmental cues and to realise optimal concentrations of heparan sulfate chains at the required times and sites. Glypican e.g., well equipped but not unique in its antithrombin III-binding properties, may allow endothelial cells to express anticoagulant heparan sulfate chains at their apical surfaces. Moreover, through its constitutive shedding, glypican may act as a soluble competitor of the transmembrane forms, possibly reversing their associations. Perhaps better clues to the in vivo functions of these proteoglycans may be gained from the study of their coexpression with their potential ligands, and from the study of cells and organisms that are made specifically deficient in the different proteoglycan forms.

MATRIX-ASSOCIATED HEPARAN SULFATE PROTEOGLYCANS

In contrast to the above, other proteoglycans appear to be more strongly associated with the matrix than with the cell surface. In cultures of human lung fibroblasts e.g. the non-ionic detergent Triton X-100, used to extract the 'membrane-associated' proteoglycans, leaves one major heparan sulfate proteoglycan in association with the fibrillar 'cytoskeleton-matrix' residue[26]. The core protein of this proteoglycan is very large, with a Mr in excess of 400 kDa. Limited sequence analysis and extensive immunological cross reactivities suggest that the core of this proteoglycan is related or identical to that of the basement membrane proteoglycans, for which the low density proteoglycan isolated from the Engelbreth Holm Swarm tumor stands as a prototype[27,28]. The corresponding cDNA clones predict that this protein displays extensive structural similarities to laminin and to the neuronal cell adhesion molecule NCAM[29]. In rotary shadowing the protein has a characteristic beads-on-a-string aspect, hence the name 'perlecan', with the heparan sulfate chains clustered at one end of the molecule[28]. Northern and western blot analyses indicate that this proteoglycan is produced by a large variety of cells, including fibroblasts, suggesting that it is a nearly ubiquitous component of cell-matrix interfaces. Immunohistochemical analyses reveal that this proteoglycan is very abundant or obvious in basement membranes, and confirm that it also accumulates around fibroblasts in situ[30].

The mechanisms that are responsible for the association of this proteoglycan with the extracellular matrix are not known, but once incorporated in a fibroblast matrix, do not seem to call upon its

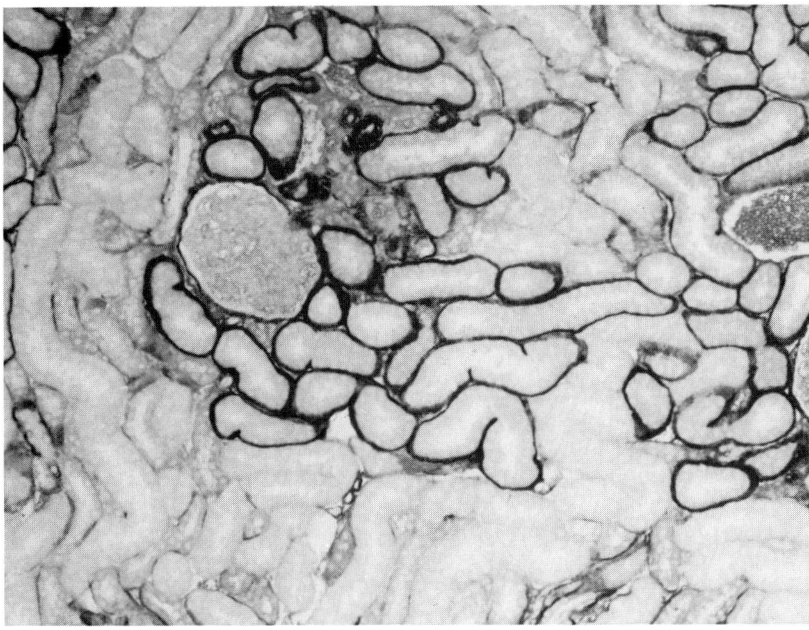

Fig. 5. Heterogeneity of the heparan sulfate composition of basement membranes
Section of the kidney of an adult hamster stained with a monoclonal antibody that reacts specifically with heparan sulfate, showing differences in the abundance of this epitope between basement membranes of proximal and distal tubules. Since anti-Δglucuronosyl-antibodies on heparitinase-treated sections and anti-core protein antibodies reveal no staining differences, these results suggest that the structure of the heparan sulfate chains of the proteoglycans rather than their number or exposure varies at these sites.

continued substitution with heparan sulfate chains as heparitinase does not solubilize the proteoglycan[26]. It may be relevant, in this context, that the core protein of this proteoglycan binds to fibronectin with high affinity and specificity. This binding interaction seems to engage a protease-sensitive fibronectin segment which maps close to the gelatin-binding domain, possibly in the second type III repeat, and a glycosaminoglycan-free core protein fragment which may map in proximity of a heparin-binding site in this protein[31]. Interestingly, studies on the topological organisation of this proteoglycan in the matrix, using polyclonal and core protein-specific monoclonal antibodies, suggest that the protein core may span the full thickness of the basement membranes and contact the interstitial matrix by one of its extremities[30]. These findings imply that fibronectin may function as a docking protein for the anchorage of the proteoglycan in the interstitial matrix.

As for the cell surface proteoglycans, the specific functions of this and potentially still other matrix-associated proteoglycans are unknown. Moreover, as reported for syndecan, the glycosaminoglycan composition of these matrix proteoglycans seems to vary depending on their source (fig. 5). Yet, their heparan sulfate chains appear to be able to bind several of the components identified as ligands for the cell surface proteoglycans, varying from fibrillar type I collagen[26], to antithrombin III[32] and bFGF[33]. Studies which have identified the heparan sulfate chains of the basement membrane proteoglycans as major binding sites for antithrombin III[34] and as repositories for the tissue stores of bFGF[33] lend support to the physiological relevance of these in vitro findings.

It is tempting then to speculate that the different heparan sulfate proteoglycans may be important modulators of the biological activities of their ligands, not only by acting as their cofactors, but also by acting on their compartmentalisation, shed-, surface- and matrix-associated forms entering in mutual competition for the same reactants. The 'mobile' cell surface proteoglycans like fibroglycan, syndecan and glypican may be involved in their recruitment and clearance, the 'stationary' matrix-associated proteoglycans in their storage in metabolically quiescent but readily available forms. In that instance the rather ubiquitous distribution of matrix proteoglycans, but specific and fluctuating patterns of cell surface proteoglycan expression suggest that regulatory mechanisms that affect this balance likely act upon the cell surface forms.

In conclusion, differential expression of the various heparan sulfate proteoglycans in different cell types, membrane domains and developmental stages suggests that each of these molecules assumes a specific task. The heparan sulfate chains may thereby be considered as the effector parts of the molecules, involved in the binding, activation and sequestration of growth factors, adhesion receptors, extracellular matrix proteins and proteinase inhibitors. The core proteins, in contrast, harbor the structures for targeting the proteoglycans to their final destinations, directing them to specific membrane domains, endocytotic compartments or extracellular docking proteins, and provide means for controlling the abundance and temporal expression of heparan sulfate chains at these sites.

ACKNOWLEDGEMENTS

The personal investigations cited in this summary were supported by grants 9001486 and 3006687 of the Nationaal Fonds voor Geneeskundig Wetenschappelijk Onderzoek, Belgium, by the Interuniversity Network for Fundamental Research, and by a grant 'Geconcerteerde Acties' from the Belgian Government.
Guido David is a 'research director' of the Nationaal Fonds voor Wetenschappelijk Onderzoek, Belgium.

REFERENCES

1. U. Lindahl, L. Thunberg, G. Bäckstrom, J. Riesenfeld, K. Nordling, and I. Björk, Extension and structural variability of the antithrombin-binding sequence in heparin, J. Biol. Chem. 259: 12368 (1984).
2. A. Yayon, M. Klagsbrun, J.D. Esko, P. Leder, and D.M. Ornitz, Cell surface, heparin-like molecules are required for binding of basic fibroblast growth factor to its high affinity receptor, Cell 64:841 (1991).

3. A. Rapraeger, A. Krufka, B.B. Olwin, Requirement of heparan sulfate for bFGF-mediated fibroblast growth and myoblast differentiation, Science 252:1705 (1991).
4. I. Thesleff, M. Jalkanen, S. Vainio, and M. Bernfield, Cell surface proteoglycan expression correlates with epithelial-mesenchymal interaction during tooth morphogenesis, Dev. Biol. 129:565 (1988).
5. S. Vainio, M. Jalkanen, and I. Thesleff, Syndecan and tenascin expression is induced by epithelial-mesenchymal interactions in embryonic tooth mesenchyme, J. Cell Biol. 108:1945 (1989).
6. M. Bernfield, and R.D. Sanderson, Syndecan, a morphogenetically regulated cell surface proteoglycan that binds extracellular matrix and growth factors, Phil. Trans. R. Soc. Lond. 327: 171 (1990).
7. L. Kjellén, I. Pettersson, and M. Höök, Cell-surface heparan sulfate: an intercalated membrane proteoglycan, Proc. Natl. Acad. Sci. 78:5371 (1981).
8. A.C. Rapraeger, and M. Bernfield, Heparan sulfate proteoglycans from mouse mammary epithelial cells, J. Biol. Chem. 258:3632 (1983).
9. V. Lories, G. David, J.J. Cassiman, and H. Van den Berghe, Heparan sulfate proteoglycans of human lung fibroblasts, Eur. J. Biochem. 158:351 (1986).
10. V. Lories, J.J. Cassiman, H. Van den Berghe, and G. David, Multiple distinct membrane heparan sulfate proteoglycans in human lung fibroblasts, J. Biol. Chem. 264:7009 (1989).
11. M. Ishihara, N.S. Fedarko, and H.E. Conrad, Involvement of phosphatidylinositol and insulin in the coordinate regulation of proteoheparan sulfate metabolism and hepatocyte growth, J. Biol. Chem. 262:4708 (1987).
12. D.J. Carey, and R.C. Stahl, Identification of a lipid-anchored heparan sulfate proteoglycan in Schwann cells, J. Cell Biol. 111:2053 (1990).
13. P. Marynen, J. Zhang, J.J. Cassiman, H. Van den Berghe, and G. David, Partial primary structure of the 48- and 90-Kilodalton core proteins of cell surface-associated heparan sulfate proteoglycans of lung fibroblasts, J. Biol. Chem. 264:7017 (1989).
14. A. Rapraeger, M. Jalkanen, E. Endo, J. Koda, and M. Bernfield, The cell surface proteoglycan from mouse mammary epithelial cells bears chondroitin sulfate and heparan sulfate glycosaminoglycans, J. Biol. Chem. 260:11046 (1985).
15. G. David, and H. Van den Berghe, Heparan-sulfate chondroitin sulfate hybrid proteoglycan of the cell surface and basement membrane of mouse mammary epithelial cells, J. Biol. Chem. 260:11067 (1985).
16. S. Saunders, M. Jalkanen, S. O'Farrell, and M. Bernfield, Molecular cloning of syndecan, an integral membrane proteoglycan, J. Cell Biol. 108:1547 (1989).
17. M. Mali, P. Jaakkola, A.-M. Arvilommi, and M. Jalkanen, Sequence of human syndecan indicates a novel gene family of integral membrane proteoglycans, J. Biol. Chem. 265:6884 (1990).
18. K. Hayashi, M. Hayashi, M. Jalkanen, J.H. Firestone, R.L. Trelstad, and M. Bernfield, Immunocytochemistry of cell surface heparan sulfate proteoglycan in mouse tissues, A light and electron microscopic study, J. Histochem. Cytochem. 35:1079 (1987).
19. G. David, V. Lories, B. Decock, P. Marynen, J.J. Cassiman, and H. Van den Berghe, Molecular cloning of a phosphatidylinositol-anchored membrane heparan sulfate proteoglycan from human lung fibroblasts, J. Cell Biol. 111:3165 (1990).

20. N.T. Ktistakis, D'Nette Thomas, and M.G. Roth, Characteristics of the tyrosine recognition signal for internalization of transmembrane surface glycoproteins, J. Cell Biol. 111:1393 (1990).
21. K.G. Rothberg, Y. Ying, J.F. Kolhouse, B.A. Kamen, and R.G.W. Anderson, The glycophospholipid-linked folate receptor internalizes folate without entering the clathrin-coated pit endocytic pathway, J. Cell Biol. 110:637 (1990).
22. K. Elenius, M. Salmivirta, P. Inki, M. Mali, and M. Jalkanen, Binding of human syndecan to extracellular matrix proteins, J. Biol. Chem. 265:17837 (1990).
23. M.C. Kiefer, J.C. Stephans, K. Crawford, K. Okino, and P.J. Barr, Ligand-affinity cloning and structure of a cell surface heparan sulfate proteoglycan that binds basic fibroblast growth factor, Proc. Natl. Acad. Sci. USA 87:6985 (1990).
24. R.D. Sanderson, and M. Bernfield, Molecular polymorphism of a cell surface proteoglycan: distinct structures on simple and stratified epithelia, Proc. Natl. Acad. Sci. USA 85:9562 (1988).
25. M. Salmivirta, K. Elenius, S. Vainio, U. Hofer, R. Chiquet-Ehrismann, I. Thesleff, and M. Jalkanen, Syndecan from embryonic tooth mesenchyme binds tenascin, J. Biol. Chem. 266:7733 (1991).
26. A. Heremans, J.J. Cassiman, H. Van den Berghe, and G. David, Heparan sulfate proteoglycan from the extracellular matrix of human lung fibroblasts, J. Biol. Chem. 263:4731 (1988).
27. J.R. Hassell, W.C. Leyshon, S.R. Ledbetter, B. Tyree, S. Suzuki, M. Kato, K. Kimata, and H.K. Kleinman, Isolation of two forms of basement membrane proteoglycans, J. Biol. Chem. 260:8098 (1985).
28. M. Paulsson, P.D. Yurchenco, G.C. Ruben, J. Engel, and R. Timpl, Structure of low density heparan sulfate proteoglycan isolated from a mouse tumor basement membrane, J. Mol. Biol. 197:297 (1987).
29. D.M. Noonan, E.A. Horigan, S.R. Ledbetter, G. Vogeli, M. Sasaki, Y. Yamada, and J.R. Hassell, Identification of cDNA clones encoding different domains of the basement membrane heparan sulfate proteoglycan, J. Biol. Chem. 263:16379 (1988).
30. A. Heremans, B. Van der Schueren, B. De Cock, M. Paulsson, J.J. Cassiman, H. Van den Berghe, and G. David, Matrix-associated heparan sulfate proteoglycan: core protein-specific monoclonal antibodies decorate the pericellular matrix of connective tissue cells and the stromal side of basement membranes, J. Cell Biol. 109:3199 (1989).
31. A. Heremans, B. De Cock, J.J. Cassiman, H. Van den Berghe, and G. David, The core protein of the matrix associated heparan sulfate proteoglycan binds to fibronectin, J. Biol. Chem. 265:8716 (1990).
32. G. Pejler, G. Backstrom, and U. Lindahl, Structure and affinity for antithrombin of heparan sulfate chains derived from basement membrane proteoglycans, J. Biol. Chem. 262:5036 (1987).
33. A.-M. Gonzalez, M. Buscaglia, M. Ong, and A. Baird, Distribution of basic fibroblast growth factor in the 18-day rat fetus: localization in the basement membranes of diverse tissues, J. Cell Biol. 110:753 (1990).
34. A.I. de Agostini, S.C. Watkins, H.S. Slayter, H. Youssoufian, and R.D. Rosenberg, Localization of anticogulantly active heparan sulfate proteoglycans in vascular endothelium: antithrombin binding on cultured endothelial cells and perfused rat aorta, J. Cell Biol. 111:1293 (1990).

SYNDECAN - A CELL SURFACE PROTEOGLYCAN THAT SELECTIVELY BINDS EXTRACELLULAR EFFECTOR MOLECULES

Markku Jalkanen[1,2], Klaus Elenius[1] and Markku Salmivirta[1]

Department of Medical Biochemistry[1], University of Turku, SF-20520 Turku, Finland
Turku Biotechnology Center[2], P.O. Box 123, SF-20521 Turku, Finland

INTRODUCTION

Syndecan is a cell surface proteoglycan (Saunders et al., 1989), which was originally isolated from mouse mammary epithelial (NMuMG) cells as a hybrid proteoglycan containing both heparan sulfate and chondroitin sulfate glycosaminoglycan (GAG) chains (Rapraeger et al., 1985). The ratio and total amount of GAG bound to syndecan is known to vary in different tissues (Sanderson & Bernfield, 1988; Salmivirta et al., 1991) and cells (Rapraeger 1989). Syndecan binds several extracellular matrix molecules such as fibrillar collagens of interstitial matrix (Koda et al., 1985), fibronectin (Saunders & Bernfield, 1988) and thrombospondin (Sun et al., 1989) but shows no binding to some other heparin binding matrix molecules, like vitronectin and laminin (Koda et al., 1985; Elenius et al., 1990). Syndecan also binds growth factors, like bFGF (Kiefer et al., 1990), and could therefore play a major role in the regulation of cell adhesion, growth and differentiation (for a review see Jalkanen et al., 1991).

The conserved structure of the core protein of syndecan (Mali et al., 1990) could be involved in the signal transduction through the plasma membrane but the GAG-chains provide both specificity and affinity for the ligand binding. Syndecan is known to alter its glycosylation in different tissues and cells, suggesting that this variation in glycosylation could result also in the selective recognition of matrix molecules by syndecan of different cell types. This tissue specificity in syndecan glycosylation is an example of specific use of GAG in the interactions resulting in the regulation of cell proliferation, differentiation, organ formation and tissue maintenance.

Heparin and Related Polysaccharides
Edited by D.A. Lane *et al.*, Plenum Press, New York, 1992

SYNDECAN AND TENASCIN COLOCALIZE IN THE DIFFERENTIATING MESENCHYME OF TOOTH

During tooth formation syndecan expression follows morphogenetic rather than histological boundaries (Thesleff et al., 1988). It is expressed by budding epithelium but soon after epithelial penetration, the neural crest-derived mesenchyme next to epithelium shows strong positive staining for syndecan. Indeed, our recent results indicate that syndecan expression by differentiating tooth mesenchyme is regulated by epithelial contact, both at protein (Vainio et al., 1989) and mRNA (Vainio et al., 1991) level. Although studied for decades, no indication of the chemical nature of a possible "morphogen" exists (Thesleff et al., 1991). Hence, syndecan expression could be used as a marker to study these interactions and components involved in a more comprehensive manner.

Similar to syndecan, also tenascin - a matrix glycoprotein, is expressed by condensing tooth mesenchyme (Vainio et al., 1989). Tenascin is an adhesive matrix glycoprotein with several interesting structural domains (Spring et al., 1990), including binding domain(s) for heparin/heparan sulfate (Chiquet et al., 1991). Furthermore, it has alternatively spliced variants and its expression is developmentally regulated (for review see Chiquet-Ehrismann, 1990). Our recent interest has focused on the possible interaction between syndecan and tenascin. By this interaction syndecan and tenascin could participate in the regulation of mesenchymal differentiation during organ formation by providing a spatially and temporally regulated cell surface anchorage of to the matrix.

SYNDECAN FROM TOOTH MESENCHYME SELECTIVELY BINDS TENASCIN

We have previously observed that syndecan isolated from condensing mesenchyme of tooth reveals selective recognition of tenascin not observed for syndean from NMuMG cells (Salmivirta et al., 1991). If the binding of syndecan to tenascin is compared to binding of syndecan to fibronectin, the mesenchymal syndecan reveals five-fold higher tendency to bind tenascin than syndecan from NMuMG epithelial cell line (Fig. 1A). Syndecan lacking GAG chains does not bind tenascin, indicating that the specific but so far unknown characteristics of heparan sulfate of this syndecan must be responsible for tenascin recognition (Salmivirta et al. 1991). Syndecan from tooth mesenchyme contains only heparan sulfate and is not a hybrid containing also chondroitin sulfate, like syndecan from NMuMG cells. We have not been able to increase the binding of NMuMG-derived syndecan to tenascin after enzymatic removal of chondroitin sulfate from syndecan, suggesting that the role of chondroitin sulfate is not preventive for tenascin binding. Our focus is more now in the analysis of heparan sulfate immunoisolated from tooth mesenchyme and, as described later, from 3T3 cells.

It is also interesting to note that the changes observed in tenascin expression during tooth formation, has not been observed for other matrix proteins, including fibronectin and collagen Type III fibrils (Vainio et al., 1989). It is therefore conceivable that the intriguing expression patterns of tenascin and syndecan may regulate cellular differentiation and proliferation by providing a proper cell-matrix interaction, which can also modulate growth factor influence, as described in the next paragraph.

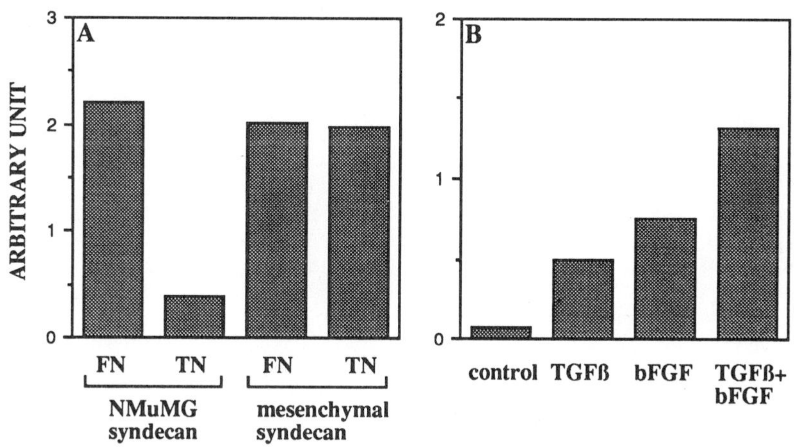

Fig 1. The binding properties of mesenchymal syndecan to matrix and the inductivity of syndecan expression in mesenchymal 3T3 fibroblasts. (**A**) 50 µg of fibronectin and tenascin were loaded on nitrocellulose filters which were subsequently incubated in the presence of sulfate-labeled syndecan ectodomain isolated either from a mammary epithelial cell line (NMuMG) or from the condensing mesenchyme of developing tooth. The amount of radioactive syndecan bound was detected with autoradiography and densitometry. (**B**) 3T3 cells were cultured for 24 h in the absence or presence of 2.5 ng/ml TGFß1, 10 ng/ml bFGF or their combination and the amount of syndecan at the cell surface was quantitated by a radioimmunoassay and densitometry.

GROWTH FACTORS - PUTATIVE REGULATORS OF SYNDECAN EXPRESSION IN MESENCHYMAL CELLS

During the last decade the information on locally secreted growth factors with developmentally important influence on growth and differentiation of adjacent cells has increased significantly. Two best known such growth factor families are transforming growth factos (TGF; Massague, 1990) and fibroblast growth factors (FGF; Burgess & Maciag, 1989). Recently these growth factors were also shown to be expressed around and within the condensing and proliferating tooth mesenchyme. TGF-ß1 is first expressed on the tips of penetrating epithelia but early during mesenchymal condensation, the mesenchymal cells adjacent to penetrating epithelia, also produce TGF-ß (Vaahtokari et al., 1991). Int-2, which is an embryonal member of bFGF-family, is expressed by the condensing mesemchyme so that its expression temporally correlates to the morphogenetic activation of tooth mesenchyme (Wilkinson et al., 1989).

Our recent results indicate that syndecan expression by mesenchymal cells (3T3) can be enhanced in vitro by simultaneous exposure of cells to bFGF and TGF-ß (Elenius et al., 1991b). The influence of these growth factors is synergistic (Fig. 1B), indicating that they both can participate in the regulation of syndecan expression by mesenchymal cells. Furthermore, the mesenchymal syndecan binds bFGF (Elenius et al., 1991b), suggesting that the expression of syndecan on the surfaces of mesenchymal cells could be a regulatory requirement for the action of growth factor action, as recently described for heparin and heparan sulfate (Yayon et al., 1991; Rapraeger et al., 1991). The fact that cultured mesenchymal cells can be induced to synthesize syndecan, will provide an alternative source of syndecan for comparison of its GAG composition and structure with syndecan from NMuMG cells. By this comparison it will be possible to reveal initial information regarding the specificity of ligand recognition and structural requirement of GAG chains for individual syndecan.

Syndecan expression is also enhanced by regenerating tissues, e.g. of cutaneous wounds (Elenius et al., 1991a). Increased syndecan expression can be observed close to wound site by keratinocytes of skin epidermis and hair follicles, and suggest participation of soluble factors in the regulation. Also endothelial cells of sprouting capillaries of granulation tissue express tempo-spatially syndecan that could indicate a role for syndecan as a growth factor promoter in angiogenesis (Elenius et al., 1991a). Therefore, growth factors can regulate in many situations syndecan expression, and on the other hand, enhanced syndecan expression can influence the action of growth factors.

SYNDECAN CAN LOCALIZE GROWTH FACTOR PROMOTION TO THE SITE OF CELL-MATRIX INTERFACE

Our recent results suggest that syndecan could have tissue specific interactions, which could provide selective binding of syndecan to different extracellular effector molecules during different stages of cellular differentiation and organ formation. All the information of syndecan interactions available today indicate that they are mediated via GAG-chains of syndecan. Thus, it seems likely that cells have capability to regulate GAG composition attached to the syndecan core protein. They could also regulate the composition of oligosaccharide organization within one GAG-chain, because of the known diversity of heparan sulfate structure (for review see Gallagher 1989; Kjellen & Lindahl, 1991).

During morphogenetic movement and proliferation in the embryo, ECM components in defined patterns can serve as substrates for motile and differentiating cells. The fact that syndecan can bind both matrix components and growth factors, suggests that it can immobilize the growth factor promotion to the site where syndecan interacts with matrix. One such situation is the condensing tooth mesenchyme, which shows expression of syndecan, tenascin and growth factors. The dualistic binding capacity of syndecan provides one explanation of how the known influence of extracellular matrix (Stoker et al., 1990) can be translated into cellular behaviour. It also promotes the importance of GAGs in the regulation of cell proliferation, differentiation and morphogenesis leading to organ formation and maintenance.

ACKNOWLEDGEMENTS

The original work summarized in this review has been supported by The Academy of Finland, The Finnish Cancer Union, The National Institute of Health (NIH) Grant DE09399-01, The Paulo Foundation and The Research and Science Foundation of Farmos.

REFERENCES

Burgess, W.H., Maciag, T., 1989, The heparin-binding (fibroblast) growth factor family of proteins, Annu. Rev. Biochem., 58: 575-606.
Chiquet-Ehrismann, R., 1990, What distinquishes tenascin from fibronectin, FASEB J., 4: 2598-2604
Chiquet, M., Vrucinic-Filipi, N., Schenk, S., Beck, K., and Chiquet-Ehrismann, R., 1991, Isolation of chick teneascin variants and fragments: A c-terminal heparin-binding fragment produced by cleavage of the extra domain from the largest subunit splicing variant, Eur. J. Biochem., in press.
Elenius, K., Salmivirta, M., Inki, P., Mali, M., and Jalkanen, M., 1990, Binding of human syndecan to extracellular matrix, J. Biol. Chem., 265: 17837-17843.
Elenius, K., Vainio, S., Laato, M., Salmivirta, M., Thesleff, I., and Jalkanen, M., 1991a, Induced expression of syndecan in healing wounds, J. Cell Biol., 114: 585-596.
Elenius, K., Määttä, A., Salmivirta, M., and Jalkanen, M., 1991b, Growth factors induce 3T3 cells to express bFGF-binding syndecan, EMBO J., submitted.
Gallagher, J.T., 1989, The extended family of proteoglycans: social residents of the pericellular zone, Curr. Opin. Cell Biol., 1: 1201-1218.
Jalkanen, M., Jalkanen, S., and Bernfield, M., 1991, Binding of extracellular effector molecules by cell surface proteoglycans, in: Receptors for extracellular matrix, J. MacDonald, and R. Mecham, Eds., Academic Press, San Diago, in press.

Kiefer, M.C., Stephans, J.C., Crawford, K., Okino, K. and Barr, P.J., 1990, Ligand-affinity cloning and structure of a cell surface heparan sulfate proteoglycan that binds basic fibroblast growth factor, Proc. Natl. Acad. Sci. USA, 87: 6985-6989.

Kjellen, L., and Lindahl, U., 1991, Proteoglycans: Structure and interactions, Annu. Rev. Biochem., 60: 443-75.

Koda, J., Rapraeger, A., and Bernfield, M., 1985, Heparan sulfate proteoglycans from mouse mammary epithelial cells. Cell surface proteoglycan as a receptor for interstitial collagens, J. Biol. Chem.,260: 8157-8162.

Mali, M., Jaakkola, P., Arvilommi, A.-M., and Jalkanen, M., 1990, Sequence of human syndecan indicates a novel gene family of integral membrane proteoglycans, J. Biol. Chem., 265: 6884-6889.

Massague, J., 1990, The transforming growth factor-ß family, Annu. Rev. Cell Biol., 6: 597-642.

Rapraeger, A., Jalkanen, M., Endo, E., Koda, J., and Bernfield, M., 1985, The cell surface proteoglycan from mouse mammary epithelial cells bears chondroitin sulfate and heparan sulfate glycosaminoglycans, J. Biol. Chem., 260: 11046-11052.

Rapraeger, A., 1989, Transforming growth factor (type ß) promotes the eaddition of chondroitin sulfate chains to the cell surface proteoglycan (syndecan) of mouse mammary epithelia, J. Cell Biol., 109: 2509-2518.

Rapraeger, A., Krufka, A., Olwin, B.B., 1991, Requirement of heparan sulfate for bFGF-mediated fibroblast growth and myoblast differentiation, Science, 252: 1705-1708.

Salmivirta, M., Elenius, K., Vainio, S., Hofer, U., Chiquet-Ehrismann, R., Thesleff, I., and Jalkanen, M., 1991, Syndecan from tooth mesenchyme binds tenascin, J. Biol. Chem., 266: 7733-7739.

Sanderson, R., and Bernfield, M., (1988) Molecular polymorphism of a cell surface proteoglycan: distinct structures on simple and stratified epithelia, Proc. Natl. Acad. Sci. USA, 85: 9562-9566.

Saunders, S., and Bernfield, M., 1988, Cell surface proteoglycan binds mouse mammary epithelial cells to fibronectin and behaves as a receptor for interstitial matrix, J. Cell Biol., 106: 423-430.

Saunders, S., Jalkanen, M., O'Farrell, S., and Bernfield, M., 1989, Molecular cloning of syndecan, an integral membrane proteoglycan, J. Cell Biol., 108: 1547-1556.

Spring, J., Beck, K., and Chiquet-Ehrismann, R., 1989, Two contrary functions of tenascin: dissection of the active sites by recombinant tenascin fragments, Cell, 59: 325-334.

Stroker, A.W., Streuli, C.H., Martins-Green, M., and Bissell, M.J., 1990, Designer microenvironments for the analysis of cell and tissue function. Curr. Opin. Cell Biol., 2: 864-874.

Sun, X., Mosher, D.F., and Rapraeger, A., 1989, Heparan sulfate-mediated binding of epithelial cell surface proteoglycan to thrombospondin, J. Biol. Chem., 264: 2885-2889.

Thesleff, I., Jalkanen, M., Vainio, S., and Bernfield, M., 1988, Cell surface proteoglycan expression correlates with epithelial-mesenchymal interaction during tooth morphogenesis, Dev. Biol., 129: 565-572.
Thesleff, I., Jalkanen, M., Partanen, A.-M., and Vainio, S., 1991, Molecular changes in dental mesenchyme during tooth development, in: "Aspects of Oral Molecular Biology. Frontiers of Oral Physiology", Ferguson, D.B., Ed., S. Karger, Basel, vol. 8, pp 42-56.
Vaahtokari, A., Vainio, S., and Thesleff, I., 1991, Associations between transforming growth factor ß1 RNA expression and epithelial-mesenchymal interactions during tooth morphogenesis, Development, in press.
Vainio, S., Jalkanen, M., and Thesleff, I., 1989, Syndecan and tenascin expression is induced by epithelial-mesenchymal interactions in embryonic tooth mesenchyme. J. Cell Biol., 108: 1945-1954.
Vainio, S., Jalkanen, M., Vaahtokari, A., Sahlberg, C., Mali, M., Bernfield, M., and Thesleff, I., 1991, Expression of syndecan gene is induced early, is transient, and correlates with changes in mesenchymal cell proliferation during tooth morphogenesis, Dev. Biol., 147: in press.
Wilkinson, D.G., Bailes, J.A., and McMahon, A.P., 1989, Expression pattern of the FGF-related proto-oncogene int-2 suggests multiple roles in fetal development, Development, 105: 131-136.
Yayon, A., Klagsbrun, M., Esko, J.D., Leder, P., and Ornitz, D.M., 1991, Cell surface, heparin-like molecules are required for binding of basic fibroblast growth factor to its high affinity receptor, Cell, 64: 841-848.

HEPARAN SULFATE PROTEOGLYCANS

AND SIGNALLING IN CELL ADHESION

Anne Woods and John R. Couchman

Department of Cell Biology
University of Alabama at Birmingham
Birmingham, AL 35294-0019, USA.

INTRODUCTION

Studies on cell adhesion to isolated and purified extracellular matrix molecules in vitro clearly show that substratum-bound matrix molecules can influence cytoskeletal architecture, through some form of transmembrane signalling process. Fibronectin is one such matrix molecule, and its biological properties in terms of the induction of cell adhesion are now known to be mediated by several domains of the molecule, including the "cell"-binding domain which binds to integrin receptors of the β_1 family (reviewed in Hynes, 1990; Ruoslahti, 1988). Fibronectin has, in addition, both an amino-terminal 29-kD and a 31-kD heparin-binding domain, which we showed previously were involved in the organization of spread cells (Woods et al., 1986). Cells not only attach to fibronectin, but also spread and can form specialized adhesion structures, focal adhesions (reviewed in Burridge et al. 1988; Hynes, 1990; Woods & Couchman, 1988; and see Fig. 1). These are highly specialized foci of cell-matrix interaction where concentrations of specific membrane and cytoskeletal components form a continuum from the substrate to which the cells are attached to the internal microfilament system (Woods & Couchman, 1988). These structures are not found in highly motile cells (Couchman & Rees, 1979), and form concomitantly as migratory cells stop moving and start to export endogenously synthesized extracellular matrix. Thus, fibroblasts in vitro can exhibit either a locomotory phenotype or a stationary phenotype characterized by the ability of the cells to alter their environment. This is of great importance in many disease processes, since loss of anchorage, and changes in matrix deposition are often causative events in various connective tissue diseases.

Focal adhesions have been studied with respect to their morphology and components. For example, interference reflection microscopy (IRM), a technique that indicates the proximity of the cell underside to the substrate (Izzard & Lochner, 1976, 1980), has been used to monitor their formation and dissolution in living cells (Fig. 1). Electron microscopy has also shown that focal adhesions are the areas of the cell which are closest to the substrate, and that they contain a submembraneous electron dense plaque to which actin-containing microfilament bundles (stress fibers) are attached (reviewed in Burridge et al., 1988; Woods & Couchman, 1988). Some of the plaque components have now been determined, mainly through the use of immunological methods

Heparin and Related Polysaccharides
Edited by D.A. Lane *et al.*, Plenum Press, New York, 1992

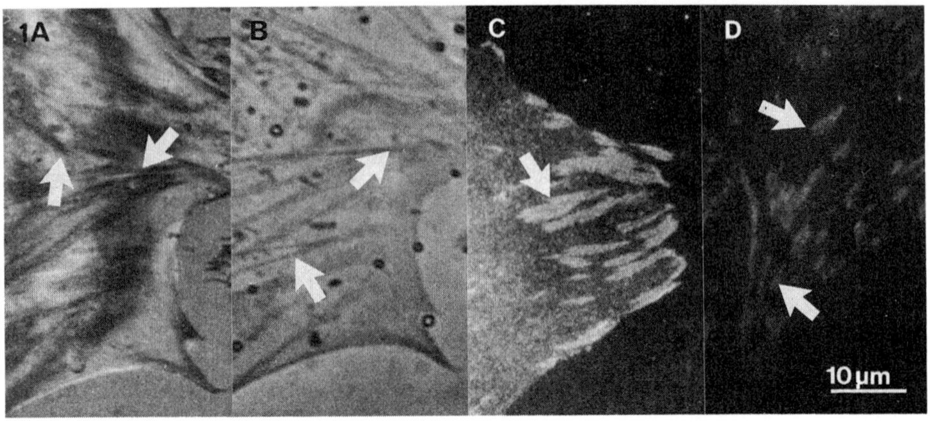

Figure 1. IRM (A), phase contrast microscopy (B), and immunofluorescent labeling of talin (C) and integrin β_1 subunits in human embryo fibroblasts spread on fibronectin substrates for 3 hours. Focal adhesions and stress fibers are arrowed.

(reviewed in Burridge et al., 1988, Woods & Couchman, 1988). In addition to actin, myosin and other components of the microfilaments, they contain increased concentrations of vinculin and talin, which are often used to specifically label focal adhesions, and α-actinin (Burridge et al., 1988; Woods & Couchman, 1988: Fig. 1). These are the structural components of the focal adhesion, but other components which may have regulatory roles are now being identified. These include calpain II (Beckerle et al., 1987) and, at least in some cells, protein kinase C (PKC) type 3 (Jaken et al., 1989). Their role in the formation or stability of focal adhesions have not, however, yet been elucidated, and the possible role of PKC is the subject of our present studies.

Membrane components of focal adhesions are also specialized. When cells adhere to fibronectin, $\alpha_5\beta_1$ integrins are found in increased amounts in focal adhesions whereas $\alpha_v\beta_3$ integrins may be used by cells on vitronectin (Fath et al., 1989; Hynes, 1990; Ruoslahti, 1988). In addition, cells can form focal adhesions on substrates of laminin or type I collagen, which both utilize different receptors. Thus cells can form similar structures using different ligand-receptor systems, indicating some common mechanism of formation. The plaque components tested so far appear to be similar regardless of the substrates and another common feature is the concentration of cell surface heparan sulfate proteoglycans in the membrane overlying internal stress fibers and at focal adhesions (Woods et al., 1984), indicative of some role for these membrane components in focal adhesion structure or formation.

The cascade of events underlying focal adhesion formation is not determined, although it is known that integrins may interact with both talin and α-actinin, and that talin, vinculin, α-actinin and actin can form a multicomplex (Burridge & Mangeat, 1984; Burridge et al., 1988; Horwitz et al., 1986; Otey et al., 1990). The stimuli for these interactions are not known, but they may depend on the concentration of these individual components, so that a mechanism which increases their concentration in a specific area of the cell may induce their interactions.

ROLE OF HEPARAN SULFATE IN FOCAL ADHESION FORMATION

Our earlier studies monitoring the distribution of cell surface heparan sulfate proteoglycans by immunofluorescence and immunoelectron microscopy indicated that they were concentrated along the membrane overlying stress fibers, at focal adhesions and at cell-cell contact areas (Woods et al., 1984). Later studies showed that a hydrophobic heparan sulfate proteoglycan population was retained when live cells were treated with detergent (Woods et al., 1985). This technique has been extensively used to determine whether membrane components are linked to the cytoskeleton of cells since free membrane components should be solubilized. In addition, biochemical analysis of cell-matrix adhesion zones has indicated the presence of a hydrophobic heparan sulfate proteoglycan which is capable of binding fibronectin (Lark & Culp, 1984).

Cell biological studies have indicated that heparan sulfate proteoglycans may be important in focal adhesion formation (Couchman et al., 1988; LeBaron et al., 1988; Woods et al., 1986). As stated above, fibronectin has several sites which can interact with cell surfaces. The best studied interaction has been that of the RGD sequence of the 105-kD "cell"-binding domain of fibronectin binding to cell surface integrins (reviewed in Hynes, 1990; Ruoslahti, 1988). Our previous studies (Woods et al., 1986), and that of others (Beyth & Culp, 1984; Izzard et al., 1986; Streeter & Rees, 1987) indicated, however, that these interactions were insufficient for focal adhesion formation. Normal fibroblasts will attach, spread and form focal adhesions when plated on substrates of fibronectin, even when treated with cycloheximide to prevent protein synthesis (Woods et al., 1986: Fig. 1). Thus fibronectin can provide all the 'signals' needed by cells to cause the requisite rearrangement of membrane and cytoskeletal components into these specialized areas. In contrast, although cells will attach and spread on substrates of the 105-kD fragment (Woods et al., 1986; Fig. 2), they do not progress to focal adhesion formation in the absence of protein synthesis, and remain in a locomotory phenotype. The later phase of focal adhesion formation appears to require interactions of the cells with heparin-binding fibronectin fragments, and can be promoted in cells prespread on 105-kD fragments by the addition of low amounts (0.2 nM) of either the 29- or 31-kD heparin-binding fibronectin fragments (Woods et al., 1986). This occurs within 30 minutes, even in the continued presence of cycloheximide, and thus represents a rearrangement of preexisting components, rather than the induction or alteration of protein synthesis. Two other studies using mutant cells have highlighted the possible role of heparan sulfate proteoglycans in focal adhesion formation. First, mutant CHO cells, which lack the ability to synthesize proteoglycans, cannot form focal adhesions, even when plated on substrates of intact fibronectin (LeBaron et al., 1988). Second, mutant BHK cells, which do not form focal adhesions in response to fibronectin, have a cell surface heparan sulfate of reduced half-life and sulfation, and reduced capacity to bind fibronectin (Couchman et al., 1988).

SIGNALLING MECHANISMS INVOLVED IN FOCAL ADHESION FORMATION

The signalling mechanisms involved in cell adhesion processes have not been well studied to date. It is indeed difficult to monitor individual signals from matrix molecules when normal cells seeded in serum-containing medium are undergoing a continuum of events such as attachment, spreading and focal adhesion formation. First, not all cells are at the same stage simultaneously. Second, endogenous matrix synthesis is occurring and we have previously shown that the stage of adhesion can determine

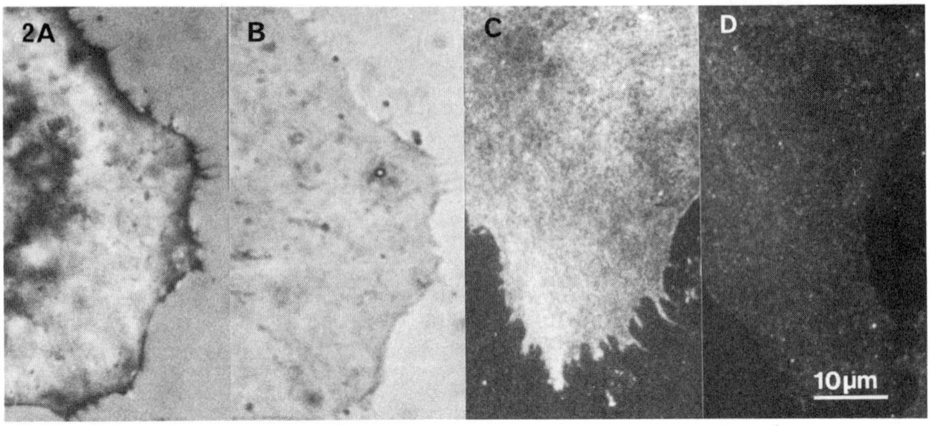

Figure 2. IRM (A), phase contrast microscopy (B), and immunofluorescent labeling of talin (C) and integrin β_1 subunits in human embryo fibroblasts spread for 3 hours on a substrate of 105-kD fibronectin fragment.

export of matrix components (Couchman & Rees, 1979; Woods et al., 1983). Endogenously synthesized matrix molecules could, in turn, affect the metabolism of the cells. We therefore prefer to use the system of cycloheximide-treated cells adhering to a defined substrate of fibronectin or its fragments in serum-free medium to investigate specific 'signals' given to the cells at different adhesion stages. In particular, the system described above where cells prespread on substrates of 105-kD fibronectin are treated with soluble heparin-binding fragments allows investigation of the processes specifically accompanying the final phase of cell adhesion, that of focal adhesion formation. Our experiments were designed to determine three things.
1). Is any kinase system specifically stimulated during the induction of focal adhesion formation by fibronectin?
2). Does the signal transduced to the cell on addition of heparin-binding fibronectin fragments involve this same kinase system?
3). Are there artificial means to stimulate these systems and thus promote focal adhesion formation in the absence of heparin-binding activity of fibronectin?

Here, we present evidence that PKC activation is needed for focal adhesion formation, rather than that of cAMP- or cGMP-dependent kinases, and that this activation can be promoted by the binding of heparin-binding domains of fibronectin.

Prevention of Focal Adhesion Formation by Kinase Inhibitors

We used human embryo fibroblasts, treated with cycloheximide to ensure we were monitoring only the response to the substrate onto which the cells were seeded, and plated these onto fibronectin substrates for 1 hour to allow attachment and spreading to be completed (Woods & Couchman, submitted). These cultures were then incubated for a further 2 hours in the presence or absence of two kinase inhibitors to monitor the effects of inhibition of different kinases on the ability of the cells to form focal adhesions. The kinase inhibitors used were H7 and HA1004, both available commercially. These are isoquinolinesulfonamides with varying specificities for PKC, cAMP- or cGMP-dependent kinases, and myosin light chain kinase (Asano & Hidaka, 1984;

Hidaka et al., 1984). H7 inhibits the first three kinases at similar levels (Ki=3.0 - 6.0 μM), but has little effect on myosin light chain kinase until higher concentrations are used (Ki=97 μM). This was used to determine the effect of inhibiting all three kinases. In contrast, another inhibitor, HA1004, affects cAMP- and cGMP-dependent kinases at similar concentration to that of H7 (Ki=2.3 and 1.3 μM respectively), but has little effect on PKC until higher concentration (Ki=40 μM). Thus, if H7 affected focal adhesion formation at one concentration, it could be due to inhibition of PKC, or cAMP- or cGMP-dependent kinase. Use of HA1004, at the same level should determine if the effects were due to cAMP- or cGMP-dependent kinase inhibition, and use of a higher concentration of HA1004, where PKC is additionally inactivated should confirm any effects seen with H7 which are PKC-dependent. We monitored focal adhesion formation by IRM, and found that H7 and HA1004 did indeed prevent focal adhesion formation, but in both cases, only at levels where PKC was inhibited, not at levels where cAMP- and cGMP-dependent kinases should be inactivated (Woods & Couchman, submitted). Thus activation of PKC appears to be needed for focal adhesion formation. As a further check on focal adhesion formation we monitored the distribution of talin, vinculin and integrin β_1 subunits in untreated cells spread on fibronectin for 3 hours and those treated with 20 μM H7, 20 μM HA1004 or 80μM HA1004. Both talin (Fig. 1) and vinculin (not shown) were found in focal adhesions in untreated cells and in those treated with 20 μM HA1004, but in cells treated with H7 or HA1004 at the higher concentration, these two components were diffusely distributed. In contrast, integrin β_1 subunits appeared to be in focal adhesion-like plaques (Fig. 1) under all circumstances. This was initially surprising, but further studies indicated that this was due to the experimental procedure. To ensure that attachment and spreading were complete, we allowed the cells to spread for 1 hour prior to addition of the inhibitors. When cells spread for 1 hour were monitored, we found that talin and integrin β_1 subunits were already condensed into small focal adhesion-like plaques, although vinculin was still diffuse in most cells and IRM indicated an absence of focal adhesions. Thus, the effect of the inhibitors was to cause an apparent dispersal of talin from these plaques, i.e. to allow dispersal of any talin-integrin interactions. This need not however, be a specific effect on either talin or the integrin per se; it could be that this initial interaction, since it is of low affinity (Akiyama et al., 1985; Hynes, 1990; Ruoslahti, 1988), may be transient unless stabilized by further interactions. This stabilization could be provided by other focal adhesions components such as vinculin, alpha-actinin, or an as yet unidentified molecule. This is supported by our finding that long term focal adhesions (24 hours) were not susceptible to disruption by kinase inhibitors.

Promotion of Focal Adhesion Formation by Heparin-Binding Fibronectin Fragments

Our previous study (Woods et al., 1986) had used IRM to monitor the presence of focal adhesions and labeling of actin or phase contrast microscopy (Fig. 1) to detect stress fibers in cells spread on fibronectin substrates, their absence in cells on substrates of the 105-kD "cell"-binding fibronectin fragment, and the induction of their formation by addition of heparin-binding fragments. As noted above, this did not, however, automatically allow determination of the distribution of focal adhesion components involved in the cellular response to individual fibronectin domains. We therefore monitored the distribution of talin, vinculin and β_1 subunits in cells spread on substrates of 105-kD fragments before and after stimulation with heparin-binding fragments. As seen in figure 2, the IRM image of cells spread for 3 hours on 105-kD substrates without heparin-binding fragment stimulation showed an absence of focal adhesions and stress fibers, in contrast to cells spread on intact fibronectin. Immunofluorescent labeling of talin (Fig. 2), vinculin (not shown) and integrin β_1 subunits (Fig. 2) also indicated a diffuse distribution in the majority of cells. In contrast, cells treated with heparin-binding fragments for 30 mins prior to fixation showed focal adhesions by IRM, and by labeling for talin, vinculin and integrin β_1 subunits (Table 1). Thus, four independent methods of monitoring indicated specific induction of focal adhesion formation. In addition, it appears that the binding of heparin-binding fibronectin fragments, presumably to cell surface heparan sulfate proteoglycans, can induce a redistribution of cellular components with which they do not directly interact.

Since focal adhesion formation in response to intact fibronectin substrates appeared to be dependent on activation of PKC (Woods & Couchman, submitted), we next determined whether inhibition of PKC with 20 μM H7 or 80 μM HA1004 could prevent the induction of focal adhesion formation by heparin-binding fragments. We allowed cells to spread for 2 hours on substrates of 105-kD fibronectin fragments, added kinase inhibitors for 30 mins and then added heparin-binding fragments for a further 30 mins. The cells were then monitored by IRM and labeling for talin, vinculin and integrin β1 subunits (Table 1). The majority of cells treated with 29- or 30-kD heparin-binding fragments formed focal adhesions when no inhibitors were present, or when HA1004 was added at 20 μM (denoted '+' in Table 1 to indicate availability of PKC for activation). In contrast, those cells treated with heparin-binding fragments in the presence of 20 μM H7 or 80 μM HA1004 did not form focal adhesions ('-' in Table 1). Thus it appears as though the 'signal' from heparin-binding fragments may be transduced through activation of PKC.

Promotion of Focal Adhesion Formation by Activation of PKC with Phorbol Esters

Since inhibition of PKC appeared to prevent focal adhesion formation, we wondered whether artificial stimulation of PKC with phorbol esters could induce focal adhesion formation under conditions where they would not normally form. We allowed cells to spread for 2.5 hours on substrates of 105-kD fibronectin fragment and then added phorbol myristyl acetate (PMA) or 4-ß-phorbol 12,13-didecanoate (PDDa) for 30 mins. The inactive phorbol ester 4-α-phorbol 12,13-didecanoate (PDDi) was used as a control. IRM indicated that the active phorbol esters, but not the inactive ester, induced focal adhesion formation in a dose-dependent manner, with maximal activity at 200 nM. Immunofluorescent labeling for talin, vinculin and integrin β_1 subunits confirmed the redistribution of these components into focal adhesions in the majority of cells treated with 200 nM PMA or PDDa, but their diffuse distribution in cells treated with 200 nM PDDi (Table 2). Thus, the requirement for the stimulation of cells by heparin-binding activity of fibronectin to promote focal adhesion formation can be fulfilled by artificial activation of PKC.

POSSIBLE MECHANISM OF ACTION OF PKC ACTIVATION BY HEPARIN-BINDING FIBRONECTIN ACTIVITY

Focal adhesion formation is a complex process. Of the known components, many can interact with each other and also self associate at high concentration. However, individual interactions appear to be of low affinity, inconsistent with the formation of highly stable structures such as focal adhesions. This may reflect the need for cells to alter their adhesion type. Under normal circumstances in vivo, fibroblasts tend to be relatively inactive, replacing matrix slowly to maintain homeostasis. Under conditions of

Table 1. Percentage of cells with focal adhesions measured by IRM, or labeling for talin (T), integrin β_1 subunits (B) or vinculin (V) after spreading on 105-kD substrates, and treatment with heparin-binding fragments in the presence or absence of kinase inhibitors. '+' = PKC available.

INHIBITOR		ADDITIVE None	29-kD	31-kD
	IRM	7	69	77
	T	15	68	81
None	B	15	64	82
+	V	17	65	78
	IRM	1	3	2
	T	1	10	7
20μM H7	B	0	2	4
-	V	6	7	1
	IRM	25	54	69
	T	25	70	76
20μM HA1004	B	24	63	77
+	V	36	78	77
	IRM	7	10	11
	T	11	13	21
80μM HA1004	B	8	15	33
-	V	19	18	24

wound repair, or in some connective tissue diseases, however, the fibroblasts become more active in remodeling the tissue and exporting matrix. We have previously shown that in vitro, the formation of focal adhesions appears to correlate with the ability of cells to export and organize matrix in a controlled fashion. Although fibroblast focal adhesions are not widely seen in vivo, similar entities known as fibronexus structures do exist, particularly in wound beds (Singer, 1979). It may be that through subtle alterations in the cytoskeleton as these adhesions change, cellular metabolism can be greatly modified. Thus, the capacity to 'fine tune' the cellular adhesion state may be important. Having dual or multiple interactions necessary for stable adhesion may be one control mechanism. For example, fibronectin is highly susceptible to proteases, and during inflammation or wounding the heparin-binding domains may be cleaved allowing fibroblasts to adopt a more locomotory phenotype as is seen in the initial phase of wound repair. In addition, cells can vary their proteoglycan metabolism, and this could be reflected in their adhesive behavior. For example, transformed cells, which tend to have a locomotory phenotype and lack focal adhesions, often have heparan sulfate proteoglycans of reduced sulfation (Robinson et al., 1984), which may decrease their interaction with fibronectin (Couchman et al., 1988).

Table 2. Percentage of cells containing focal adhesions measured by IRM, or labeling for talin (T), integrin β_1 subunits (B) or vinculin (V) on 105-kD substrates after treatment with PMA, PDDa or PDDi.

ADDITIVE	IRM	T	B	V
None	18	15	18	12
PMA	77	84	48	55
PDDa	86	82	61	45
PDDi	23	16	16	22

The interactions of integrins with fibronectin are low affinity interactions, compatible with labile adhesion to allow for locomotion. When focal adhesions form, it may be additional interactions of cell surface heparan sulfate proteoglycans which stabilize these interactions and increase the avidity of the interaction. This need not be direct, but could be by phosphorylation of other membrane or cytoplasmic components by PKC. There are analogous situations recently described in lymphocytes. For example, when antibodies are used to cause patching of lymphocyte surface integrins, talin does not subpatch unless PKC is also activated (Burn et al., 1988). In addition, dual interactions appear to be needed in lymphocytes to achieve higher avidity adhesion, and direct activation of PKC can substitute for one of these interactions (reviewed in O'Rourke & Mescher, 1990). T cells do not normally adhere to fibronectin or laminin, even though they have cell surface integrins, unless treated with phorbol esters or antibodies to the T-cell receptor (Shimizu et al., 1990). Similarly, activation of PKC in monocytes and macrophages can also augment integrin-mediated adhesion to laminin and to endothelium (Gladwin et al., 1990; Shaw et al., 1990). Thus, there may be common mechanisms between different cell types.

The cellular components phosphorylated by PKC to promote this stable adhesion are not yet known, and are the subject of our current investigations. Similarly, the type of membrane heparan sulfate proteoglycans to which the heparin-binding fibronectin fragments bind is under study. Recent advances in determining the variety and distribution of membrane proteoglycans, and elucidation of the sequence of the core proteins of these membrane components will be invaluable in determining how the adhesive signal is transduced into PKC activation.

ACKNOWLEDGEMENTS

We thank Dr. K. Burridge (Univ. of North Carolina, USA) and Dr. S. Johansson (Univ. of Uppsala, Sweden) for generous gifts of antibodies against talin and integrin β_1 subunits respectively. John Couchman and Anne Woods were supported by NIH grants AR39741, and AR20614 to the Multipurpose Arthritis Center at UAB. J.R.C. is an Established Investigator of the American Heart Association.

REFERENCES

Akiyama, S. K., Hasegawa, E., Hasegawa, T., and Yamada, K. M., 1985, The ineraction of fibronectin fragments with fibroblastic cells, J. Biol. Chem., 260:13256.

Asano, T., and Hidaka, H., 1984, Vasodilatory action of HA1004 [N-(2-guanidinoethyl)-5-isoquinolinesulfonamide], a novel calcium antagonist with no effects on cardiac function, J. Pharmacol. Exp. Ther, 231:141..

Beckerle, M. C., Burridge, K., DeMartino, G. N., and Croall, D. E., 1987, Colocalization of calcium-dependent protease II and one of its substrates at sites of cell adhesion, Cell, 51:569.

Beyth, R. C., and Culp, L. A., 1984, Complementary adhesive responses of human skin fibroblasts to the cell-binding domain of fibronectin and the heparan sulfate-binding protein platelet factor 4, Exp. Cell Res., 155:537.

Burn, P., Kupfer, A., and Singer, S. J., 1988, Dynamic membrane cytoskeletal interactions: specific association of integrin and talin arises in vivo after phorbol ester treatment of peripheral blood lymphocytes, Proc. Natl. Acad. Sci. USA, 85:497.

Burridge, K., and Mangeat, P., 1984, An interaction between vinculin and talin, Nature (Lond.), 308:744.

Burridge, K., Fath, K., Kelly, T., Nuckolls, G., and Turner, C., 1988, Focal adhesions: Transmembrane junctions between the extracellular matrix and the cytoskeleton, Annu. Rev. Cell Biol., 4:487.

Couchman, J. R., and Rees, D. A., 1979, The behaviour of fibroblasts migrating from chick heart explants: changes in adhesion, locomotion and growth, and in the distribution of actomyosin and fibronectin, J. Cell Sci., 39:149.

Couchman, J. R., Austria, R., Woods, A., and Hughes, R. C., 1988, An adhesion defective BHK cell mutant has cell surface heparan sulfate proteoglycans of altered properties, J. Cell Physiol., 136:226.

Fath, K. R., Edgell, C.-J. S., and Burridge, K., 1989, The distribution of distinct integrins in focal contacts is determined by the substratum composition, J. Cell Sci., 92:67.

Gladwin, A.-M., Hassall, D. G., Martin, J. F., and Booth, R. F. G., 1990, MAC-1 mediates adherence of human monocytes to endothelium via a protein kinase C dependent mechanism, Biochem. Biophys. Acta., 1052:166.

Hidaka, H., Inagaki, M., Kawamoto, S., and Sasaki, Y., 1984, Isoquinolinesulfonamides, novel and potent inhibitors of cyclic nucleotide dependent protein kinase and protein kinase C, Biochemistry, 23:5036.

Horwitz, A., Duggan, K., Buck, C., Beckerle, M. C., and Burridge, K., 1986, Interactions of plasma membrane fibronectin receptor with talin - a transmembrane linkage, Nature (Lond.), 320:531.

Hynes, R. O., 1990, "Fibronectins," Springer-Verlag, New York.

Izzard, C. S., and Lochner, L. R., 1976, Cell to substrate contacts in living fibroblasts: an interference relection study with an evaluation of the technique, J. Cell Sci., 21:128.

Izzard, C. S., and Lochner, L. R., 1980, Formation of cell-to-substrate contacts during fibroblast motility: an interference-reflexion study, J. Cell Sci., 42:81.

Izzard, C. S., Radinsky, R., and Culp, L. A., 1986, Substratum contacts and cytoskeletal reorganization of Balbc/3T3 cells on a cell-binding fragment and heparin-binding fragments of plasma fibronectin, Exp. Cell Res., 165:320.

Jaken, S., Leach, K., and Klauck, T., 1989, Association of Type 3 protein kinase C with focal contacts in rat embryo fibroblasts, J. Cell Biol., 109:697.

Lark, M. W., and Culp, L. A., 1984, Multiple classes of heparan sulfate proteoglycans from fibroblast substratum adhesion sites. Affinity fractionation on columns of platelet factor 4, plasma fibronectin and octyl-Sepharose, J. Biol. Chem., 259:6773.

LeBaron, R. G., Esko, J. D., Woods, A., Johansson, S., and Höök, M., 1988, Adhesion of glycosaminoglycan-deficient Chinese Hamster Ovary cell mutants to fibronectin substrata, J. Cell Biol., 106:945.

O'Rourke, A. M., and Mescher, M. F., 1990, T-cell receptor activated adhesion systems, Curr. Opinion Cell Biol., 2:888.

Otey, C. A., Pavalko, F. M., and Burridge, K., 1990, An interaction between α-actinin and the β_1 integrin subunit in vitro, J. Cell Biol., 111:721.

Robinson, J., Viti, M., and Höök, M., 1984, Structure and properties of an undersulfated heparan sulfate proteoglycan synthesized by a rat hepatoma cell line, J. Cell Biol., 98:946.

Ruoslahti, E., 1988, Fibronectin and its receptors, Annu. Rev. Biochem., 57:375.

Shaw, L. M., Messier, J. M., and Mercurio, A. M., 1990, The activation dependent adhesion of macrophages to laminin involves cytoskeletal anchoring and phosphorylation of the $\alpha_6\beta_1$ integrin, J. Cell Biol., 110:2167.

Shimizu, Y., van Seventer, G. A., Horgan, K. J., and Shaw, S., 1990, Regulated expression and binding of three VLA (β_1) integrin receptors on T cells, Nature (Lond.), 345:250.

Singer, I. I., 1979, The fibronexus: a transmembrane association of fibronectin-containing fibers and bundles of 5nm microfilaments in hamster and human fibroblasts, Cell 16:675.

Streeter, H. B., and Rees, D. A., 1987, Fibroblast adhesion to RGDS shows novel features compared with fibronectin, J. Cell Biol., 105:507.

Woods, A., and Couchman, J. R., 1988, Focal adhesions and cell-matrix interactions, Collagen Rel. Res., 8:155.

Woods, A., Couchman, J. R., Johansson, S., and Höök, M., 1986, Adhesion and cytoskeletal organisation of fibroblasts in response to fibronectin fragments, EMBO (Eur. Mol. Biol. Org.) J., 5:665.

Woods, A., Höök, M., Kjellén, L., Smith, C. G., and Rees, D. A., 1984, Relationship of heparan sulfate proteoglycans to the cytoskeleton and extracellular matrix of cultured fibroblasts, J. Cell Biol., 99:1743.

Woods, A., Smith, C. G., Rees, D. A., and Wilson, G., 1983, Stages in specialization of fibroblast adhesion and deposition of extracellular matrix, Eur. J. Cell Biol., 32:108.

ANIMAL CELL MUTANTS DEFECTIVE IN HEPARAN SULFATE POLYMERIZATION

Jeffrey D. Esko

Department of Biochemistry
Schools of Medicine and Dentistry
University of Alabama at Birmingham
Birmingham, Alabama 35294

INTRODUCTION

Proteoglycans mediate diverse cellular processes by interacting with a variety of protein ligands. Some of these interactions depend on the proteoglycan core protein, but most involve electrostatic interactions with the glycosaminoglycan chains attached to the core protein (Jackson et al., 1991; Kjellén and Lindahl, 1991; Esko, 1991). These latter interactions depend on the composition and arrangement of monosaccharide residues in the chains, which in turn depend on the glycosyltransferases, sulfotransferases, and epimerases that catalyze chain polymerization and modification. Thus, the biological activity of proteoglycans is intimately related to the regulation of glycosaminoglycan biosynthesis.

Mutational analysis provides a way to study proteoglycan biosynthesis and provides an empirical method for determining the biological function of proteoglycans in intact cells. This approach has been applied successfully to Chinese hamster ovary (CHO) cells (Esko, 1991). CHO cells have many advantages over other cell lines. They produce both cell-associated and secreted proteoglycans containing chondroitin sulfate and heparan sulfate chains. They readily clone from single cells, grow rapidly in monolayer and suspension cultures, and form tumors in athymic mice. Screening and selection procedures have been developed for identifying mutants altered in a variety of cellular processes (Esko, 1989), and a large collection of proteoglycan-deficient mutants has been obtained (Table 1). In many instances, the biochemical deficiency that alters proteoglycan biosynthesis has been elucidated as well.

The CHO mutants have been sorted into genetic complementation groups through cell hybridization studies. Each strain is designated by a three letter acronym, *pgs*, which

Heparin and Related Polysaccharides
Edited by D.A. Lane *et al.*, Plenum Press, New York, 1992

Table 1. Proteoglycan-deficient mutants of CHO cells

Complementation Group	Biochemical Defect	Phenotype
*pgs*A	Xyl Transferase	Proteoglycan-deficient
*pgs*B	Gal Transferase I	Proteoglycan-deficient
*pgs*C	Anion-carrier	Normal proteoglycans
*pgs*D	GlcNAc and GlcA Transferases	Heparan sulfate-deficient & accumulates chondroitin sulfate
*pgs*E	N-Sulfotransferase	Undersulfated heparan sulfate

indicates a deficiency in *p*roteo*g*lycan *s*ynthesis (Table 1). The capital letter refers to a specific genetic complementation group, and the number refers to an individual strain. For example, mutant *pgs*A-745 fails to attach D-xylose (Xyl) to specific serine residues within core proteins, whereas *pgs*B-761 fails to attach D-galactose (Gal) in β1→4 linkage to Xyl. These mutants contain decreased amounts of the Xyl transferase and Gal transferase I, respectively, as measured with synthetic substrates in cell-free extracts. Both heparan sulfate and chondroitin sulfate synthesis are reduced in vivo, providing genetic evidence that the reactions measured in vitro are responsible for glycosaminoglycan synthesis in intact cells (Esko et al., 1985; 1987). Sulfate uptake is altered in *pgs*C mutants, thus defining a sulfate carrier on the cell surface (Esko et al.,1986; Knurr et al., 1988). Mutants belonging to *pgs*E express less N-sulfotransferase activity and produce undersulfated heparan sulfate in vivo (Bame and Esko, 1989; Bame et al., 1991a; 1991b).

*pgs*D DEFINES A NEW CLASS OF HEPARAN SULFATE-DEFICIENT MUTANTS

We recently described a set of mutants that by cell hybridization studies defines a new complementation group designated *pgs*D (Lidholt et al. 1991). Unlike other mutants, *pgs*D strains fail to produce heparan sulfate due to deficiencies of enzymes involved in heparan sulfate polymerization. As discussed below, *pgs*D mutants also accumulate chondroitin sulfate compared to wild-type cells, suggesting some form of coordinate control may exist over the enzymes involved in chain polymerization reactions.

The general behavior of *pgs*D mutants is typified by *pgs*D-677. As shown in Fig. 1, *pgs*D mutants make only chondroitin sulfate, whereas wild-type cells typically produce about 70%

heparan sulfate (filled bars) and 30% chondroitin sulfate (darkened hatched bars). Since, the total amount of glycosaminoglycan/μg of cell protein is unaltered in the mutant, chondroitin sulfate accumulates by a factor of 3 compared to the wild-type. The extent of chondroitin sulfate accumulation varies from 2 to 4-fold among different *pgs*D strains, and appears to depend on the amount of residual heparan sulfate produced. No alteration was found in the distribution of material between the cell layer and the growth medium.

To test if heparan sulfate proteoglycan core protein expression was altered in *pgs*D-677, cells were incubated with 30 μM estradiol-β-D-xyloside (EDX), a primer for heparan sulfate and chondroitin sulfate (Lugemwa and Esko, 1991). [^{35}S]glycosaminoglycans made in the presence and absence of primer were analyzed by chondroitinase and nitrous acid digestion (Fig. 2). When EDX was fed to *pgs*A-745 cells (Xyl transferase-deficient), about 30% of the primed glycosaminoglycan chains consisted of heparan sulfate and about 70% was chondroitin sulfate. In contrast, when EDX was fed to *pgs*D-677, only chondroitin sulfate was made. EDX actually stimulated chondroitin sulfate synthesis, causing a 3-fold accumulation of glycosaminoglycan chains. These findings suggested that the mutation in *pgs*D-677

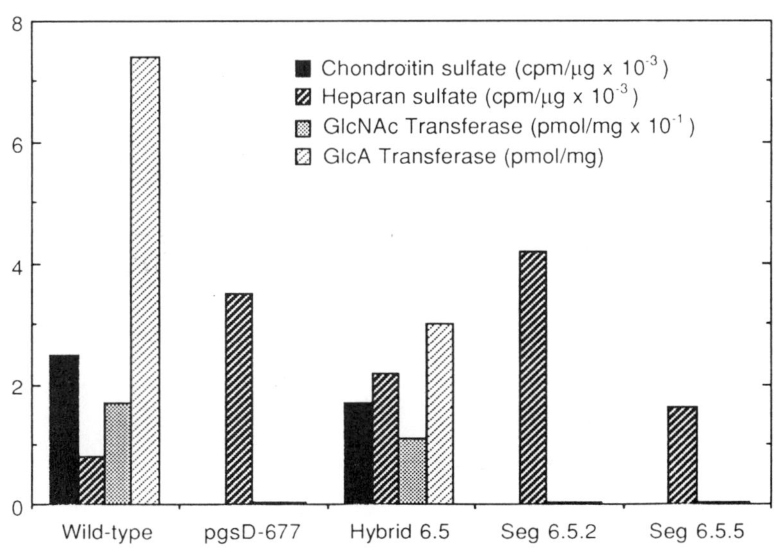

Fig. 1. Phenotypic traits of wild-type, mutant, and hybrids. [^{35}S]Glycosaminoglycans were isolated from the indicated cell lines and subjected to chondroitinase ABC and nitrous acid degradation to determine the proportion of [^{35}S]chondroitin sulfate and [^{35}S]heparan sulfate. Extracts of the cells were assayed for GlcNAc and GlcA transferases using defined oligosaccharides as acceptors (Lidholt et al., 1991).

selectively affected heparan sulfate synthesis. Thus, it seemed likely that one of the glycosyltransferases involved in heparan sulfate polymerization might be altered.

To test this possibility, GlcNAc and GlcA transferase activities were measured in CHO cell extracts using as acceptors defined oligosaccharides from E. coli K5 capsular N-acetylheparosan (Lidholt et al., 1991). The extent of transfer was measured instead of the rate in order to potentiate any differences between mutant and wild-type enzymes. As shown in Fig. 1, *pgs*D-677 lacked both enzyme activities. In contrast, wild-type extracts transferred about 2 pmols of GlcNAc (stippled bars) and about 75 pmols of GlcA (open hatched bars) to oligosaccharide acceptors per mg of cell protein. The absolute values for the transferase activities in the wild-type do not reflect the actual amount of enzyme activity since the concentration of UDP-[^{3}H]GlcNAc in the GlcNAc transferase reaction and the concentration of oligosaccharide acceptors in the GlcA transferase reaction were below their respective K_m values. Mixtures of wild-type and mutant extracts transferred one-half the amount transferred when the wild-type was assayed alone, indicating that the mutant did not produce a soluble inhibitor.

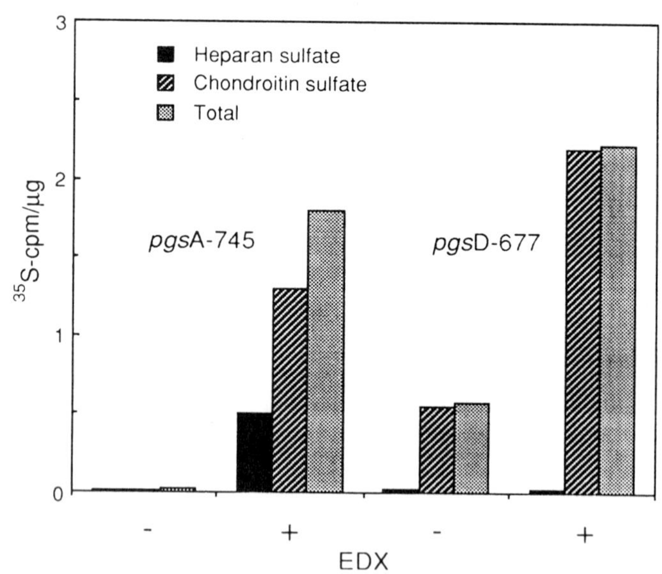

Fig. 2. Priming of heparan sulfate on estradiol-β-D-xyloside. Cells were incubated with 30 µM estradiol-β-D-xyloside (EDX) and pulse-labelled with $^{35}SO_4$ for 4 hr. Newly made [^{35}S]glycosaminoglycans were treated with chondroitinase ABC and nitrous acid to determine the amount of chondroitin sulfate and heparan sulfate generated in the (+) presence and (-) absence of xyloside.

The dual enzyme deficiency in *pgs*D-677 suggested the possibility that more than one mutation might be present. However, several pieces of evidence indicated that only one mutation was responsible for the dual enzyme deficiency. First, N-sulfotransferase activity in the mutant was normal, suggesting that other enzymes involved in heparan sulfate synthesis were unaltered. Second, another independent isolate belonging to *pgs*D complementation group (*pgs*D-803) also showed the dual enzyme deficiency and diminution of heparan sulfate synthesis. Since these strains were isolated from independent stocks of mutagen-treated cells, it seems unlikely that the same two mutations occurred in both strains.

We also have tested by segregation analysis whether the deficiencies were genetically linked. In this experiment, mutant and wild-type cells were fused with poly(ethylene)glycol and hybrid clones were selected. Analysis of glycosaminoglycan synthesis in the hybrids showed that heparan sulfate synthesis was restored and that both enzyme activities were present in cell extracts. Hybrid 6.5 typifies the behavior of these strains (Fig. 1). Cell hybrids spontaneously lose chromosomes during mitosis. During this process, unlinked traits residing on different chromosomes can segregate. Thus, if more than one mutation were responsible for the dual enzyme deficiency and failure to produce heparan sulfate and they resided on different chromosomes, then the traits might separate during cell division. When ~10^4 colonies of hybrid 6.5 were assayed for proteoglycan synthesis by colony autoradiography (Esko, 1989), two clones were found that did not produce heparan sulfate (strains 6.5.2 and 6.5.5, Fig. 1). Both clones re-expressed the dual enzyme deficiency noted in the original mutant (Table 1). Interestingly, both strains accumulated chondroitin sulfate by a factor of 3, like the original mutant. Together, these findings showed that the failure to produce heparan sulfate and the dual enzyme deficiency were recessive traits and that all of the traits behaved as though they were linked. A single mutation that alters two enzymes suggests that *pgs*D may define a shared subunit of the glycosyltransferases or a protein possessing both catalytic activities.

PROTEOGLYCAN INTERMEDIATES IN *pgs*D MUTANTS

Structural studies of core proteins that accumulate in *pgs*D cells could provide insight into the mechanism by which heparan sulfate synthesis occurs. Core proteins might contain linkage region tetrasaccharides (-GlcA-β1,3-D-Gal-β1,3-D-Gal-β1,4-D-Xyl-β-O-L-[Ser]) at sites where heparan sulfate synthesis prematurely terminated. Alternatively, the cores might contain a greater number of chondroitin sulfate chains if glycosaminoglycan attachment sites normally used for heparan sulfate assembly were used for chondroitin sulfate synthesis. If the latter hypothesis were correct, it could explain the accumulation of chondroitin sulfate in the mutant.

To test these possibilities, we examined the composition of betaglycan, a cell surface proteoglycan that binds TGF-β. Betaglycan was affinity labelled by crosslinking ^{125}I-TGF-β to cells, and crosslinked material was subjected to SDS-PAGE (Cheifetz and Massagué, 1986). Three bands of material were labelled in wild-type cells corresponding to Type-I, Type-II and Type-III receptors. Betaglycan, the Type III proteoglycan-form of the receptor, contains about 90% heparan sulfate and about 10% chondroitin sulfate in wild-type cells. When *pgs*D-677 was affinity labelled, Type I and Type II receptors labelled normally, but the high molecular weight Type III receptor collapsed to a band of about 110 kDa (data not shown).

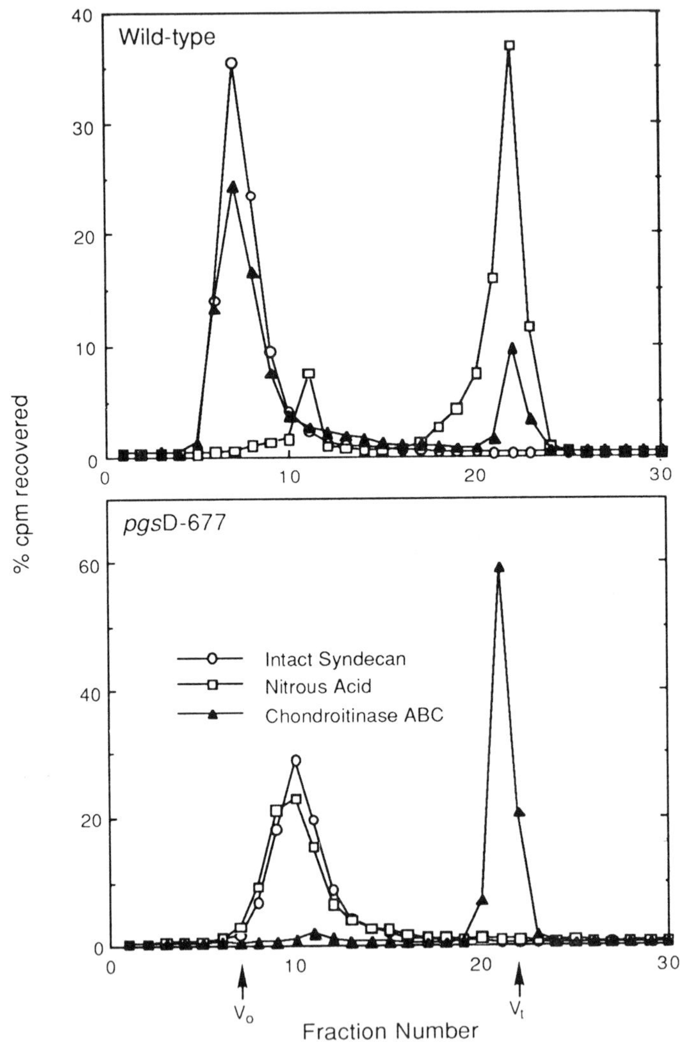

Fig. 3. Gel filtration of CHO syndecan. Cells were labelled with $^{35}SO_4$ and [^{35}S]proteoglycans were isolated by DEAE-chromatography. The preparation was then separated by affinity chromatography using anti-syndecan monoclonal antibody 2E9 conjugated to Sepharose (Marynen et al., 1989). A portion of material was analyzed by gel filtration HPLC before and after chondroitinase ABC and nitrous acid treatment.

This material resembled betaglycan from wild-type cells after digestion with heparitinase and chondroitinase ABC (Cheifetz and Massagué, 1986). A similar band was obtained when *pgs*A-745 (Xyl transferase-deficient) was affinity labelled. These findings suggested that betaglycan accumulated in *pgs*D as core proteins lacking glycosaminoglycan chains.

Betaglycan is most likely a minor proteoglycan in CHO cells and may not exemplify the behavior of all cell surface proteoglycans. Recent studies have shown that the major proteoglycan in CHO cells resembles syndecan, a hybrid proteoglycan containing both heparan sulfate and chondroitin sulfate. [^{35}S]Syndecan was affinity purified from mutant and wild-type cells using an anti-syndecan monoclonal antibody (2E9, a generous gift from G. David). Analysis by gel filtration HPLC showed a single peak of material that eluted near the V_o of the column (Fig. 3). When treated with chondroitinase ABC, a small amount of radioactivity shifted from the V_o fraction and eluted as disaccharides in the V_t fraction, suggesting that about 20% of the glycosaminoglycan was chondroitin sulfate. When syndecan was treated with nitrous acid, the majority of counts shifted towards the V_t of the column, and a small amount of material migrated as core proteins containing only chondroitin sulfate chains and heparan sulfate stubs left from the nitrous acid digestion. When syndecan from *pgs*D-677 was analyzed, it eluted later than syndecan from the wild-type, suggesting that it might be smaller in size. Nitrous acid treatment had no effect, but treatment with chondroitinase ABC shifted virtually all of the counts to the V_t fractions. Analysis of the residual oligosaccharides on syndecan is currently underway to determine if the cores contain linkage fragments or whether chondroitin sulfate chains assembled at sites normally used for heparan sulfate synthesis.

ALTERED BINDING OF BASIC FIBROBLAST GROWTH FACTOR

The availability of mutants like *pgs*D-677 has allowed us to examine the biological consequences of altered glycosaminoglycan synthesis (Esko et al., 1988; LeBaron et al., 1988, 1989; Murphy-Ullrich et al., 1988; Kaesberg et al., 1989; Yayon et al., 1991; Shieh et al., 1991). For example, recent studies by Yayon et al. (1991) and Repraeger et al. (1991) show that altered heparan sulfate synthesis affects the binding of basic fibroblast growth factor (bFGF). Binding of bFGF to CHO cells shows typical saturation behavior (Fig. 4). When ^{125}I-bFGF was incubated with *pgs*D-677 cells, essentially background binding was observed, suggesting that cell surface heparan sulfate accounts for most of the binding measured in wild-type cells. Yayon et al. (1991) have shown that the initial interaction with the heparan sulfate is required for subsequent binding to high affinity bFGF receptors. Binding to heparan sulfate also depends on the degree of sulfation, since mutant *pgs*E-606, altered in N-sulfation of heparan sulfate, showed intermediate levels of binding (Fig. 4).

These observation suggested a novel screening method for isolating mutants defective in heparan sulfate synthesis. Most of the existing CHO mutants were isolated through a replica plating strategy in which animal cell colonies were allowed to form between the plastic surface of a tissue culture plate and an overlying disk of woven polyester cloth (Esko, 1989). Some cells from each developing colony grow into the fibers of the disk and some remain attached to the plate. After 12-14 days, the disk contains a faithful replica of the colony pattern on the bottom of the dish. Multiple polyester replicas can be generated by stacking several disks in each plate. Stacking allows each colony to be assayed for more than one alteration.

To screen for mutants in bFGF binding, cells were treated with a chemical mutagen (ethylmethane sulfonate) to produce random mutations throughout the genome. Single cells were cloned under stacks of polyester disks. One of the disks was incubated with ^{125}I-bFGF and rinsed with saline to remove unbound bFGF. A second disk was incubated with $^{35}SO_4$ in order to monitor [^{35}S]glycosaminoglycan synthesis in the colonies. Both disks were exposed to X-ray film and after autoradiography they were stained with Coomassie Brilliant Blue in order to visualize all colonies. Mutants were identified by comparing the stained disks to the corresponding autoradiograms (Esko, 1989). Putative mutants, defined as blue colonies that lacked a corresponding image on the X-ray film, were retrieved from the original master dishes. Only those strains that failed to bind ^{125}I-bFGF *and* that incorporated normal amounts of $^{35}SO_4$ were selected. Some of these strains behaved like *pgs*D mutants described above, but others did not. These latter strains may contain subtle alterations of heparan sulfate sequences that define binding sites for bFGF. The characterization of these strains is currently underway.

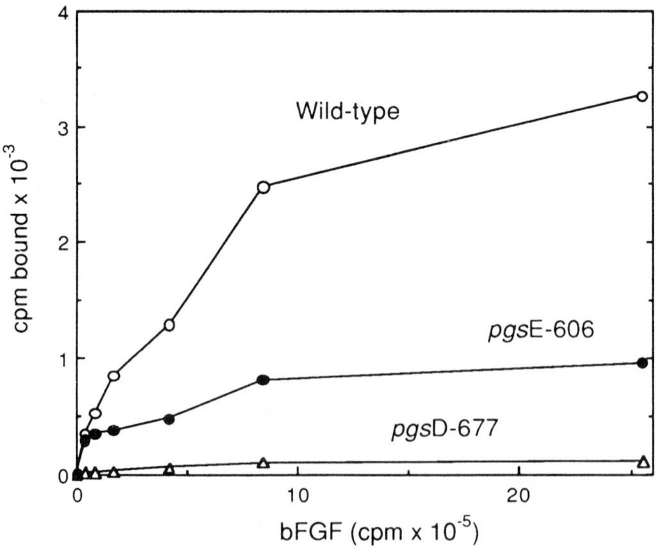

Fig. 4. Binding of bFGF depends on heparan sulfate. Cells were incubated with ^{125}I-bFGF for one hour at 4 °C and bound material was quantitated by liquid scintillation counting.

ACKNOWLEDGEMENTS

This study was made possible by the work of J. Weinke, K. Bame, L. Zhang, and S. Ribeiro (USA) on the isolation of mutants, bFGF binding studies, and isolation of syndecan, K. Lidholt and U. Lindahl (Sweden) for assay of GlcNAc and GlcA transferases, S. Cheifetz and J. Massagué (USA) for studies on betaglycan, and G. David (Belgium) for anti-syndecan antibodies. I also thank Ulf Lindahl and his organizing committee for an excellent meeting on heparin and related polysaccharides.

REFERENCES

Bame, K.J., and Esko, J.D., 1989, Undersulfated heparan sulfate in a Chinese hamster ovary cell mutant defective in heparan sulfate N-sulfotransferase, J. Biol. Chem., 264:8059.

Bame, K.J., Lidholt, K., Lindahl, U., and Esko, J.D., 1991a, Biosynthesis of heparan sulfate. Coordination of polymer-modification reactions in a Chinese hamster ovary cell mutant defective in N-sulfotransferase, J. Biol. Chem., 266:10287.

Bame, K.J., Reddy, R.V., and Esko, J.D., 1991b, Coupling of N-deacetylation and N-sulfation in a Chinese hamster ovary cell mutant defective in heparan sulfate N-sulfotransferase, J. Biol. Chem., 266:12461.

Cheifetz, S., and Massagué, J., 1989, Transforming growth factor-beta (TGF-beta) receptor proteoglycan. Cell surface expression and ligand binding in the absence of glycosaminoglycan chains, J. Biol. Chem., 264:12025.

Elgavish, A., Esko, J.D., and Knurr, A., 1988, Chinese hamster ovary cell mutants deficient in an anion exchanger functionally similar to the erythroid band 3, J. Biol. Chem., 263:18607.

Esko, J.D., Stewart, T.E., Taylor, W.H., 1985, Animal cell mutants defective in glycosaminoglycan biosynthesis, Proc. Natl. Acad. Sci. U S A, 82:3197.

Esko, J.D., Elgavish, A., Prasthofer, T., Taylor, W.H., and Weinke, J.L., 1986, Sulfate transport-deficient mutants of Chinese hamster ovary cells. Sulfation of glycosaminoglycans dependent on cysteine, J. Biol. Chem., 261:15725.

Esko, J.D., Weinke, J.L., Taylor, W.H., Ekborg, G., Rodén, L., Anantharamaiah, G., and Gawish. A., 1987, Inhibition of chondroitin and heparan sulfate biosynthesis in Chinese hamster ovary cell mutants defective in galactosyltransferase I, J. Biol. Chem., 262:12189.

Esko, J.D., Rostand, K.S., and Weinke, J.L., 1988, Tumor formation dependent on proteoglycan biosynthesis, Science, 41:1092.

Esko, J.D., 1989, Replica plating of animal cells, Meth. in Cell Biol., 32:387.

Esko, J.D., 1991, Genetic analysis of proteoglycan structure, function and metabolism, Curr. Opin. Cell Biol., 3:805.

Jackson, R.L., Busch, S.J., and Cardin, A.D., 1991, Glycosaminoglycans: molecular properties, protein interactions, and role in physiological processes, Physiol. Rev., 71:481.

Kaesberg, P.R., Ershler, W.B., Esko, J.D., and Mosher, D.F., 1989, Chinese hamster ovary cell adhesion to human platelet thrombospondin is dependent on cell surface heparan sulfate proteoglycan, J. Clin. Invest., 83:994.

Kjellén, L. and Lindahl, U., 1991, Proteoglycans: structure and interactions, Ann. Rev. Biochem., 60:443.

LeBaron, R.G., Esko, J.D., Woods, A., Johansson, S., and Höök, M., 1988, Adhesion of glycosaminoglycan-deficient Chinese hamster ovary cell mutants to fibronectin substrata, J. Cell Biol., 106:945.

LeBaron, R.G., Höök, A., Esko, J.D., Gay, S., and Höök, M., 1989, Binding of heparan sulfate to type V collagen. A mechanism of cell-substrate adhesion, J. Biol. Chem.,264:7950.

Lidholt, K., Weinke, J.L., Kiser, C.S., Lugemwa, F.N., Bame, K.J., Cheifetz, S., Massagué, J., Lindahl, U., and Esko, J.D., 1992, Chinese hamster ovary cell mutants defective in heparan sulfate biosynthesis, Proc. Natl. Acad. Sci. USA, in press.

Lugemwa, F.N. and Esko, J.D., 1991, Estradiol-β-D-xyloside, an efficient primer of heparan sulfate biosynthesis, J. Biol. Chem., 266:6674.

Marynen, P., Zhang, J., Cassiman, J.J., Van den Berghe, H., and David, G., 1989, Partial primary structure of the 48- and 90-kilodalton core proteins of cell surface-associated heparan sulfate proteoglycans of lung fibroblasts. Prediction of an integral membrane domain and evidence for multiple distinct core proteins at the cell surface of human lung fibroblasts, J. Biol. Chem., 264:7017.

Murphy-Ullrich, J.E., Westrick, L.G., Esko, J.D., and Mosher D.F., 1988, Altered metabolism of thrombospondin by Chinese hamster ovary cells defective in glycosaminoglycan synthesis, J. Biol. Chem., 263:6400.

Repraeger, A.C., Krufka, A., and Olwin, B.B., 1991, Requirement for heparan sulfate for bFGF-mediated fibroblast growth and myoblast differentiation, Science, 252:1705.

Shieh, M.-T., WuDunn, D., Montgomery, R.I., Esko, J.D., and Spear, P.G., 1991, Heparan sulfate proteoglycans are cell surface receptors for Herpes simplex virus, J. Cell Biol., submitted.

Yayon, A., Klagsbrun, M., Esko, J.D., Leder, P., and Ornitz, D.M., 1991, Cell surface, heparin-like molecules are required for binding of basic fibroblast growth factor to its high affinity receptor, Cell, 64:841.

TWO ENZYMES IN ONE: N-DEACETYLATION AND N-SULFATION IN HEPARIN BIOSYNTHESIS ARE CATALYZED BY THE SAME PROTEIN

Lena Kjellén, Inger Pettersson, Erik Unger and Ulf Lindahl[1)]

Dept. of Veterinary Medical Chemistry, Swedish University of Agricultural Sciences and [1)]Dept. of Medical and Physiological Chemistry University of Uppsala, The Biomedical Center Box 575, S-751 23 Uppsala, Sweden

HEPARIN AND HEPARAN SULFATE BIOSYNTHESIS

The biosynthesis of heparin and heparan sulfate (see Lindahl et al., 1986; Lindahl & Kjellén, 1987; Lindahl, 1989) involves polymerization of a polysaccharide backbone consisting of alternating glucuronic acid (GlcA) and N-acetylglucosamine (GlcNAc) residues, and modification of the polymerization product into a sulfated polysaccharide with complex structure (Fig. 1). While the polymerization and modification processes may be artificially segregated, e.g. in isolated microsomes, these processes are likely to occur concomitantly in the intact cell (Lidholt et al., 1989; Lidholt 1991). The first modification reaction, N-deacetylation, converts GlcNAc into glucosamine (GlcN) residues. Essentially all GlcN residues are then N-sulfated. The subsequent modification reactions, C-5 epimerization of GlcA to iduronic acid (IdoA) residues and O-sulfation in various positions, occur exclusively in the vicinity of previously incorporated N-sulfate groups. In contrast to N-sulfation, these reactions are incomplete, leaving a fraction of the potential target units unmodified.

REGULATORY ROLE OF N-DEACETYLATION

Since N-deacetylation is prerequisite to N-sulfation, the N-deacetylase enzyme has a key role in regulating the overall modification process. Hence, a polysaccharide with a high content of N-sulfate groups, such as mast-cell heparin, can be expected to be more extensively O-sulfated and to contain a larger proportion of IdoA units than heparan sulfate, which generally is N-sulfated to a lower degree. Further, the distribution of N-acetyl and N-sulfate groups is of importance for the generation of specific saccharide sequences. For instance, a functional antithrombin-binding region requires that two of the three GlcN residues of the pentasaccharide sequence shown in Fig. 2 be N-sulfated. In fact, the N-substitution pattern of this region may contribute toward the generation of the appropriate sequence of hexuronic

Heparin and Related Polysaccharides
Edited by D.A. Lane *et al.*, Plenum Press, New York, 1992

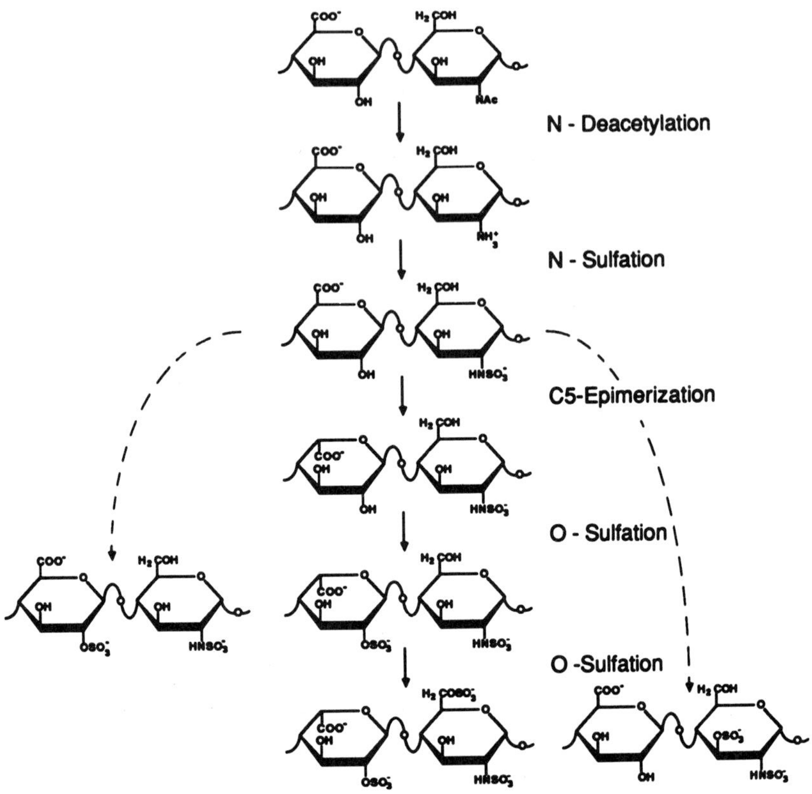

Fig. 1. Polymer modification reactions in heparin/heparan sulfate biosynthesis. Major reactions are indicated by the continuous sequence in the middle, whereas the dashed arrows represent less common O-sulfation reactions.

acid units, an N-sulfate group at *unit 3* rendering the GlcA *unit 4* susceptible to epimerization, whereas an N-acetylated *unit 1* will ensure retention of the *D-gluco* configuration of *unit 2*. Finally, the N-deacetylase may possibly be involved in regulating the polysaccharide chain length, since the N-sulfate group at the penultimate GlcN residue appears to promote the addition of GlcA, a potential rate-limiting step, to the growing polysaccharide chain (Lidholt, 1991).

CHARACTERIZATION OF THE N-DEACETYLASE

Recent studies on the N-deacetylase in mouse mastocytoma tissue indicated that at least two proteins were required for the expression of enzyme activity (Pettersson et al., 1991). These proteins could be separated by affinity chromatography on wheat germ agglutinin-Sepharose into a lectin-binding and a nonbinding component (designated E and F, respectively, see Fig. 3). The lectin-binding Component E was purified to homogeneity and

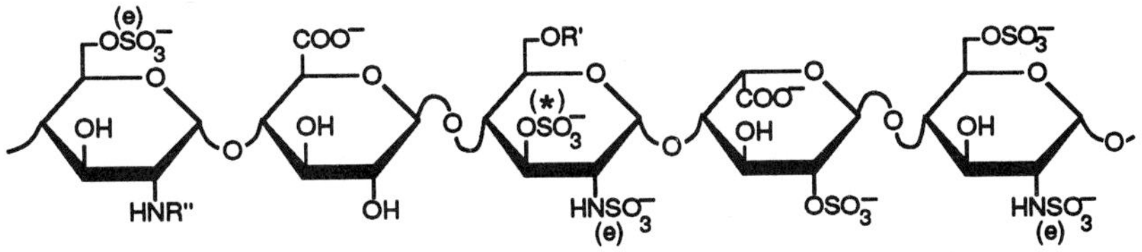

Fig. 2. The antithrombin-binding region in heparin. The pentasaccharide sequence is composed of three GlcN (*units 1,3* and *5*), one GlcA (*unit 2*) and one IdoA unit (*unit 4*). Structural variants are indicated by -R' (-H or SO_3^-) or -R" ($-COCH_3$ or $-SO_3^-$). The 3-O-sulfate group (asterisk), a marker component for the antithrombin-binding region, and sulfate groups indicated by (e) are essential for high-affinity binding to antithrombin.

showed a single band, M_r ~110,000, on SDS-PAGE. Analysis of the 110 kDa protein demonstrated that it contained also N-sulfotransferase activity. However, the latter activity was detectable also in the absence of Component F. While Component E thus appeared to harbour the active site for the N-sulfotransferase, it could not be immediately concluded whether the N-deacetylase site was located in Component E or F (Pettersson et al., 1991). However, more recent results indicate that Component F can be replaced by strongly basic proteins, such as histones, and also by Polybrene, a synthetic polycation, without loss of N-deacetylase activity (I. Pettersson, E. Malmport, J.-p. Li, U. Lindahl & L. Kjellén, unpublished observation). The active site for N-deacetylation therefore must reside in Component E, along with the N-sulfotransferase site (Fig. 3); in the absence of Component F (or other exogenous polycationic macromolecules) only the latter site generates detectable activity. A protein with N-sulfotransferase activity has previously been purified from rat liver (Brandan & Hirschberg, 1988). The mastocytoma Component E is most likely an analogous murine protein.

REDUCED SULFATION OF HEPARAN SULFATE IN DIABETES

Several of the complications of diabetes have been attributed to a lowered production of heparan sulfate (Sternberg et al., 1985). For instance, a decreased amount of basement membrane heparan sulfate proteoglycans is believed to perturb the electrostatic filtration barrier in the kidneys (Kanwar et al., 1980, 1983; Deckert et al., 1989). However, in addition, the heparan sulfate produced in diabetic tissues appears to be undersulfated, as shown for rat liver (Kjellén et al., 1983) and kidney heparan sulfate (Cohen et al., 1988).

We recently demonstrated that the activity of the N-deacetylase was ~40% lower in hepatocytes from diabetic rats than in control cells (Unger et al. 1991), thus explaining the reduced sulfation of liver heparan sulfate in diabetes. In contrast, the activity of the GlcA C-5 epimerase, which converts GlcA to IdoA in the polymer-modification reaction immediately subsequent to N-deacetylation/N-sulfation (see Fig. 1), was unaffected. We could also show

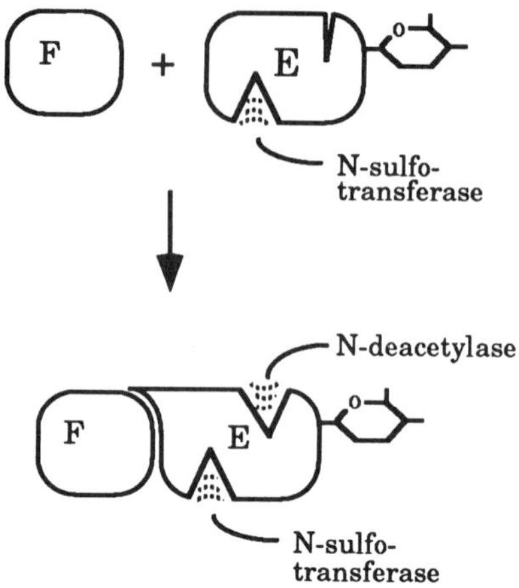

Fig. 3. Model of the interaction between the components of the N-deacetylase/N-sulfotransferase complex. Component E contains the active sites for both N-deacetylase and N-sulfotransferase activity. In the absence of Component F, only the N-sulfotransferase site is active. Binding of Component F induces a conformational change in E, resulting in an activation of the N-deacetylase site.

that the decreased N-deacetylase activity was due to a reduction in Component E activity, while Component F was present in excess amounts in both control and diabetic cells (Unger et al., 1991).

To study if the activities of biosynthetic enzymes other than the N-deacetylase/N-sulfotransferase were also altered due to the diabetic state, a detailed structural comparison of heparan sulfate from control and diabetic hepatocytes was performed (E. Unger, A. Stark & L. Kjellén, unpublished results). In addition to a lowered N-sulfation, also the contents of O-sulfate groups and IdoA residues were reduced in heparan sulfate from diabetic cells. Since the activity of the C-5 epimerase was the same in diabetic and control hepatocytes (Unger et al., 1991), the decreased formation of N-sulfated GlcN residues, required for substrate recognition by this enzyme, may alone be responsible for the decreased level of IdoA. In addition, similar ratios of N-sulfate/O-sulfate groups were found for control and diabetic heparan sulfate, indicating that also the lowered O-sulfation was due to a reduced level of susceptible target units for the corresponding enzymes. The structural alterations in heparan sulfate due to diabetes thus can all be ascribed to a reduced N-sulfation of the polysaccharide during biosynthesis, in accord with the postulated regulatory key role of the N-deacetylase/N-sulfotransferase system.

REFERENCES

Brandan, C., and Hirschberg, C.B., 1988, Purification of rat liver N-heparan sulfate sulfotransferase, J. Biol. Chem., 263:2417-2422

Cohen, M.P., Klepser, H., and Wu, V.-Y., 1988, Undersulfation of glomerular basement membrane heparan sulfate in experimental diabetes and lack of correction with aldose reductase inhibition, Diabetes, 37:1324-1327

Deckert, T., Feldt-Rasmussen, B., Borch-Kohnsen, K., Jensen, T., and Kofoed-Enevoldsen, A., 1989, Albuminuria reflects widespread vascular damage. The Steno hypothesis, Diabetologia 32:219-226

Kanwar, Y.S., Linker, A., and Farquhar, M.G., 1980, Increased permeability of the glomerular basement membrane to ferritin after removal of glycosaminoglycans (heparan sulfate) by enzyme digestion, J. Cell Biol. 86:688-693

Kanwar, Y.S., Rosenzweig, L.J., Linker, A., and Jakubowski, M.L., 1983, Decreased de novo synthesis of glomerular proteoglycans in diabetes: Biochemical and autoradiographic evidence, Proc. Natl. Acad. Sci. U.S.A. 80:2272-2275

Kjellén, L., Bielefeld, D., and Höök, M., 1983, Reduced sulfation of liver heparan sulfate in experimentally diabetic rats, Diabetes 32:337-342

Lidholt, K., Kjellén, L., and Lindahl, U., 1989, Biosynthesis of heparin. Relationship between the polymerization and sulphation processes, Biochem. J. 261:999-1007

Lidholt, K., 1991, A new model for the biosynthesis of heparin, Ph.D. thesis, The Swedish University of Agricultural Sciences, Uppsala, Sweden

Lindahl, U., Feingold, D.S., and Rodén, L., 1986, Biosynthesis of heparin, TIBS 11:221-225

Lindahl, U., and Kjellén, L., 1987, Biosynthesis of heparin and heparan sulfate, in: "Biology of proteoglycans", Wight, T.N., and Mecham, R.P. eds., Academic Press, New York, pp.59-104

Lindahl, U., 1989, Biosynthesis of heparin and related polysaccharides, in: "Heparin: Chemical and Biological Properties, Clinical Applications", Lane, D.A., and Lindahl, U., eds., Edward Arnold, London, pp. 159-189

Pettersson, I., Kusche, M., Unger, E., Wlad, H., Nylund, L., Lindahl, U., and Kjellén, L., 1991, Biosynthesis of heparin. Purification of a 110-kDa mouse mastocytoma protein required for both glucosaminyl N-deacetylation and N-sulfation, J. Biol. Chem. 266:8044-8049

Sternberg, M., Cohen-Forterre, L., and Peyroux, J., 1985, Connective tissue in diabetes mellitus: Biochemical alterations of the intercellular matrix with special reference to proteoglycans, collagens and basement membranes, Diab. Metabol. 11:27-50

Unger, E., Pettersson, I., Eriksson, U.J., Lindahl, U., and Kjellén, L., 1991, Decreased activity of the heparan sulfate-modifying enzyme glucosaminyl N-deacetylase in hepatocytes from Streptozotocin-diabetic rats, J. Biol. Chem. 266:8671-8674

METABOLISM OF PLASMA MEMBRANE-ASSOCIATED HEPARAN SULFATE PROTEOGLYCANS

Masaki Yanagishita

Bone Research Branch
National Institute of Dental Research
National Institutes of Health
Bethesda, Maryland 20892, U.S.A.

INTRODUCTION

Heparan sulfate (HS) proteoglycans are widely distributed throughout animal tissues in two main localized areas; in association with the plasma membrane and in the extracellular matrix (especially in basement membranes). Discussions on the metabolism of plasma membrane-associated HS proteoglycans is the subject of this chapter.

Biosynthetic processes of HS proteoglycans follow those of general glycoproteins: N-Linked oligosaccharide precursors are transferred to the core protein cotranslationally in the rough endoplasmic reticulum, and undergo extensive carbohydrate modification in the Golgi apparatus. These modifications include glycosaminoglycan and O-linked oligosaccharide synthesis on the core protein in addition to processing of N-linked oligosaccharides. After the completion of the HS proteoglycan in the trans Golgi network, they are targeted to their final destination based on the distinct structure of core proteins.[1-4] Metabolic state of the cells can also influence this process.[5] On the plasma membrane, three types of interaction between HS proteoglycan and membrane lipid bilayer have been demonstrated; (i) direct intercalation of the core protein in the plasma membrane,[2,3] (ii) through glycosylphosphatidylinositol (GPI)-anchor covalently linked to core protein,[4,6-10] and (iii) through specific or non-specific binding of the HS proteoglycans to other cell surface molecules.[11] Reasons for selective expression of HS proteoglycan species reflecting precise function of the molecules in different cells, embryonic stages or metabolic status have not been understood well. Based on the cell surface localization, interaction of HS proteoglycans with both extracellular and intracellular molecules, their involvement in cell-cell or cell-extracellular matrix interaction, substrate adhesion etc. have been speculated. However, the results of studies have been mostly suggestive but not conclusive of these concepts.

Catabolic stages of HS metabolism involve removal of molecules either by shedding from the cell surface or endocytosis followed by eventual degradation in lysosomes. Shedding of protein-intercalated HS proteoglycans would require proteolytic cleavage of the core protein while GPI-anchored HS proteoglycans can be shed either by proteolytic cleavage or by the cleavage of their GPI anchor. Mechanisms involved in

Heparin and Related Polysaccharides
Edited by D.A. Lane *et al.*, Plenum Press, New York, 1992

shedding of HS proteoglycans and their regulation have not been elucidated in detail. Endocytosed HS proteoglycans are eventually degraded to their constituent monosaccharides and sulfate in lysosomes. Lysosomal degradation of glycosaminoglycans including HS has been extensively studied and the enzymes responsible for the final degradation of HS, a group of specific exoglycosidases and sulfatase, have been well studied.[12]

Metabolism of HS proteoglycans between biosynthesis and lysosomal degradation is a reflection of their biological functions and naturally vary considerably. Discussions in this chapter focus on underlying cellular processes in the metabolism of plasma membrane-associated HS proteoglycans in variety of cell systems. These include endocytosis, recycling, prelysosomal degradation and targeting of HS proteoglycans to specific cellular compartments. Most data presented in this chapter are based on *in vitro* studies using rat ovarian granulosa cells as a model system.[13-16] Most metabolic processes elucidated in this system can be seen among many different cell types with some variations. There are also many other cell types which either show or lack certain metabolic processes. These differences among cell types are also discussed briefly.

GENERAL EXPERIMENTAL DESIGN FOR METABOLIC STUDY OF MEMBRANE-ASSOCIATED HS PROTEOGLYCAN

The metabolic study of plasma membrane-associated HS proteoglycan invariably involves isotopic pulse labeling of proteoglycans using cell or tissue cultures and chase protocols to study the metabolic fate of these molecules. Appropriate isotope precursors and the timing of pulse-chase protocols are designed according to the cell system and the aspect of metabolism to study. The cell surface localization of labeled proteoglycans can be generally examined by the accessibility of proteoglycans to various enzymes exogenously added to cell cultures; e.g. proteases which efficiently remove most HS proteoglycans from the cell surface, and a phosphatidylinositol-specific phospholipase C (PI-PLC) which specifically removes GPI-anchored HS proteoglycan. After the appropriate pulse labeling-chase protocol is complete, then subcellular fractionation techniques can be used to determine the localization of proteoglycans. Standard chromatography techniques including ion exchange chromatography and gel filtration, in combination with various chemical and enzymatic treatments, are used to determine the structures of HS proteoglycans.

TRANSPORT OF MEMBRANE-ASSOCIATED HS PROTEOGLYCANS FROM GOLGI APPARATUS TO THE CELL SURFACE

Glycosaminoglycan synthesis on the core protein is the last step of posttranslational modification of HS proteoglycans. Sulfation, the last step of glycosaminoglycan synthesis, occurs in the trans Golgi network. Transit time required for the transport of HS proteoglycans from the trans Golgi network to the cell surface can be measured by a short [^{35}S]sulfate pulse-labeling (e.g. 2 min) followed by monitoring the appearance of proteoglycans on the exterior cell surface (measured by accessibility of ^{35}S-labeled HS proteoglycan to exogenously added trypsin or PI-PLC). In the rat granulosa cells and in most cell systems we have tested (except for parathyroid cells), the majority (>90%) of the HS proteoglycans are transferred onto the cell surface. Both protein-intercalated HS proteoglycan and GPI-anchored HS proteoglycan appear on the cell surface with indistinguishable transit times (12-13 min), suggesting that the anchoring mechanism has little influence on the transport of the HS proteoglycan from the Golgi to the cell surface,

Fig.1. Minor amounts of HS proteoglycan (<10% of the completed HS proteoglycan) are not transferred to the cell surface but instead remained in an intracellular compartment (a relatively large proportion in the parathyroid cell under normal extracellular calcium conditions).

TURNOVER OF HS PROTEOGLYCANS ON THE CELL SURFACE

Turnover of HS proteoglycans on the cell surface occurs by two basic routs; by endocytosis and by shedding into medium. The mechanism involved in shedding of proteoglycans is not well understood. HS proteoglycans anchored to the plasma membrane through their core protein seem to require a proteolytic cleavage of the core proteins at or near the plasma membrane. It is not known whether this proteolysis occurs at the exterior surface of the cells, or after the endocytosis of the molecule. Shedding of GPI-anchored HS proteoglycan has been a subject of debate because it has a potential of being specifically regulated by endogenous PI-PLC. Variable data have been reported in different systems; in human lung fibroblast culture virtually all GPI-anchored HS proteoglycan are shed in basal culture conditions[4] while in ovarian granulosa cell and osteoblast (UMR 106-01) cultures very little, if any, of GPI-anchored HS proteoglycan are shed in both basal conditions and after stimulation with insulin,[16,17] which is known to cause shedding of GPI-anchored molecule in some other systems.[18] There has been no report of a system which shows a regulated shedding of GPI-anchored HS proteoglycan.

In most systems we studied, the majority of plasma membrane-associated HS proteoglycans (~70% of the core protein-intercalated and nearly 100% of GPI-anchored species) are endocytosed. Half life ($T_{1/2}$) values for the endocytosis differ among the systems studied; 4-24 h for core protein intercalated HS proteoglycans, and shorter and less variable (3-4 h) $T_{1/2}$ for a few cases of GPI-anchored HS proteoglycan studied.[16,17] The anchoring mechanism of HS proteoglycan seems to critically influence the endocytotic mechanisms.

Receptor mediated endocytosis of peripherally cell-associating HS proteoglycans has not been clearly demonstrated as has been the case for the decorin (chondroitin/dermatan sulfate containing proteoglycan).[19]

PRELYSOSOMAL, INTRACELLULAR DEGRADATION PROCESSES OF HS PROTEOGLYCAN

Pulse labeling with [^{35}S]sulfate and chase experiments performed using rat granulosa cells have clearly defined the presence of two kinetically distinct intracellular degradation pathways for HS proteoglycan. Similar experiments using various inhibitors of proteases and lysosomal enzymes also indicated that some degradation steps occur in prelysosomal compartments. A summary of these experiments using rat ovarian granulosa cell culture is presented in Fig. 1.

Degradation pathway 1 (rapid)

Endocytosed GPI-anchored HS proteoglycans are quickly ($T_{1/2}$ ~25 min) transferred to lysosomes and undergo rapid, complete degradation, Fig. 1, left panel. No appreciable degradation intermediates can be demonstrated. Lysosomal degradation is totally inhibited by lysosomotropic drugs (such as chloroquine, NH_4Cl or monensin).

Degradation pathway 2 (slow)

Endocytosed protein-intercalated HS proteoglycans undergo waves of distinct degradation steps which occur over an extended period of time,

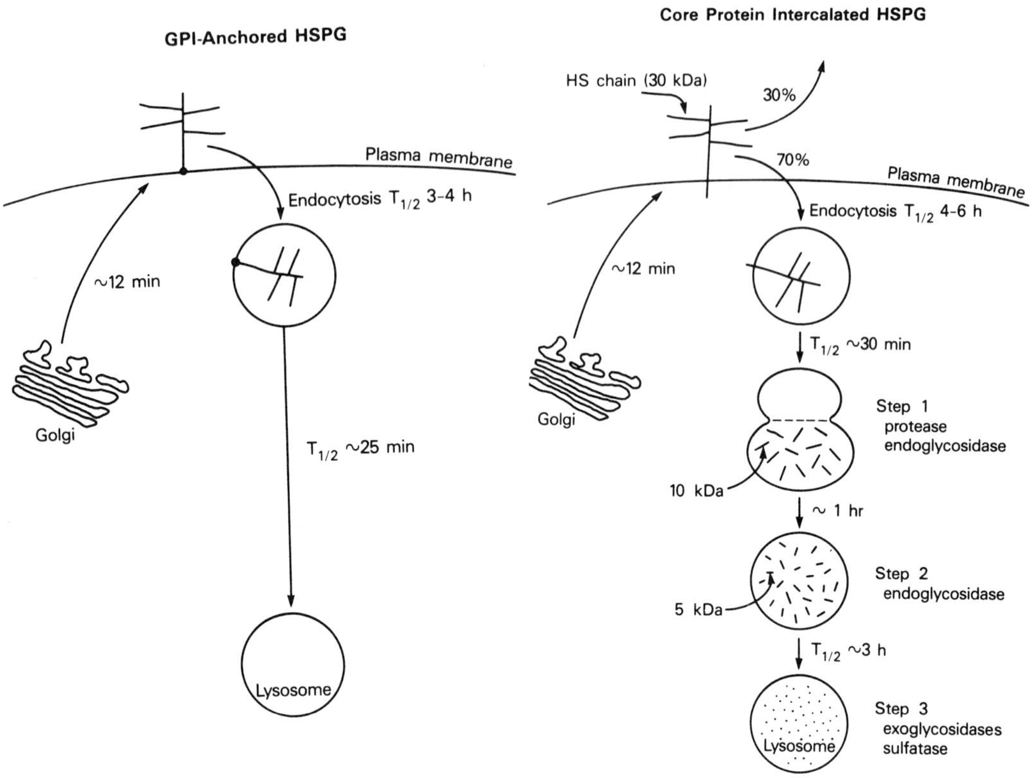

Fig. 1. Schematic models for transport and degradation pathways for plasma membrane-associated HS proteoglycans in rat ovarian granulosa cells. Left panel: GPI-anchored HS proteoglycan, and right panel: protein-intercalated HS proteoglycan

Fig. 1, right panel. The first stage of degradation occurs soon after endocytosis ($T_{1/2}$ ~30 min). Initially, the core protein is extensively digested liberating free HS chains (average 30 kDa), which is closely followed by an endoglycosidic degradation generating HS chains having an average size of 10 kDa. In normal culture conditions, these processes of proteolytic degradation and endoglycosidic degradation are closely coordinated and cannot be separated. These two processes were only partially separated by the use of a protease inhibitor, leupeptin, which slowed down the initial proteolytic process. Analysis of the intermediate product accumulated after leupeptin treatment suggested that proteolysis is a required step for the subsequent endoglycosidic cleavage to occur. This implies that the proteolytic and endoglycosidic degradation steps are functionally segregated, and that proteolysis is required for the transfer of liberated HS chains to the next compartment containing the endoglycosidic enzyme. Step 1 degradation, Fig. 1, proceeds irrespective of the presence of lysosomotropic agents (chloroquine, monensin etc.), suggesting that these degradation processes occur in an environment with neutral pH such as early endosomes. These 10 kDa HS fragments stay unchanged for approximately 1 h before they undergo the next wave of degradation (Step 2), indicating the storage of these fragments in a functionally separate compartment from the next degradation compartment. The second wave of degradation involves further endoglycosidic degradation of the HS fragments resulting in the generation of HS oligosaccharides with an average size of 5 kDa. This step is completely inhibited by lysosomotropic agents indicating it occurs in an acidic environment. Alternatively, the transfer of HS fragments from the preceding degradation compartment to this degradation environment may also require acidification of the compartment to proceed. Endoglycosidase activity demonstrated in Steps 1 and 2 degradation stages may be similar to those found in other cells with a characteristic pH optimum between 5-7.[20-27] The two distinct endoglycosidase activities in Step 1 and Step 2 may indicate either the presence of two enzymes with different substrate requirements or the same enzyme working on the conformationally altered HS chains in a different pH environment (Step 1 in a neutral and Step 2 in an acidic environment). HS oligosaccharides generated by the second step of degradation stay unchanged for a relatively long time ($T_{1/2}$ ~3 h). The final stage of degradation (Step 3) occurs in a compartment which can be described as a classic lysosome; ^{35}S-labeled HS oligosaccharides are rapidly and completely degraded to free sulfate. There are no appreciable intermediates detected between the preceding 5 kDa HS oligosaccharides and the free sulfate generated, indicating that the degradation of HS oligosaccharides is completed in such a very short time. This in turn indicates that the compartment which stores the 5 kDa HS oligosaccharide is functionally segregated from the final lysosomal compartment. This final step is also completely inhibited by lysosomotropic drugs, confirming the lysosomal nature of the final degradation compartment and/or the requirement of an acidified compartment for the transfer of 5 kDa HS oligosaccharide from the preceding storage compartment to the lysosomes. Interestingly, a serine protease inhibitor, leupeptin, seems to inhibit this final degradation step in lysosomes. Since the majority of the protein components of the HS proteoglycan should have been removed from the 5 kDa HS oligosaccharides at this stage, this observation implies the requirement of proteolytic process for the final transfer of HS oligosaccharides to lysosomes or for the activation of key exoglycosidases.

In granulosa cells, GPI-anchored HS proteoglycan and protein-intercalated HS proteoglycan are clearly segregated and targeted to distinct degradation pathways, therefore, the anchoring mechanism seems to critically determine the routing of HS proteoglycans to different degradation pathways. As described above, the kinetics of endocytosis of HS proteoglycan also seem to be determined by the anchoring mechanism. Elucidation of regulatory mechanisms involved in the endocytotic and degradation pathways of HS proteoglycans especially in relation to their defined functional roles will be an interesting subject for future study.

Metabolic processes found in other systems

The basic endocytotic, and intracellular degradation schemes elucidated for the granulosa cell system seem to be consistent with the metabolic scheme observed in other cell culture systems that have been used for the study of HS proteoglycan metabolism. However, there are several metabolic behavior of HS proteoglycan reported in other systems which have not been clearly identified in the granulosa cells.

Recycling of HS proteoglycan between the cell surface and an intracellular compartment has been demonstrated in a rat parathyroid cell line.[5] Recycling of HS proteoglycan is observed only when the extracellular calcium concentration is reduced below physiological levels. Recycling is as rapid (average cycling time is ~9 min) as other recycling molecules reported. In the granulosa cell, recycling of plasma membrane-associated HS proteoglycan cannot be demonstrated within the resolution of pulse-chase experiments (2 min).

Release of free HS chains into the medium compartment is observed in some cell cultures including hepatocytes[27,28] and parathyroid cells.[5]

In other cell systems such as the granulosa culture, free HS chains generated through the degradation pathway 2 are not exocytosed and all of them undergo lysosomal degradation. Release of HS chains into the medium may suggest the presence of an endoglycosidase near the plasma membrane or the presence of a recycling mechanism involving free HS chains.

An endocytotic mechanism of GPI-anchored HS proteoglycans after their cleavage from the plasma membrane by PI-PLC using an inositol-recognizing receptor has been proposed for a hepatoma cell line.[6] Uptake of shed HS proteoglycan (either GPI-anchored, PI-PLC-released, or protein-intercalated HS proteoglycans) is minimal in other cell culture systems.

Nuclear localization of HS chains has been demonstrated in a hepatoma cell line.[29,30] HS chains associated with the nuclear compartment are enriched with oversulfated disaccharide components and the amount of HS chains accumulating in the nucleus seems to correlate with proliferative activity of the cells.

CONCLUSION

Basic mechanisms involved in the metabolism of plasma membrane-associated HS proteoglycans have been identified in several model experimental systems. Future studies will elucidate regulation of specific biological functions of the plasma membrane-associated HS proteoglycans by the basic cellular metabolic activities.

REFERENCES

1. D. M. Noonan, E. A. Horigan, S. R. Ledbetter, G. Vogeli,M. Sasaki, Y. Yamada, and J. R. Hassell, Identification of cDNA clones encoding different domains of the basement membrane heparan sulfate proteoglycan, J. Biol. Chem. 263:16379 (1988).
2. S. Saunders, M. Jalkanen, S. O'Farrell and M. Bernfield, Molecular cloning of syndecan, an integral membrane proteoglycan, J. Cell Biol. 18:1547 (1989).
3. P. J. Marynen, J. Zhang, J. J. Cassiman, H. Van den Berghe and G. David, Partial primary structure of 48- and 90-kilodalton core proteins of cell surface-associated heparan sulfate proteoglycans of lung fibroblasts, J. Biol. Chem. 264:7017 (1989).
4. G. David, V. Lories, B. Marynen, J.-J. Cassiman, and H. Van den Berghe, Molecular cloning of a phosphatidylinositol-anchored membrane heparan sulfate proteoglycan from human lung fibroblasts, J. Cell Biol. 111:3165 (1990).
5. Y. Takeuchi, K. Sakaguchi, M. Yanagishita, G. D. Aurbach and V. C. Hascall, Extracellular calcium regulates transport and distribution of proteoglycans in a rat parathyroid cell line, J. Biol. Chem. 265:13661 (1990).
6. M. Ishihara, N. S. Fedarko, and H. E. Conrad, Involvement of phosphatidylinositol and insulin in the coordinated regulation of proteoheparan sulfate metabolism and hepatocyte growth, J. Biol. Chem. 262:4708 (1987).
7. M. Yanagishita and D. J. McQuillan, Two forms of plasma membrane-intercalated heparan sulfate proteoglycans in rat ovarian granulosa cells: labeling of proteoglycans with a photoactivatable hydrophobic probe, and effect of membrane-anchor specific phospholipase C, J. Biol. Chem. 264:17551 (1989)
8. M. Yanagishita, Metabolic labeling of glycosylphosphatidylinositol-anchor of heparan sulfate proteoglycans in rat ovarian granulosa cells, submitted (1991).

9. D. J. Carey, D. M. Crumbling, R. C. Stahl, and D. M. Evans, Association of cell surface heparan sulfate proteoglycans of Schwann cells with extracellular matrix proteins, J. Biol. Chem. 265:20627 (1990).
10. A. Schmidtchen, R. Sundler, and L.-Å. Fransson, A fibroblast heparan sulphate proteoglycan with a 70 kDa core protein is linked to membrane phosphatidylinositol, Glycoconjugate J. 7:563, (1990).
11. L. Kjellén; Å. Oldberg; M. Höök, Cell-surface heparan sulfate: Mechanisms of proteoglycan-cell association, J. Biol. Chem. 255:10407 (1980)
12. E. F. Neufeld and J. Muenzer, The mucopolysaccharidoses, *in*: "The Metabolic Basis of Inherited Disease," C. R. Scriver, A. L. Beaudet, W. S. Sly and D. Valle, eds. McGraw-Hill, New York, pp. 1565-1587 (1989).
13. M. Yanagishita, and V. C. Hascall, Proteoglycan synthesized by rat ovarian granulosa cell culture: multiple intracellular degradation pathways and the effect of chloroquine, J. Biol. Chem. 259:10270 (1984).
14. M. Yanagishita, and V. C. Hascall, Effects of monensin on the synthesis,transport and intracellular degradation of proteoglycans in rat ovarian granulosa cells in culture, J. Biol. Chem. 260:5445 (1985).
15. M. Yanagishita, Inhibition of intracellular degradation of proteoglycans by leupeptin in rat ovarian granulosa cells, J. Biol. Chem. 260:11075 (1985).
16. M. Yanagishita, Glycosylphosphatidylinositol-anchored and core protein intercalated heparan sulfate proteoglycans in rat ovarian granulosa cells have distinct secretory, endocytotic and intracellular degradation pathways submitted (1991).
17. D. J. McQuillan, R. J. Midura, V. C. Hascall and M. Yanagishita, Plasma membrane-intercalated heparan sulfate proteoglycans in osteoblastic cell line (UMR 106-01, BSP), submitted
18. B. L. Chan, M. P. Lisanti, E. Rodriguez-Boulan and A. R. Saltiel, Insulin-stimulated release of lipoprotein lipase by metabolism of its phosphatidylinositol anchor, Science 241:1670 (1988)
19. H. Hausser and H. Kresse, Binding of heparin and of the small proteoglycan decorin to the same endocytosis receptor proteins leads to different metabolic consequences, J. Cell Biol. 114:45 (1991)
20. G. M. Oosta, L. V. Favreau, D. L. Beeler and R. D. Rosenberg, Purification and properties of human platelet heparitinase, J. Biol. Chem. 257:11249, (1982).
21. L. Thunberg, G. Bäckström, Å. Wasteson, H. C. Robinson, S. Ögren and U. Lindahl, Enzymatic depolymerization of heparin-related polysaccharides: substrate specificities of mouse mastocytoma and human platelet endo-β-D-glucuronidases, J. Biol. Chem. 257:10278 (1982).
22. Å. Oldberg, C.-H. Heldin, Å. Wasteson, C. Busch and M. Höök, Characterization of a platelet endoglycosidase degrading heparin-like polysaccharides, Biochemistry 19:5755 (1980).
23. U. Klein and K. von Figura, Substrate specificity of a heparan sulfate-degrading endoglucuronidase from human placenta, Hoppe-Seyler's Z. Phyiol. Chem. 360:1465 (1979).
24. M. Höök, Å. Wasteson and Å. Oldberg, A heparan sulfate-degrading endoglycosidase from rat liver tissue, Biochem. Biophys. Res. Commun. 67:1422 (1975).
25. U. Klein, H. Kresse and K. von Figura, Evidence for degradation of heparan sulfate by endoglycosidases: Glucosamine and hexuronic acid are reducing terminals of intracellular heparan sulfate from human skin fibroblasts, Biochem. Biophys. Res. Commun. 69:158 (1976).
26. L. Kjellén, H. Pertoft, Å. Oldberg and M. Höök, Oligosaccharides generated by an endoglucuronidase are intermediates in the intracellular degradation of heparan sulfate proteoglycans, J. Biol. Chem. 260:8416 (1985).

27. J. T. Gallagher, A. Walker, M. Lyon, and W. H. Evans, Heparan sulphate-degrading endoglycosidase in liver plasma membranes, Biochem. J. 250:719 (1988).
28. M. Piepkorn, P. Hovingh, and A. Linker, Glycosaminoglycan free chains, external plasma membrane component from the membrane proteoglycans, J. Biol. Chem. 264:8662 (1989).
29. N. S. Fedarko and H. E. Conrad, A unique heparan sulfate in the nuclei of hepatocytes: structural changes with growth state of the cells, J. Cell Biol. 102:587 (1986)
30. M. Ishihara, N. S. Fedarko and H. E. Conrad, Transport of heparan sulfate into the nuclei of hepatocytes, J. Biol. Chem. 261:13575 (1986)

LYSOSOMAL DEGRADATION OF HEPARIN AND HEPARAN SULPHATE

Craig Freeman and John Hopwood

Lysosomal Diseases Research Unit
Department of Chemical Pathology
Adelaide Children's Hospital
North Adelaide South Australia 5006

Introduction

This review is a limited update of a previous review of enzymes that degrade heparan sulphate (HS) and heparin (Hopwood, 1989). Newly-synthesised HS proteoglycan (HSPG) is internalised from the cell surface and catabolised with a half-time of 4 to 6 h in rat ovarian granulosa cells (Yanagishita and Hascall, 1984), and by more than 28 h in human colon carcinoma cells (Iozzo, 1987). Studies of both cell types have shown the existence of preliminary protease and an initial endoglycosidase activity in non-lysosomal (chloroquine-insensitive compartments) to generate HS intermediates of Mr 10 kDa, and further chloroquine-sensitive endoglycosidic activity to produce HS fragments of Mr 5 kDa which are rapidly degraded in the lysosome by a series of exohydrolases to monosaccharides and sulphate without the generation of intermediates. A number of distinct lysosomal membrane transporters are involved in the efflux of the monomeric products GlcNAc, GlcA and IdoA and sulphate ions from the lysosome (Jonas et al., 1989; Jonas and Jobe, 1990a, 1990b; Mancini et al., 1989) which can be reutilised in biosynthetic pathways (Rome and Hill, 1986).

Endoglycosidic Activity

To date, each of the endoglycosidase activities studied have been endo-ß-glucuronidases. Although Yanagishita and Hascall (1984) and Iozzo (1987) reported that the endoglycosidase acted on internalised HS, studies have not excluded the possibility that HSPG is hydrolysed on the cell surface by these activities and the products endocytosed following receptor-mediated endocytosis (Krüger and Kresse, 1986; Barzu et al., 1987). The secretion of endo-ß-glucuronidase by B16 melanoma cells has been proposed to be an important factor for the invasive properties of those malignant cells (Nakajima et al., 1988). Gallagher et al. (1988) reported the presence of a HS-degrading endoglycosidase activity in rat liver plasma membranes that was active towards hydrophobic membrane-bound HSPG but had little activity towards HSPG displaced from the membranes by NaCl. This may indicate either substrate specificity towards different populations of HSPG or perhaps inaccessibility of the enzyme to the extrinsic HSPG. Enzyme activity was maximal at pH 7.5 to 8.0, was absent below pH 5.5, and was retained in membranes solubilised in 1% Triton X-

Heparin and Related Polysaccharides
Edited by D.A. Lane *et al.*, Plenum Press, New York, 1992

100. The high pH optimum and cell surface location distinguish the enzyme from other endoglycosidases which have more acidic pH optima and may be lysosomal or endosomal in origin (see Hopwood, 1989). The plasma membrane endoglycosidase may modulate cellular interactions mediated by HS and/or release biologically active HS fragments from the cell periphery. Alternatively, the endoglycosidase activity could be co-internalised with HSPG into primary endosomes to act at the higher pH before endosomal acidification, which is compatible with the dual endoglycosidase activities proposed by Yanagishita and Hascall (1984), and Iozzo (1987), to produce HS fragments with Mr of 5 and 10 kDa in separate compartments. The endoglycosidase activity may be a source of nuclear HS which occurs independently of endosome acidification (Ishihara et al., 1986).

Sewell et al. (1989) showed human mononuclear cells contained a cell-associated HS-degrading endoglycosidase activity which was not secreted into the medium following cell culture. The substrate used for detection of the activity was a xyloside-initiated ^{35}S-labelled HS fraction isolated from the medium of cultured bovine glomeruli, and was used to construct a solid phase substrate following coupling to Sepharose 4B. The enzyme, partially purified from mononuclear cells derived from human spleen, had an apparent native protein Mr of 50 kDa and activity towards the solid phase substrate with a broad pH optimum from pH 4 to 7. The practice of using heparin rather than HS to detect endo-ß-glucuronidase activity may have selectively identified activities with a preference for highly sulphated sequences rather than activities toward the HS common GlcA-GlcNAc linkages that are less common in heparin (for example, B16 melanoma and human platelet activities) (Oostra et al., 1982; Nakajima et al., 1984). Based on limited substrate specificity there appears to be at least three different types of mammalian cell endo-ß-glucuronidases: B16 melanoma activities degrade HS rather than heparin (Nakajima et al., 1988), human platelet activities cleave both heparin and a heparin precursor devoid of O-sulphate groups and the GlcA-GlcNS3S linkage in the antithrombin-binding sequence, while mouse mastocytoma activities act toward heparin rather than the heparin precursor and do not cleave the antithrombin regions of heparin (Oldberg et al., 1980; Thunberg et al., 1982). Jin et al. (1990) purified an endo-ß-glucuronidase from cultured murine melanoma cells. Polyclonal antibodies raised against a synthetic N-terminal peptide derived from the 97 kDa polypeptide were used to localise the antigen in the cytoplasm or at the cell surface of tumour cells, but not in normal tissue.

Exo-Enzyme Activities

The final stage of HS degradation occurs when the fragments of endo-ß-glucuronidase activity are processed by the concerted action of nine lysosomal exo-enzymes (see Hopwood, 1989) to yield inorganic sulphate and monosaccharide products. These enzyme activities include five sulphatases: glucosamine-3-sulphatase, glucosamine-6-sulphatase (G6S), sulphamate sulphohydrolase, iduronate-2-sulphatase (IDS) and glucuronate-2-sulphatase; 3 glycosidases: α-N-acetylglucosaminidase, ß-D-glucuronidase and α-L-iduronidase (IDUA); and a bond making enzyme: acetyl CoA:α-glucosamine-N-acetyltransferase (N-acetyltransferase), an integral membrane enzyme required for the transfer of acetyl groups from cytosolic-derived acetyl CoA to the non-reducing terminal GlcN-residues exposed by sulphamate sulphohydrolase activity. Apart from glucosamine-3-sulphatase and glucuronate-2-sulphatase, a deficiency in humans of any one of these enzyme activities is known to lead to the accumulation of HS fragment substrates for the deficient enzyme, and disorders known as the Mucopolysaccharidoses (MPS).

Since the previous review (Hopwood, 1989) most reports concerning the purification, characterisation, catalytic properties and molecular biology of the HS-degradative enzymes have concentrated on the uronic acid-acting enzymes IDS and IDUA.

Iduronate-2-Sulphatase

Iduronate-2-sulphatase (IDS) de-O-sulphates non-reducing terminal IdoA2S residues in heparin, HS and dermatan sulphate (DS). In humans, a deficiency of the enzyme results in an MPS-II, (or Hunter syndrome) phenotype, an X-linked recessive trait which is characterised by mild to severe skeletal dysmorphism, coarse facies, hepatosplenomegaly, cardiovascular problems and frequently, though not always, mental retardation (Neufeld and Muenzer, 1989; Hopwood and Morris, 1990).

IDS has been purified more than 500,000-fold with 5% recovery of activity from human liver (Bielicki et al., 1990). Two major forms were separated by chromatofocusing chromatography in approximately equal amounts of recovered enzyme activity. Form A (pI 4.5) and Form B (pI <4.0) each consisted of polypeptides of Mr 42 and 14 kDa when analysed by SDS-PAGE under both reducing and non-reducing conditions. Form A had a native Mr in the range of 42 to 65 kDa determined by gel permeation chromatography. IDS was purified from human kidney, placenta and lung and shown to have similar native Mr and subunit components to that observed for liver enzyme. The observed behaviour of IDS during gel permeation was dependent upon the nature of the gel matrix and the pH and ionic strength of the elution buffer, and a range of native protein apparent Mr values of 14 to 130 kDa were obtained for Form A. The dependence of apparent Mr on buffer pH and ionic strength was also observed with purified human liver IDUA (Clements et al., 1985a). Thus, it would be unwise to make comparisons between reports of the native size of IDS prepared from different sources using different analytical methods and conditions (see Hopwood, 1989). Bielicki et al. (1990) observed that both forms of human liver IDS were active towards a variety of substrates derived from heparin and DS. The pH optima and kinetic parameters (K_m and k_{cat}) for human liver IDS activity toward each substrate was dependent upon the substrate structure and the buffer composition.

The addition of a C-6 sulphate ester to IdoA2S-anM to produce the disaccharide IdoA2S-anM6S gave a 63-fold increase in catalytic efficiency, the largest increase for any single structural change and involved both a 5-fold and 13-fold increase in binding affinity and turnover number respectively (Table 1). IdoA2S-anT4S had a 5.7-fold higher affinity for enzyme but was turned over 4-fold less efficiently such that it was hydrolysed only 1.4-times more efficiently than IdoA2S-anM6S. Activities toward tetrasaccharide substrates which differ only in their GlcN-substituent enabled a comparison of the effect of aglycone N-substituted GlcN-residues. The effect of the GlcNS6S substituent was to increase the binding affinity by up to 2-fold compared to substrates bearing aglycone GlcNAc6S or GlcNH6S residues and the disaccharide with anM6S residue. The dominating influence of the penultimate residue C-6 sulphate ester upon both substrate binding and turnover was demonstrated in that the K_m and k_{cat} values for activity toward IdoA2S-anM6S were only 2.8-fold and 2.3-fold less respectively than the strongest binding and most reactive tetrasaccharide substrates. IdoA2S-anM6S was acted upon 3-fold less efficiently than either of the N-sulphated or N-acetylated tetrasaccharides which were hydrolysed twice as efficiently as IdoA2S-GlcNH6S-IdoA2S-anM6S, therefore demonstrating the lack of influence by the GlcN substitution upon enzyme activity in contrast to the observations for purified G6S (Freeman and Hopwood, 1987). The pentasaccharide substrate

IdoA2S-GlcNS-UA-GlcNAc-GlcOA, which lacks a penultimate C-6 sulphate, was bound with similar affinity to the tetrasaccharide substrates, but with a 10-fold higher affinity than IdoA2S-anM. The turnover number for IdoA2S-anM was 4.7-fold less than for the pentasaccharide which was hydrolysed 48-times more efficiently than IdoA2S-anM, demonstrating the influence of aglycone structures further away from the site of hydrolysis . Form A removed only the non-reducing terminal sulphate ester from each of the tetrasaccharide substrates. Neither Form A nor Form B were active toward Ido2S-anM6S, the carboxy-reduced derivative of IdoA2S-anM6S, or the C-6 epimer GlcA2S-anM6S, which demonstrated the important influence upon enzyme specificity of the presence and configuration of the C-6 carboxy group. However, both GlcA2S-anM6S and Ido2S-anM6S were potent competitive inhibitors of IDS activity toward IdoA2S-anM6S with Ki values of 1 μM each, which suggest that they bind with greater affinity than the disaccharide substrate IdoA2S-anM6S. IdoA-anM6S and IdoA-anM, the products of IDS activity toward IdoA2S-anM6S and IdoA2S-anM respectively, along with anM6S, the product of IDS activity toward IdoA-anM6S, were all strongly binding competitive inhibitors of IDS activity with Ki values of 1.7, 11.7 and 0.25 μM respectively, however the monosaccharides GlcNAc, GlcNS, GlcNAc6S and anM did not inhibit IDS activity toward IdoA2S-anM6S.

TABLE 1. Human liver IDS (Form A) activity toward a variety of substrates (Bielicki et al., 1990)

	pH optimum	Km (μM)	kcat[a]	10^{-6} x kcat/Km (catalytic efficiency)
IdoA2S-anM	5.4	19.2	161	8 (1.0)[b]
IdoA2S-anM6S	4.0	4.0	2114	529 (63)[b]
IdoA2S-anT4S	5.0	0.7	507	724 (86)[b]
IdoA2S-GlcNS6S-IdoA2S-anM6S	5.7	1.4	2177	1568 (187)[b]
IdoA2S-GlcNAc6S-IdoA2S-anM6S	5.7	3.1	4858	1568 (187)[b]
IdoA2S-GlcNH6S-IdoA2S-anM6S	5.4	2.5	1925	770 (92)[b]
IdoA2S-GlcNS-UA-GlcNAc-GlcOA	5.4	1.9	756	399 (48)[b]

a, kcat, turnover number (mol/min per mol of enzyme); b, kcat/Km calculated relative to a value for IdoA2S-anM = 1; c, GlcOA is C-1 reduced GlcA

A 2.3 kb cDNA clone coding for human IDS has been isolated and sequenced and used in Northern analysis of RNA from human placenta to identify four species of 5.8, 5.4, 2.1 and 1.4 kb (Wilson et al., 1990). The IDS gene is located on the X chromosome at the q27/28 boundary, with its 5′ end toward the telomere (Roberts et al., 1989; Wilson et al., 1991). Analysis of the deduced 550-amino acid precursor sequence indicated that IDS has a 25-amino acid amino-terminal signal sequence, followed by 8-amino acids that are removed from the proprotein. Proteolytic cleavage occurs to produce mature enzyme that contains a 42 kDa polypeptide N-terminal to a 14 kDa polypeptide. The amino acid sequence has strong sequence homology with other sulphatases, suggesting that IDS arose through a process of gene duplication and divergent evolution (Wilson et al., 1990). Approximately 20% of MPS-II patients have structural alterations and gross deletions in their IDS gene. All of these patients have a clinically severe Hunter phenotype (Wilson et al., 1991).

α-L-Iduronidase

Hydrolysis of the non-reducing end α-L-iduronide glycosidic bond in heparin, HS and DS results from α-L-iduronidase (IDUA) activity. A deficiency of IDUA activity in humans results in MPS-I (Hurler or Scheie syndromes depending on clinical severity). Severely affected (Hurler) patients may have mental retardation, severe skeletal deformities, coarse hirsute facies, corneal clouding and early death, while mildly affected (Scheie) patients may have both a normal intelligence and life-span with mild skeletal deformities (Neufeld and Muenzer, 1989). The considerable variation of clinical presentation is probably a result of different mutations in the IDUA gene (Neufeld and Muenzer, 1989; Hopwood and Morris, 1990).

A monoclonal antibody against human liver IDUA was used to immunopurify IDUA 170,000-fold from human liver, lung, kidney and urine in yields of greater than 50% (Clements et al., 1989). SDS-PAGE analysis of IDUA isolated showed the major polypeptide present in enzyme from all sources had a Mr of 65 kDa with polypeptides with Mr 74, 60, 49, 44, 18 and 13 kDa also being present. Two forms of human liver, lung and kidney IDUA were eluted from the affinity column. Although composed of different proportions of some of the seven polypeptides, the two liver forms showed no difference in their activities towards the heparin and DS-derived substrates IdoA-anM6S and IdoA-anT4S. The native protein Mr determined by gel permeation for both of the liver forms was 65 kDa. Six of the polypeptides are probably derived by three proteolytic clips of the 74 kDa polypetide (Clements et al., 1989; Scott et al., 1991). Taylor et al. (1991) studied the synthesis and maturation of IDUA in cultured human skin fibroblasts using a monoclonal antibody. Pulse-chase labelling of the fibroblasts showed IDUA was synthesised as an 81 kDa precursor and was processed within 24 h via intermediates of 76 and 70 to a mature 69 kDa species, while an 82 kDa precursor was secreted into the culture medium. Myerowitz and Neufeld (1981) found a similar sequence of maturation in cultured skin fibroblasts using a polyclonal antibody.

Substantially different catalytic properties (pH optima, K_m and V_{max} values) have been reported for various preparations of purified enzyme, and for IDUA in leucocyte and cultured human skin fibroblast homogenates. For example, IDUA activity from different tissue sources has been reported to have apparent K_m values ranging from 0.06 to 2.2 mM for activity towards IdoA-MU, and from 0.04 to 9.0 mM for activity towards IdoA-anM (see Hopwood, 1989). These differences may be due to the tissue source, the method of purification, the presence of modifier proteins or different assay protocols which have made the comparison of kinetic data between the laboratories difficult. Freeman and Hopwood (1991c) determined the catalytic properties of purified human liver IDUA activity toward heparin-derived substrates. Enzyme activity at low pH was stimulated in the presence of BSA and potently inhibited by NaCl, previously shown to be a competitive inhibitor of IDUA (Clements et al., 1985b), therefore the data reported in Table 2 for activity toward the disaccharide substrates have lower pH optima and K_m values, and higher k_{cat} values than reported by Clements et al. (1985b). The major influence on substrate binding substrate and turnover by IDUA resulted from the addition of a C-6 sulphate ester to IdoA-anM to give IdoA-anM6S, resulting in a 138-fold increase in catalytic efficiency. A similar observation was made for purified human liver IDS activity toward IdoA2S-anM6S, compared to IdoA2S-anM (see above). IdoA-MU was a poorly bound substrate with a K_m value of 5.5-fold and 50-fold higher than IdoA-anM and IdoA-anM6S respectively, but it was turned over 121- and 8-times faster, respectively. Extension of the disaccharide IdoA-anM6S to the tetrasaccharide substrates permitted a comparison of the effect of the N-substitution and aglycone structures further away from the site of catalysis upon substrate binding and catalysis (Table 2). Each of the tetrasaccharides was bound to the enzyme with more than twice the affinity than observed for IdoA-anM6S. Whereas binding of the substrate to the enzyme was independent of the type of N-substitution, the rate of substrate turnover was influenced by the N-substitution on the adjacent GlcN-residue. Maximal hydrolysis of IdoA-GlcNH6S-IdoA2S-anM6S was 8.2-fold and 29-fold slower than when the substrate possessed an aglycone 6-sulphated GlcNS or GlcNAc residue. However, the most catalytically efficient substrate, the tetrasaccharide IdoA-GlcNAc6S-IdoA2S-anM6S was hydrolysed only 1.9-times more efficiently than the disaccharide IdoA-anM6S. IDS similarly hydrolysed its most efficient substrate IdoA2S-GlcNAc6S-IdoA2S-anM6S only 3-times more efficiently than IdoA2S-anM6S (Table 1). The small change in catalytic efficiency was unexpected compared to sulphamate sulphohydrolase and G6S which each acted towards their most efficient substrate GlcNS6S-IdoA2S-anM6S 3,300- and 130-fold more efficiently than for their most efficient disaccharide substrates (see Hopwood, 1989).

TABLE 2. Human liver IDUA activity toward a variety of substrates (Freeman and Hopwood, 1991c)

	pH optimum	Km (μM)	kcat	10^{-6} x kcat/Km (catalytic efficiency)
IdoA-MU	2.4	110	2132.0	19.4 (22)[b]
IdoA-anM	3.0	20	17.9	0.9 (1)[b]
IdoA-anM6S	2.7	2.2	273.6	124.4 (138)[b]
IdoA-GlcNS6S-IdoA2S-anM6S	3.6	1.0	65.0	65.0 (72)[b]
IdoA-GlcNAc6S-IdoA2S-anM6S[a]	3.3	1.0	229.5	229.5 (255)[b]
IdoA-GlcNH6S-IdoA2S-anM6S	3.0	0.9	7.9	8.8 (10)[b]

a, kcat, turnover number (mol/min per mol of enzyme); b, kcat/Km efficiencies calculated relative to a value for IdoA-anM = 1.0; c, this is an artificial sequence as GlcNAc is not naturally found between IdoA residues; Abbreviations, MU = 4-methylumbelliferyl

When human liver IDUA activity towards IdoA-MU and IdoA-anM6S was determined at pH 4.8 over a wide substrate concentration range, Lineweaver-Burk plots revealed the existence of two apparent K_m values of 37 μM and 1.92 mM for IdoA-MU, and 10 μM and 475 μM for activity towards IdoA-anM6S (Freeman and Hopwood, 1991c). For both substrates, the lower K_m form acted 23-times more efficiently towards its respective substrate than did the high K_m form. The observation of two apparent K_m values for human liver IDUA activity towards IdoA-MU and IdoA-anM6S may offer an explanation for the range of K_m values that have been reported for IDUA activity toward IdoA-MU and IdoA-anM. IDUA activity toward IdoA-anM6S and IdoA-GlcNAc6S-IdoA2S-anM6S was inhibited by the addition of oligosaccharides which were substrates for other HS-degrading enzymes, but which IDUA itself did not hydrolyse, for example, the substrate analogues IdoA2S-anM6S and GlcA2S-anM6S. Oligosaccharides with non-reducing terminal GlcN residues GlcNAc6S-IdOA, GlcNAc6S-IdoA2S-anM6S and GlcNS-IdoA2S-anM6S, possibly by binding to the aglycone binding sites of IDUA, inhibited IDUA activity toward IdoA-anM6S by 20, 52 and 89% respectively. The potency of their inhibition also reflected the affinity of the analogues for their own enzymes (K_m values of 11.1, 0.76 and 0.07 μM respectively).

Scott et al. (1991) reported the isolation and sequence of cDNA coding for human IDUA. Northern analysis of RNA from human placenta identified a single 2.3 kb in RNA species. Scott et al. (1990) also reported the relocation of the gene for IDUA from chromosome 22 to 4p16.3. The deduced precursor 653-amino acid IDUA polypeptide has a 26-amino acid signal sequence that is cleaved immediately prior to the amino terminus of the 74 kDa polypeptide identified in human liver IDUA (Clements et al., 1989; Scott et al., 1991).

Glucosamine-6-Sulphatase

The hydrolysis of non-reducing end α-linked glucosamine-6-sulphate residues on heparin or HS and ß-linked GlcNAc6S residues on keratan sulphate has been shown to occur by the same G6S activity. A deficiency of the enzyme in humans has been shown to result in the expression of a Sanfilippo phenotype (MPS-IIID). Four forms of human G6S activities which differed in their pI values were isolated from human liver (Freeman et al., 1987) and were shown to have similar substrate specificities towards oligosaccharides derived from heparin and keratan sulphate (Freeman and Hopwood, 1987). cDNA coding for the full sequence of the mature G6S has been isolated and used in Northern analysis to identify a 4.2 and 5.0 kb mRNA species in human skin fibroblasts RNA (Robertson et al., 1988a and unpublished results). The chromosomal location of the gene for G6S is 12q14 (Robertson et al., 1988b). The amino acid sequence has strong sequence homology with other sulphatases (see IDS above and Robertson et al., 1988*a*; Wilson et al., 1990). Western blot analysis, using polyclonal antibodies to G6S in fibroblast lysates from normal controls, revealed two bands at 82 and 78 kDa (Siciliano et al., 1991). Endoglycosidase H treatment of the fibroblast extracts resulted in a single Mr 74 kDa species indicating that the two forms result from different carbohydrate processing of the same polypeptide. Urine and skin fibroblast G6S activities were also observed to be composed of activities which differed in the relative amounts of the different pI forms compared to the liver enzyme, however their kinetic parameters (pH optimum, K_m and V_{max}) and influence of aglycone substrate structure upon the observed activity were similar to that observed for the human liver enzyme (Freeman and Hopwood, 1987, 1991b).

Influence of Aglycone Substrate Structure Upon HS-Degradation

A comparison can be made of the effect of aglycone structural features upon those enzyme activities from purified human liver, which act towards the uronic acid and GlcN residues in the highly sulphated regions of HS. The absolute requirement for the presence of a C-6 carboxy group on the IdoA residue being attacked was observed for both IDUA and IDS activities (Clements et al., 1985b; Bielicki et al., 1990). Carboxy reduction of IdoA-anM6S and IdoA2S-anM6S abolished activity towards either substrate (Clements et al., 1985b; Bielicki et al., 1990), however the carboxy-reduced substrates were potent competitive inhibitors of activities towards the parent compounds. Disaccharide substrates for both sulphamate sulphohydrolase and G6S activities where the aglycone C-6 carboxy group was reduced were able to bind strongly to the respective enzyme, however substrate turnover in each case was poor compared to activity toward the intact substrate (Freeman and Hopwood, 1986, 1987). These results suggest that for each of the HS-degrading enzymes, the presence of C-6 carboxy groups are not critical for binding of the substrate to each of the enzymes, but are very important for the turnover of substrate.

The presence of an aglycone C-6 sulphate ester on the penultimate residue gave the greatest single influence upon both the binding and the rate of substrate turnover for both of the enzymes which act towards IdoA residues (Bielicki et al., 1990; Clements et al., 1985a; Freeman and Hopwood, 1991c). The influence of a C-6 sulphate ester on the GlcNS residue being acted upon by sulphamate sulphohydrolase was dependent upon the rest of the substrate structure. The substrates GlcNS-IdoA2S-anM6S, GlcNS-Ido and GlcNS-Ido2S were hydrolysed from 1.6- to 10.6-fold less efficiently, while GlcNS6S-IdOA was hydrolysed 6-fold more efficiently than the 6-sulphated GlcNS analogues. The influence of a penultimate residue possessing a 2-O-sulphate ester, for example GlcNS6S-Ido2S, increased the catalytic efficiency for sulphamate sulphohydrolase and G6S activities by 10- and 3-fold respectively, compared to GlcNS6S-Ido. However, substrates possessing a penultimate residue with an IdoA2S residue were hydrolysed at rates much greater than the sum of the separate influences from aglycone C-6 carboxy and C-2 sulphate ester residues, for example, GlcNS6S-IdoA2S-anM6S were hydrolysed by sulphamate sulphohydrolase 21,000- and 3,500-fold more efficiently, and by G6S 1,000-fold more efficiently than either GlcNS6S-IdOA or GlcNS6S-Ido2S respectively. The influence of aglycone structures further away from the site of hydrolysis was evident in that sulphamate sulphohydrolase hydrolysed the tetrasaccharide GlcNS-IdoA-GlcNS-IdOA 20-fold more efficiently than the disaccharide GlcNS-IdOA, while IDUA hydrolysed the pentasaccharide substrate 48-fold more efficiently than IdoA-anM. However, IDS and IDUA each hydrolysed their most efficient tetrasaccharide substrates less than 3-fold more efficiently than their most reactive disaccharide substrates. (Abbreviations: Ido, idose; Ido2S, idose-2-sulphate; IdOA, C-1 reduced IdoA).

The binding of complex tetrasaccharide and trisaccharide substrates to IDS and G6S was influenced by up to 3-fold by the N-substituent on the aglycone GlcN-residue while IDUA activity was independent of N-substitution (Freeman and Hopwood, 1987, 1991c; Bielicki et al., 1990). However, the effect of the N-substituent upon the rate of substrate turnover was clearly more influential for the enzymes as they acted in turn along the sequence IdoA2S-GlcNX6S-IdoA2S-anM6S. Whereas IDS activity towards the tetrasaccharide IdoA2S-GlcNH6S-IdoA2S-anM6S was 2-fold less efficient than toward either the N-sulphated, N-acetylated substrates (Bielicki et al., 1990), IDUA hydrolysed IdoA-GlcNAc6S-IdoA2S-anM6S 3.6-fold and 26-fold more efficiently than the N-sulphated or non-substituted substrates respectively (Freeman and Hopwood, 1991c). G6S, however, acted towards the trisaccharide substrates GlcNS6S-IdoA2S-anM6S 10-fold and 300-fold more efficiently than towards GlcNAc6S-IdoA2S-anM6S and GlcNH6S-IdoA2S-anM6S respectively (Freeman and Hopwood, 1987).

Therefore, each of the HS-degradative enzymes acting either toward uronic acid or GlcN-residues appear to be influenced by similar aglycone substrate structures. In effect, each enzyme binds the same substrate which differ only in the non-reducing end terminal residue, which may explain why each of the enzymes studied to date are potently inhibited by substrates for other HS-degrading enzymes. The Ki values for IdoA-anM6S and IdoA-anM inhibition of IDS activity are similar to the K_m values for IDUA activity towards either substrate (Bielicki et al., 1990; Freeman and Hopwood, 1991c). Models have been proposed for both IDUA and sulphamate sulphohydrolase (Clements et al., 1985b; Freeman and Hopwood, 1986, 1991c) to account for the catalytic requirements. Following binding of the substrate to the enzyme via hydrogen bonding with the ring hydroxy groups, the ring oxygens and interaction with the C-2 sulphate (sulphamate sulphohydrolase and G6S) or C-6 sulphate ester (IDUA and IDS), cationic groups of the enzyme that may interact with the carboxy group and other sulphate esters lie just beyond reach. Following a substrate binding-induced conformational change of the enzyme, strong interactions, for example ion pairing, may occur between the sulphate and carboxy groups and cationic amino acids such as arginine and histidine residues.

The tetrasaccharide substrate, IdoA2S-GlcNS6S-IdoA2S-anM6S can be used as a model substrate for comparing the activities of the HS-degradative enzymes which act in sequence to degrade the highly sulphated regions of HS. The activities of N-acetyltransferase and α-N-acetylglucosaminidase towards more complex heparin-derived substrates have yet to be determined. The substrate IdoA2S-GlcNS6S-IdoA2S-anM6S represents the major repeating structure (-IdoA2S-GlcNS6S-) in human lung heparin (Lindahl, 1989). After IDS and IDUA activities, there are two possible pathways for the degradation of GlcNS6S-IdoA2S-anM6S to proceed (Table 3). In the first pathway IdoA2S-GlcNS6S-IdoA2S-anM6S is acted upon sequentially by IDS, IDUA, G6S and sulphamate sulphohydrolase, which act towards their respective substrates with increased binding activity (Table 3). However, purified sulphamate sulphohydrolase acts towards GlcNS6S-IdoA2S-anM6S with both a K_m value 4-fold lower and a catalytic efficiency 10-fold higher compared to G6S activity towards the same substrate K_m. Human skin fibroblast homogenate activities towards GlcNS6S-IdoA2S-anM6S have shown sulphamate sulphohydrolase preferentially de-N-sulphates the substrate to produce GlcNH6S-IdoA2S-anM6S which is both a relatively inefficient substrate for subsequent human liver G6S activity and is not a substrate for N-acetyltransferase activity (Freeman and Hopwood, 1987, 1991b; Hopwood, 1989). Therefore, the efficient degradation of HS requires the facilitated passage of GlcNS6S-IdoA2S-anM6S from IDUA to G6S.

Table 3. Kinetics of enzymes involved in the sequence degradation of the highly sulphated (IdoA2S-GlcNS6S-) region of heparin (Freeman and Hopwood, 1991c).

	Enzyme[a]	**Km (μM)**	**10^{-6}x kcat/Km (catalytic efficiency)**
IdoA2S-GlcNS6S-IdoA2S-anM6S[b]	IDS	1.4	1568
IdoA-GlcNS6S-IdoA2S-anM6S	IDUA	1.0	66
GlcNS6S-IdoA2S-anM6S	G6S	0.25	91
GlcNS-IdoA2S-anM6S	NS	0.07	665
GlcNH-IdoA2S-anM6S	NAT	-	-
GlcNAc-IdoA2S-anM6S	NAG	-	-
IdoA2S-anM6S	IDS	4.0	529
IdoA-anM6S	IDUA	2.2	123
anM6S			
IdoA2S-GlcNS6S-IdoA2S-anM6S[c]	IDS	1.4	1568
IdoA-GlcNS6S-IdoA2S-anM6S	IDUA	1.0	66
GlcNS6S-IdoA2S-anM6S	NS	0.08	942
GlcNH6S-IdoA2S-anM6S	G6S	0.35	0.3
GlcNH-IdoA2S-anM6S	NAT	-	-

a, Abbreviations: -, not done; NS, sulphamate sulphohydrolase; NAT, acetyl-CoA:glucosamine-N-acetyltransferase; NAG, α-N-acetylglucosaminidase. b, Pathway for initial G6S activity toward GlcNS6S-IdoA2S-anM6S. c, Pathway for initial NS activity toward GlcNS6S-IdoA2S-anM6S

Lysosomal Efflux of the Products of HS Degradation

Rome and Hill (1986) reported that the efflux of sulphate and N-acetylhexosamines from lysosomes, isolated from cultured human skin fibroblasts, occured at a rate that paralleled their production. The existence of genetic disorders which result from deficiencies in specific transport mechanisms for cysteine (Gahl et al., 1982) and sialic acid (Renlund et al., 1986) supported the notion that the egress of products of HS degradation would involve specific lysosomal transport systems. Jonas and Jobe (1990a) demonstrated the presence of a specific pH-regulated sulphate transporter using membrane vesicles prepared from rat liver lysosomes where sulphate and protons exit lysosomes in exchange for chloride ions. Sulphate transport was increased at lowered buffered pH, the K_m for transport at pH 5.0 was 0.16 mM, while at pH 7.0 a lower affinity system with a K_m of 1.4 mM was present. The specific transport of both GlcNAc and GalNAc was shown in rat liver lysosomes (Jonas et al., 1989) and in rat liver lysosomal membrane vesicles (Jonas and Jobe, 1990b). Characteristics of GlcNAc transport in the vesicles (K_m 1.3 mM) were similar to those observed in intact lysosomes (K_m 4.4 mM). Sulphation or phosphorylation of the substrate resulted in a loss of recognition by the carrier as did removal of the acetyl group to form GlcNH. Transport of GalNAc and GlcNAc was competitive, but not affected by Glc, GlcN, GlcA or N-acetylneuraminic acid, or by neutral and cationic amino acids. GlcNAc transport was not dependent upon KCl, $ATP/MgCl_2$ and was unaffected by variation in buffer pH between 6.0 and 8.0. A proton-driven carrier for acidic monosaccharides including N-acetylneuraminic acid and GlcA was observed in rat liver lysosomal membrane vesicles (Mancini et al., 1989). Neither glucuronolactone nor GlcNAc were recognised by the carrier. Transport was strongly influenced by a pH gradient across the membrane.

A Hypothetical Model for HS Degradation

The lysosomal degradation of HS has been proposed to proceed by a highly organised and coupled process such that the HS-degradative enzymes function as a multienzyme complex in close proximity to the lysosomal membrane (Hopwood, 1989). There are several observations supporting this concept shown in Figure 1; (1) Yanagishita and Hascall (1984) have shown lysosomal turnover of HS in cultured cells is extremely rapid and intermediate fragments are not observed; (2) There are several reaction sequences in the degradative pathway where the product of one activity always becomes the substrate for the next, for example -IdoA2S- (IDS -> IDUA), -GlcA2S- (glucuronate-2-sulphatase -> ß-D-glucuronidase) and -GlcNS- (sulphamate sulphohydrolase -> N-acetyltransferase -> α-N-acetylglucosaminidase), providing potential pairing of enzymes for channeling of substrates; (3) Each enzyme activity has similar aglycone structure requirements and has been shown to display both product and substrate inhibition as well as inhibition by non-substrates, therefore requiring organisation to modulate substrate availability and product removal and to ensure an efficient pathway for the degradation of GlcNS6S residues, since initial sulphamate sulphohydrolyase activity would produce poor substrates for subsequent G6S activity; (4) Buckmaster et al. (1986) and Freeman and Hopwood (unpublished) have shown a pH-dependent association of lysosomal enzymes with each other and the lysosomal membrane with maximal association occurring at pH 4.8 to 5.1, the physiological pH. High salt concentrations are needed to solubilise many of the enzymes, which is consistent with their physical association with integral membrane proteins, perhaps the N-acetyltransferase or one of the transport proteins, especially since monosaccharides and sulphate are potent inhibitors of enzyme activities; (5) The degradation of

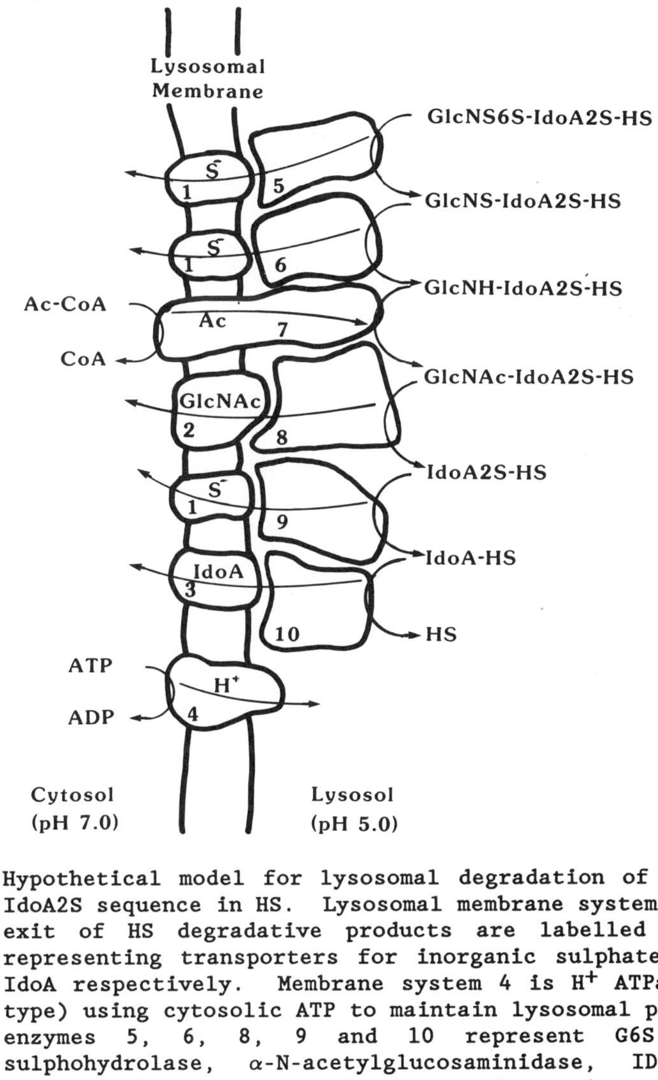

FIGURE 1 Hypothetical model for lysosomal degradation of the GlcNS6S-IdoA2S sequence in HS. Lysosomal membrane systems involved in exit of HS degradative products are labelled 1, 2 and 3 representing transporters for inorganic sulphate, GlcNAc and IdoA respectively. Membrane system 4 is H^+ ATPase (vacuolar type) using cytosolic ATP to maintain lysosomal pH. Lysosomal enzymes 5, 6, 8, 9 and 10 represent G6S, sulphamate sulphohydrolase, α-N-acetylglucosaminidase, IDS and IDUA respectively. Integral lysosomal membrane protein 7 represents N-acetyltransferase involved in the transfer of acetyl groups from the cytosolic substrate acetyl-CoA to the lysosolic substrate GlcNH-IdoA2S-HS.

glycosaminoglycans in intact lysosomes is stimulated by ATP (proton pump) and N-acetyltransferase (Rome and Hill, 1986); (6) Co-purification of enzyme activities has occurred between consecutively-acting enzymes, for example glucuronate-2-sulphatase and ß-D-glucuronidase (Freeman and Hopwood, 1989; 1991a); (7) Enzyme activities may be modulated by the presence of other proteins. BSA and poly-L-lysine stimulated and stabilised glucuronate-2-sulphatase activity, and modulated the kinetic parameters, for example, the combination of poly-lysine and $MnCl_2$ decreased the K_m value at pH 4.8 for activity towards GlcA2S-anM6S by 10-fold and increased the V_{max} value 4-fold (Freeman and Hopwood, 1989, 1991a, unpublished). A multienzyme complex in close proximity to the lysosomal membrane near N-acetyltransferase and transport proteins may permit lysosolic control of degradation, removal of inhibitory sulphate and monosaccharide products and reducing exposure to protease activities.

REFERENCES

BARZU, T., van RIJN, J. L. M. L., PETITOU, M., TOBELEM, G. and CAEN, J. P., 1987, Heparin degradation in the endothelial cells, Thromb. Res., 47:601.

BIELICKI, J., FREEMAN, C., CLEMENTS, P. R., and HOPWOOD, J. J., 1990, Human liver iduronate-2-sulphatase: purification, characterisation and catalytic properties, Biochem. J., 271:75.

BUCKMASTER, M. J., FERRIS, A. L., and STORRIE, B., 1988, Effects of pH, detergent and salt on aggregation of Chinese-hamster-ovary-cell lysosomal enzymes, Biochem. J., 249:921.

CLEMENTS, P. R., BROOKS, D. A., McCOURT, P. A. G., and HOPWOOD, J. J., 1989, Immunopurification and characterisation of human α-L-iduronidase using monoclonal antibodies, Biochem. J., 259:199.

CLEMENTS, P. R., BROOKS. D. A., SACCONE, G. T. P., and HOPWOOD, J. J., 1985a, Human α-L-iduronidase. Purification, monoclonal antibody production, native and subunit molecular mass, Eur. J. Biochem., 152:21.

CLEMENTS, P. R., MULLER, V., and HOPWOOD, J. J., 1985b, Human α-L-iduronidase. 2. Catalytic properties, Eur. J. Biochem., 152:29.

FREEMAN, C., and HOPWOOD, J. J., 1986, Human liver sulphamate sulphohydrolase: determination of native protein and subunit Mr values and influence of substrate aglycone structure on catalytic properties, Biochem. J., 234:83.

FREEMAN, C., and HOPWOOD, J. J., 1987, Human liver N-acetylglucosamine 6-sulphate sulphatase: catalytic properties, Biochem. J., 246:355.

FREEMAN, C., and HOPWOOD, J. J., 1989, Human liver glucuronic acid-2-sulphatase: purification, characterization and catalytic properties, Biochem. J., 259:209.

FREEMAN, C., and HOPWOOD, J. J., 1991a, The estimation of glucuronate-2-sulphatase activity in skin fibroblast homogenates, Biochem. J., 278:In press.

FREEMAN, C., and HOPWOOD, J. J., 1991b, Human glucosamine-6-sulphatase deficiency: diagnostic enzymology towards heparin-derived trisaccharide substrates, Biochem. J., In press.

FREEMAN, C., and HOPWOOD, J. J., 1991c, Human liver α-L-iduronidase: catalytic activity toward heparin-derived oligosaccharide substrates, Biochem. J., In press.

FREEMAN, C., CLEMENTS, P. R., and HOPWOOD, J. J., 1987, Human liver N-acetylglucosamine 6-sulphate sulphatase: purification and characterization, Biochem. J., 246:347.

GAHL, W. A., TIETZE, F., BASHAN, N., STEINHERZ, J. D., and SCHULMAN, J. D., 1982, Defective cysteine exodus from isolated lysosomal rich fractions of cystinoic leucocytes, Science, 257:9570.

GALLAGHER, J. T., WALKER, A., LYON, M., and EVANS, W. M., 1988, Heparan sulphate-degrading endoglycosidase in liver plasma membranes, Biochem. J., 250:719.

HOPWOOD, J. J., 1989, Enzymes that degrade heparin and heparan sulphate, In: 'Heparin: Chemical and Biological Properties, Clinical Applications', (Lane D. W., and Lindahl, U., eds), Edward Arnold, London, pp 191.

HOPWOOD, J. J., and MORRIS, C. P., 1990, The mucopolysaccharidoses: diagnosis, molecular genetics and treatment, Mol. Biol. Med., 7:381.

IOZZO, R. V., 1987, Turnover of heparan sulfate proteoglycan in human colon carcinoma cells, J. Biol. Chem., 262:1888.

ISHIHARA, M., FEDARKO, N. S., and CONRAD, H. E., 1986, Transport of heparan sulphate into nuclei of hepatocytes, J. Biol. Chem., 261:13575.

JIN, L., NAKAJIMA, M., and NICOLSON, G. L., 1990, Immunochemical localization of heparanase in mouse and human melanomas. Int. J. Cancer, 45:1088.

JONAS, A. J., and JOBE, H., 1990a, Sulfate transport by rat liver lysosomes, J. Biol. Chem., 265:17545.

JONAS, A. J., and JOBE, H., 1990b, N-Acetyl-D-glucosamine countertransport in lysosomal membrane vesicles. Biochem. J., 268:41.

JONAS, A. J., SPELLER, R. J., CONRAD, P. B., and DUBINSKY, W. P., 1989, Transport of N-acetyl-D-glucosamine and N-acetyl-D-galactosamine by rat liver lysosomes, J. Biol. Chem., 264:15247.

KRÜGER, U., and KRESSE, H., 1986, Endocytosis of proteoheparan sulfate by cultured skin fibroblasts, Biol. Chem. Hoppe-Seyler, 367:465.

LINDAHL, U., 1989, Biosynthesis of heparin and related polysaccharides, In: 'Heparin: Clinical and Biological Properties, Clinical Applications', (Lane, D., and Lindahl, U., eds), Edward Arnold, London, pp 159.

MANCINI, G. M. S., de JONG, H. R., GALJAARD, H. and VERHEIJEN, F. W., 1989, Characterisation of a proton-driven carrier for sialic acid in the lysosomal membrane, J. Biol. Chem., 264:15247.

MYEROWITZ, R., and NEUFELD, E. F., 1981, Maturation of α-L-iduronidase in cultured human fibroblasts, J. Biol. Chem., 256:3044.

NAKAJIMA, M., IRIMURA, T., and NICOLSON, G. I., 1988 Heparanase and tumor metastasis, J. Cell Biochem., 36:157.

NAKAJIMA, M., IRIMURA, T., Di FERRANTE, N., and NICOLSON, G. L., 1984, Metastatic melanoma cell heparanase: characterization of heparan sulfate degradation fragments produced by B16 melanoma endoglucuronidase, J. Biol. Chem., 259:2283.

NEUFELD, E. F., and MUENZER, J., 1989, The mucopolysaccharidoses, In: 'Metabolic Basis of Inherited Disease', (Scriver, C. R., Beaudet, A. L., Sly, W. S., and Valle, D., eds), McGraw-Hill, New York, 6th Ed., pp 1565.

OLDBERG, A., HELDIN, C-H., WASTESON, Å., BUSCH, C., and HÖÖK, M., 1980, Characterization of a platelet endoglycosidase degrading heparin-like polysaccharides, Biochem., 19:5755.

OOSTA, G. M., FAVREAU, L. V., BELLER, D. L., and ROSENBERG, R. D., 1982, Purification and properties of human platelet heparitinase, J. Biol. Chem., 257:11249.

RENLUND, M., TIETZE, F., GAHL, W. A., 1986, Defective sialic acid egress from isolated fibroblast lysosomes of patients with Salla disease, Science, 232:759.

ROBERTS, S. H., UPADHYAYA, M., SARFARAZI, M., and HARPER, P. S., 1989, Further evidence localising the gene for Hunter's syndrome to the distal region of the X chromosome, J. Med. Genet., 26:309.

ROBERTSON, D. A., CALLEN, D. F., BAKER, E. G., MORRIS, C. P., and HOPWOOD, J. J., 1988b, Chromosomal location of the gene for human glucosamine-6-sulphatase to 12q14, Hum. Genet., 79:175.

ROBERTSON, D. A., FREEMAN, C., NELSON, P. V., MORRIS, C. P., and HOPWOOD, J. J., 1988a, Human glucosamine 6-sulphatase cDNA reveals homology with steroid sulphatase, Biochem. Biophys. Res. Commun., 157:218.

ROME, L. H., and HILL, D. F., 1986, Lysosomal degradation of glycoproteins and glycosaminoglycans: efflux and recycling of sulphate and N-acetylhexosamines, Biochem. J., 235:707.

SCOTT, H. S., ANSON, D. S., ORSBORN, A. M., NELSON, P. V., CLEMENTS, P. R., MORRIS, C. P., and HOPWOOD, J. J., 1991, Human α-L-iduronidase: cDNA isolation and expression. Proc. Natl. Acad. Sci. USA, In press.

SCOTT, H. S., ASHTON, L. J., EYRE, H. J., BAKER, E., BROOKS, D. A., CALLEN, D. F., MORRIS, C. P., and HOPWOOD, J. J., 1990, α-L-Iduronidase is isolated to chromosome 4p16.3, Am. J. Hum. Genet., 47:802.

SEWELL, R. F., BRENCHLEY, P. E. C., and MALLICK, N. P., 1989, Human mononuclear cells contain an endoglycosidase specific for heparan sulphate glycosaminoglycan demonstrable with the use of a specific solid-phase metabolically radiolabelled substrate. Biochem. J., 264:777.

SICILIANO, L., FIUMARA, L. P., PAVONE, L., FREEMAN, C., ROBERTSON, D., MORRIS, C. P., HOPWOOD, J. J., DiNATALE, P., MUSUMECI, S., and HORWITZ, A. L., 1990, Sanfilippo syndrome type D in two adolescent sisters, J. Med. Genet., 28:402.

TAYLOR, J. A., GIBSON, G. J., BROOKS, D. A., and HOPWOOD, J. J., 1991, α-L-Iduronidase in normal and mucopolysaccharidosis type I human skin fibroblasts, Biochem. J., 274:263.

THUNBERG, L., BÄCKSTRÖM, G., WASTESON, Å., ROBINSON, H. C., ÖGREN, S., and LINDAHL, U., 1982, Enzymatic depolymerization of heparin-related polysaccharides: substrate specificities of mouse mastocytoma and human platelet endo-ß-D-glucuronidase, J. Biol. Chem., 257:10278.

WILSON, P. J., BIELICKI, J., CLEMENTS, P. R., OCCHIODORO, T., ANSON, D., MORRIS, C. P., and HOPWOOD, J. J., 1990, Hunter syndrome: isolation of a cDNA clone encoding human iduronate-2-sulphatase and analysis of patient DNA, Proc. Natl. Acad. Sci. USA, 87:8531.

WILSON, P. J., SUTHERS, G. K., CALLEN, D. F., BAKER, E., NELSON, P., COOPER, A., WRAITH, E. J., SUTHERLAND, G. R., MORRIS, C. P., and HOPWOOD, J. J., 1991, Frequent deletions at Xq28 indicate genetic heterogeneity in Hunter syndrome, Hum. Genet., 86:505.

YANAGISHITA, M., and HASCALL, V. C., 1984, Metabolism of proteoglycans in rat ovarian granulosa cell culture: multiple intracellular degradative pathways and the effect of chloroquine, J. Biol. Chem., 259:10270.

HEPARIN BINDING PROPERTIES OF THE CARBOXYL TERMINAL DOMAIN OF [$A^{103,106,108}$] ANTISTASIN 93-119

George D. Manley, Thomas J. Owen, John L. Krstenansky, Robert G. Brankamp and Alan D. Cardin

Marion Merrell Dow Research Institute
2110 E. Galbraith Road
Cincinnati, Ohio 45215

SUMMARY

Antistasin is a 119 amino acid protein with anticoagulant, antimetastatic and heparin-binding properties derived from the salivary glands of the leech Haementaria officinalis (1). This protein contains a specific consensus sequence for heparin binding at its carboxyl terminal end and a region between residues 32 and 48 putatively involved in glycosaminoglycan interactions. The cyclic peptide antistasin 37-48 (C-P-H-G-F-Q-R-S-R-Y-G-C) and the carboxyl terminal fragment [$A^{103,106,108}$] antistasin 93-119 (P-N-G-L-K-R-D-K-L-G-A-E-Y-A-E-A-R-P-K-R-K-L-I-P-R-L-S) were synthesized by solid-phase peptide chemistry and their interactions with ^{125}I-labeled heparin were investigated. Heparin binding to [$A^{103,106,108}$] antistasin 93-119 was specific and saturable as binding was blocked by addition of the unlabeled glycosaminoglycan. The rank order of potency of various glycosaminoglycans in blocking ^{125}I-labeled heparin binding to [$A^{103,106,108}$] antistasin 93-119 was dextran sulfate > heparin >> dermatan sulfate ≥ chondroitin sulfate A and C indicating a specificity of the peptide for the glycosaminoglycan structure. Moreover, heparin binding increased linearly with increasing salt and was optimal at 0.15 M NaCl and physiological pH. In contrast, binding of heparin to the basic peptide antistasin 37-48 decreased linearly as the ionic strength of the medium was increased to physiological concentration (0.15 M) thus showing a greater specificity of heparin for [$A^{103,106,108}$] antistasin 93-119. These studies indicate that residues 93-119 of antistasin mediate this inhibitor's interaction with heparin.

INTRODUCTION

Antistasin is a potent anticoagulant-antimetastatic protein (1) that is produced in the salivary gland cells of the proboscis leech, Haementeria officinalis, a predatory annelid indigenous to Mexico and surrounding regions of South America (2,3). Antistasin is sequence-related to the more potent inhibitor, ghilanten from Haementeria ghilianii (4). These proteins block the active site function of Factor Xa (1,5), a key enzyme in the regulation of the hemostatic mechanism. In addition, these inhibitors exhibit a high affinity for heparin (1,5). Based on limited homology of antistasin with properdin and thrombospondin it was proposed that the region encompassing residues 32-48 was involved in heparin-binding (6,7). We have previously identified a unique heparin-binding consensus sequence present in a variety of glycosaminoglycan-binding proteins including the coagulation inhibitor antithrombin III (8), antistasin and ghilanten (4). The present study reports a structure-activity investigation of the heparin-binding regions of antistasin.

Heparin and Related Polysaccharides
Edited by D.A. Lane *et al.*, Plenum Press, New York, 1992

METHODS AND MATERIALS

Peptides were synthesized on a 0.5 mmol scale by solid-phase methods using an Applied Biosystems model 430A peptide synthesizer and the appropriate Nα-t-Boc-amino acid Pam resin (Applied Biosystems, Foster City, CA). All Nα-t-Boc protected amino acids were double coupled as their symmetrical anhydrides first in dimethylformamide and then in dichloromethane except for arginine and glutamine which were double coupled using dicyclohexylcarbodiimide and 1-hydroxybenzotriazole. The side chain protection of the amino acids was as follows: Arg (Tos), Lys (2-ClZ), Glu (Bzl), Tyr (2-Br-Z), Ser (Bzl). The peptides were cleaved from the resin and deprotected in anhydrous hydrogen fluoride (HF) (containing 5% anisole) at -5°C for 30 min. After removal of HF in vacuo, the peptides were extracted from the resin with 30% aqueous acetic acid and 25% aqueous acetonitrile. Cyclization of the 37-48 fragment was accomplished by oxidation of the Cys side chains with $K_3Fe(CN)_6$. The peptides were purified by preparative HPLC on a Dynamax C_{18} column (21.4 x 250 mm, Rainin Instruments, Woburn, MA) using various gradients of acetonitrile in 0.1% aqueous trifluoroacetic acid. The purity and identity of the peptides were assessed by analytical high pressure liquid chromatography, quantitative amino acid analysis (9) and fast atom bombardment mass spectrometry.

Heparin was radioiodinated as described previously (10). Briefly, 5 mg of heparin were reacted with 94 µmoles of cyanogen bromide in 0.2 ml H_2O and the solution maintained at pH 11 for 5 min with 0.2 M NaOH. The sample was then desalted on a PD-10 column (Pharmacia, Uppsala, Sweden) equilibrated in 0.2 M sodium borate, pH 8.0 and then reacted with 1 mg of fluoresceinamine overnight at 4°C. The fluoresceinamine-labeled heparin (FL-Hep) was desalted as above yielding a stoichiometry of 0.2 fluoresceinamine groups per heparin of 13,000 average molecular weight. The FL-Hep (200 mg in 0.4 ml of borate buffer, pH 8.0) was added to a 1.5 ml screw-cap vial (Alltech Associates, Inc., Deerfield, IL) precoated with 50 mg of Iodogen (Pierce, Rockford, IL). To this reaction vessel was added 18.5 mBq of carrier-free Na ^{125}I (Amersham, Arlington Heights, IL). Radioiodination was performed at 0°C on ice for 2 min and the reaction was terminated by gel filtration of the sample on a 0.5 x 16 cm BIOGEL™ P2 column equilibrated and eluted in phosphate buffered saline (PBS); 1 ml fractions were collected. The specific radioactivity was 250 dpm/ng.

Binding studies were performed with Immulon 4 Removawell polystyrene microtitre plates (Dynatech Laboratories, Inc., Chantilly, VA). Peptide, 5 µg in 50 µl of 50 mM HEPES, pH 7.4, was added to each well. The wells were incubated for 1.5 hours at room temperature and were then aspirated and washed four times with PBS to remove the free peptide. For saturation binding studies, increasing concentrations of ^{125}I-labeled heparin were added to each well in a final volume of 50 µl PBS. Incubation was for 2 h at room temperature. The wells were then washed four times with 250 µl of PBS to remove unincorporated counts and the amount of ^{125}I-labeled-heparin bound to peptide was determined by gamma counting. In other binding studies, either 50 µl of salt or glycosaminoglycans at various concentrations were added to each well followed by the addition of 5 µl of ^{125}I-labeled heparin. Incubation, washing and counting were performed as described above.

RESULTS

Figure 1 shows that the binding of ^{125}I-labeled heparin to [A^{103,106,108}] antistasin 93-119 increased linearly and then plateaued indicating a saturable interaction. The binding of ^{125}I-labeled heparin was blocked by excess unlabeled glycosaminoglycan and did not occur in the absence of the peptide or in bovine serum albumin coated wells thus demonstrating a specific interaction with [A^{103,106,108}] antistasin 93-119. Table I shows the amino acid sequence of the carboxyl terminal regions (residues 93-119) of antistasin and ghilanten and the corresponding synthetic fragment; alanine for cysteine substitutions were incorporated at positions 103, 106 and 108 to prevent peptide oligomerization for the studies.

Figure 2 shows that the binding of ^{125}I-labeled-heparin to [$A^{103,106,108}$] antistasin 93-119 occurred maximally at physiological salt concentration and pH. In fact, from 0.0 to 0.15 M NaCl binding was increased by 100%. For the amino terminal basic peptide 37-48 increasing the salt to physiological concentration decreased heparin binding by 70% and binding decreased to nearly zero at 0.5 M NaCl.

The inhibition of ^{125}I-labeled-heparin binding to [$A^{103,106,108}$] antistasin 93-119 by various sulfated polysaccharides is shown in Figure 3. The dextran sulfates (5,000 and 500,000)

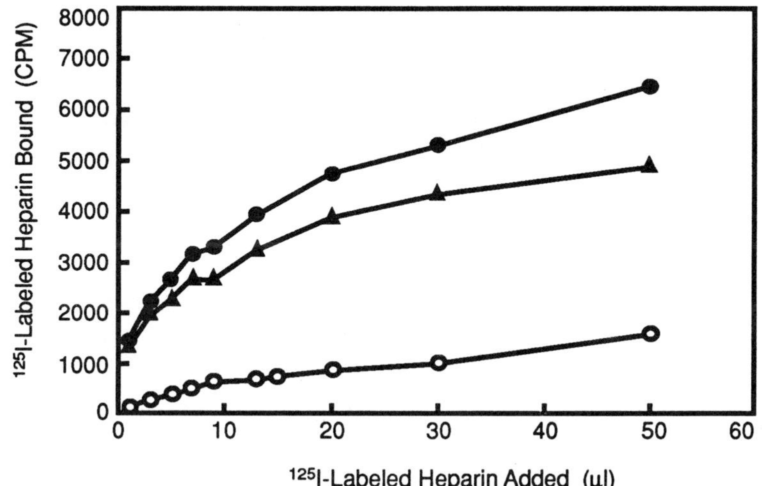

FIG. 1 The binding of ^{125}I-labeled-heparin to peptide 93-119 is saturable and specific. ^{125}I-labeled-heparin binding in the presence (●) and absence (O) of peptide 93-119 represents total and nonspecific binding respectively; specific binding (▲) represents total binding-nonspecific binding. Figure reproduced by permission from: Brankamp RG, Manley GG, Blankenship DT, Bowlin TL and Cardin AD, "Studies on the anticoagulant, antimetastatic and heparin-binding properties of ghilanten-related inhibitors." Blood Coagulation and Fibrinolysis 2:161-166, 1991 (Rapid Communications of Oxford, Ltd.).

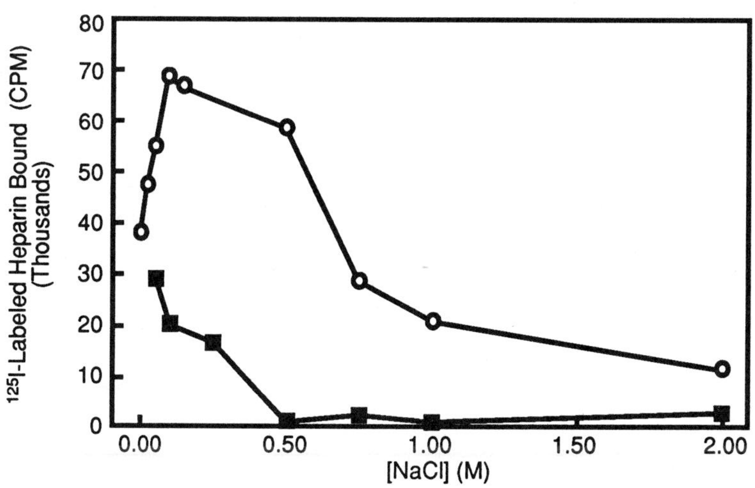

FIG. 2 Effects of salt concentration on the binding of ^{125}I-labeled heparin to the antistasin peptides: (O) 93-119 and (■) 37-48.

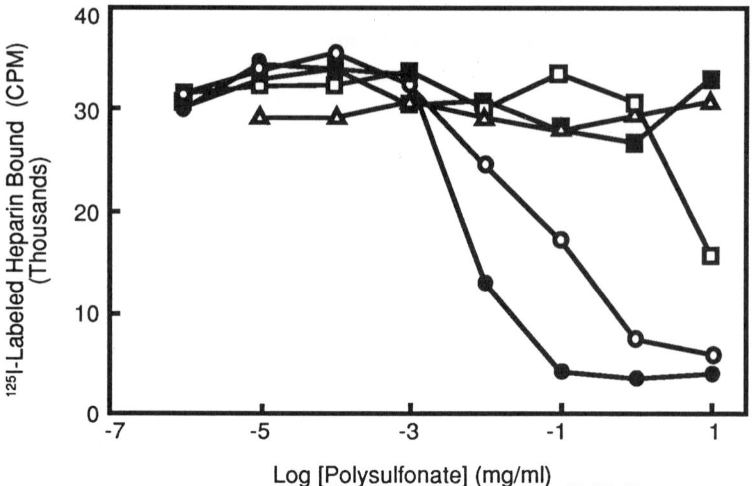

FIG. 3 Inhibition of binding of ^{125}I-labeled heparin to [$A^{103,106,108}$] antistasin 93-119 by the following sulfated polysaccharides: (●) dextran sulfate, (○) heparin, (■) chondroitin sulfate A, (□) chondroitin sulfate B, and (▲) chondroitin sulfate C.

TABLE 1

SEQUENCE ORGANIZATION OF BASIC RESIDUE CLUSTERS IN GHILANTEN, ANTISTASIN AND [$A^{103,106,108}$] ANTISTASIN 93-119

Protein	[B-B-X-B][a]	[B-B-B-X-X-B][a]
Ghilanten	P-N-G-L-K-R-D-K-L-G-C-E-Y-C-E-C-K-P-K-R-K-L-V-P-R-L-S	
Antistasin	P-N-G-L-K-R-D-K-L-G-C-E-Y-C-E-C-R-P-K-R-K-L-I-P-R-L-S	
Peptide 93-119	P-N-G-L-K-R-D-K-L-G-A-E-Y-A-E-A-R-P-K-R-K-L-I-P-R-L-S	

[a]B = probability of a basic residue; X = a hydropathic residue.

TABLE 2

BINDING ACTIVITY OF VARIOUS SULFATED POLYSACCHARIDES TO [$A^{103,106,108}$] ANTISTASIN 93-119

Sulfated Polysaccharide	IC_{50} (μg/ml)
Heparin	100
Dextran Sulfate 5000	7
Dextran Sulfate 500,000	8
Dermatan Sulfate	>1000
Chondroitin Sulfate A	>1000
Chondroitin Sulfate C	>1000

were the most potent inhibitors followed by heparin. The chondroitin sulfates A-C showed little or no inhibition of the binding indicating a selectivity of the carboxyl terminal fragment for the heparin and dextran sulfate glycosaminoglycan structures. The relative binding potencies (IC_{50}'s) of the sulfated polysaccharides are shown in Table 2.

DISCUSSION

Antistasin (1) and ghilanten (4) are sequence-related heparin-binding proteins with potent anticoagulant and antimetastatic activities (5,11). These highly basic proteins have a two-fold repeated internal homology with ten disulfide bonds stabilizing their folded conformations (4,12) and a carboxyl terminal consensus motif for the recognition of heparin-like mucopolysaccharides (4). To investigate the putative heparin-binding properties of this region, a synthetic fragment of antistasin (residues 93-119) was synthesized and a heparin-binding assay was developed. The carboxyl terminal fragment showed optimal affinity for heparin at physiological conditions of salt and pH. It has been speculated that the region 32-48 is important in the binding of antistasin to various sulfated glycoconjugates including heparin (6). This region contains two cysteines at residues 37 and 48 that potentially cyclize to form a highly basic loop domain. In this study the cyclized peptide 37-48 showed limited affinity for heparin under physiological conditions. We have also demonstrated potent heparin binding activity with the more limited fragment antistasin 109-119 (data not shown). Thus the peptide composition and not the peptide length of antistasin 37-48 accounts for its reduced binding activity. Specificity of binding of ^{125}I-labeled heparin was indicated further by the saturable binding characteristics, the requirement for peptide in the assay and its blockade by heparin and dextran sulfate but not by dermatan sulfate or chondroitin sulfates A and C. That dextran sulfate and heparin have a higher affinity for the carboxyl terminal fragment than do the other glycosaminoglycans is consistent with previous reports (6,7) that show the same rank order of affinities of these sulfated polysaccharides for whole antistasin. Further studies with protein mutants lacking the carboxyl terminal region should clarify the role of these residues in the binding of heparin.

ACKNOWLEDGEMENTS

We wish to thank Debbra Wagner for preparing this manuscript, Steven Biedenbach, Andrew Stein and David Davis for graphics, Elaine Semancik for obtaining copyright permission to reproduce Figure 1 and Phil Smith for critical evaluation of this manuscript.

REFERENCES

1. G. P. Tuszynski, T. B. Gasic, and G. J. Gasic, Isolation and characterization of antistasin: an inhibitor of metastasis and coagulation, J. Biol. Chem. 262:9718-9723 (1987).
2. F. Filippi, de Sopra un nuovo genre (Haementeria) di Annelidi della famiglia della Sanquisughe, Mem. Acad. Sci. Torino 10:1-14 (1849).
3. R. T. Sawyer, Leech biology and behaviour. Vol. 2. Feeding biology, ecology, and systematics, Clarendon Press, Oxford (1986).
4. D. T. Blankenship, R. G. Brankamp, G.D. Manley, et al., Amino acid sequence of ghilanten: anticoagulant-antimetastatic principle of the South American leech, Haementeria ghilianii, Biochem. Biophys. Res. Comm. 166:1384-1389 (1990).
5. R. G. Brankamp, D. T. Blankenship, P. S. Sunkara, et al., Ghilantens: anticoagulant-antimetastatic proteins from the South American leech, Haementeria ghilianii, J. Lab. Clin. Med. 115:89-97 (1990).

6. G. D. Holt, H. C. Krivan, G. J. Gasic, et al., Antistasin, an inhibitor of coagulation and metastasis, binds to sulfatide (Gal(3-SO_4)β1-1Cer) and has a sequence homology with other proteins that bind sulfated glycoconjugates, J. Biol. Chem. 264:12138-12140 (1989).
7. G. D. Holt, M. K. Pangburn, V. Ginsburg, Properdin binds to sulfatide [Gal(3-SO_4)β1-1Cer] and has a sequence homology with other proteins that bind sulfated glycoconjugates, J. Biol. Chem. 265:2852-2855 (1990).
8. A. D. Cardin, H. J. R. Weintraub, Molecular modeling of protein-glycosaminoglycan interactions, Arteriosclerosis 9:21-32 (1989).
9. D. T. Blankenship, M.A. Krivanek, B. L. Ackermann, et al., High-sensitivity amino acid analysis by derivatization with 0-phthalaldehyde and 9-fluorenylmethyl chloroformate using fluorescence detection: applications in protein strucutre determination, Anal. Biochem. 178:227-232 (1989).
10. A. D. Cardin, C. J. Randall, N. Hirose, et al., Physical - chemical interaction of heparin and human plasma low-density lipoproteins, Biochemistry 26:5513-5518 (1987).
11. G. J. Gasic, A. Iwakawa, T. B. Gasic, et al., Leech salivary gland extract from Haementeria officinalis, a potent inhibitor of cyclophosphamide- and radiation-induced artificial metastasis enhancement, Cancer Res. 44:5670-5676 (1984).
12. E. Nutt, T. Gasic, J. Rodkey, et al., The amino acid sequence of antistasin; a potent inhibitor of factor Xa reveals a repeated internal structure, J. Biol. Chem. 263:10162-10167 (1988).

HEPARIN PROTEIN INTERACTIONS

Fanyu Zhou, Teresia Höök, John A. Thompson
Magnus Höök

Department of Biochemistry
University of Alabama at Birmingham
Birmingham, AL 35294

INTRODUCTION

To exert their various biological activities, glycosaminoglycans such as heparin often bind to specific proteins; over the years, a number of heparin-binding proteins have been identified. In Table 1, we have listed some of the heparin-binding proteins. These are grouped according to the functions they perform. For example, the heparin-binding growth factors presumably have a common function, e.g. all act as mitogens and have very similar sequences. The matrix proteins fibronectin, laminin, vitronectin, and thrombospondin have very different structures, but all can mediate the substrate adhesion of eucaryotic cells. These proteins contain binding sites for cellular receptors of the integrin type and also glycosaminoglycan-binding

TABLE 1

HEPARIN BINDING PROTEINS

Adhesive Matrix Proteins:
Fibronectin
Vitronectin
Laminin
Collagens
Thrombospondin

Serine Protease Inhibitors:
Antithrombin III
Heparin co-factor II
Protease nexins

Growth Factors:

Acidic-FGF	HBGF-1
Basic-FGF	HBGF-2
int-2	HBGF-3
hst/ks	HBGF-4
FGF-5	HBGF-5
FGF-6	HBGF-6
KGF	HBGF-7

Other Proteins:
Superoxide dismutase
Elastase
Platelet factor 4
N-CAM
Viral coat protein
Transcription factors
Coagulase enzymes

Proteins Involved in Lipid Metabolism:
Lipoprotein lipase
Hepatic triglyceride lipase
Apolipoprotein B
Apolipoprotein E

Heparin and Related Polysaccharides
Edited by D.A. Lane *et al.*, Plenum Press, New York, 1992

domains which indicates that an interaction with glycosaminoglycans such as cell surface associated heparan sulfate proteoglycans may be of importance for the cell adhesion mediating function of the proteins. In a third category, here represented by proteins involved in lipid metabolism, different proteins have been grouped which clearly fulfill different functions, but since all bind glycosaminoglycans, this interaction may play a central role in a common process involving the different proteins. For example, the apoproteins, B1 and A1, which are components of the LDL and HDL particles, respectively, have glycosaminoglycan binding sites as do several lipolytic enzymes such as lipoprotein lipase. It is tempting to speculate that binding to glycosaminoglycans is of importance for the function of these molecules. Perhaps a cell surface proteoglycan with several glycosaminoglycan chains could serve as an anchor for both the lipoproteins and the lipoprotein lipase allowing the substrate and the enzyme to remain in close contact for a number of catalytic cycles.

There are a number of heparin-binding proteins where we do not see the protein as a member of a group or family of functionally or structurally related molecules. In many cases, the importance of the glycosaminoglycan binding properties of these proteins is unclear. Although it is possible that further studies will reveal functional roles for the heparin binding properties of these proteins, it should be pointed out that the most noticeable characteristic of heparin is its strong negative charge. When the polysaccharide is coupled to Sepharose, the matrix will serve as a cation-exchange column and any positively charged protein could conceivably bind to such a column and hence demonstrate heparin-binding activity. For example, when the amino acid lysine is polymerized to polylysine, this "protein" will now bind to heparin-Sepharose. One should be cautious in ascribing a functional significance to a heparin binding property for a protein when this interaction is weak and the protein is eluted from a Heparin-Sepharose by lower then physiological salt concentrations.

SPECIFIC AND POSITIVE, COOPERATIVE BINDING

Traditionally, we have recognized two forms of heparin/protein interactions - a positive, cooperative binding and a specific binding. Perhaps one should not necessarily regard these types of interactions as mutually exclusive. In the case of positive, cooperative binding, the polysaccharide may bind to a protein at different sites; and the affinities shown at the individual interactive sites contribute to the overall affinity of the polysaccharide for its protein ligand. An example of a positive, cooperative interaction is that between the polysaccharide and a lipoprotein particle with several apolipoprotein molecules. The longer the polysaccharide chain is, the more binding sites will fit along the chain and the higher the apparent affinity that the polysaccharide shows for the lipoprotein particle. Conversely in this model, oligosaccharides of decreasing size will bind to the ligand with decreasing affinity. In the simplest model of positive cooperativity, we can consider a theoretical protein with several well defined heparin binding sites spaced along the polypeptide and the affinity of an oligosaccharide for this protein would depend on if it contained structural units recognized by the protein and if the spacing of these units will match that of the heparin binding sites in the protein.

Heparin binding sites in many proteins seem not to be well defined, which often is reflected by a lack of specificity in its glycosaminoglycan interaction. These types of proteins typically bind a variety of glycosaminoglycans where the apparent affinities differ dependent on the composition of uronic acid units and sulfate group contents.

The specific binding of heparin to protein has been considered as a second type of polysaccharide protein interaction. The classical example of a specific heparin protein interaction is that between heparin and antithrombin III (AT-III). In this case, other glycosaminoglycans cannot substitute for heparin in its binding nor in exerting a biological activity. The biological activity of heparin seems to be mediated by a conformational change in the AT-III molecule induced primarily by its binding to one specific structural unit in the heparin molecule.

In conclusion, earlier studies of the interaction of heparin with protein suggest two types of interaction: those involving one well defined binding site and those involving multiple binding sites resulting in a positive cooperativity. These binding sites may be specific for certain glycosaminoglycan "sequences", or be less well defined and allow a binding to several glycosaminoglycans (for more detailed discussions of glycosaminoglycan protein interaction, see References 1,2,3). So far only AT-III has been shown to bind to a specific structural unit in heparin. Are there other protein heparin interactions that involve specific heparin sequences? Obviously, this is an area where further work is needed. With the recent technical advances in terms of glycosaminoglycan structural analysis, genetic manipulation of the glycosaminoglycan biosynthetic machinery and the availability of recombinant versions of specific heparin-binding proteins or protein segments to be used in studies of oligosaccharide interaction, substantial progress can be predicted. Recent work in our laboratory has indicated that other proteins in addition to AT-III may also bind to specific sequences in heparin. We would like to point out that the data presented in this communication is of a preliminary nature and that the conclusions made have to be confirmed in future studies.

PROTEINS RECOGNIZING THE AT-III BINDING SITE IN HEPARIN

Is the AT-III binding pentasaccharides uniquely recognized by this protease inhibitor, or is it also recognized by other proteins which may be structurally or functionally related? To address these questions, we have initiated a series of experiments to try to isolate heparin-binding proteins that show the same specificity as that of AT-III. The strategy we have chosen for this work is based on the fact that the AT-III binding sequence is only present in some heparin molecules but not in others. Initially we therefore made some AT-III Sepharose and used this affinity matrix to fractionate heparin into two populations with high affinity and low affinity for AT-III, respectively. To be assured that the population of low affinity heparin did not contain residual molecules with affinity to AT-III, this population was passed over the affinity matrix several times. The two populations of heparin; high affinity and low affinity heparin, showed an expected dramatic difference in their ability to potentiate the AT-III activity. Whereas the high affinity heparin efficiently activated AT-III, the low affinity heparin was essentially without effect (Figure 1). The two populations of heparin were then coupled to Sepharose to provide new affinity matrixes which showed a dramatic difference in their capacity to bind AT-III. The columns were used in attempts to isolate proteins other than AT-III that show the same specificity for a binding site in heparin as the protease inhibitor. Initially we have focused on plasma proteins and used the following protocol: First, heparin binding proteins are isolated by affinity chromatography on a column containing unfractionated heparin. These heparin binding proteins are then passed over a low affinity heparin-Sepharose. Breakthrough material are applied to a high affinity heparin-Sepharose. We

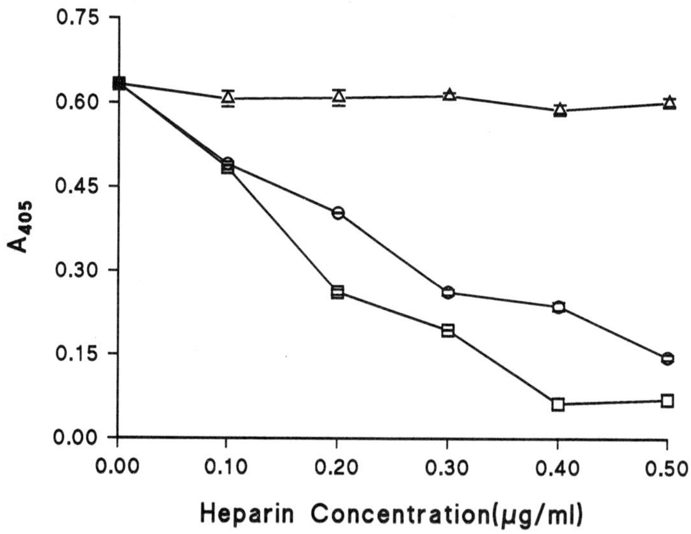

Figure 1. AT-III activating properties of different populations of heparin eluted from AT III-Sepharose (○) unfractionated heparins: (△) low affinity heparin: eluted by 0.15 M NaCl; (□) high affinity heparin: eluted by 3.5 M NaCl. Samples (200μl) of AT III solution (10μg/ml) and of heparin solutions of different concentrations were mixed and incubated for 3 minutes at 37° C. 120μl thrombin (10 NIH Units/ml) was then added, and after additional incubation of 30 minutes, the remaining thrombin activity was measured by monitoring the release of *p*-nitroanilline (absorbing at 405 *nm)* from the synthetic thrombin substrate S2238 (AB Kabi, Sweden).

would expect AT-III like proteins not to bind to the low affinity heparin column but be retained on the high affinity heparin column. SDS-PAGE of the different fractions shows that the principle of this approach appears to be correct since the material retained on the high affinity heparin column contained large amounts of AT-III. However, it seems that AT-III is such a dominant component in normal plasma that on SDS-PAGE analysis it hides other heparin binding proteins of similar molecular weight. Fortunately, we had access to a patient that had a deficiency in AT-III. When plasma from this patient was analyzed, according to the outlined scheme, we could identify a second heparin binding component that does not bind to low affinity heparin and is retained on high affinity Heparin-Sepharose. This component has a somewhat higher molecular weight than AT-III (Figure 2). Furthermore, this component could be separated cleanly from anti-thrombin III by ion-exchange chromatography on a Mono Q column.

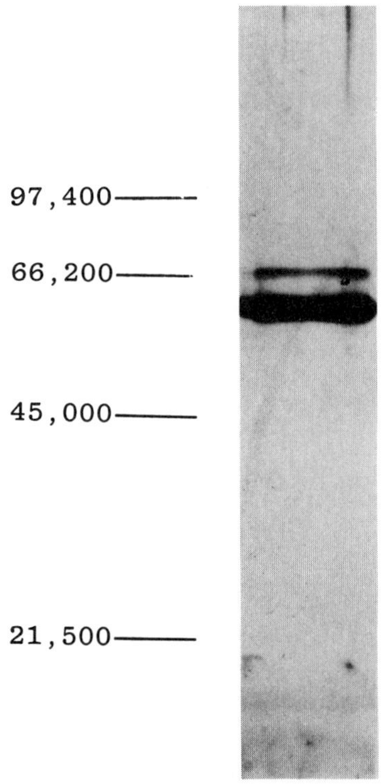

Figure 2. Heparin binding plasma proteins from a patient with an AT-III deficiency were further fractionated on a column of LA-heparin-Sepharose. The non-binding fraction was passed over the LA-heparin-Sepharose a second time and finally absorbed to a HA-heparin-Sepharose. The figure shows a PAGE analysis of proteins bound to the immobilized HA-Heparin. Lines indicate the migration of molecular weight standards.

We have not, at this point, identified this new heparin binding component or determined its relationship to AT-III. However, this study does show that this approach may be useful in the future to isolate heparin-binding proteins that show the same specificity as that demonstrated by AT-III.

HEPARIN BINDING GROWTH FACTORS

The family of heparin-binding growth factors (HBGF) presently consists of seven structurally related polypeptides. Acidic fibroblast growth factor and basic fibroblast growth factor were the first two members of this family to be described.[4,5] These growth factors are now referred to as HBGF-1 and HBGF-2. Subsequently, four sequence-related oncogene products, Int-2, Hst, FGF-5 and FGF-6, have been identified and designated as HBGF-3, HBGF-4, HBGF-5 and HBGF-6. More recently, an epithelial cell specific growth factor named keratinocyte growth factor (KGF or FGF-7) has been characterized as the seventh member of the HBGF family. (For recent reviews on the heparin-binding growth factors, see References 6,7).

HBGF-1 and HBGF-2 both have a M_r of about 18 kd and consist of non-glycosylated single chain polypeptides with 53% sequence identity. The three dimensional structures of both HBGF-1 and HBGF-2 have recently been determined by x-ray crystallographic analysis and appear to consist primarily of 12 anti-parallel β-strands.[8] It is noteworthy that no specific heparin binding sites were obvious from these analyses. Neither HBGF-1 nor HBGF-2 contains signal sequences, and it is presently unclear how these growth factors find their way to the extracellular matrix. The other identified members of this family of proteins contain signal sequence. The HBGF-1 and -2 polypeptides are encoded by separate genes which occur as single copies and are similar in their overall organization containing three exons separated by two introns.

HBGF-1 and HBGF-2 are potent mitogens for a large number of established cell lines, as well as for a variety of normal mammalian mesoderm and neuroectoderm derived cells. The biological effects of HBGF-1 and HBGF-2 are mediated by specific high affinity cell surface receptors. So far, at least four different high affinity receptors for the HBGF's have been identified. These include the *flg* and *bek* gene products,[9,10] fibroblast growth factor receptor 3 (FGFR-3)[11] and FGFR-4.[12] These receptors are structurally related and all contain glycosylated immunoglobulin-like motifs in the extracellular domains, and a conserved tyrosine kinase domain split by a short kinase insert in the cytoplasmic segment. The *flg* and the *bek* gene products have been shown to bind both HBGF-1 and 2, whereas, FGFR-4 appears to bind HBGF-1 but not HBGF-2.

The transmembrane signalling pathway of HBGF-mediated cellular proliferation has not been fully characterized. The interaction of HBGFs with their high affinity receptors has been reported to stimulate tyrosine kinase activity. In addition, HBGFs have been shown to induce expression of proto-oncogenes such as *c-fos* and *c-myc*. However, other cellular events seem to be necessary for the stimulation of DNA synthesis and cell proliferation. HBGF-1 seems to contain a putative nuclear translocation sequence located near the N-terminus of the protein. Imamuar *et al.* demonstrated that a mutant of HBGF-1 lacking this sequence failed to induce DNA synthesis and cell proliferation, whereas the intracellular receptor-mediated tyrosine phosphorylation and proto-oncogene expression proceeded as with the native growth factor.[13]

The HBGFs bind tightly to heparin. HBGF-1, which has an isoelectric point of 5.6, is eluted from a heparin affinity column with 1-1.2 M NaCl. HBGF-2, which has an isoelectric point of 9.6, is eluted from a similar heparin column at a higher salt concentration (1.5-1.8 M NaCl).

The site in HBGF-1 which is responsible for its binding to heparin is presently unclear, and available experimental data appear somewhat conflicting. Consensus sequence analyses of the glycosaminoglycan-protein interaction predicted that the sequence ^{124}GLKKNGRSK^{132} in bovine HBGF-1 might be responsible for heparin binding.[14] The lysine residue at position 132 has been suggested to be of particular importance. Harper and Lobb demonstrated that limited methylation of lysine 132 in bovine HBGF-1 resulted in a significantly reduced affinity of the modified growth factor for heparin. The mitogenic and receptor binding activities were also reduced several fold.[15] Furthermore, substitution of lysine 132 with glutamic acid through site-directed mutagenesis resulted in a protein with drastically reduced specific mitogenic activity and apparent affinity for immobilized heparin. The binding of this mutant to the high affinity cell surface receptor appeared unaltered. Similarly, mutant HBGF-1 can stimulate tyrosine kinase activity and induce proto-oncogene expression as the native protein does.[16]

However, other regions in HBGF-1 have also been implicated in its heparin binding. Cleavage of bovine HBGF-1 by thrombin produces a 16 kd fragment with reduced but appreciable affinity for heparin. The mitogenic activity of this fragment appears to be 50-fold less than that of intact HBGF-1. The peptide bonds cleaved by thrombin to generate this fragment were identified as Arg 136 and Thr 157, implying that amino acid residues distal to lysine 132 contribute to the heparin binding site of HBGF-1.[17] Mehlman and Brugess have also shown that a synthetic peptide corresponding to residues 49-72 in HBGF-1 is able to compete with HBGF-1 in a heparin binding assay.[18]

The heparin binding sites in HBGF-2 have also not been defined. In one study, more than 25 overlapping peptides covering the entire sequence of HBGF-2 were synthesized. Two peptides corresponding to residues 32-76 and to residues 114-123 were found to bind to ^{3}H-labeled heparin.[19] In another study, the heparin binding activity of truncated versions of recombinant HBGF-2 were examined. Using this technique, the major heparin binding region was located to residues 115-151 in the carboxyl terminus of the growth factor.[20]

BIOLOGICAL EFFECTS OF HEPARIN ON THE HBGF'S

Heparin strongly potentiates the growth promoting activity of HBGF-1 (Figure 3), suggesting a functionally important relationship between heparin and/or heparan sulfate proteoglycans and the HBGF polypeptide. This effect has been correlated with the high affinity of HBGF-1 to heparin and with the capacity of heparin to stabilize the protein, which in complex with heparin is less sensitive to proteolytic degradation, and acid or heat denaturation. Although exogenous heparin does not potentiate the activity of HBGF-2, this growth factor

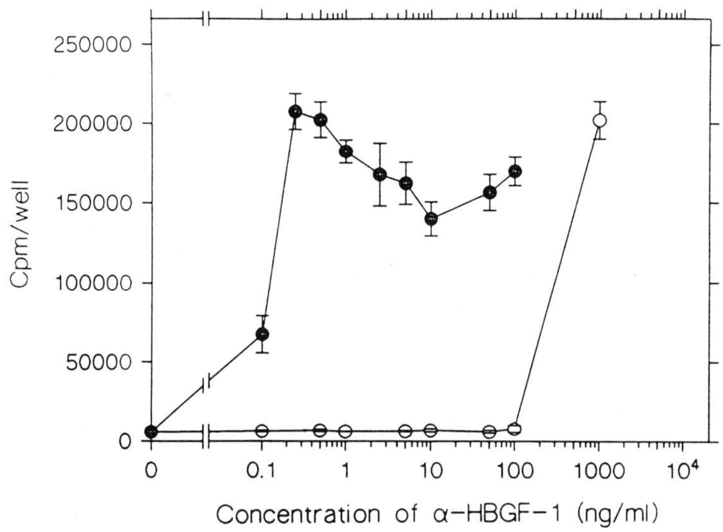

Figure 3: Stimulation of DNA synthesis in NIH 3T3 cells by HBGF-1 in the presence or absence of heparin. Cells were seeded into 24 well plates and grown to near confluence in DME containing 10% fetal bovine serum (FBS). The cells were serum starved (DME, 0.5% FBS) for 36 hr, incubated with indicated concentrations of HBGF-1 in the presence (●) or absence (○) of heparin (150 μg/ml) for 16 hr, then pulsed with 1.0 μCi [^{3}H]thymidine for 4 hr. The cells were harvested and incorporation of radioactivity was determined.

is also stabilized by complexing to heparin. Recent data suggest that a binding of HBGF-2 to cell surface associated heparan sulfate is required before it can bind to its high affinity receptors.[21,22]

DOES HBGF-1 BIND TO A SPECIFIC SEQUENCE IN HEPARIN?

In our search for a protein which could potentially bind to a specific heparin sequence, we have focused on HBGF-1 for the following reasons:

A) The affinity of the growth factor for heparin is apparently high as indicated by the high salt concentration needed to elute the protein from a heparin affinity column.

B) The isoelectric point of the protein suggests that the protein does not bind to heparin via nonspecific cation-anion interactions.

C) The relative short size of the polypeptide (155 amino acid residues) suggests that extensive positive cooperativity is unlikely.

The approach we have chosen for these studies is to fractionate heparin oligosaccharides on a column of Sepharose substituted with HBGF-1. The protein is produced as a procaryotic recombinant using the vector pKK233.2.[23] The synthesized protein is purified by affinity chromatography on heparin-Sepharose and itself coupled to Sepharose in complex with acetylated heparin; a technique originally developed in Ulf Lindahl's lab for preparing an AT-III affinity column for fractionation of heparin.[24] The presence of heparin will protect the heparin binding sites during the coupling reaction. After the coupling is completed, the acetylated heparin which does not contain any primary amino groups and therefore is not coupled to the matrix is eluted by 2M NaCl and the column is ready for use.

Heparin was chemically depolymerized by partial nitrous acid treatment. Generated oligosaccharides were separated by gel chromatography on a column of Sephadex G-50. The fractions correlated to tetra-, hexa-, octa- and decasaccharides were pooled separately and labelled with sodium [^{3}H]-borohydride, yielding products with terminal 2,5-anhydro[^{3}H]mannitol units. The size of the oligosaccharides in each pool was confirmed by analysis on size-exclusion HPLC using chondroitin sulfate derived oligosaccharides as internal standards. The affinity of the heparin derived oligosaccharide fragments for immobilized HBGF-1 was analyzed and compared with that of intact heparin using the affinity column described above, eluted with a continuous sodium chloride gradient. The results of these studies (Figure 4) show that the different oligosaccharides were heterogeneous in their apparent affinity for immobilized HBGF-1. Whereas the tetra- and the hexasaccharides showed low to intermediate affinity for the immobilized HBGF-1, the octasaccharide pool contained components that required the same salt concentration to be eluted from the HBGF-1 affinity column as did intact heparin. We, therefore, focused our continuing analysis on these octasaccharides. The pool of octasaccharides was divided into low affinity, medium affinity and high affinity with respect to apparent affinity to immobilized HBGF-1, which can be eluted from HBGF-1 affinity matrix by 0.62-0.78 M, 0.78-0.86 M, 0.93-1.03 M NaCl respectively. The relative affinities were confirmed by rechromatography of the different octasaccharide pools. Preliminary structural analysis of the octasaccharides were then conducted using an ion-exchange chromatography on a Mono Q column fitted onto a FPLC system. The ion-exchange column was eluted with a linear salt gradient, and the results of these analyses demonstrated (Figure 5) that the different octasaccharide pools were heterogenous with respect to overall charge density. However, there was no clear relationship between apparent affinity for HBGF-1 and overall charge density as revealed by the ion-exchange chromatography. In fact, the

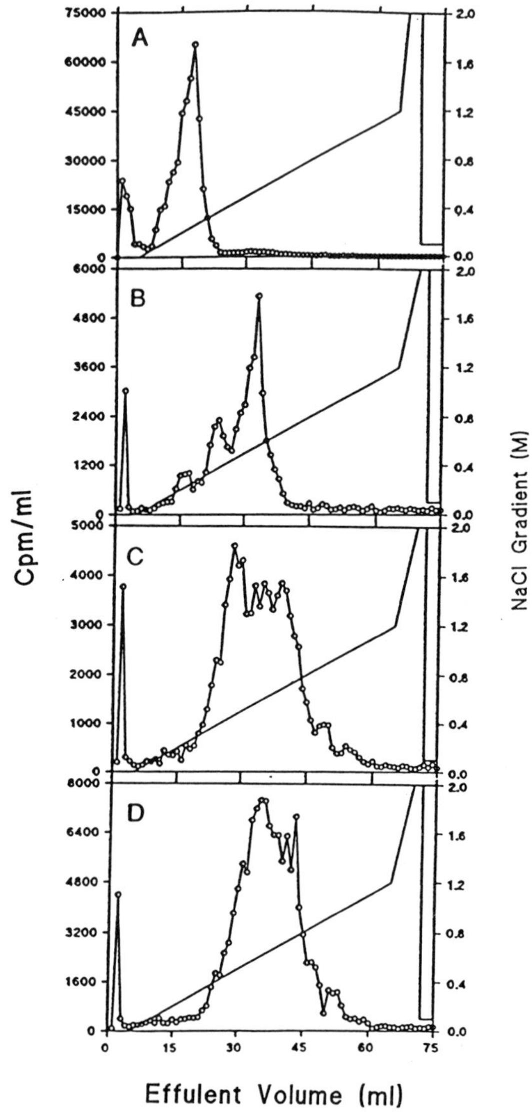

Figure 4. HBGF-1 affinity chromatographic analysis of ^{3}H-labeled heparin-derived oligosaccharides. The tetra-, hexa-, octa- and decasaccharides separated from gel filtration chromatography were pooled and labeled with $NaB[^{3}H]_4$ individually. The labeled products were applied to a column (0.5x5 cm) of immobilized HGBF-1 equilibrated with 20 mM Tris-HCl, 1 mM EDTA, pH 7.2. The column was eluted with the indicated NaCl gradient at a rate of 0.5 ml/min. 1.0 ml fractions were collected and analyzed for radioactivity. Figure 4A, 4B, 4C, 4D represent tetra-, hexa-, octa- and decasaccharide respectively.

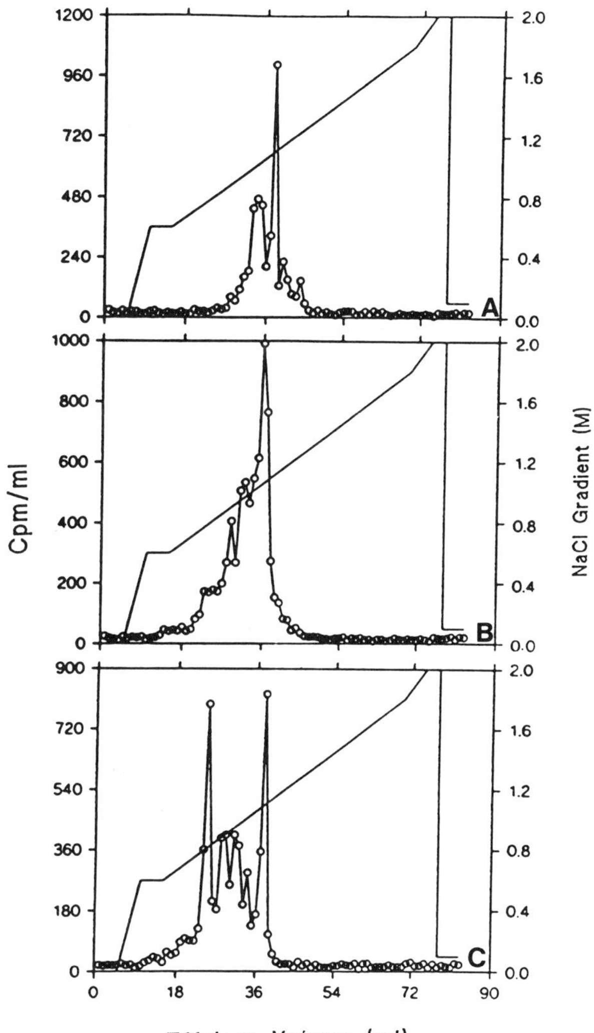

Figure 5. Anion-exchange chromatographic patterns of octasaccharides with different affinity for HBGF-1. Octasaccharides with low affinity(A), medium affinity (B) and high affinity (C) for HBGF-1 were applied to a Mono Q HR5/5 column individually and eluted with a NaCl gradient as indicated. The flow rate of elution was 0.5 ml/min and 1.0 ml fraction were collected for radioactivity measurements.

high affinity octasaccharide also contained molecules of intermediate charge density indicating that the distribution of the charged groups on the octasaccharide is of greater importance than the overall charge of the molecule for interacting with acidic FGF. These data would tend to suggest that a specific sequence can be found in heparin which is required for the high affinity binding of the polysaccharide to HBGF-1.

Although the data so far are preliminary, the results presented in this study suggest that, in addition to the binding of heparin to antithrombin III, other proteins may bind to specific regions in the polysaccharide molecule and these specific sequences will, undoubtedly, be the topic of future studies.

REFERENCES

1. R. L. Jackson, S. J. Busch, and A.D. Cardin, Glycosaminoglycans: Molecular properties, protein interactions and role in physiological processes, *Physiological Reviews*, 71:481 (1991).
2. J. T. Gallagher, M. Lyon, and W.P. Steward, Structure and function of heparan sulfate proteoglycans, *Biochem. J.* 236:313 (1986).
3. L. Kjellén, and U. Lindahl, Proteoglycans: Structures and interactions, *Annu. Rev. Biochem.* 60:443 (1991).
4. K. A. Thomas, M. C. Riley, S. K. Lemmon, N. C. Baglan, and R. A. Bradshaw, Brain fibroblast growth factor, *J. Biol. Chem.*, 255:5517 (1980).
5. D. Godpodarowicz, Growth factors and their action *in vivo* and *in vitro*, *J. Pathology*, 141:201 (1983).
6. W. H. Burgess, and T. Maciag, The heparin-binding (fibroblast) growth factor family of proteins, *Annu. Rev. Biochem.* 58:575 (1989).
7. M. Klagsbrun, Fibroblast growth factor family, *Current Opinion in Cell Biology* 2:857 (1990).
8. X. Zhu, H. Komiya, A. Chirino, S. Faham, G. M. Fox, T. Arakawa, B.T. Hsu, D. C. Rees, Three-dimensional structures of acidic and basic fibroblast growth factors, *Science* 251:90 (1991).
9. P. L. Lee, D. E. Johnson, L. S. Cousens, V. A. Fried, and L. T. Williams, Purification and complementary DNA cloning of a receptor for basic fibroblast growth factor, *Science* 245:57 (1989).
10. S. Kornbluth, K.E. Paulson, and H. Hanafusa, Novel tyrosine kinase identified by phosphotyrosine antibody screening of cDNA libraries, *Mol. Cell. Biol.* 8:5541 (1988).
11. K. Keegan, D.E. Johnson, L. T. Williams, and M. J. Hayman, Isolation of an additional member of the fibroblast growth factor receptor family, FGFR-3, *Proc. Natl. Acad. Sci. USA* 88:1095 (1991).
12. J. Partanen, T.P. Mäkelä, E. Erola, J. Korhonen, H. Hirvonen, L. Claesson-Welsh, and K. Alitalo, FGFR-4, a novel acidic fibroblast growth factor receptor with a distinct expression pattern, *EMBO Journal* 10:1347 (1991).

13. T. Imamura, K. Engleka, X. Zhan, Y. Tokita, R. Forough, D. Roeder, A. Jackson, J. A. M. Maier, T. Hla, T. Maciag, Recovery of mitogenic activity of a growth factor mutant with a nuclear translocation sequence, *Science* 249:1567 (1990).
14. A. D. Cardin, and H. J. R. Weintraub, Molecular modeling of protein-glycosaminoglycan interactions, *Arteriosclerosis* 9:21 (1989).
15. J. W. Harper, and R. R. Lobb, Reductive methylation of lysine residues in acidic fibroblast growth factor: Effect on mitogenic activity and heparin affinity, *Biochemistry* 27:671 (1988).
16. W. H. Burgess, A. M. Shaheen, M. Ravera, M. Jaye, P. J. Donohue, and J. A. Winkles, Possible dissociation of the heparin-binding and mitogenic activities of heparin-binding (acidic fibroblast) growth factor-1 from its receptor-binding activities by site-directed mutagenesis of a single lysine residue, *J. Cell Biol.* 111:2129 (1990).
17. R. R. Lobb, Thrombin inactivates acidic fibroblast growth factor but not basic fibroblast growth factor, *Biochemistry* 27:2572 (1988).
18. T. Mehlman, and W. H. Burgess, Detection and characterization of heparin-binding proteins with a gel overlay procedure, *Anal. Biochem.* 188:159 (1990).
19. A. Baird, D. Schubert, N. Ling, and R. Guillemin, Receptor- and heparin-binding domains of basic fibroblast growth factor, *Proc. Natl. Acad. Sci. USA* 85:2324 (1986).
20. M. Seno, R. Sasada, T. Kurokawa, and K. Igarishi, Carboxyl-terminal structure of basic fibroblast growth factor significantly contributes to its affinity for heparin, *Eur. J. Biochem.*, 188:239 (1990).
21. A. Yayon, M. Klagsbrun, J. D. Esko, P. Leder, and D. M. Ornitz, Cell surface, heparin-like molecules are required for binding of basic fibroblast growth factor to its high affinity receptor, *Cell* 64:841 (1991).
22. A. C. Rapraeger, A. Krufka, and B. Olwin, Requirement of heparan sulfate for bFGF-mediated fibroglast growth and myoblast differentiation, *Science* 252:1705 (1991).
23. R. Forough, K. Engleka, J. A. Thompson, A. Jackson, T. Imamura, and T. Maciag, Differential expression in *Eschericha coli* of the α and β forms of heparin-binding acidic fibroblast growth factor-1: potential role of RNA secondary structure, Biochim. Biophys. Acta. (in press).
24. M. Höök, I. Björk, J. Hopwood, and U. Lindahl, Anticoagulant activity of heparin: Separation of high-activity and low-activity heparin species by affinity chromatography on immobilized antithrombin, *FEBS Lett.* 66:90 (1976).

ROLE OF PROTEIN CONFORMATIONAL CHANGES, SURFACE APPROXIMATION AND PROTEIN COFACTORS IN HEPARIN-ACCELERATED ANTITHROMBIN-PROTEINASE REACTIONS

Steven T. Olson and Ingemar Björk

Division of Biochemical Research, Henry Ford Hospital, Detroit, MI 48202, U.S.A. and Department of Veterinary Medical Chemistry, Swedish University of Agricultural Sciences, S-751 23 Uppsala, Sweden

Antithrombin functions as the principal plasma protein inhibitor of most blood coagulation proteinases.[1,2] The essential role of this inhibitor in regulating the activity of these proteinases in vivo is indicated from the well-established link between inherited or acquired deficiencies of antithrombin and the tendency to develop thrombotic disease. Antithrombin is a member of the serpin superfamily of protein proteinase inhibitors and its main target enzymes include the blood coagulation factors IXa, Xa and thrombin. This and other serpins are distinguished from other family members in that their reactions with target enzymes are greatly accelerated by the binding of heparin or heparan sulfate glycosaminoglycans. This property is chiefly responsible for the anticoagulant activity of heparin and has suggested a role for endogenous heparin and heparan sulfate in the regulation of blood coagulation proteinases by antithrombin. In this article, we will review our present understanding of the relationship between antithrombin structure and function based on currently available evidence. Our discussion will focus principally on two areas: 1) the mechanism by which antithrombin and other serpins inhibit their target proteinases; and 2) the molecular basis of heparin's accelerating effect on antithrombin-proteinase reactions.

MECHANISM OF ANTITHROMBIN INHIBITION OF PROTEINASES

Antithrombin inhibits serine proteinases by forming tight, equimolar complexes which involve an interaction between a specific reactive bond of the inhibitor and the active-site of the enzyme.[1,2] The resistance of these complexes to dissociation has suggested that they represent stable tetrahedral or acyl-intermediates formed during the cleavage of the reactive bond as a normal substrate. The existence of a normal substrate pathway in which the proteinase cleaves the inhibitor reactive bond instead of forming a stable complex is suggested by the observation that small amounts of a free reactive-site cleaved form of the inhibitor is produced during the reaction of antithrombin with proteinases.[3,4] These findings have suggested that antithrombin is activated to trap proteinases at an intermediate stage of cleaving the reactive bond, but that a fraction of the proteinase can escape this trapping by completing the cleavage of the reactive bond before trapping occurs.

The reactive bond of antithrombin has been identified as the Arg 393-Ser 394 bond near the C-terminus of the inhibitor based on the observations that 1) the inhibitor is cleaved at this site following dissociation of antithrombin-proteinase complexes and that 2) similar cleavage sites exist in other members of the serpin family at homologous positions.[1,2] The essential role

Heparin and Related Polysaccharides
Edited by D.A. Lane *et al.*, Plenum Press, New York, 1992

of this bond in the inhibition of proteinases by antithrombin is further indicated from studies of natural variants of antithrombin in which the amino acid residues of this bond have been mutated. Three distinct variants in which the Arg 393 residue is mutated to His,[5-8] Cys[5,9] or Pro[10] have been isolated and shown to be completely inactive as inhibitors of proteinases, consistent with the Arg 393 residue being the P1 recognition site for a target enzyme. This assignment is also consistent with the specificity of the target proteinases of antithrombin for cleaving Arg-X bonds. The importance of the P1' Ser 394 residue of the reactive bond is also indicated from another variant, antithrombin-Denver, in which the P1' residue is mutated to Leu.[11] This variant is only weakly active as an inhibitor of thrombin. Stephens et al. have produced further variants at the P1' position by site-directed mutagenesis of a recombinant antithrombin and shown that Gly, Ala and Thr can substitute for Ser in this position with only minimal effects on inhibitor function.[12] Examination of a large number of variants suggested that the size and hydrophobicity of the P1' residue were the important determinants for inhibitor function with thrombin as the target enzyme. In collaborative studies, we have further shown that the importance of the P1' Ser residue is highly dependent on the target proteinase.[13] Second-order inhibition rate constants for antithrombin-Denver reactions with thrombin, factor Xa and plasmin were thus found to be reduced 430-fold, 7-fold and 45-fold relative to antithrombin, which resulted in an increased selectivity of antithrombin for inhibiting factor Xa. Similar reductions in heparin-accelerated inhibition rate constants were observed when native and mutant inhibitor reactions with these enzymes were compared, consistent with the P1' mutation affecting proteinase binding and not heparin binding. These results suggest that the substrate specificity of the target enzyme in large part determines the importance of the P1' residue in inhibitor function.

Studies of other antithrombin variants have demonstrated that the P8-P12 region on the N-terminal side of the reactive bond also is critical for antithrombin to function as an inhibitor of proteinases. The P8, P10, P11 and P12 residues are highly conserved in serpins which function as inhibitors.[14] Natural antithrombin variants in which the P10 Ala is mutated to Pro[15,16] and P12 Ala to Thr[17] have been isolated and shown to be inactive as inhibitors of proteinases. In contrast to the complete unreactivity of the P1 variants toward proteinases, however, these variants are excellent substrates of their target enzymes, which efficiently cleave antithrombin at the reactive bond. A similar conversion from an inhibitor to a substrate of proteinases occurs with natural P10[18] and P12[19] variants of C1-inhibitor as well as with a variant of α_2-antiplasmin in which an Ala is inserted at the P8 position[20]. Asakura et al.[21] have further shown that a monoclonal antibody that binds to the P8-P12 region also transforms antithrombin from an inhibitor to a substrate of thrombin.

An understanding of how antithrombin and other serpins can react either as inhibitors or substrates of their target enzymes has emerged from the elucidated three-dimensional structures of several homologous serpins by X-ray crystallography, although the only known structures of serpins which function as inhibitors , namely, α_1-antitrypsin[22] and antithrombin,[23] are of the reactive-bond cleaved forms. Surprisingly, cleavage of the reactive bond was found to result in a large separation of the P1 and P1' residues to opposite ends of the molecule suggesting that a large structural change accompanies this cleavage. The P1-P16 region, amino-terminal to this bond, was further found to comprise the central strand of a six-membered β-sheet in the core of the protein. Engh et al.[24] have shown by molecular dynamics simulations that a reconstruction of the native inhibitor structure is possible simply by extracting the central P1-P16 strand of the major β-sheet and reconnecting the P1 residue at the C-terminus of this strand to the P1' residue at the opposite end of the molecule to form an exposed peptide loop that extends away from the protein core (fig. 1). The closing of the resulting 5-membered β-sheet in the protein core can then be accomplished with minimal perturbations in structure. The validity of this reconstruction is suggested from the recently determined structure of the noninhibitory serpin, ovalbumin, in which an exposed peptide loop containing the homologous position of the reactive bond was directly observed.[25] The insertion of the P1-P16 region of the reactive-site loop into the major β-sheet of the protein that accompanies cleavage of the reactive bond has suggested that a partial insertion of this loop into the β-sheet may be involved in trapping a proteinase in a stable complex.[19] According to this idea, mutations in the P8-P12 region of the loop induce structural alterations that prevent the insertion of the loop into the β-sheet and thereby allow the exposed reactive bond to be cleaved as a normal substrate. Support for this idea has come from studies of Schulze et al.[26] in which a synthetic peptide corresponding to the P1-P14 residues of α_1-antitrypsin was shown to bind to native antitrypsin and result in a complete loss in inhibitor activity, presumably due to the peptide binding to the site required for the insertion of the reactive-site loop and the consequent trapping of a proteinase.

We have extended these studies by testing a further prediction of the proposed reactive-site loop insertion model for serpin trapping of proteinases in stable complexes.[27] Thus, blocking of the insertion site for the reactive loop by the binding of the P1-P14 peptide should not only prevent the inhibitor from trapping a proteinase in a stable complex but should also allow the

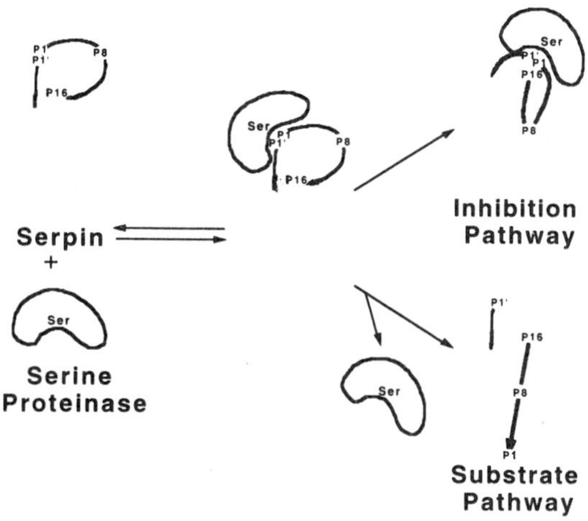

Fig. 1. Reaction model for serpin inhibition of serine proteinases.

exposed reactive bond to be cleaved as a normal substrate. Addition of a synthetic peptide corresponding to the P1-P14 region of antithrombin to the inhibitor was found to result in a time-dependent loss in the ability of antithrombin to inhibit thrombin which did not occur in a control reaction lacking the peptide. Chromatography of a mixture of antithrombin and a radiolabelled peptide on heparin-Sepharose after nearly complete inhibition (~24 hr) resulted in the elution of a labelled protein-peptide complex at physiological salt, i.e., at a considerably lower salt concentration than native antithrombin, indicating that peptide binding had markedly decreased the affinity of antithrombin for heparin, similar to the behavior of reactive-site cleaved antithrombin.[3] The stoichiometry of peptide binding to antithrombin was 1:1 based on the incorporation of the radiolabelled peptide and the observation of 14 additional residues, corresponding to the composition of the peptide, in the amino acid composition of the antithrombin-peptide complex as compared to that of antithrombin. The isolated antithrombin-peptide complex was completely inactive as an inhibitor of thrombin, as indicated from activity measurements in the absence or presence of heparin as well as from the failure to observe the formation of a stable thrombin-antithrombin complex by SDS-gel electrophoresis. Incubation of the inhibitor-peptide complex with thrombin did, however, result in proteolytic cleavage of the reactive bond of the inhibitor, as evidenced by SDS-gel electrophoresis and N-terminal sequence analyses. These results thus confirmed the prediction of the reactive-site loop insertion model that binding of the P1-P14 peptide does convert antithrombin from an inhibitor to a substrate of thrombin.

Together, the available data are therefore consistent with the reaction model of fig. 1 for the mechanism of serpin inhibition of proteinases, which is a modified version of the scheme of Skriver et al.[19] The reactive bond of antithrombin initially binds at the active-center of the proteinase as in a normal substrate, but the subsequent attack of the active-center serine residue of the proteinase on the reactive bond activates the inhibitor to partially insert the amino-terminal region of the reactive-site loop into the β-sheet of the protein core. This insertion of the loop arrests the cleavage of the reactive bond and thereby traps the proteinase in a stable complex. However, a fraction of the proteinase escapes this trapping by completing the cleavage of the reactive bond before trapping occurs. This liberates the P1-P16 strand of the reactive-site loop so that it can be completely inserted into the β-sheet. According to this model, mutations in the P1-P1' reactive bond residues interfere with the recognition of the inhibitor by the target proteinase and its ability to react either as an inhibitor or a substrate of a proteinase. In contrast, mutations in the P8-P12 region of the reactive-site loop or the binding of the P1-P14 peptide to the insertion site for this loop both have the effect of blocking the insertion of the loop into the β-sheet. This prevents the trapping of a proteinase in a stable complex and forces the inhibitor to react solely as a substrate.

MECHANISM OF HEPARIN ACCELERATION OF ANTITHROMBIN-PROTEINASE REACTIONS

Binding of Heparin to Antithrombin

Heparin accelerates the reactions of antithrombin with its main target enzymes up to several thousand-fold.[1,2] The magnitude of these rate enhancements reduces the in vivo lifetime of these proteinases from minutes to milliseconds at plasma concentrations of antithrombin and is the basis for heparin anticoagulant activity. An essential component of the mechanism of heparin's accelerating effect on antithrombin-proteinase reactions is the binding of antithrombin to a unique pentasaccharide sequence in the heparin molecule.[1,2] Heparin molecules which contain this sequence bind antithrombin with a high affinity that is at least 1000 times greater than heparin molecules lacking this sequence and are responsible for the bulk of the anticoagulant activity of commercial heparin preparations. Four sulfate groups in this sequence, including a unique 3-O-sulfate marker of this sequence, have been shown to be essential for the high-affinity interaction, consistent with the number of ionic interactions shown to be involved in the binding of the pentasaccharide to antithrombin from the salt-dependence of this binding.

The site on antithrombin to which the heparin pentasaccharide binds has been mapped by studies of antithrombin variants and chemically modified derivatives in which heparin binding is decreased or abolished. Antithrombin variants with defective heparin binding have implicated an N-terminal region in this binding including Ile 7,[28] Arg 24,[29] Pro 41,[1,2] and Arg 47[1,2] and a region further upstream in the sequence including Arg 129[30] and Asn 135.[1,2] Chemical modification and NMR studies have further implicated His 1[31] and Trp 49[1,2] in the N-terminal region and 5 basic residues including Lys 107, Lys 114, Lys 125, Lys 136 and Arg 145 in the upstream region.[1,2,32,33] A disulfide bond connecting these two regions is also essential for heparin binding.[34] These residues may either directly or indirectly contribute to heparin binding by participating in ionic interactions with the polysaccharide or by maintaining the structural integrity of the binding site. Carrell and coworkers[22] have shown that the two regions of the inhibitor implicated in heparin binding map to the A and D α-helices on the surface of a three-dimensional model of antithrombin based on the cleaved α_1-antitrypsin structure. The basic residues in this model were observed to form a band of positive charge which is of an appropriate size to bind the pentasaccharide. Other evidence implicating this region of the inhibitor in heparin binding includes: 1) the basic residues in this region are conserved in several other heparin-activatable serpins;[22] 2) peptide fragments of antithrombin containing this region bind heparin;[1,2] and 3) a monoclonal antibody that recognizes one of these peptides blocks heparin binding to antithrombin.[35] The putative heparin binding site lies underneath the major β-sheet involved in the insertion of the reactive-site loop and is adjacent to the reactive-bond region.

Spectroscopic, chemical modification and kinetic studies of the interaction of heparin molecules containing the pentasaccharide with antithrombin have shown that this interaction is accompanied by a conformational change in the inhibitor.[1,2] In collaborative studies with Jean Choay, we have compared the fluorescence, ultraviolet and circular dichroism spectroscopic changes which accompany the binding of pentasaccharide and full-length heparins containing this specific sequence.[36] Highly similar spectroscopic changes were found to be induced by the two heparins, consistent with the two heparins inducing the same conformational change in antithrombin. The pentasaccharide site in the full-length heparin therefore appears to be solely responsible for inducing this conformational change. Binding and kinetic studies of the interactions of pentasaccharide and full-length heparins are also consistent with this conclusion. Rapid kinetic studies have shown that both heparins bind antithrombin in a two-step process consisting of an initial weak heparin interaction that is essentially identical for the two heparins, followed by a conformational change that is responsible for producing the high-affinity interaction. This conformational change is stabilized to a somewhat greater extent by the full-length heparin than by the pentasaccharide. Binding studies further indicate that the major portion of the binding energy of the full-length heparin interaction (93%) is accounted for by the pentasaccharide interaction. Analysis of the salt dependence of these binding interactions suggests that the small additional binding energy of the full-length heparin interaction can be accounted for by a single electrostatic interaction made by the larger heparin outside the pentasaccharide binding region. Together, the results are consistent with the two heparins inducing the same high heparin affinity conformation in antithrombin but with an additional electrostatic interaction of the full-length heparin further stabilizing, but not otherwise affecting this conformation.

<u>Role of Antithrombin Conformational Change and Surface Approximation Mechanisms</u>

The conformational change induced in antithrombin by the binding of the pentasaccharide region of heparin has lead to the hypothesis, first proposed by Rosenberg and Damus,[38] that this conformational change is responsible for activating antithrombin to be a better inhibitor of its target proteinases. This conformational activation is thought to make antithrombin more reactive toward proteinases by causing the reactive bond of the inhibitor to be more accessible or complementary to the active-site of these proteinases (fig. 2).[22,37] This hypothesis predicts that the heparin pentasaccharide or small oligosaccharides containing this sequence should produce the same rate-enhancing effect on antithrombin-proteinase reactions as a full-length heparin. However, early studies of the dependence of heparin's accelerating effect on heparin chain-length did not support this prediction for all target proteinases.[1,2] In keeping with this conformational change hypothesis, heparin oligosaccharides as small as the pentasaccharide were observed to be as effective as larger heparin chains in accelerating factor Xa inhibition by antithrombin, indicating that heparin chains just large enough to bind antithrombin and induce the conformational change are sufficient to account for heparin's rate-enhancing effect on this antithrombin-proteinase reaction. In contrast, small heparin oligosaccharides were essentially inactive in accelerating thrombin inhibition by antithrombin. Instead, significant accelerating activity was not observed until the chain-length was 18 saccharides long, indicating that the antithrombin conformational change could not account for heparin's rate-enhancing effect on this latter antithrombin-proteinase reaction. A second mechanism postulated to explain the unusual chain-length dependence of heparin rate enhancement with thrombin as the proteinase was that heparin was acting as a surface or bridge to approximate antithrombin and the enzyme by the binding of both proteins to the same heparin chain (fig. 2). According to this mechanism, the larger heparin chain required to accelerate the antithrombin-thrombin reaction is necessary to accomodate both the inhibitor and proteinase on the same polysaccharide chain. While a substantial body of evidence supports the importance of the surface approximation or bridging mechanism for heparin acceleration of the antithrombin-thrombin reaction, other investigators have suggested that the bridging mechanism plays only a secondary role and that activation of antithrombin through the conformational change is the primary basis for heparin acceleration of all antithrombin-proteinase reactions.[1,2,22,38]

To determine whether the antithrombin conformational change mechanism contributes to heparin's rate-enhancing effect on the antithrombin-thrombin reaction, we have exploited the distinguishing features of the conformational change and bridging mechanisms. In the case of the bridging mechanism, heparin binding to antithrombin promotes the interaction of antithrombin with a proteinase indirectly through the additional binding energy provided by the proteinase-heparin interaction, whereas for the conformational change mechanism, heparin binding to antithrombin directly enhances the binding energy of the antithrombin-proteinase

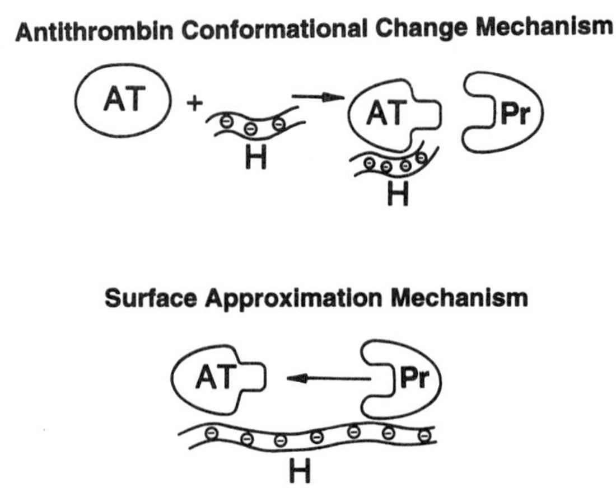

Fig. 2 Alternative mechanisms of heparin acceleration of antithrombin-proteinase reactions.

interaction (fig. 2). Because proteinase-heparin interactions are ionic and therefore strongly dependent on salt, the bridging mechanism predicts that the heparin-enhanced rate of association of the proteinase with antithrombin bound to heparin should show a strong salt dependence that parallels the salt dependence of proteinase binding to heparin. Moreover, this rate enhancement should be quantitatively accounted for by the binding energy of the proteinase-heparin interaction. In contrast, the conformational change mechanism predicts a weaker salt dependence of the heparin-enhanced rate of association of the proteinase with antithrombin bound to heparin which should parallel the salt dependence of proteinase binding to antithrombin.

To test these predictions, we compared pentasaccharide and full-length heparin rate enhancements of antithrombin reactions with thrombin and factor Xa at physiological salt and at double this salt concentration (fig. 3).[36] To insure that we were measuring maximal rate enhancements at each salt concentration, conditions were chosen under which heparin chains were fully complexed with antithrombin. For the antithrombin-factor Xa reaction, the pentasaccharide enhancement of the second-order inhibition rate constant (270-fold) was comparable in magnitude to the full-length heparin enhancement (570-fold), consistent with the minimal dependence of heparin's accelerating effect on the polysaccharide chain-length that was previously observed. Moreover, these rate enhancements were either unaffected or only weakly affected by doubling the salt concentration, as was the unaccelerated reaction rate, indicating that the pentasaccharide and full-length heparin-enhanced interactions of antithrombin with factor Xa were either completely or mostly nonionic and therefore minimally involved the approximation of antithrombin and factor Xa on the heparin surface. Instead, these results were consistent with heparin rate enhancement being primarily due to the antithrombin conformational change enhancing the binding energy of a nonionic antithrombin-factor Xa interaction. In contrast, for the antithrombin-thrombin reaction, the pentasaccharide enhancement of the second-order inhibition rate constant was minor (1.7-fold) relative to the full-length heparin rate enhancement (4300-fold), confirming the requirement for a larger chain-length for significant heparin acceleration of this reaction. Consistent with this chain-length requirement being due to the need

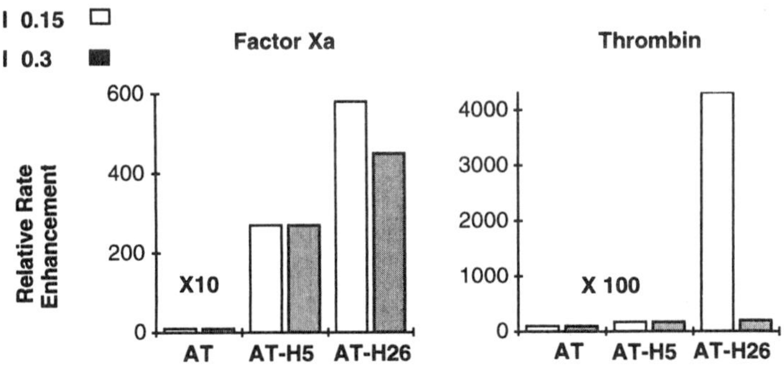

Fig. 3 Chain-length and salt dependence of heparin enhancements of second-order rate constants for antithrombin reactions with thrombin and factor Xa.

for heparin to bind both antithrombin and thrombin, only the rate enhancement by the larger heparin was substantially reduced by doubling the salt concentration (to 200-fold), whereas the pentasaccharide accelerated and unaccelerated inhibition rate constants were unchanged at the higher salt concentration. A detailed study of the salt dependence of the full-length heparin acceleration revealed it to be indistinguishable from the salt dependence of thrombin binding to heparin and completely accounted for by the binding energy of the thrombin-heparin interaction.[39] These results indicate that heparin approximation of antithrombin and thrombin bound to the polysaccharide quantitatively accounts for heparin's rate-enhancing effect on the antithrombin-thrombin reaction and that the antithrombin conformational change makes only a minor contribution to the rate enhancement. Together, these results suggest that both the antithrombin conformational change and surface approximation mechanisms contribute to heparin's accelerating effect on antithrombin-proteinase reactions, but that the relative contribution of each mechanism is dependent on the target proteinase.

Protein Cofactor Involvement in Heparin's Accelerating Effect

Further support for the involvement of two distinct mechanisms in heparin's rate-enhancing effect on antithrombin-proteinase reactions has come from studies of the reactions of antithrombin with the proteinases, plasma kallikrein and factor XIa,[40,41] which are only modestly accelerated by heparin (10-50-fold).[1,2] A possible reason for this low enhancement was suggested from the fact that the proenzyme forms of these proteinases require a cofactor protein, high molecular weight kininogen (H-kininogen), to mediate their binding to a negatively charged surface, where they can be activated by surface-bound factor XIIa. We therefore hypothesized that this cofactor protein might similarly be required to promote the binding of these enzymes to the negatively charged surface of heparin, where their reaction with heparin-bound antithrombin could be further enhanced by a bridging mechanism. Support for this hypothesis initially came from our previous demonstration that H-kininogen binds heparin comparably to the major heparin binding protein of plasma, histidine-rich glycoprotein.[42] To test our hypothesis, we examined the effect of heparin and H-kininogen on the rate of antithrombin inhibition of kallikrein and factor XIa. H-kininogen was indeed found to greatly stimulate the heparin-accelerated rate of kallikrein and factor XIa inactivation by antithrombin, in sharp contrast to the antagonizing effect of the cofactor on the heparin-accelerated antithrombin-thrombin reaction previously shown to result from cofactor binding to heparin. This stimulating effect was maximal below plasma concentrations of the cofactor protein (~0.7 uM), which were expected to largely saturate the interaction of these enzymes with the cofactor based on

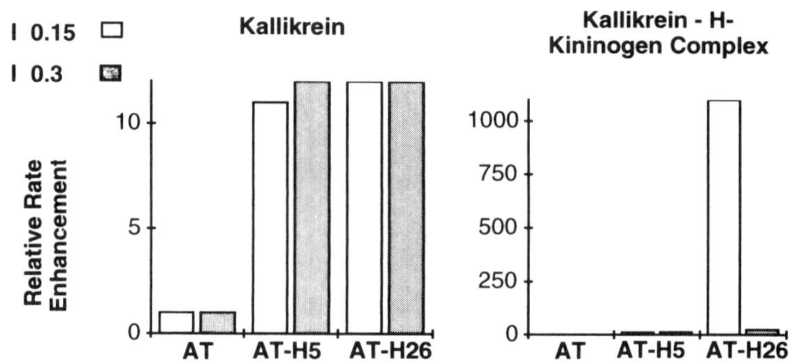

Fig. 4 Chain-length and salt dependence of heparin enhancements of second-order rate constants for antithrombin-kallikrein reactions.

measured dissociation constants. The rate-enhancing effect of H-kininogen was dependent on the presence of heparin, since the cofactor alone had no effect on the slow rate of antithrombin inhibition of these enzymes when added at the same levels required for maximal rate enhancement in the presence of heparin. Further kinetic studies of the reactions of antithrombin and antithrombin-heparin complex with kallikrein and factor XIa in the absence and presence of saturating cofactor over a range of inhibitor concentrations indicated that H-kininogen maximally increased the heparin-enhancement of the second-order inhibition rate constants for these reactions to about 1000-fold for both enzymes; i.e., comparable to heparin rate enhancements of other antithrombin-proteinase reactions. Consistent with the prediction of these H-kininogen-dependent heparin rate enhancements, antithrombin was observed to be a major inhibitor of kallikrein and factor XIa in plasma containing therapeutic levels of heparin (~1 unit/ml). This was demonstrated by addition of the enzymes to normal and deficient plasmas and detection of the proteinase-inhibitor complexes formed and separated by SDS-gel electrophoresis using radiolabelled enzyme or Western blotting.

To determine whether H-kininogen was acting to stimulate the heparin-accelerated reactions of antithrombin with kallikrein and factor XIa by promoting the binding of these enzymes to heparin and thereby allowing the bridging of the enzymes with heparin-bound antithrombin, we compared pentasaccharide and full-length heparin accelerations of the antithrombin-kallikrein reaction in the absence and presence of the cofactor protein at physiological salt and at double this salt concentration (fig. 4).[43] The maximal rate enhancement was again obtained by insuring that heparin was fully complexed with antithrombin and that kallikrein was saturated with H-kininogen at each salt concentration. For the antithrombin-kallikrein reaction in the absence of H-kininogen, the pentasaccharide and full-length heparins produced equivalent enhancements of second-order inhibition rate constants (11-12-fold) which were unaffected by doubling the salt concentration, as was the unaccelerated reaction rate constant. These results indicated that, like the heparin-enhanced antithrombin-factor Xa reaction, heparin binding to just antithrombin was required for heparin's rate-enhancing effect and that this rate enhancement could be explained by the antithrombin conformational change induced by heparin binding enhancing the binding energy of a nonionic antithrombin-proteinase interaction. Contrasting these results, the pentasaccharide enhancement of the second-order inhibition rate constant for the reaction of antithrombin with kallikrein-H-kininogen complex was minor (12-fold) relative to the full-length heparin rate enhancement (1100-fold), due to the failure of the cofactor to stimulate the pentasaccharide acceleration. This indicated that, like the heparin-enhanced antithrombin-thrombin reaction, a heparin chain-length larger than that necessary to bind antithrombin was required for cofactor stimulation of the heparin rate enhancement. Consistent with this larger chain-length requirement again being due to the need for heparin to bind both antithrombin and the H-kininogen-kallikrein complex, the H-kininogen-dependent full-length heparin rate enhancement was substantially reduced by doubling the salt concentration (to 24-fold) whereas the cofactor-dependent pentasaccharide rate enhancement was unaffected at the higher salt concentration.

Together these results indicate that the predominant mechanism of heparin acceleration of the antithrombin-kallikrein reaction shifts from a factor Xa-type conformational change mechanism in the absence of H-kininogen to a thrombin-type surface approximation mechanism in the presence of the cofactor. Moreover, they support the conclusion that heparin's accelerating effect on all antithrombin-proteinase reactions results from additive contributions of both antithrombin conformational change and surface approximation mechanisms, with the relative contributions of these two mechanisms depending on the target proteinase and, in the case of kallikrein and probably also factor XIa, other plasma proteins that act to promote the binding of the proteinase to heparin. This suggests the intriguing possibility that for proteinases such as factor Xa, which mostly utilize the conformational change mechanism, as yet undiscovered plasma proteins may exist to promote the binding of these proteinases to heparin so as to produce an additional surface approximation mechanism component to heparin rate enhancement that assures their proper regulation by antithrombin.

REFERENCES

1. I. Björk, S. T. Olson, and J. D. Shore, Molecular mechanisms of the accelerating effect of heparin on the reactions between antithrombin and clotting proteinases, in: "Heparin. Chemical and Biological Properties. Clinical Applications", D. A. Lane and U. Lindahl, eds., Edward Arnold, London (1989).
2. S. T. Olson and I. Björk, Regulation of thrombin by antithrombin and heparin cofactor II, in: "Thrombin: Structure and Function", L. J. Berliner, ed., Plenum, New York City (1991).
3. I. Björk and W. W. Fish, Production in vitro and properties of a modified form of bovine antithrombin, cleaved at the active site by thrombin, J. Biol. Chem. 257:9487 (1982).
4. S. T. Olson, Heparin and ionic strength-dependent conversion of antithrombin III from an inhibitor to a substrate of a-thrombin, J. Biol. Chem. 260:10153 (1985).
5. H. Erdjument, D. A. Lane, M. Panico, V. DiMarzo, and H. R. Morris, Single amino acid substitutions in the reactive site of antithrombin leading to thrombosis. Congenital substitution of arginine 393 to cysteine in antithrombin Northwick Park and to histidine in antithrombin Glasgow, J. Biol. Chem. 263:5589 (1988).

6. M. C. Owen, C. H. Beresford, and R. W. Carrell, Antithrombin Glasgow, 393 Arg to His: a P_1 reactive site variant with increased heparin affinity but no thrombin inhibitory activity, FEBS Lett. 231:317 (1988).
7. D. A. Lane, H. Erdjument, A. Flynn, V. DiMarzo, M. Panico, H. R. Morris, M. Greaves, G. Dolan, and F. E. Preston, Antithrombin Sheffield: amino acid substitution at the reactive site (Arg 393 to His) causing thrombosis, Brit. J. Haematol. 71:91 (1989).
8. H. Erdjument, D. A. Lane, M. Panico, V. DiMarzo, H. R. Morris, K. Bauer, and R. D. Rosenberg, Antithrombin Chicago, amino acid substitution of arginine 393 to histidine, Thromb. Res. 54:613 (1989).
9. H. Erdjument, D. A. Lane, H. Ireland, V. DiMarzo, M. Panico, H. R. Morris, A. Tripodi, and P. M. Manucci, Antithrombin Milano, single amino acid substitution at the reactive site, Arg 393 to Cys, Thromb. Hemostas. 60:471 (1988).
10. D. A. Lane, H. Erdjument, E. Thompson, M. Panico,, V. DiMarzo,, H. R. Morris, G. Leone, V. DeStefano, and S. L. Thein, A novel amino acid substitution in the reactive site of a congenital variant antithrombin. Antithrombin Pescara, Arg^{393} to Pro, caused by a CGT to CCT mutation, J. Biol. Chem. 264:10200 (1989).
11. A. W. Stephens, B. S. Thalley, and C. H. W. Hirs, Antithrombin III-Denver, a reactive site variant, J. Biol. Chem. 262:1044 (1987).
12. A. W. Stephens, A. Siddiqui, and C. H. W. Hirs, Site-directed mutagenesis of the reactive center (serine 394) of antithrombin III, J. Biol. Chem. 263:15849 (1988).
13. S. T. Olson, R. Sheffer, A. W. Stephens, and C. H. W. Hirs, Molecular basis of the reduced reactivity of antithrombin-Denver with thrombin and factor Xa. Role of the P1' residue, Thromb. Haemostas. 65:670 (1991).
14. S. C. Bock, Antithrombin III genetics, structure and function, in: "Recombinant Technology in Hemostasis and Thrombosis", L. W. Hoyer and W. N. Drohan, eds., Plenum, New York City (1990).
15. P. Molho-Sabatier, M. Aiach, I. Gaillaird, J. N. Fiessinger, A. M. Fischer, G. Chadeuf, and E. Clause, Molecular characterization of antithrombin III (AT III) variants using polymerase chain reaction. Identification of the AT III Charleville as an Ala 384 →Pro mutation, J. Clin. Invest. 84:1236 (1989).
16. R. Caso, D. A. Lane, E. A. Thompson, R. J. Olds, S. L. Thein, M. Paqnico, I. Blench, H. R. Morris, J. M. Freyssinet, M. Aiach, F. Rodeghiero, and G. Finazzi, Antithrombin Vicenza, Ala 384 to Pro (GCA to CCA) mutation, transforming the inhibitor into a substrate, Brit. J. Haematol. 77:87 (1991).
17. R. Devraj-Kizuk, D. H. K. Chui, E. V. Prochownik, C. J. Carter, F. A. Ofosu, M. A. Blajchman, Antithrombin-III-Hamilton: A gene with a point mutation (guanine to adenine) in codon 382 causing impaired serine protease reactivity, Blood 72:1518 (1988).
18. N. J. Levy, N. Ramesh, M. Cicardi, R. A. Harrison, A. E. Davis, Type II hereditary angioneurotic edema that may result from a single nucleotide change in the codon for alanine-436 in the C1 inhibitor gene, Proc. Natl. Acad. Sci., U.S.A. 87:265 (1990).
19. K. Skriver, W. R. Wikoff, P. A. Patston, F. Tausk, M. Schapira, A. P. Kaplan, and S. C. Bock, Substrate properties of C1 Inhibitor Ma (alaninie 434→glutamic acid). Genetic and structural evidence suggesting that the P12-region contains critical determinants of serine protease inhibitor inhibitor/substrate status, J. Biol. Chem. 266:9216 (1991).
20. W. E. Holmes, H. R. Lijnen, L. Nelles, C. Kluft, H. K. Nieuwenhuis, D. C. Rijken, and D. Collen, α_2-antiplasmin Enschede: alanine insertion and abolition of plasmin inhibitory activity, Science 238:209 (1987).
21. S. Asakura, H. Hirata, H. Okazaki, T. Hashimoto-Gotoh, and M. Matsuda, Hydrophobic residues 382-386 of antithrombin III, ala-ala-ala-ser-thr, serve as an epitope for an antibody which facilitates hydrolysis of the inhibitor by thrombin, J. Biol. Chem. 265:5135 (1990).
22. R. Huber and R. W. Carrell, Implications of the three-dimensional structure of α_1-antitrypsin for structure and function of serpins, Biochemistry 28:8951 (1989).
23. L. Mourey, J. P. Samama, M. Delarue, J. Choay, J. C. Lormeau, M. Petitou, and D. Moras, Antithrombin III: structural and functional aspects, Biochimie 72:599 (1990).

24. R. A. Engh, H. T. Wright, and R. Huber, Modeling of the intact form of the α_1-proteinase inhibitor, Protein Engng. 3:469 (1990).
25. P. E. Stein, A. G. W. Leslie, J. T. Finch, W. G. Turnell, P. J. McLaughlin, and R. W. Carrell, Crystal structure of ovalbumin as a model for the reactive centre of serpins, Nature 347:99 (1990).
26. A. J. Schulze, U. Baumann, S. Knof, E. Jaeger, R. Huber, and C.-B. Laurell, Structural transition of α_1-antitrypsin by a peptide sequentially similar to β-strand s4A, Eur. J. Biochem. 194:51 (1990).
27. I. Björk, K. Ylinenjarvi, S. T. Olson, and P. E. Bock, Conversion of antithrombin from an inhibitor of thrombin to a substrate with reduced heparin affinity and enhanced conformational stability by binding of a tetradecapeptide corresponding to the P_1 to P_{14} region of the putative reactive-bond loop of the inhibitor, J. Biol. Chem. in press (1991).
28. S. O. Brennan, J. Y. Borg, P. M. George, C. Soria, J. Soria, J. Caen, and R. W. Carrell, New carbohydrate site in mutant antithrombin (7 Ile→Asn) with decreased heparin affinity, FEBS Lett. 237:118 (1988).
29. J. Y. Borg, S. O. Brennan, R. W. Carrell, P. George, D. J. Perry, and J. Shaw, Antithrombin Rouen-IV, 24 Arg→Cys. The amino terminal contribution to heparin binding, FEBS Lett. 266:163 (1990).
30. S. Gandrille, M. Aiach, D. A. Lane, D. Vidaud, P. Molho-Sabatier, R. Caso, P. deMoerloose, J. N. Fiessinger, and E. Clauser, Crucial role of Arg 129 in heparin binding site of antithrombin III: Identification of a novel mutation Arg 129 to Gln, J. Biol. Chem. 265:18997 (1990).
31. P. Gettins and E. W. Wooten, On the domain structure of antithrombin III. Localization of the heparin-binding region using ^{1}H NMR spectroscopy, Biochemistry 26:4403 (1987).
32. J. Y. Chang, Binding of heparin to human antithrombin III activates selective chemical modification at lysine 236. Lys-107, lys-125, and lys-136 are situated within the heparin-binding site of antithrombin III, J. Biol. Chem. 264:3111 (1989).
33. X. J. Sun and J. Y. Chang, Evidence that arginine-129 and arginine-145 are located within the heparin binding site of human antithrombin III, Biochemistry 29:8957 (1990).
34. X. J. Sun and J. Y. Chang, Heparin binding domain of human antithrombin III inferred from the sequential reduction of its three disulfide linkages. An efficient method for structural analysis of partially reduced proteins, J. Biol. Chem. 264:11288 (1989).
35. J. W. Smith, N. Dey, and D. J. Knauer, Heparin binding domain of antithrombin III: Characterization using a synthetic peptide directed polyclonal antibody, Biochemistry 29:8950 (1990).
36. S. T. Olson, I. Björk, P. A. Craig, J. D. Shore, and J. Choay, Role of the high-affinity pentasaccharide in heparin acceleration of antithrombin III inhibition of thrombin and factor Xa, Thromb. Haemostas. 58:8 (1987).
37. R. D. Rosenberg and P. S. Damus, The purification and mechanism of action of human antithrombin-heparin cofactor, J. Biol. Chem. 248:6490 (1973).
38. C. H. Beresford and M. C. Owen, Minireview. Antithrombin III, Int. J. Biochem. 22:121 (1990).
39. S. T. Olson and I. Björk, Predominant contribution of surface approximation to the mechanism of heparin acceleration of the antithrombin-thrombin reaction. Elucidation from salt concentration effects, J. Biol. Chem. 266:6353 (1991).
40. S. T. Olson, High molecular weight-kininogen enhancement of the heparin-accelerated rate of plasma kallikrein inactivation by antithrombin III, J. Cell Biol. 107:827a (1989).
41. S. T. Olson and J. D. Shore, High molecular weight-kininogen and heparin acceleration of factor XIa inactivation by plasma proteinase inhibitors, Thromb. Haemostas. 62:381 (1989).
42. I. Björk, S. T. Olson, R. G. Sheffer, and J. D. Shore, Binding of heparin to human high molecular weight kininogen, Biochemistry 28:1213 (1989).
43. S. T. Olson and J. Choay, Mechanism of high molecular weight-kininogen stimulation of the heparin-accelerated antithrombin/kallikrein reaction, Thromb. Haemostas. 62:326 (1989).

THE INTERACTION OF GLYCOSAMINOGLYCANS WITH HEPARIN COFACTOR II: STRUCTURE AND ACTIVITY OF A HIGH-AFFINITY DERMATAN SULFATE HEXASACCHARIDE

Douglas M. Tollefsen

Division of Hematology-Oncology
Department of Internal Medicine
Washington University, St. Louis, Missouri 63110

INTRODUCTION

The anticoagulant activities of glycosaminoglycans are mediated by antithrombin III (ATIII) and heparin cofactor II (HCII), members of the "serpin" family that are present in plasma at micromolar concentrations. ATIII inhibits several of the proteases involved in coagulation, particularly thrombin and factor Xa, whereas HCII specifically inhibits thrombin.[1] Both dermatan sulfate and heparin increase the rate of inhibition of thrombin by HCII more than 1000-fold.[2] In contrast, ATIII is stimulated only by heparin. HCII functions as a pseudosubstrate for thrombin, forming a stable 1:1 complex. Glycosaminoglycans accelerate complex formation in part by providing a template to which both the inhibitor and the protease bind.[3]

STRUCTURE OF HCII

HCII is composed of a single polypeptide chain of 480 amino acid residues[4,5] (Fig. 1A). Near the N-terminus is an acidic domain which contains two tyrosine residues that become sulfated during biosynthesis in hepatocytes.[6] Peptides cleaved from the N-terminal portion of HCII by neutrophil proteases have potent chemotactic activity.[7]

Reactive site

The reactive site peptide bond attacked by thrombin is located near the C-terminus and contains a leucine in the P1 position[8], which is atypical of thrombin substrates. Because of the P1 leucine, HCII inhibits thrombin very slowly in the absence of a glycosaminoglycan. In fact, HCII inhibits chymotrypsin more rapidly than thrombin.[9] Substitution of an arginine for leucine at the P1 position increases the rate of thrombin inhibition about 100-fold[10], producing a rate similar to that of α1-antitrypsin Pittsburgh. Therefore, it appears that HCII has evolved to be essentially inactive in the absence of a glycosaminoglycan.

Heparin and Related Polysaccharides
Edited by D.A. Lane *et al.*, Plenum Press, New York, 1992

Glycosaminoglycan-binding site

HCII Oslo. The middle portion of HCII contains a cluster of basic amino acid residues that comprise the glycosaminoglycan-binding site. This region was identified by analysis of a variant of HCII discovered in a healthy Norwegian blood donor. Based on crossed immunoelectrophoresis, Abildgaard and coworkers reported that this variant (which we have termed "HCII Oslo") binds heparin but does not bind dermatan sulfate.[11] We amplified a portion of the HCII gene from this individual using the polymerase chain reaction and identified a nucleotide substitution in the codon for arginine-189, resulting in a histidine at this position in HCII Oslo.[12] To confirm that this mutation was responsible for the observed phenotype, we constructed the same point mutation in the normal HCII cDNA and expressed the recombinant protein in *E. coli*. Native recombinant HCII and the Oslo variant required similar concentrations of heparin to stimulate thrombin inhibition. In contrast, recombinant HCII Oslo required about 100 times more dermatan sulfate than native HCII to produce an equivalent degree of thrombin inhibition. These experiments suggested that the positive charge on arginine-189 may be involved in binding dermatan sulfate but not heparin and provided the first evidence that the binding sites for the two glycosaminoglycans are not identical.

Site-directed mutagenesis of the glycosaminoglycan-binding site. To define further the glycosaminoglycan binding sites in HCII, we have mutated nearby arginine and lysine residues to eliminate their positive charge.[12-14] Mutations of arginine-189, -192, and -193 specifically decrease the interaction with dermatan sulfate but have no effect on heparin binding. In contrast, mutations of lysine-173 decrease binding to heparin but not dermatan sulfate. Mutations of arginine-184 or lysine-185 in

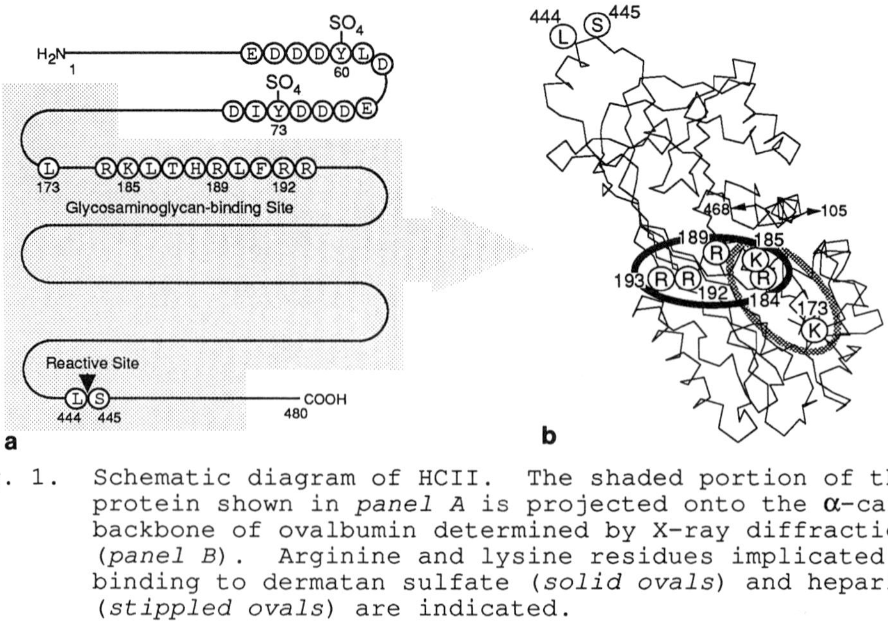

Fig. 1. Schematic diagram of HCII. The shaded portion of the protein shown in *panel A* is projected onto the α-carbon backbone of ovalbumin determined by X-ray diffraction (*panel B*). Arginine and lysine residues implicated in binding to dermatan sulfate (*solid ovals*) and heparin (*stippled ovals*) are indicated.

the middle of this domain affect the interaction of HCII with both glycosaminoglycans. Similar results have been obtained by Ragg and coworkers.[15,16] Thus, the binding sites for heparin and dermatan sulfate in HCII appear to be overlapping but not identical (Fig. 1B).

BINDING OF DERMATAN SULFATE AND HEPARIN OLIGOSACCHARIDES TO HCII

Dermatan sulfate consists of alternating uronic acid and N-acetyl galactosamine residues. The structure is heterogeneous as a result of incomplete biosynthetic modifications. For example, glucuronic acid may be epimerized to iduronic acid, and sulfation may occur at the 2 position of iduronic acid or at the 4 or 6 position of N-acetyl galactosamine. In heparin, the amino sugar is glucosamine, and heparin is generally more highly sulfated than dermatan sulfate. ATIII binds to a specific pentasaccharide structure in heparin which contains a 3-O-sulfate group in the middle position.

We were interested in determining whether HCII also recognizes specific oligosaccharide sequences within the heparin or dermatan sulfate polymers. To isolate oligosaccharides with high affinity for HCII, we subjected heparin and dermatan sulfate to partial chemical depolymerization with nitrous acid. Gel filtration chromatography was then used to obtain fragments of various chain lengths. Each size fraction was tested for the ability to bind to HCII-Sepharose, and the low-affinity and high-affinity molecules were analyzed further.

HCII-Sepharose chromatography. As indicated in Fig. 2, the smallest dermatan sulfate oligosaccharides that bound to HCII were hexasaccharides.[17] About 6% of the hexasaccharides bound under conditions in which the column was not overloaded. As the chain length increased to 14 monosaccharide units, the bound fraction increased to 25%. About 80% of intact porcine skin dermatan sulfate molecules bound, indicating that most of these chains contained at least one HCII-binding site. In contrast to dermatan sulfate, the HCII column bound at least 75% of each heparin oligosaccharide greater than two sugars in length.[18]

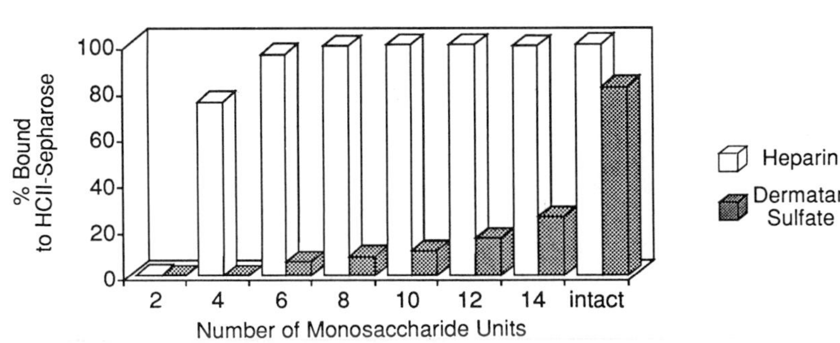

Fig. 2. HCII-Sepharose chromatography of heparin and dermatan sulfate oligosaccharides. Size-fractionated oligosaccharides were applied to the affinity column under non-saturating conditions in 50 mM NaCl, 50 mM Tris-HCl, pH 7.4. Data from refs. 17 & 18.

These results suggest that HCII binds non-specifically to heparin, but that it preferentially binds to a subpopulation of dermatan sulfate oligosaccharides containing at least six residues.

Structure of the high-affinity dermatan sulfate hexasaccharide. To analyze the high-affinity hexasaccharide, material from the HCII column was fractionated by anion-exchange chromatography. Fig 3A shows a chromatograph of dermatan sulfate hexasaccharides prior to affinity chromatography on HCII-Sepharose. Most of the material eluted as a cluster of peaks containing 3 sulfates per hexasaccharide, but there were also minor clusters containing 4, 5 or 6 sulfates per hexasaccharide. The flow-through material from the HCII-Sepharose column was similar to unfractionated hexasaccharides

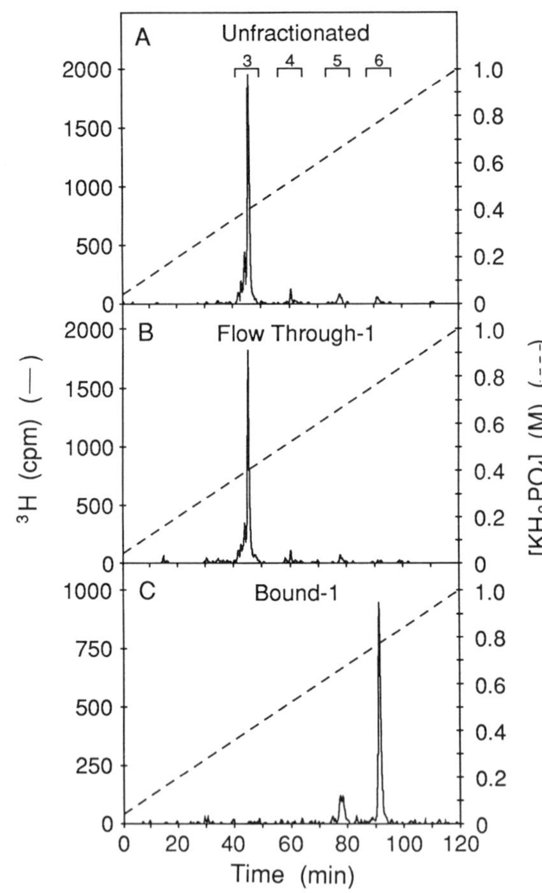

Fig. 3. Ion-exchange chromatography of ^{3}H-labeled dermatan sulfate hexasaccharides. Unfractionated hexasaccharides (*panel A*) and the flow through (*panel B*) and bound (*panel C*) fractions from the HCII-Sepharose column were chromatographed on a Partisil 10 SAX HPLC column eluted with a gradient of KH_2PO_4. The *number* above each peak represents the apparent degree of sulfation of the hexasaccharides contained in that peak. Data from ref. 17.

except for a reduction in the peak containing 6 sulfates (Fig. 3B). In agreement with this finding, most of the material bound to HCII-Sepharose eluted as a peak containing 6 sulfate groups (Fig. 3C). Repeated application of the flow- through material to the affinity column indicated that hexasaccharides containing 4 or 5 sulfate groups bound to HCII with lower affinity than the hexasaccharide containing 6 sulfate groups. However, none of the hexasaccharides containing only 3 sulfate groups bound to HCII.

To determine the structure of the high-affinity hexasaccharide, we pooled the major peak shown in Fig. 3C and determined its disaccharide composition. Six disaccharides are

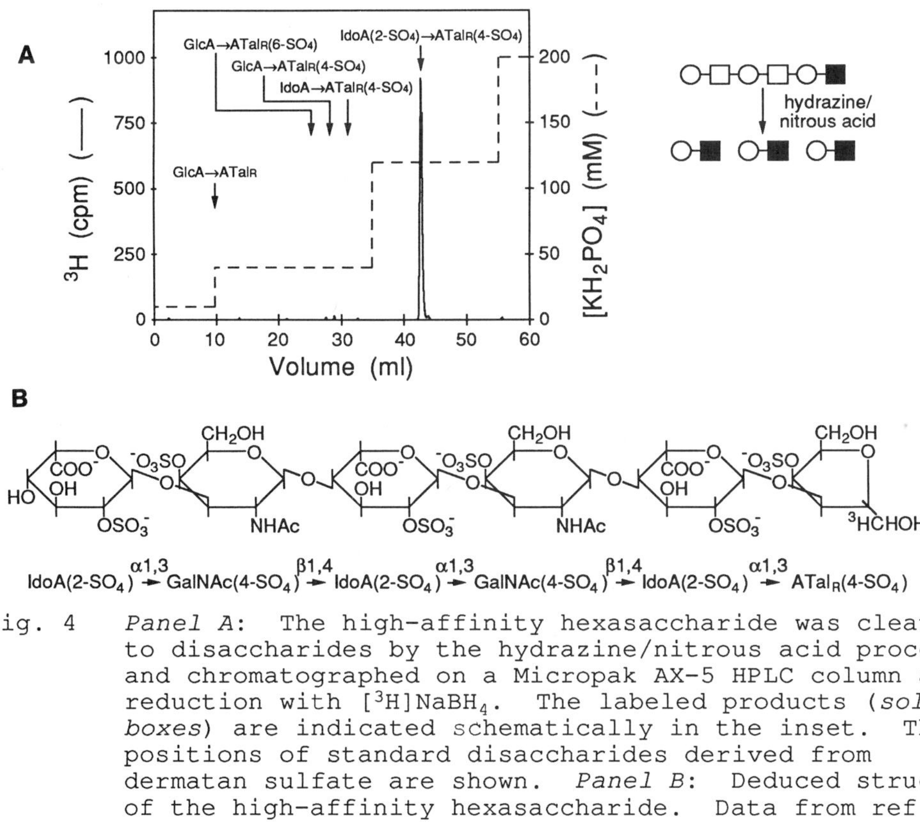

Fig. 4 *Panel A*: The high-affinity hexasaccharide was cleaved to disaccharides by the hydrazine/nitrous acid procedure and chromatographed on a Micropak AX-5 HPLC column after reduction with [^{3}H]NaBH$_4$. The labeled products (*solid boxes*) are indicated schematically in the inset. The positions of standard disaccharides derived from dermatan sulfate are shown. *Panel B*: Deduced structure of the high-affinity hexasaccharide. Data from ref. 17.

obtained from porcine skin dermatan sulfate by exhaustive depolymerization with hydrazine and nitrous acid. In each disaccharide, the reducing terminal N-acetyl galactosamine is converted to anhydrotalitol by the cleavage procedure. Exhaustive depolymerization of the high-affinity hexasaccharide yielded a single peak that coeluted with the disulfated disaccharide standard, iduronic acid 2-sulfate → anhydrotalitol 4-sulfate (Fig. 4A). Identification of this product was confirmed by paper chromatography, paper electrophoresis, and analysis of the products obtained by mild acid hydrolysis.[17]

Based on these experiments, the high-affinity hexasaccharide consists of three disulfated disaccharides, iduronic acid 2-sulfate alternating with N-acetyl galactosamine 4-sulfate, with the reducing terminal sugar converted to anhydrotalitol 4-sulfate by the cleavage procedure (Fig. 4B). The repeating disaccharide unit is relatively uncommon in the intact dermatan sulfate polymer. About 75% of the disaccharides derived from intact dermatan sulfate are iduronic acid → anhydrotalitol 4-sulfate. In contrast, only 5% of the disaccharides are iduronic acid 2-sulfate → anhydrotalitol 4-sulfate, the repeating disaccharide unit found in the high-affinity hexasaccharide. If this disaccharide were distributed randomly in the dermatan sulfate polymer, the probability of finding three next to each other would be about 1 in 10,000. In fact, we found that about 2% of the hexasaccharides derived from dermatan sulfate have the structure of the high-affinity hexasaccharide. Therefore, the disulfated disaccharides must be clustered in the dermatan sulfate polymer to form the high-affinity binding sites for HCII.

FUNCTION OF THE N-TERMINAL ACIDIC DOMAIN OF HCII

The N-terminal acidic domain contains two repeated structures (Fig. 1A) that can be aligned with the C-terminal portion of hirudin. X-ray crystallographic studies of the thrombin-hirudin complex indicate that amino acids in the C-terminal portion of hirudin form ionic and hydrophobic contacts with the surface of thrombin; many of these residues appear to be conserved in the acidic domain of HCII. Hortin et al.[19] found that a synthetic peptide representing residues 54-75 of HCII competes with hirudin for binding to thrombin, suggesting that thrombin may bind to this site in the intact HCII molecule.

Kinetics of protease inhibition by N-terminal deletion mutants. To investigate the function of the acidic domain, we constructed a series of N-terminal deletions in recombinant HCII.[20] Fig. 5 summarizes the second-order rate constants for inhibition of thrombin or chymotrypsin by each variant. None of the deletions had a significant effect on the rate of inhibition of thrombin in the absence of a glycosaminoglycan, nor did they affect the rate of inhibition of chymotrypsin. These experiments indicate that deletion of the acidic domain does not disrupt the tertiary structure of the reactive site of HCII. More importantly, the acidic domain is not required for inhibition of thrombin in the absence of a glycosaminoglycan.

Deletion of the first 31 or 52 amino acid residues had no effect on the ability of dermatan sulfate or heparin to increase the rate of thrombin inhibition. Both glycosaminoglycans increased the rate constants about 5000-fold. However, deletion of 67 amino acids, which includes a portion of the acidic domain, reduced the ability of dermatan sulfate to stimulate thrombin inhibition by about three orders of magnitude and had slightly less of an effect on heparin.

Binding to heparin-Sepharose. To rule out the possibility that the results shown in Fig. 5 could be explained by decreased binding of the glycosaminoglycans to the deletion mutants, we performed heparin-Sepharose chromatography of the various deletion mutants mixed with intact recombinant HCII as an internal standard. We found that deletion of the acidic domain actually *increased* the affinity of HCII for heparin-Sepharose.[20] Thus, both native HCII and the variant lacking the first 52 amino acid residues eluted at about 0.35 M NaCl, while elution of forms missing part or all of the acidic domain required 0.5-0.8 M NaCl.

Model for inhibition of thrombin. Our experiments, and similar studies performed by Ragg and coworkers[15,16], suggest a new model for thrombin inhibition by HCII (Fig. 6). According to this model, the N-terminal acidic domain of native HCII binds intramolecularly to the glycosaminoglycan-binding site, and covalent complex formation with thrombin occurs at the basal rate of about 2×10^4 $M^{-1}min^{-1}$. Glycosaminoglycans displace the hirudin-like acidic domain from the glycosaminoglycan-binding site, allowing it to interact with the anion-binding exosite of thrombin. The glycosaminoglycan also serves as a template, binding HCII and thrombin simultaneously. These *non-covalent* binding reactions accelerate thrombin inhibition 5000-fold. Deletion of the acidic domain has no effect on the basal rate of inhibition of thrombin, since the acidic domain in native HCII is not free to interact with thrombin in the absence of a glycosaminoglycan. However, deletion of the acidic domain increases the affinity for glycosaminoglycans by uncovering the glycosaminoglycan-binding site. Glycosaminoglycans modestly stimulate thrombin inhibition by the deletion mutant, perhaps through the template mechanism.

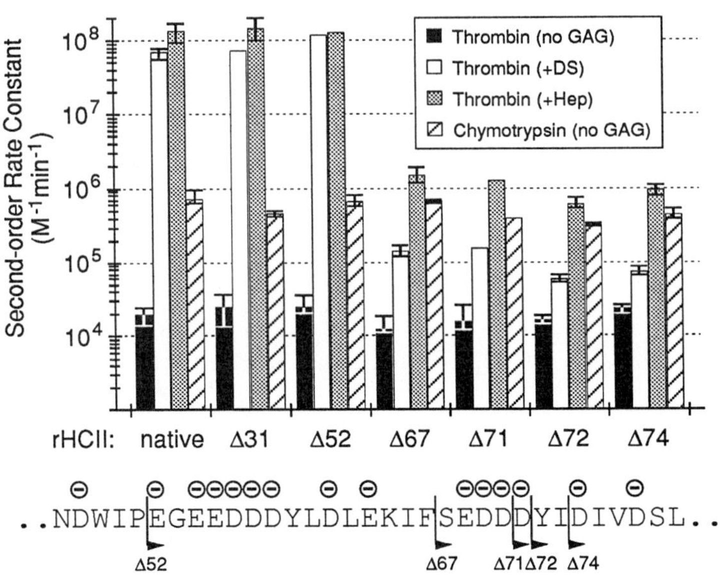

Fig. 5 Second-order rate constants for inhibition of α-thrombin and chymotrypsin by native and N-terminal deletion mutants of recombinant HCII. Inhibition of α-thrombin was determined in the absence (*no GAG*) or presence of 25 μg/ml of heparin (*+Hep*) or dermatan sulfate (*+DS*). Data from ref. 20.

Further evidence for this model comes from studies with the high-affinity dermatan sulfate hexasaccharide. The hexasaccharide stimulates thrombin inhibition by native HCII about 50 to 100-fold[20], even though it is presumably too small to form a template. The stimulatory effect seems to be mediated entirely by the acidic domain, since deletion of the acidic domain reduces thrombin inhibition to the basal rate in the presence of the hexasaccharide. Ultimately, confirmation of this model will require crystallization of native HCII and complexes of HCII with thrombin and glycosaminoglycans.

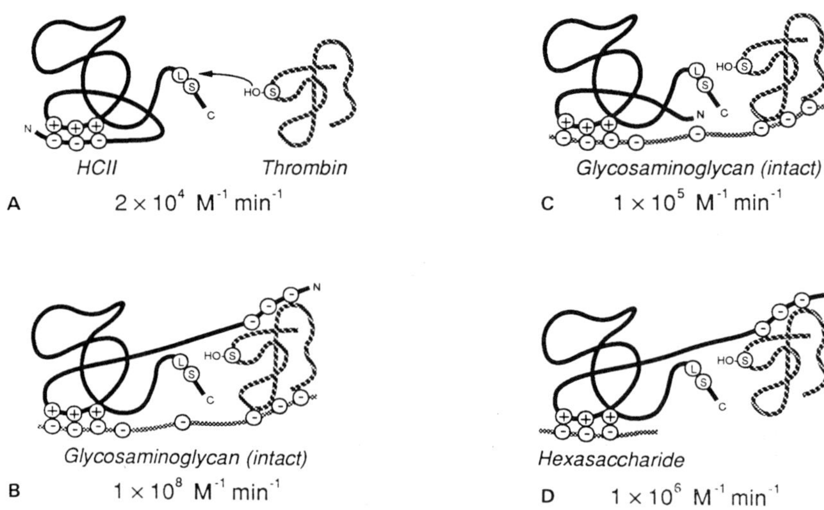

Fig. 6 Model for inhibition of thrombin by HCII. *Panel A*: The active site serine (S) hydroxyl group of thrombin attacks the reactive site leucyl-serine (LS) peptide bond of HCII to form a covalent complex. In the absence of a glycosaminoglycan, the N-terminal acidic domain of HCII (-) forms ionic bonds with the glycosaminoglycan-binding site (+) and is unable to interact with thrombin. *Panel B*: A glycosaminoglycan chain displaces the N-terminal acidic domain of HCII from the glycosaminoglycan-binding site. The acidic domain then interacts with the hirudin-binding site of thrombin. Binding of thrombin both to the N-terminal acidic domain of HCII and to the glycosaminoglycan template greatly increases the rate of covalent complex formation. *Panel C*: A deletion mutant of HCII lacking the acidic domain is shown. Complex formation with thrombin can be stimulated only by the glycosaminoglycan template mechanism. *Panel D*: The dermatan sulfate hexasaccharide displaces the N-terminal acidic domain of HCII but is of insufficient length to bind thrombin simultaneously. Complex formation can be stimulated only by the displacement mechanism. Approximate second-order rate constants are indicated. Data from ref. 20.

SUMMARY

The binding sites for dermatan sulfate and heparin in HCII overlap but are not identical. This may explain the observation that HCII binds non-specifically to heparin oligosaccharides but preferentially binds to a minor hexasaccharide isolated from dermatan sulfate having the structure shown in Fig. 4B. The tissue distribution of dermatan sulfate molecules containing the high-affinity HCII binding site may regulate HCII activity *in vivo*. Finally, in the presence of dermatan sulfate or heparin, the N-terminal acidic domain of HCII may interact with the hirudin-binding site of thrombin to produce maximal stimulation of the thrombin-HCII reaction.

REFERENCES

1. K. A. Parker, and D. M. Tollefsen, The protease specificity of heparin cofactor II. Inhibition of thrombin generated during coagulation, J. Biol. Chem. 260:3501 (1985).
2. D. M. Tollefsen, C. A. Pestka, and W. J. Monafo, Activation of heparin cofactor II by dermatan sulfate, J. Biol. Chem. 258:6713 (1983).
3. M. J. Griffith, Heparin-catalyzed inhibitor/protease reactions: kinetic evidence for a common mechanism of action of heparin, Proc. Natl. Acad. Sci. U.S.A. 80:5460 (1983).
4. H. Ragg, A new member of the plasma protease inhibitor gene family, Nucleic Acids Res. 14:1073 (1986).
5. M. A. Blinder, J. C. Marasa, C. H. Reynolds, L. L. Deaven, and D. M. Tollefsen, Heparin cofactor II: cDNA sequence, chromosome localization, restriction fragment length polymorphism, and expression in Escherichia coli, Biochemistry 27:752 (1988).
6. G. Hortin, D. M. Tollefsen, and A. W. Strauss, Identification of two sites of sulfation of human heparin cofactor II, J. Biol. Chem. 261:15827 (1986).
7. C. W. Pratt, R. B. Tobin, and F. C. Church, Interaction of heparin cofactor II with neutrophil elastase and cathepsin G, J. Biol. Chem. 265:6092 (1990).
8. M. J. Griffith, C. M. Noyes, J. A. Tyndall, and F. C. Church, Structural evidence for leucine at the reactive site of heparin cofactor II, Biochemistry 24:6777 (1985).
9. F. C. Church, C. M. Noyes, and M. J. Griffith, Inhibition of chymotrypsin by heparin cofactor II, Proc. Natl. Acad. Sci. U.S.A. 82:6431 (1985).
10. V. M. Derechin, M. A. Blinder, and D. M. Tollefsen, Substitution of arginine for Leu444 in the reactive site of heparin cofactor II enhances the rate of thrombin inhibition, J. Biol. Chem. 265:5623 (1990).
11. T. R. Andersson, M. L. Larsen, and U. Abildgaard, Low heparin cofactor II associated with abnormal crossed immunoelectrophoresis pattern in two Norwegian families, Thromb. Res. 47:243 (1987).
12. M. A. Blinder, T. R. Andersson, U. Abildgaard, and D. M. Tollefsen, Heparin cofactor IIOslo. Mutation of Arg-189 to His decreases the affinity for dermatan sulfate, J. Biol. Chem. 264:5128 (1989).
13. M. A. Blinder, and D. M. Tollefsen, Site-directed mutagenesis of arginine 103 and lysine 185 in the

proposed glycosaminoglycan-binding site of heparin cofactor II, J. Biol. Chem. 265:286 (1990).
14. H. C. Whinna, M. A. Blinder, M. Szewczyk, D. M. Tollefsen, and F. C. Church, Role of lysine 173 in heparin binding to heparin cofactor II., J. Biol. Chem. 266:8129 (1991).
15. H. Ragg, T. Ulshöfer, and J. Gerewitz, On the activation of human leuserpin-2, a thrombin inhibitor, by glycosaminoglycans, J. Biol. Chem. 265:5211 (1990).
16. H. Ragg, T. Ulshöfer, and J. Gerewitz, Glycosaminoglycan-mediated leuserpin-2/thrombin interaction. Structure-function relationships, J. Biol. Chem. 265:22386 (1990).
17. M. M. Maimone, and D. M. Tollefsen, Structure of a dermatan sulfate hexasaccharide that binds to heparin cofactor II with high affinity, 265:18263 (1990).
18. M. M. Maimone, Characterization of heparin and dermatan sulfate molecules that bind and activate heparin cofactor II, Ph.D. thesis (1990).
19. G. L. Hortin, D. M. Tollefsen, and B. M. Benutto, Antithrombin activity of a peptide corresponding to residues 54-75 of heparin cofactor II, J. Biol. Chem. 264:13979 (1989).
20. V. M. D. Van Deerlin, and D. M. Tollefsen, The N-terminal acidic domain of heparin cofactor II mediates the inhibition of α-thrombin in the presence of glycosaminoglycans, J. Biol. Chem., in press (1991).

THROMBOMODULIN: AN ANTICOAGULANT CELL SURFACE PROTEOGLYCAN WITH PHYSIOLOGICALLY RELEVANT GLYCOSAMINOGLYCAN MOIETY

John F. Parkinson[1], Takatoshi Koyama[2], Nils U. Bang[1] and Klaus T. Preissner[3]

[1] Lilly Research Laboratories, Indianapolis, Indiana, USA
[2] Tokyo Medical and Dental University, Tokyo, Japan
[3] Haemostasis Research Unit, Kerckhoff-Klinik, Max-Planck-Gesellschaft, Bad Nauheim, Germany

INTRODUCTION

The vessel wall forms the barrier between the circulating blood and the underlying tissues and it provides a continuous luminal endothelial cell layer which is actively and dynamically involved in the balance of the haemostatic system. Different coagulation and fibrinolytic components are synthesized, secreted or bound by the vascular endothelium, but the intact vessel wall prevents the activation of the haemostatic system by provision of its thrombo-resistant surface. At least two major anticoagulant mechanisms contribute to this control and regulation of blood coagulation, once it is initiated by injury of the vessel wall: (a) Endothelial cell surface-expressed thrombin receptors, in particular thrombomodulin (TM) (Esmon and Owen, 1981; Owen and Esmon, 1981), and (b) endothelial cell surface-associated reactions involving serine protease inhibitors such as antithrombin III (AT III), heparin cofactor II (HC II), or protease nexin I and their possible interaction with heparan sulfate proteoglycans (Preissner, 1990). When bound to intact TM, thrombin effectively invokes the onset of the protein C pathway, involving activation of the circulating pro-enzyme protein C, subsequent complex formation of activated protein C with cofactor protein S, and the ultimate proteolytic degradation of coagulation cofactors Va and VIIIa (Esmon, 1987). Thus, the central role of TM as anticoagulant cofactor thereby converts thrombin from a pro- into an anticoagulant enzyme that initiates a negative feedback loop which leads to inhibition of thrombin formation by eliminating the critical cofactors. The central physiological function of the protein C pathway has been demonstrated by a strong correlation between the occurence of recurrent thromboembolic complications and congenital deficiencies of protein C or protein S (Bertina, 1988). In the following, structure/function relationships of TM will be summarized indicating that a major O-linked glycan moiety of the thrombin receptor acts as main modulator of receptor function.

Heparin and Related Polysaccharides
Edited by D.A. Lane *et al.*, Plenum Press, New York, 1992

THROMBOMODULIN REPRESENTS A PROTEOGLYCAN

TM is an endothelial cell-specific membrane receptor present on the endothelial cell surface of arteries, veins, and capillaries as well as on lymphatics and trophoblasts of human placenta (Maruyama et al., 1985; DeBault et al., 1986a). About 50% of thrombin-binding sites at the endothelial cell surface are related to TM (Maruyama and Majerus, 1985). The presence of TM is not only responsible for the rapid clearance of low doses of the enzyme from the circulation (Lollar and Owen 1980) but increases the rate of protein C activation by thrombin, the only physiologically relevant activator, by about 20,000-fold (Esmon and Owen, 1981; Esmon et al., 1982b).

An appreciable pool of intracellular TM is found associated with the Golgi-endoplasmatic reticulum complex in endothelial cells (DeBault et al., 1986b). Initial immunocytochemical staining revealed the absence of TM from human brain (Ishii et al., 1986), however, cryosections of capillaries in the brain stained positive for TM, which was also expressed by those cells _in vitro_ (Wong et al., 1991). Lung and placenta tissues contain the highest concentration of TM, indicating that the relative density of TM is high in the microcirculation (Owen and Esmon, 1981). Upon cultivation also smooth muscle cells express TM indicating that _in vitro_ conditions may lead to altered expression of the receptor (Soff et al., 1991) as demonstrated for a variety of cells. In particular, transcriptional control of TM as studied in various cells in culture appears to guide the expression of the intron-deleted TM gene. Interestingly, non-vascular distribution of a surface marker, designated fetomodulin, has been found in mouse embryos (Imada et al., 1987) and this protein has been shown to be identical with TM (Imada et al., 1990b) indicating that TM may play a role to regulate thrombin's function during fetal development as well. TM has been isolated from rabit lung (Esmon et al., 1982c), human placenta (Salem et al., 1984; Kurosawa and Aoki, 1985) as well as from bovine lung (Jakubowski et al., 1986; Suzuki et al., 1986) and mouse tissue (Kumada et al., 1988), and comparison revealed a high similarity in structure and function.

Different stages of physiologically important reactions are initiated by high affinity (K_D = 0.5 nM), active-site independent binding of thrombin to TM: formation of a 1:1 molar complex with the receptor which primarily is calcium-independent abrogates virtually all of the thrombin procoagulant activities such as fibrin formation, factor V activation (Esmon et al., 1982a), activation of platelets (Esmon et al. 1983), and activation of factor XIII (Polgar et al., 1986). While thrombin is not blocked at its active site, its specificity for physiological macromolecular substrates is changed and this change is greatly dependent on acidic domain(s) of rabbit TM (Bourin et al., 1986; Preissner et al., 1987). Initial comparison of rabbit lung and human placenta TM revealed that both receptors were not equivalent in their direct anticoagulant activity which also involves the 4-8 fold acceleration of thrombin inhibition by AT III. Human placenta TM was shown to contain additional contaminations by basic protein(s) which may mask the functionally relevant glycosaminoglycan moiety (Preissner et al., 1990). In fact, TM may be designated as proteoglycan bearing a major O-linked glycosaminoglycan, and in the following we will present evidence that most of TM's functions depend on this major carbohydrate moiety.

STRUCTURE-FUNCTION RELATIONSHIPS IN THROMBOMODULIN

The complex structure of TM was elucidated by cloning of the genes for the human, bovine, and mouse proteins (Wen et al., 1987; Jackman et al., 1986, 1987; Suzuki et al., 1987; Dittman et al., 1988). An 18-residue signal peptide is followed by a domain with structural homology to the animal lectin fold lectin domains (Patthy, 1988), six tandemly repeated domains homologous to epidermal growth factor (EGF domains), a serine/threonine-rich domain, a hydrophobic transmembrane domain and a cytoplasmic tail. The protein is highly conserved between species: 67% identical amino acids between mouse and bovine, 68% between mouse and human (Dittman et al., 1988). The human protein encoded by the TM cDNA sequence has a calculated M_r of 60,300 (Wen et al., 1987) compared to 75,000 (unreduced) and about 100,000 (reduced) of human TM isolated from placenta (Salem et al., 1984), suggesting that the mature protein may be extensively glycosylated.

Limited proteolysis of native rabbit TM and deletion mutagenesis of recombinant human TM have shown that the primary thrombin-binding site in TM lies within the 5th and 6th EGF domains (Kurosawa et al., 1988; Zushi et al., 1989). Peptides homologous to the acidic C-loop of the 5th EGF domain (residues 426-444) block thrombin binding to human TM (Hayashi et al., 1990). The primary thrombin-binding site in TM thus appears to be the 5th EGF domain. The minimal functional unit of TM with cofactor activity (i. e. the ability to bind thrombin and accelerate protein C activation) appears to be a segment comprising the 4th, 5th and 6th EGF domains (Hayashi et al., 1990; J.F. Parkinson, unpublished observations). Fragments smaller than this have extremely low cofactor activity which is difficult to distinguish kinetically from background activity in the absence of TM.

Glycosylation of Thrombomodulin

N-Linked Carbohydrates. Human TM has consensus sequences for five N-glycosylation sites at Asn^{47}, Asn^{115} and Asn^{116} in the lectin domain, Asn^{382} in the 4th EGF domain and Asn^{409} in the 5th EGF domain. The presence of carbohydrate at each of these sites was recently confirmed by tryptic peptide analysis of a secretable mutant of recombinant human TM (Parkinson et al., 1991a). The composition of these N-linked carbohydrates has recently been analyzed by its recognition with selective lectins (Koyama et al., 1991a), but their structure and function is not known, although a carbohydrate-dependent adhesive function for TM expressed at intercellular junctions during embryonal development has been proposed (Imada et al., 1990a; 1990b).

The O-Glycosylation Domain. Human TM also has eight serine and threonine residues within a putative "O-glycosylation" domain of 34 amino acid residues that separates the EGF domains from the transmembrane portion. Residues Ser^{492}, Ser^{498}, Thr^{500}, Thr^{504} and Thr^{506} were recently confirmed as being glycosylated in a secretable mutant of recombinant human TM (Parkinson et al., 1991a). A number of other cell surface glycoproteins have similar O-glycosylation domains: the low-density lipoprotein receptor (Südhof et al., 1985), decay accelerating factor (Medof et al., 1987) and platelet glycoprotein Ib (Lopez et al., 1987). Although the function of these O-glycosylation domains is not clear, some evidence suggests that they are important for protection against proteolysis (Kozarsky et al., 1988). These domains have also been proposed to adopt rigid,

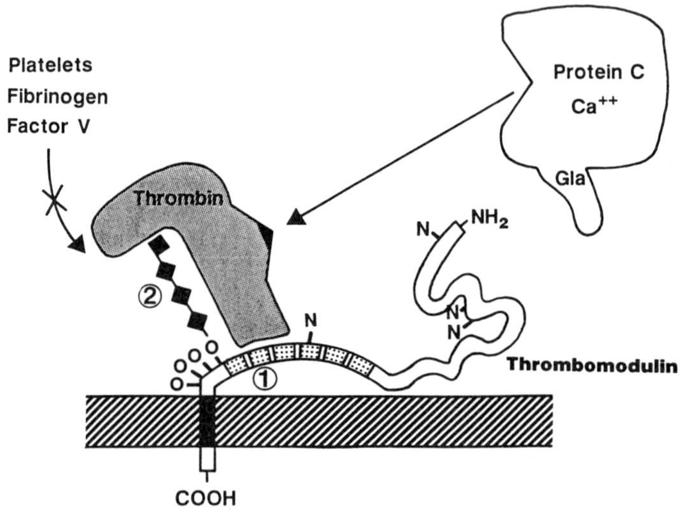

Fig. 1. Hypothetical structural model of thrombomodulin and its interaction with thrombin. The structure of intact (high M_r form) TM containing the EGF-like repeats as primary thrombin binding site (1) and the O-linked chrondroitin sulphate, representing the secondary binding site for the enzyme (2) is depicted. The acidic chondroitin sulphate moiety, which is responsible for TM direct anticoagulant function and which is absent in the low M_r form of TM, apparently interacts with the exo-site protein-binding region in thrombin, such that the enzyme may not reach its procoagulatory macromolecular substrates (e. g. platelets, fibrinogen, factor V). Rather, protein C is recognized as an anticoagulant substrate, which may also interact with the EGF-like repeats of TM via its γ-carboxy glutaminc acid domain (Gla) in a calcium-dependent manner. Residual O-linked glycans (O) in-accessible to chondroitinases and N-linked glycans (N) contain terminally linked sialic acid and other carbohydrate moieties whose spatial positioning may lead to internal stabilization of TM. (Taken from Koyama et al., 1991a)

extended conformations that can project the globular, ligand-binding domains of cell surface glycoproteins above the cell-surface glycocalyx (Jentoft, 1990). A length of 0.25 nm for each residue was proposed, indicating that for human TM the distance between the transmembrane domain and the 6th EGF domain could be as much as 8.5 nm. Fluorescence energy transfer techniques have shown that the distance between thrombin's active site and the membrane surface is approximately 6.5 nm when complexed with rabbit TM reconstituted into phospholipid vesicles (Lu et al., 1989). These observations support the concept that the O-glycosylation domain of TM is a structural motif which maintains the primary thrombin-binding site in the EGF domains at a distance from the cell surface where ligand-binding can be achieved without substantial interference from the glycocalyx.

Thrombomodulin and Chondroitin Sulphate. One important feature contained within the O-glycosylation domain of human TM is the sequence of Ser[490]-Gly-Ser-Gly-Glu-Pro which was proposed as a potential chondroitin sulphate attachment site (Bourin et al., 1988). The presence of chondroitin sulphate has been confirmed for TM solubilised from rabbit lung (Bourin et al., 1988; Bourin & Lindahl, 1990; Bourin et al., 1990) and human TM mutants expressed in mammalian cells (Parkinson et al., 1990a; Nawa et al., 1990) but so far not in native human TM. The chondroitin sulphate chain (10,000-12,000 molecular weight) from rabbit TM has been isolated and its di- and trisaccharide composition determined (Bourin et al., 1990). The internal regions of the chain were predominantly conventional mono-sulphated disaccharides. The non-reducing termini of the chains had an unusually high amount of the oversulphated trisaccharide GalNAc(diS)-GlcA-GalNAc(diS). The structure of the chondroitin sulphate chain(s) on recombinant human TM mutants are not known. Ser[492] appears to be the site of chondroitin sulphate attachment in the O-glycosylation domain of a portion of secreted recombinant human TM expressed in human 293 cells (Parkinson et al., 1991a). This residue is conserved in human, bovine and mouse TM (Dittman et al., 1988). Another portion of soluble TM mutants lack this major chondroitin sulphate despite the presence of smaller N- and O-linked side chains (Koyama et al., 1991b).

As shown in Table 1, the presence of chondroitin sulphate has dramatic effects on the functions of the thrombin-TM complex. Studies with rabbit TM showed that direct inhibition of thrombin's procoagulant functions (fibrinogen cleavage) and acceleration of AT-III-dependent thrombin inhibition were dependent on the presence of a chondroitinase ABC-sensitive glycosaminoglycan in TM (Bourin et al., 1988). Similar results were obtained in other laboratories using rabbit TM (Preissner et al., 1987; 1990) and with a secretable deletion mutant of recombinant human TM (Koyama et al., 1991b). In contrast to accelerating the slow rate of thrombin inhibition by AT III and HC II in the absence of exogenous glycosaminoglycans, the presence of chrondroitin sulphate in human TM inhibits rapid thrombin inhibition by AT III in the presence of heparin and by HC II in the presence of dermatan sulphate (Bourin, 1989; Koyama et al., 1991b). Results from these experiments and from data mentioned above led us to propose a model for TM and its interaction with thrombin (Fig. 1). High cooperativity between both the primary (1) and secondary (2) binding sites on TM invokes conformational change(s) in thrombin that results in efficient recognition of protein C but the inability of procoagulant substrates to reach the enzyme.

Thus, the presence of chondroitin sulphate in TM has been shown to increase inhibition of other thrombin-dependent effects, which include activation of factor V (Bourin and Lindahl, 1990), platelets (Parkinson et al., 1991a) and endothelial cells (Parkinson et al., 1991b). In addition, the

Table 1. Effects of Thrombomodulin on Thrombin Function

Thrombin Function	Effect
Protein C activation	Enhanced *
Fibrinogen cleavage	Inhibited *
Factor V activation	Inhibited *
Factor VIII activation	Inhibited *
Factor XIII activation	Inhibited *
Platelet activation	Inhibited *
Endothelial cell activation	Inhibited *
Inhibition by antithrombin III (AT III)	Enhanced *
Inhibition by heparin cofactor II (HC II)	Enhanced *
Inhibition by AT III-heparin	Inhibited *
Inhibition by HC II-dermatan sulphate	Inhibited *
Single-chain urokinase inactivation	Enhanced *
Other Functions	
Factor Xa-mediated prothrombin cleavage	Inhibited
Factor Xa-mediated protein C activation	Enhanced

* Dependent on the O-linked glycan of thrombomodulin

presence of chondroitin sulphate has profound effects on the kinetics of protein C activation by the thrombin-TM complex (Parkinson et al., 1990a). Two glycoforms of a single deletion mutant of recombinant human TM, TMD1-105 and TMD1-75, which differed only in the presence of chondroitin sulphate, had different affinities for thrombin, different maximal rates of protein C activation by the thrombin-TM complex and different calcium optima for protein C activation (Parkinson et al., 1991a) (Table 2). The biochemical and functional properties of TMD1-105 can be converted to those of TMD1-75 by chondroitinase ABC digestion, suggesting that these dramatic differences are entirely due to the presence/absence of chondroitin sulphate in the extracellular domains of human TM (Parkinson et al., 1991a; Koyama et al., 1991b). The isolated chondroitin sulphate chain from rabbit TM has very poor in vitro anticoagulant activity and does not exhibit appreciable AT III-binding properties (Bourin and Lindahl, 1990; Bourin et al., 1990).

POTENTIAL ROLES FOR CHONDROITIN SULPHATE IN REGULATION OF THROMBOMODULIN FUNCTION

The studies outlined above show that TM, a novel proteoglycan of the endothelial cell surface, requires an unusually high degree of cooperation between its protein and carbohydrate constituents in order to express full biological potency as an anticoagulant. Such a degree of cooperativity is almost without

Table 2. Protein C Activation Kinetics of the Thrombin-Thrombomodulin Complex

Kinetic Parameter	TMD1-105	TMD1-75
K_d for thrombin (nM)	2.7	16.3
K_m for protein C (μM)	1.5	2.3
K_{cat} (sec^{-1})	9.2	3.3
K_{cat}/K_m ($\mu M^{-1}sec^{-1}$)	6.3	1.4
Calcium optimum (mM)	3.0	0.3

precedent in biological systems and begs the question as to why TM has been designed in this way. Since thrombin is the target of TM, perhaps some of these questions can be answered by an understanding of thrombin structure and function.

Human thrombin, a plasma serine protease, has been crystallized and its structure solved by X-ray diffraction (Bode et al., 1989; Rydel et al., 1990). In addition to its restricted active site, thrombin contains an extended hydrophobic groove which is flanked by basic arginine and lysine residues. This groove, referred to as the anion-binding exosite (Fenton, 1986), is critical for thrombin-hirudin binding interactions and for determining thrombin's macromolecular specificity (Rydel et al., 1990). Several studies have shown that TM, fibrinogen, hirudin and hirudin peptides compete for the same or overlapping binding site on thrombin (Hofsteenge and Stone, 1987). Cloning of the platelet/endothelial cell thrombin receptor has revealed structural elements also capable of interacting with the anion-binding exosite (Vu et al., 1991). Both hirudin and recombinant TM inhibit thrombin receptor-mediated platelet and endothelial cell activation (Jakubowski and Maraganore, 1990; Prescott et al., 1990; Parkinson et al., 1991a, 1991b). All of these studies serve to emphasize the critical role that the anion-binding site of thrombin plays in directing thrombin's bio-regulatory functions in thrombosis, haemostasis and cellular activation (Fenton, 1986).

The studies outlined above indicate that the ability of TM to act as a regulator of thrombin function is dependent on effective competition with ligands (e. g. fibrinogen, thrombin receptors) for binding to thrombin's anion-binding exosite. Tables 1 and 2 show that the presence of chondroitin sulphate in the O-glycosylation domain is a critical regulatory component of the thrombin-TM complex. This is true for direct inhibition of thrombin's interactions with procoagulant substrates/cellular receptors and for the kinetics of protein C activation by the thrombin-TM complex. Removal of chondroitin sulphate decreases the affinity of the thrombin-TM interaction and results in TM being a poor inhibitor of thrombin binding to its exosite-dependent substrates and a less effective cofactor for protein C activation. The glycosylation state of TM in vivo, particularly the presence or absence of chondroitin sulphate, is thus likely to have profound effects on its function. In support of this concept, recent studies with cultured human and bovine endothelial cells have shown that the thrombin affinity of cell surface TM and the kinetics of protein C activation at the cell surface are decreased after treatment of cells with ß-D-xyloside, an inhibitor of chondroitin sulphate attachment (Parkinson et al., 1990b). These studies indicate that cell surface TM is probably modified by chondroitin sulphate and its absence decreases the anticoagulant/antithrombotic potential of cell surface TM. In addition, the thrombin-binding phenotype of cell surface TM has been shown to depend on the culture conditions for endothelial cell growth (Ma et al., 1991). Cells grown in the presence of heparin had a high-affinity phenotype which was converted to a low-affinity phenotype by ß-D-xyloside treatment, and conversely, cells grown in the absence of heparin had a low-affinity phenotype which was insensitive to ß-D-xyloside treatment. The effects of heparin on these kinetic phenotypes were entirely reversible. These studies suggest that heparin stimulates attachment of chondroitin sulphate to cell surface TM in endothelial cells and are indicative for the concept that attachment of chondroitin sulphate to TM can be modified by exogenous mediators, providing a mechanism for regulating TM function in a tissue- or organ-specific manner.

An additional mechanism for regulation of TM function at the endothelial cell surface is the neutralisation of the chondroitin sulphate chain either by glycosaminoglycan-binding proteins or by its removal via glycosidase activity (e. g. chonroitinase). As yet, there is no evidence for the second of these two mechanisms but there is precedence for the first. Several glycosaminoglycan-binding proteins, polycations and basic proteins have been shown to inhibit the ability of TM to block thrombin-dependent fibrinogen cleavage (Bourin et al., 1986, 1988; Preissner et al., 1990; Koyama et al., 1991b). Physiologically, most relevant of these is likely to be platelet factor 4 (PF4) which is a highly basic protein secreted from activated platelets. The heparin-neutralising effect of PF4 has long been recognized as a potential prothrombotic event at the endothelial cell surface following local platelet activation (Lane et al., 1986). The possibility that PF4 secretion from activated platelets can also cause rapid inhibition of cell surface TM function via neutralisation of chondroitin sulphate should also be considered as an prothrombotic activity of platelet releasate.

Taken together, recent studies on the role of the major chondroitin sulphate moiety of TM revealed that this acidic part of the thrombin receptor fulfills critical functions to control the haemostasis system. It remains to be established which (cell-specific ?) mechanisms govern the post-transitional modifications (i.e. structure of the carbohydrate attached) of TM and whether always the same attachment site of the core protein is being used. Moreover, with knowledge of TM gene structure it is expected that certain dysfunctional forms of TM will be detected, that provide a correlation with possible thromboembolic risk in the affected patients.

REFERENCES

Bertina, R. M., 1988, "Protein C and related proteins", Churchill Livingstone, Edingburgh, Melbourne, New York.

Bode, W., Mayr, I., Baumann, U., Huber, R., Stone, S. T., and Hofsteenge, J., 1989, The refined 1.9 Å crystal structure of human α-thrombin: interaction with D-Phe-Pro-Arg chloromethylketone and significance of the Tyr-Pro-Pro-Trp insertion segment, EMBO J., 8:3467.
Bourin, M.-C., 1989, Effect of rabbit thrombomodulin on thrombin inhibition by antithrombin in the presence of heparin, Thromb. Res., 54:27.
Bourin, M.-C., Lundgren-Akerlund, E., and Lindahl, U., 1990, Isolation and characterization of the glycosaminoglycan component of rabbit thrombomodulin proteoglycan, J. Biol. Chem., 265:15424.
Bourin, M.-C., Boffa, M. C., Björk, I., and Lindahl, U., 1986, Functional domains of rabbit thrombomodulin, Proc. Natl. Acad. Sci. USA, 83:5924.
Bourin, M.-C., and Lindahl, U., 1990, Functional-role of the polysaccharide component of rabbit thrombomodulin proteoglycan. Effects on inactivation of thrombin by antithrombin, cleavage of fibrinogen by thrombin and thrombin-catalyzed activation of factor-V, Biochem. J., 270:419.
Bourin, M.-C., Öhlin, A. K., Lane, D. A., Stenflo, J., and Lindahl, U., 1988, Relationship between anticoagulant activities and polyanionic properties of rabbit thrombomodulin, J. Biol. Chem., 263:8044.
DeBault, L. E., Esmon, N. L., Olson, J. R., and Esmon, C. T., 1986a, Distribution of the thrombomodulin antigen in the rabbit vasculature, Lab. Invest., 54:172.
DeBault, L. E., Esmon, N. L., Smith, G. P., and Esmon, C. T., 1986b, Localization of thrombomodulin antigen in rabbit endothelial cells in culture. An immunofluorescence and immunoelectron microscope study, Lab. Invest., 54:179.
Esmon, C. T., 1987, The regulation of natural anticoagulant pathways, Science, 235:1348.
Esmon, C. T., Esmon, N. L., and Harris, K. W., 1982a, Complex formation between thrombin and thrombomodulin inhibits both thrombin-catalyzed fibrin formation and factor V activation, J. Biol. Chem., 257:7944.
Esmon, C. T., Esmon, N. L., and Saugstad, J., 1982b, Activation of protein C by a complex between thrombin and endothelial cell surface protein, in: "Pathobiology of the Endothelial Cell", H. Nossel, ed., Academic Press, New York.
Esmon, C. T., and Owen, W. G., 1981, Identification of an endothelial cell cofactor for thrombin-catalyzed activation of protein C, Proc. Natl. Acad. Sci. USA, 78:2249.
Esmon, N. L., Carroll, R. C., and Esmon, C. T., 1983, Thrombomodulin blocks the ability of thrombin to activate platelets, J. Biol. Chem., 258:12238.
Esmon, N. L., Owen, W. G., and Esmon, C. T., 1982c, Isolation of a membrane-bound cofactor for thrombin-catalyzed activation of protein C, J. Biol. Chem., 257:859.
Fenton, II, J. W., 1986, Thrombin, Ann. N. Y. Acad. Sci., 485:5.
Hayashi, T., Zushi, M., Yamamoto, S., and Suzuki, K., 1990, Further localization of binding sites for thrombin and protein C in human thrombomodulin, J. Biol. Chem., 265:20156.
Hofsteenge, J., and Stone, S. R., 1987, The effect of thrombomodullin on the cleavage of fibrinogen and fibrinogen fragments by thrombin, Eur. J. Biochem., 168:49.
Imada, M., Imada, S., Iwasaki, H., Kume, A., Yamaguchi, H., and Moore, E. E., 1987, Fetomodulin: Marker surface protein of

fetal development which is modulatable by cyclic AMP, Dev. Biol., 122:483.
Imada, S., Yamaguchi, H., and Imada, M., 1990a, Differential expression of fetomodulin and tissue plasminogen activator to charactrise parietal endoderm differentiation of F9 embryonal carcinoma cells, Dev. Biol., 141:426.
Imada, S., Yamaguchi, H., Nagumo, M., Katayanagi, S., Iwasaki, H., and Imada, M., 1990b, Identification of fetomodulin, a surface marker protein of fetal development, as thrombomodulin by gene cloning and functional assays, Dev. Biol., 140:113.
Ishii, H., Salem, H. H., Bell, C. E., Laposata, E. A., and Majerus, P. W., 1986, Thrombomodulin, an endothelial anticoagulant protein, is absent from the human brain, Blood, 67:362.
Jakubowski, H. V., Kline, M. D., and Owen, W. G., 1986, The effect of bovine thrombomodulin on the specificity of bovine thrombin.e, J. Biol. Chem., 261:3876.
Jakubowski, J. A., and Maraganore, J. M., 1990, Inhibition of coagulation and thrombin-induced platelet activities by a synthetic dodecapeptide modeled on the carboxy-terminus of hirudin, Blood, 75:399.
Jentoft, N., 1990, Why are proteins O-glycosylated ?, Trends Biochem. Sci., 15:291.
Koyama, T., Parkinson, J. F., Aoki, N., Bang, N. U., Müller-Berghaus, G., and Preissner, K. T., 1991a, Relationship between post-translational glycosylation and anticoagulant function of secretable recombinant mutants of human thrombomodulin, Br. J. Haematol., 78: (in press).
Koyama, T., Parkinson, J. F., Sié, P., Bang, N. U., Müller-Berghaus, G., and Preissner, K. T., 1991b, Different glycoforms of human thrombomodulin: Their glycosaminoglycan-dependent modulatory effects on thrombin inactivation by heparin cofactor II and antithrombin III, Eur. J. Biochem., 198:563.
Kozarsky, K., Kingsley, D., and Krieger, M., 1988, Use of mutant cell line to study the kinetics and function of O-linked glycosylation of low density lipoprotein receptors, Proc. Natl. Acad. Sci. USA, 85:4335.
Kumada, T., Dittman, W. A., and Majerus, P. W., 1988, A role for thrombomodulin in the pathogenesis of thrombin-induced thromboembolism in mice, Blood, 71:728.
Kurosawa, S., and Aoki, N., 1985, Preparation of thrombomodulin from human placenta, Thromb. Res., 37:353.
Kurosawa, S., Stearns, D. J., Jackson, K. W., and Esmon, C. T., 1988, A 10-kDa cyanogen bromide fragment from the epidermal growth factor homology domain of rabbit thrombomodulin contains the primary thrombin binding site, J. Biol. Chem., 263:5993.
Lane, D. A., Pejler, G., Flynn, A. M., Thompson, E. A., and Lindahl, U., 1986, Neutralization of heparin-related saccharides by histidine-rich glycoprotein and platelet factor 4, J. Biol. Chem., 261:3980.
Lollar, P., and Owen, W. G., 1980, Clearance of thrombin from circulation of rabbits by high-affinity binding sites on endothelium: possible role in the inactivation of thrombin by antithrombin III, J. Clin. Invest., 66:1222.
Lopez, J. A., Chung, D. W., Fujikawa, K., Hegan, F. S., Papayannopolou, T., and Roth, G. J., 1987, Cloning of the α-chain of human platelet glyocprotein Ib: A transmembrane protein with homology to leucine-rich α_2-glycoprotein,

Proc. Natl. Acad. Sci. USA, 84:5615.
Lu, R., Esmon, N. L., Esmon, C. T., and Johnson, A. E., 1989, The active site of the thrombin-thrombomodulin complex, J. Biol. Chem., 264:12956.
Ma, S.-F., Parkinson, J. F., Garcia, J. G. N., and Bang, N. U., 1991, The thrombin affinity of endothelial thrombomodulin is modulated by exogenous heparin, Thromb. Haemostas., 65:946.
Maruyama, I., Bell, C. E., and Majerus, P. W., 1985, Thrombomodulin is found on endothelium of arteries, veins, capillaries, and lymphatics, and on syncytiotrophoblast of human placenta, J. Cell Biol., 101:363.
Maruyama, I., and Majerus, P. W., 1985, The turnover of thrombin-thrombomodulin complex in cultured human umbilical vein endothelial cells and A549 lung cancer cells, J. Biol.Chem., 260:15432.
Medof, M. E., Lublin, D. M., Holers, V. M., Ayers, D. J., Getty, R. R., Leykam, J. F., Atkinson, J. P., and Tykocinski, M., 1987, Cloning and characterization of cDNAs encoding the complete sequence of decay-accelerating factor of human complement, Proc. Natl. Acad. Sci. USA, 84:2007.
Nawa, K., Sakano, K., Fujiwara, H., Sato, Y., Sugiyama, N., Teruuchi, T., Iwamoto, M., and Marumoto, Y., 1990, Presence and function of chondroitin-4-sulfate on recombinant human soluble thrombomodulin, Biochem. Biophys. Res. Commun., 17:729.
Owen, W. G., and Esmon, C. T., 1981, Functional properties of an endothelial cell cofactor for thrombin-catalyzed activation of protein C, J. Biol. Chem., 256:5532.
Parkinson, J. F., Bang, N. U., and Garcia, J. G. N., 1991b, Recombinant human thrombomodulin attenuates the inflammatory response of thrombin-stimulated human endothelial cells, Thromb. Haemostas., 65:825.
Parkinson, J. F., Garcia, J. G. N., and Bang, N. U., 1990b, Decreased thrombin affinity of cell-surface thrombomodulin following treatment of cultured endothelial cells with ß-D-xyloside, Biochem. Biophys. Res. Commun., 169:177.
Parkinson, J. F., Grinnell, B. W., Moore, R. E., Hoskins, J., Vlahos, C. J., and Bang, N. U., 1990a, Stable expression of a secretable deletion mutant of recombinant human thrombomodulin in mammalian cells, J. Biol. Chem., 265:12602.
Parkinson, J. F., Vlahos, C. J., Yan, S. C. B., and Bang, N. U., 1991a, Recombinant human thrombomodulin: Regulation of cofactor activity and anticoagulant function by chondroitin sulphate, Thromb. Haemostas., 65:874.
Patthy, L., 1988, Detecting distant homologies of mosaic proteins. Analysis of the sequences of thrombomodulin, thrombospondin complement components C9, C8α and C8ß, vitronectin and plasma cell membrane glycoprotein PC-1, J. Mol. Biol., 202:689.
Polgar, J., Lerant, I., Muszbek, L., and Machovich, R., 1986, Thrombomodulin inhibits the activation of factor XIII by thrombin, Thromb. Res., 43:585.
Preissner, K. T., 1990, Physiological role of vessel wall related antithrombotic mechanisms: Contribution of endogenous and exogenous heparin-like components to the anticoagulant potential of the endothelium, Haemostasis, 20 (Suppl. 1):30.
Preissner, K. T., Delvos, U., and Müller-Berghaus, G., 1987,

Binding of thrombin to thrombomodulin accelerates inhibition of the enzyme by antithrombin III. Evidence for a heparin-independent mechanism, Biochemistry, 26:2521.
Preissner, K. T., Koyama, T., Tschopp, J., and Müller-Berghaus, G., 1990, Domain structure of the endothelial cell receptor thrombomodulin as deduced from modulation of its anticoagulant functions. Evidence for a glycosamino-glycan-dependent secondary binding site for thrombin, J. Biol. Chem, 265:4915.
Prescott, S. M., Seeger, A. R., Zimmerman, G. A., McIntyre, T. M., and Maraganore, J. M., 1990, Hirudin-based peptide block the inflammatory effects of thrombin on endothelial cells, J. Biol. Chem., 265:9614.
Rydel, T. J., Ravichandran, K. G., Tulinsky, A., Bode, W., Huber, R., Roitsch, C., and Fenton, J. W., 1990, The structure of a complex of recombinant hirudin and human α-thrombin, Science, 249:277.
Salem, H. H., Maruyama, I., Ishii, H., and Majerus, P. W., 1984, Isolation and characterization of thrombomodulin from human placenta, J. Biol. Chem., 259:12246.
Soff, G. A., Jackman, R. W., and Rosenberg, R. D., 1991, Expression of thrombomodulin by smooth muscle cells in culture: Different effects of tumor necrosis factor and cyclic adenosine monophosphate on thrombomodulin expression by endothelial cells and smooth muscle cells in culture, Blood, 77:515.
Suzuki, K., Kusumoto, H., and Hashimoto, S., 1986, Isolation and characterization of thrombomodulin from bovine lung, Biochim. Biophys. Acta, 882:343.
Südhof, T. C., Goldstein, J. L., Brown, M. S., and Russell, D. W., 1985, The LDL receptor gene: A mosaic of exons shared with different proteins, Science, 228:815.
Vu, T.-K. V., Hung, D. T., Wheaton, V. I., and Coughlin, S. R., 1991, Molecular cloning of a functional thrombin receptor reveals a novel proteolytic mechanism of receptor activation, Cell, 64:1057.
Wen, D., Dittmann, W. A., Ye, R. D., Deaven, L. L., Majerus, P. W., and Sadler, J. E., 1987, Human thrombomodulin: complete cDNA sequence and chromosome localization of the gene, Biochemistry, 26:4350.
Wong, V. L. Y., Bready, J., Berliner, J., Cancilla, P. A., and Fisher, M., 1991, Thrombomodulin expression by bovine brain capillary endothelial cells in monolayers and capillary-like structures in vitro, Thromb. Haemostas., 65:947.
Zushi, M., Gomi, K., Yamamoto, S., Maruyama, I., Hayashi, T., and Suzuki, K., 1989, The last three consecutive epidermal growth factor-like structures of human thrombomodulin comprise the minimum functional domain for protein C-activating cofactor activity and anticoagulant activity, J. Biol. Chem., 264:10351.

THE INTERACTION BETWEEN LACI AND HEPARIN

George J. Broze, Jr., Robin Wesselschmidt, Darryl Higuchi, Thomas Girard, Karen Likert, Louise MacPhail, and T.-C. Wun

Division of Hematology/Oncology
Jewish Hospital at Washington University Medical Center
216 So. Kingshighway Blvd.
St. Louis, MO U.S.A.

Monsanto Corporate Research
Chesterfield, MO U.S.A.

INTRODUCTION

Lipoprotein-associated coagulation inhibitor (LACI) is a multivalent protease inhibitor that directly inhibits factor Xa and, in a factor Xa dependent fashion, inhibits the factor VIIa/Tissue Factor (TF) catalytic complex (1). Its structure is unique: An acidic amino-terminal domain is followed by three tandem Kunitz-type protease inhibitor domains and a basic carboxyl-terminal region (Fig. 1) (2). The Kunitz-2 domain in LACI is needed for inhibition of factor Xa, while the Kunitz-1 domain is required for the inhibition of the factor VIIa in a factor Xa-LACI-factor VIIa/TF quaternary inhibitory complex (3). The function of the Kunitz-3 domain in LACI is not known. In human blood, LACI circulates predominantly in association with the plasma lipoproteins but it is also carried in platelets which release LACI following stimulation by thrombin (4, 5). A third pool of LACI is thought to be bound to glucosaminoglycans at the surface of the endothelium and is released into plasma following the infusion of heparin in vivo (6, 7).

LACI is a slow, tight-binding, competitive, and reversible inhibitor of factor Xa (8). Typical of Kunitz-type protease inhibitors, its mechanism of action appears to conform to the kinetic model:

$$E + I \rightleftarrows EI \rightleftarrows EI^*$$

where E=enzyme, I=inhibitor, and EI represents an immediate encounter complex described by a Ki(initial), with slowly isomerizes to a final tightened complex EI* characterized by a Ki(final).

LACI binds to heparin-agarose and heparin reportedly enhances the inhibition of factor Xa by LACI (8). Recent studies in our laboratory have helped to define better the interaction between heparin and LACI.

ROLE OF THE CARBOXYL TERMINUS OF LACI

Recombinant LACI (rLACI), immunoaffinity purified from the conditioned media of mouse C127 cells, separates into two protein peaks that elute at 0.3 M and 0.6 M NaCl following heparin-agarose chromatography (Fig. 2). Western blotting, using specific antibodies against the amino-terminus of LACI (residues 1-10) and against the carboxyl-terminus of LACI (residues 265-276), shows that rLACI(0.6) is full-length, whereas rLACI(0.3) has been truncated at its carboxy-

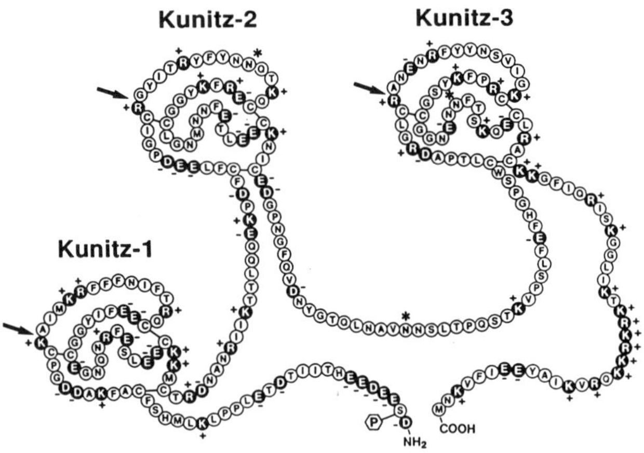

Fig. 1. Schematic diagram of the predicted secondary structure of LACI The arrows indicate the location of the P_1-P_1' residues of the active-site clefts for each Kunitz domain. Filled symbols identify charged residues (histidine is considered uncharged). Asterisks show the potential sites for N-linked glycosylation. (Reprinted by permission from Nature 338:519, copyright 1991, Macmillan Magazines, Ltd.)

terminus, presumably through proteolysis (data not shown). A rLACI, truncated following Leu252 and thus lacking the basic sequence K T K R K R K K Q R V K, was also expressed in mouse C127 cells. This rLACI(1-252) elutes from heparin-agarose at a slightly lower NaCl concentration (0.28 M) than rLACI(0.3).

Each form of LACI binds to factor Xa with 1:1 stoichiometry and they inhibit factor VIIa/TF to the same extent in an endpoint assay. Full-length rLACI(0.6), however, is a considerably more potent inhibitor than either form of truncated rLACI [rLACI(0.3) and rLACI(1-252)] of coagulation in plasma induced by factor Xa, TF, or the factor X coagulant protein from Russell's Viper venom (XCP) (Table 1). Examination of the rate of inhibition of factor Xa by rLACI(0.3) and rLACI(0.6) by coagulation assay shows that the initial degree (<5 seconds) of factor Xa

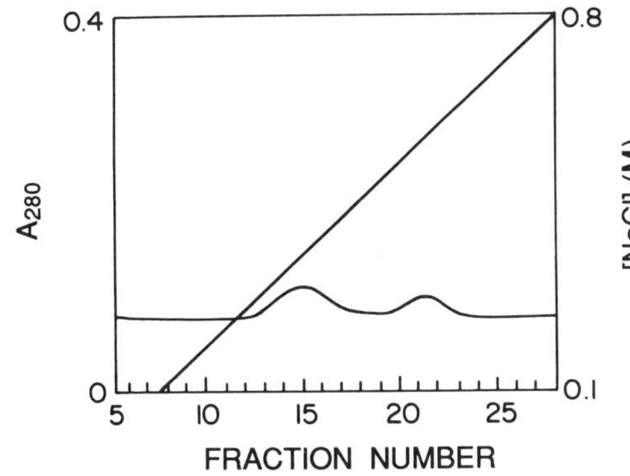

Fig. 2. Heparin agarose chromatography of rLACI. The A280 and NaCl gradient are displayed.

inhibition by rLACI(0.6) is several fold greater than that of rLACI(0.3) (Fig. 3). Studies using the continuous cleavage of a chromogenic substrate (Spectrozyme Xa) to measure remaining factor Xa activity confirm that the initial rates of factor Xa inhibition by the truncated rLACI's, rLACI(0.3) and rLACI(1-252), are inferior to that of full-length rLACI(0.6) and further demonstrate a disparate response to heparin (1 U/mL) for the inhibition of factor Xa by truncated and full-length rLACI (Fig. 4). While heparin enhances the rate of inhibition produced by rLACI(0.3) and rLACI(1-252), it appears to slow the inhibition produced by rLACI(0.6). The Ki(final)'s, determined using the method of Bieth (9), for rLACI(0.3), rLACI(1-252), and rLACI(0.6) against human factor Xa are 19.6 ± 0.8, 19.6 ± 3.0, and 3.1 ± 0.6 pM respectively.

CLEAVAGE OF LACI BY HUMAN LEUKOCYTE ELASTASE

Incubation of rLACI, containing a mixture of full-length, carboxy-terminal truncated, and possibly other forms, with a 1:60 molar concentration of human leukocyte elastase (HLE) produces limited proteolytic cleavage in the LACI molecule. SDS polyacrylamide gel electrophoretic analysis of samples from the reaction shows the time-dependent reduction in size of the major rLACI band from 42,000 to

Table 1

Inhibition of Coagulation in Plasma by rLACI (%)

	rLACI (0.6)		rLACI (0.3)		rLACI (1-252)	
	1 nM	4 nM	1 nM	4 nM	1 nM	4 nM
Factor Xa	15	46	0	6	0	5
Tissue Factor	20	59	0	6	0	5
XCP	23	61	0	8	0	6

Percent apparent inhibition of factor Xa, tissue factor, and XCP induced coagulation of plasma by the forms of rLACI (see text). Each rLACI was tested at concentrations of 1 and 4 nM.

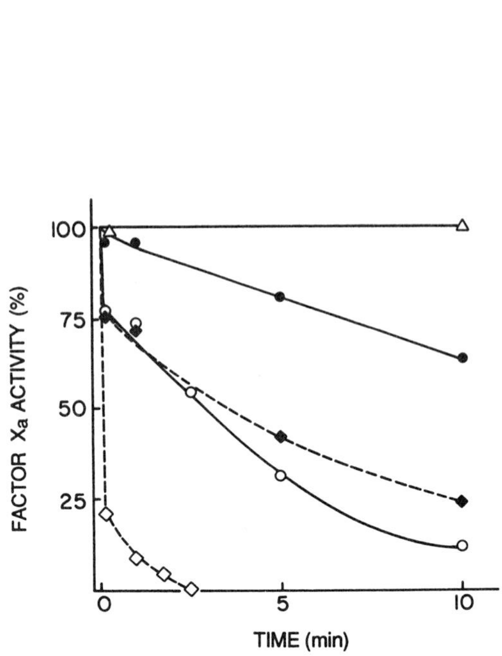

Fig. 3. Inhibition of factor Xa by rLACI. Control (no LACI), Δ – Δ; rLACI(0.3) 1 nM, ● – ●; rLACI(0.3) 4 nM O – O; rLACI(0.6) 1 nM, ◆ --- ◆; and rLACI(0.6) 4 nM, ◇ --- ◇ .

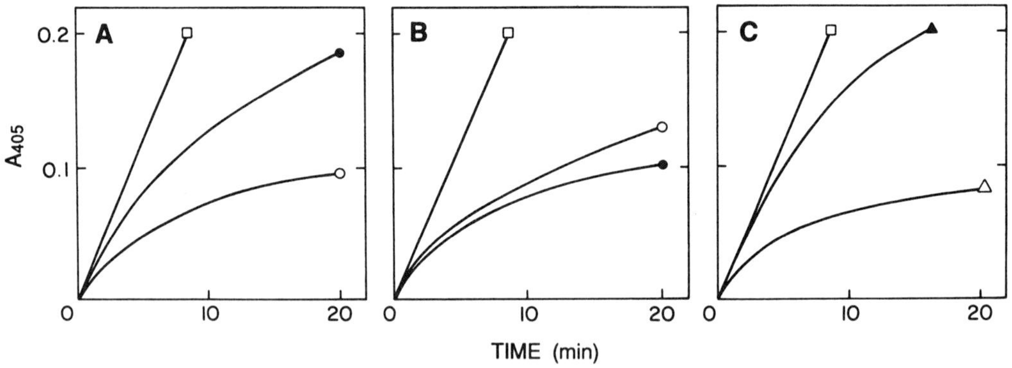

Fig. 4. Inhibition of factor Xa by rLACI in a continuous chromogenic assay. Reactions were initiated by adding human factor Xa (0.5 nM final concentration) to rLACI (2.5 nM) and Spectrozyme Xa (100 μM). In each panel □ denotes cleavage of the substrate in the absence of rLACI. A) rLACI(0.3), ●; rLACI(0.6), O. B) Same as A, except the reactions contained heparin (1 unit/ml). C) rLACI(1-252) in the absence (▲) and presence (Δ) of heparin (1 unit/ml).

34,000 MW with the simultaneous appearance of lower molecular weight bands with apparent molecular weights between 8,500 and 12,500. The polyclonal antibody raised against the amino-terminus of LACI (residues 1-10) recognizes the low molecular weight cleavage products but fails to detect the higher molecular weight product at 34,000 MW (data not shown). The amino-terminal sequence of the latter protein band is X L Q Q E K P D F X F L E E D P G I X R (X = indeterminate) which matches residues 88-107 of the predicted LACI amino acid sequence and is consistent with HLE cleavage occurring between amino acids threonine-87 and threonine-88 in LACI. As indicated by the Western blot result, the low molecular weight products have the same amino-terminal sequence as full-length LACI. Whether the different sizes of the low molecular weight bands arise from additional sites of HLE cleavage, or are related to differential post-translational modifications within the portion of the LACI molecule between residues 1-87 has not been determined.

HLE cleavage of rLACI is associated with a loss in the ability of rLACI to inhibit factor VIIa/TF in the presence of factor Xa (Fig. 5). Whereas, HLE treatment also reduces the ability of rLACI to inhibit factor Xa it had little effect on the inhibition of trypsin by rLACI, which is also mediated predominantly through the Kunitz-2 domain (Crecelius and Broze, unpublished) (Fig. 5). Progress curves of the factor Xa/rLACI interaction using the chromogenic substrate Spectrozyme Xa show that the initial degree of factor Xa inhibition produced by rLACI and HLE treated rLACI are the same (Fig. 6). This suggests that HLE induced cleavage does not effect the affinity of the initial encounter complex, EI, between factor Xa and LACI. Progressive inhibition of factor Xa by the HLE treated rLACI (EI → EI*), however, is markedly reduced as compared to that produced by untreated rLACI (Fig. 6). When progress curves are performed in the presence of 1 U/mL heparin, the initial degree of factor Xa inhibition is increased in a comparable fashion for both forms of rLACI, whereas progressive factor Xa inhibition is again markedly deficient for the HLE cleaved material.

DISCUSSION

These studies demonstrate that the carboxy-terminus of LACI is required for optimal inhibition of factor Xa. The relative affinities of rLACI(1-252), rLACI(0.3), and rLACI(0.6) for heparin-agarose parallel the rates at which they inhibit factor Xa. This suggests that both of these properties are related to the number of basic amino acids retained at the carboxy-terminus of the LACI molecule. It also implies that it is the highly basic sequence within the carboxy-terminus that may be required for the enhancement of the rate of factor Xa inhibition.

The greater factor Xa inhibition produced immediately (<5 sec.) following the addition of rLACI(0.6), as opposed to rLACI(0.3), to factor Xa (Fig. 3), suggests that the difference in the ability of these forms of rLACI to inhibit factor Xa-induced coagulation is in large part due to a lower Ki(initial) for the immediate encounter complex of full-length LACI with factor Xa. The mechanism for this effect remains to be defined, but could be related to a direct interaction between factor Xa and the carboxy-terminus of LACI. Alternatively, it may involve a particular conformation of the LACI molecule, which requires the presence of the carboxy-terminus. In the latter instance, it is tempting to speculate that an intramolecular interaction of the positively charged region of the carboxy-terminus with either the negatively charged amino-terminus of LACI, or with a sulfated oligosaccharide in LACI could be operative (10, 11).

Although the carboxy-terminus of LACI is required for high affinity binding to heparin-agarose, additional sites in LACI interact with heparin since carboxy-terminal truncated LACI also binds to heparin-agarose, albeit with lower affinity. Further, heparin enhances the inhibition of factor Xa by carboxy-terminal truncated LACI. This could be due to an effect of heparin on factor Xa, LACI, or both. Preliminary experiments suggest that heparin may increase the interaction between factor Xa and truncated LACI by serving as a template. The mechanism for the paradoxical reduction in the inhibition of factor Xa by full-length LACI induced by heparin requires further study.

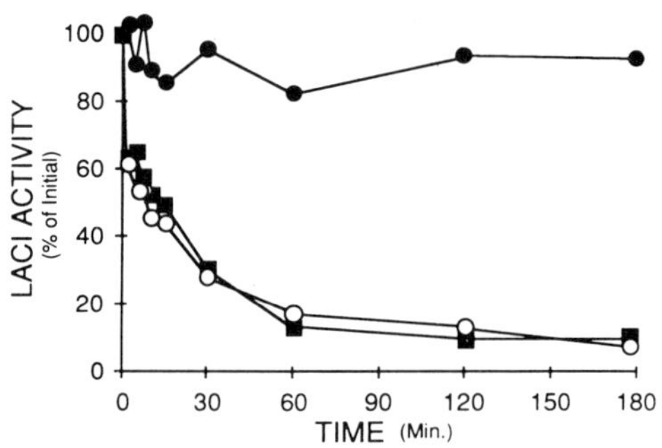

Fig. 5. Effect of HLE treatment on rLACI function. rLACI (3.3 μM) and HLE (53 nM) were incubated at room temperature. Samples were removed at the specified times, treated with methoxy-O-succinyl-Ala-Ala-Pro-Val-chloromethyl ketone (45 μM) to inactivate HLE, and then tested for LACI functional activity. Trypsin inhibition, ●; factor Xa inhibition, O; factor VIIa/TF inhibition in the presence of factor Xa, ■.

HLE separates the amino-terminal and Kunitz-1 domains of LACI from the remainder of the molecule by cleaving at threonine-87. Treatment of LACI with HLE greatly reduces the overall affinity [Ki(final)] of the factor Xa/LACI complex. The affinity [Ki(initial)] of the initial encounter complex between factor Xa and LACI is not affected by this HLE induced cleavage in LACI, and further it is enhanced in a comparable fashion by heparin for both LACI and HLE treated LACI. Thus, if an interaction between a region in LACI and heparin is involved in this enhancing effect of heparin, the epitope in LACI lies distal to threonine-88.

In sum, these studies show that more than simply the Kunitz-2 domain in LACI is required for optimal and effective inhibition of factor Xa. A variable extent of carboxy-terminal truncation has been demonstrated in the rLACI expressed by several host cell lines (unpublished data). Some degree of carboxy-terminal proteolysis is common in the rLACI present in conditioned media, and the carboxy-terminal truncated forms of rLACI frequently increase during the purification procedure itself. Whether some carboxy-terminal proteolysis of LACI occurs during the intracellular processing of the LACI molecule is not known. Whatever its source, it is clear that proteolytic truncation of the molecule significantly affects its function and must be considered when assessing the in vitro or in vivo properties of rLACI.

These studies also show that the inhibitory activity of LACI, like several other coagulation inhibitors (antithrombin III, heparin cofactor

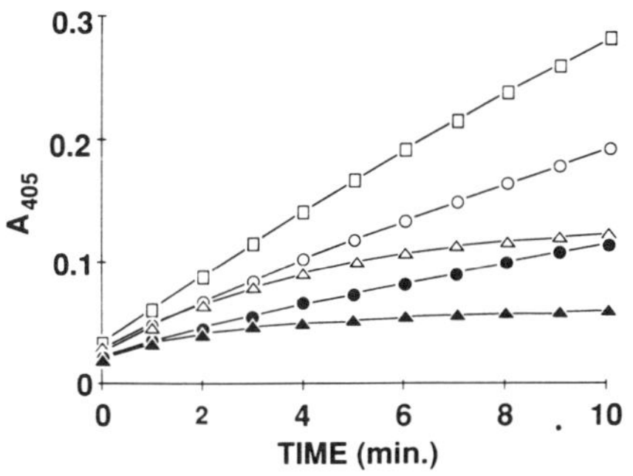

Fig. 6. Continuous chromogenic assay of factor Xa inhibition by untreated and HLE treated rLACI. Factor Xa (0.5 nM) was added to rLACI (5 nM) and Spectrozyme Xa (100 µM). Factor Xa with no added rLACI, □; HLE cleaved rLACI, O; HLE cleaved rLACI with heparin (1 unit/mL), ●; untreated rLACI, Δ; untreated rLACI with heparin (1 unit/mL), ▲.

II, C1 esterase inhibitor, and α_2-antiplasmin), is destroyed by human leukocyte elastase (12-17). Whether this effect of HLE contributes to the local activation of coagulation at sites of inflammation is not known.

REFERENCES

1. G. J. Broze, Jr., L. A. Warren, W. F. Novotny, D. A. Higuchi, J. J. Girard, and J. P. Miletich. The lipoprotein-associated coagulation inhibitor that inhibits the factor VII-tissue factor complex also inhibits factor Xa: Insight into its possible mechanism of action. Blood 71:335 (1988).

2. T.-C. Wun, K. K. Kretzmer, T. J. Girard, J. P. Miletich, G. J. Broze, Jr. Cloning and characterization of a cDNA coding for the lipoprotein-associated coagulation inhibitor shows that it consists of three tandem Kunitz-type inhibitory domains. J. Biol. Chem. 263:6001 (1988).
3. T. J. Girard, L. A. Warren, W. F. Novotny, K. M. Likert, S. G. Brown, J. P. Miletich, and G. J. Broze, Jr. Functional signficance of the Kunitz-type inhibitor domains of lipoprotein-associated coagulation inhibitor. Nature 338:518 (1989).
4. W. F. Novotny, T. J. Girard, J. P. Miletich, and G. J. Broze, Jr. Purification and characterization of the lipoprotein-associated coagulation inhibitor from huma plasma. J. Biol. Chem. 264:18832 (1989).
5. W. F. Novotny, T. J. Girard, J. P. Miletich, and G. J. Broze, Jr. Platelets secrete a coagulation inhibitor functionally and antigenically similar to the lipoprotein-associated coagulation inhibitor. Blood 72:2020 (1988).
6. P. M. Sandset, U. Abildgaard, and M. L. Larsen. Heparin induces release of extrinsic coagulation pathway inhibitor (EPI). Thromb. Res. 50:803, 1988.
7. W. F. Novotny, S. G. Brown, J. P. Miletich, D. J. Rader, and G. J. Broze, Jr. Plasma antigen levels of the lipoprotein-associated coagulation inhibitor in patient samples. Blood 78:387 (1991).
8. G. J. Broze, Jr., T. J. Girard, and W. F. Novotny. Perspectives in Biochemistry: Regulation of coagulation by a multivalent Kunitz-type inhibitor. Biochemistry 29:7539 (1990).
9. J. G. Bieth. In vivo significance of kinetic constants of protein proteinase inhibitors. Biochem. Med. 32:387 (1984).
10. P. Colburn, and V. Bounassisi. Identification of an endothelial cell product as an inhibitor of tissue factor activity. In Vitro Cell Dev. Biol. 24:1133 (1988).
11. B. J. Warn-Cramer, S. L. Maki, and S. I. Rapaport. A sulfated rabbit endothelial cell glycoprotein that inhibits factor VIIa/tissue factor is functionally and immunologically identical to rabbit extrinsic pathway inhibitor (EPI). Thromb. Res. 61:515 (1991).
12. M. Jochum, S. Lander, N. Heimburger, and H. Fritz. Effect of human granulocytic elastase on isolated human antithrombin III. Hoppe-Seyler's Physiol. Chem. 362:103 (1981).
13. R. W. Carrell, and M. C. Owen. Plakalbumin, α_1-antitrypsin, antithrombin and the mechanism of inflammatory thrombosis. Nature 317:730 (1985).
14. P. Sie, D. Dupouy, F. Dol, and B. Boneu. Inactivation of heparin cofactor II by polymorphonuclear leukocytes. Thromb. Res. 47:657 (1987).
15. C. W. Pratt, R. B. Tobin, and F. C. Church. Interaction of heparin cofactor II with neutrophil elastase and cathepsin G. J. Biol. Chem. 265:6092 (1990).
16. M. S. Brower, and P. C. Harpel. Proteolytic cleavage and inactivation of α_2-plasmin inhibitor and C1 inactivator by human polymorphonuclear leukocyte elastase. J. Biol. Chem. 257:9849 (1982).

17. P. A. Pemberton, R. A. Harrison, P. J. Lackman, and R. W. Carrell. The structural basis for neutrophil inactivation of C1 inhibitor. Biochem. J. 258:193 (1989).

TISSUE FACTOR PATHWAY INHIBITOR AND HEPARIN

Ulrich Abildgaard

Dept of Medicine
Aker University Hospital
0513 Oslo, Norway

INTRODUCTION

The anticoagulant property of heparinized blood results from interactions between heparin, protease inhibitors and clotting enzymes. The inactivation of thrombin, factor Xa and factor IXa by antithrombin (AT) is dramatically accelerated by even low concentrations of heparin. At high concentrations of heparin, inactivation of thrombin by heparin cofactor II (HC II) also contributes to the anticoagulant effect. The relative contribution of these reactions to the total anticoagulant effect is currently under debate. Moreover, their relative significance to the antithrombotic effect of heparin is not obvious. The possibility of additional anticoagulant effects of heparin tends to be ignored in recent literature.

Tissue factor (TF) is the main trigger of coagulation both in normal hemostasis and in thrombotic disease. The recognition of tissue factor pathway inhibitor (TFPI)[1] as the factor Xa dependant inhibitor of TF and factor VIIa represents a major advance in the understanding of how blood clotting is controlled. TFPI will be reviewed by George Broze in this issue. This communication will focus on some interactions between TFPI and heparin. The plasma concentration of TFPI is only about 1/1000 of the plasma concentration of AT. Yet some recent findings suggest that a significant part of the anticoagulant effect of heparin is exerted via TFPI.

THE VARYING AFFINITY OF TFPI FOR HEPARIN

When plasma is run on a heparin-Sepharose affinity column,

---

[1]The term tissue factor pathway inhibitor (TFPI) shall replace the names extrinsic pathway inhibitor (EPI), and lipoprotein associated coagulation inhibitor (LACI) as agreed upon at the ISTH sub-committee meeting in Amsterdam, June 1991.

Heparin and Related Polysaccharides
Edited by D.A. Lane *et al.*, Plenum Press, New York, 1992

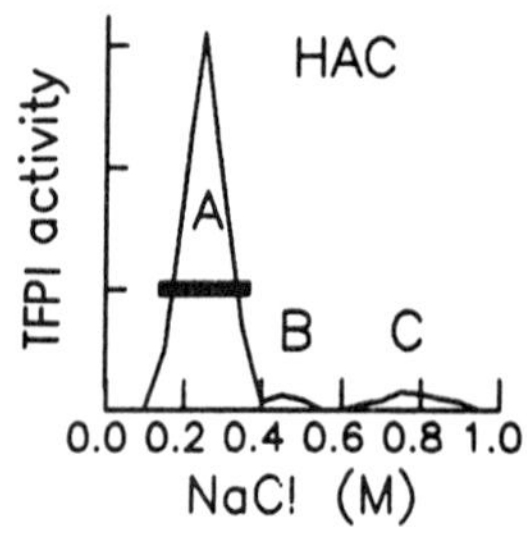

Fig. 1. Heparin affinity chromatography of normal plasma. About 25 per cent of TFPI activity (Chromogenic substrate assay) was not retained (flow through). Data from Jacobsen et al. (1990).

about 3/4 of the TFPI activity is retained. Elution with a NaCl gradient results in one major and two small activity peaks as shown in Fig. 1. The appearance of several TFPI peaks was not unexpected, since gel filtration also separated three main activity peaks (Sandset et al. 1987). TFPI fractions with no or low heparin affinity is complexed to lipoproteins and have high molecular weight. But TFPI with high heparin affinity may either be uncomplexed or complexed, illustrating the diversity of TFPI complexes in human plasma (Lindahl et al. 1990 a).

For unknown reasons, many patients with advanced cancer have TFPI activity far above that in a normal plasma. In such plasma samples, the amount of TFPI with high affinity is significantly increased (Lindahl et al. 1990 a).

RELEASE OF TFPI TO THE BLOOD BY HEPARIN

Intravenous injection of heparin produces a rapid increase in TFPI activity in plasma (Sandset et al. 1988). This TFPI response is dose dependant, but even prophylactic doses of heparin given subcutaneously are active. When a small amount of heparin was injected in a wrist vein, blood sampled from the cubital vein first showed a strong increase in heparin concentration as heparin was passing, but only a very moderate increase in TFPI activity as shown in Fig. 2. After about 2 minutes the continued sampling from the cubital vein now revealed a sustained release in TFPI activity. This finding suggested that TFPI was released from the endothelium during circulation of administered heparin as shown in Fig. 2. Kinetics of the release of TFPI by cultured endothelial cells has recently been examined by Hinz et al. (1990). The post-heparin injection increase in TFPI gradually subsides, roughly in the parallel to the heparin concentrations. Injection of low molecular weight heparin has the same effect. Following high clinical doses of heparin, more than a fourfold increase in TFPI activity may be obtained. This tells us that most of intravascular TFPI is in fact located on the intima. The equilibrium between plasma TFPI and intimal TFPI appears similar in normals and in cancer patients (Lindahl et al. 1990 b).

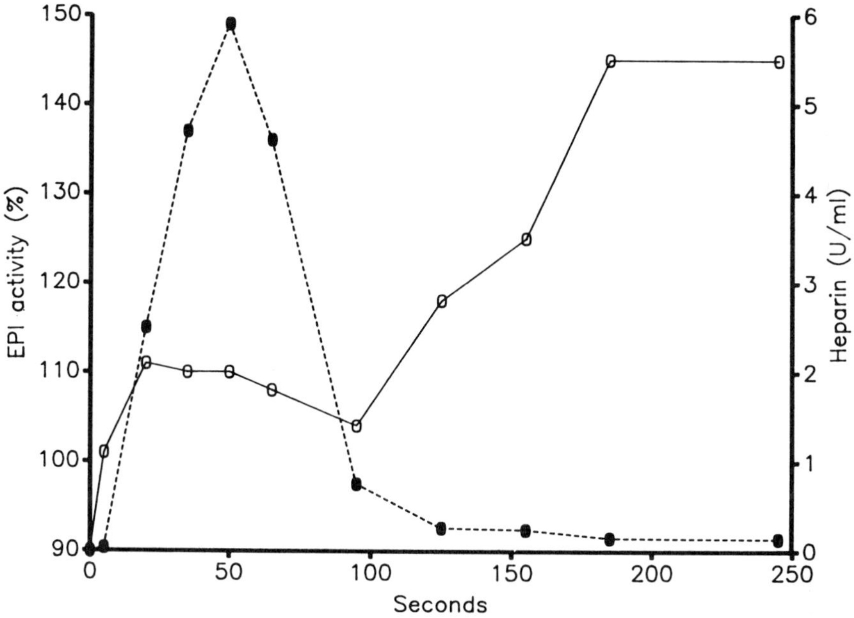

Fig.2. TFPI activity (o—o) and heparin concentration (●—●) in blood samples collected after intravenous injection of 500 U heparin. Heparin was injected in a wrist vein draining into an antecubital vein from which the blood samples were drawn. Data from Sandset et al. (1988).

Thus, the relative increase in TFPI activity after injection of heparin was the same in normals as in cancer patients, despite a significantly higher pre-injection level in the latter group.

In post-heparin plasma, particularly TFPI fractions with high affinity to heparin are increased compared to normal plasma (Lindahl et al. 1990 a). This could mean that a major part of endothelium-bound TFPI has high affinity to heparin. An alternative explanation is that heparin by replacement mainly releases TFPI that is bound to glycosaminoglycans.

TFPI AND POST-HEPARIN ANTICOAGULANT EFFECT

When heparin is added to blood or plasma in vitro, its anticoagulant effect can be neutralized by titrated amounts of either platelet factor IV, protamine or polybrene. Following injection of heparin, part of the anticoagulant effect can not be neutralized by these antagonists (Nordøy 1963, Michalski et al. 1978). Injection of heparin also induces release of lipase to the blood, and it was assumed that the post-heparin anticoagulant effect resulted from lipase action. The increase in TFPI activity suggested, however, that this inhibitor might be involved in the effect. Antibodies, specifically blocking TFPI action were kindly supplied by Dr Ole Nordfang and Thomas C Beck, Novo-Nordisk, Denmark. Pre-incubation with these antibodies abolished the post-heparin anticoagulant effect (Lindahl et al. 1991 a).

TFPI AND THE ANTICOAGULANT EFFECT INDUCED BY HEPARIN

Antibodies specifically blocking either AT, HC II or TFPI were then used in elucidating the relative role of these inhibitors to the total anticoagulant effect of heparinized plasma and blood. When coagulation was initiated by contact activation, the anticoagulant effect of heparin is reduced by about 2/3 by blocking AT. There was no effect by blocking of HC II, and only minimal effect by blocking TFPI. In plasma obtained after in vivo heparinization, the contribution of TFPI increased somewhat (Lindahl et al. 1991 b). A minor part of the inhibition of contact activation induced by heparin cannot be accounted for by any of the three inhibitors.

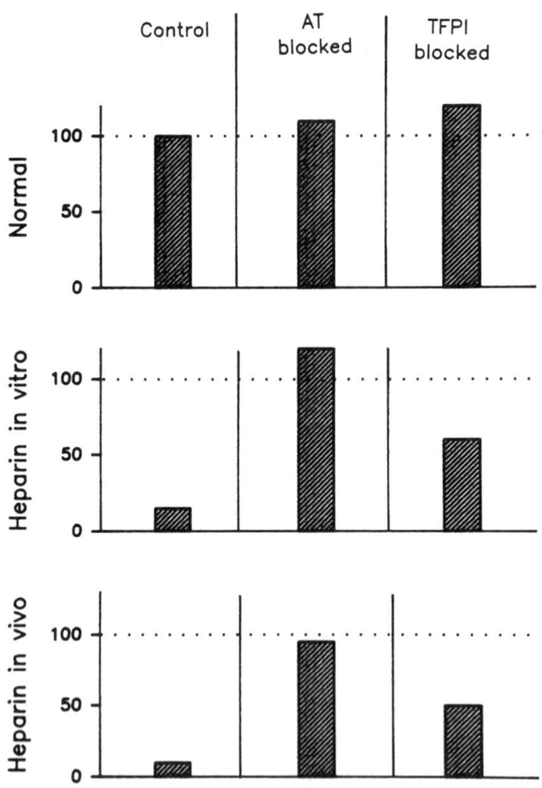

Fig. 3. The influence of blocking antibodies on the generation of thrombin following addition of tissue factor to citrated normal blood and heparinized blood. Heparin concentration 1.0 U/ml plasma. The amounts of fibrinopeptide A at 30 seconds are indicated in per cent of the value for normal blood. Data mainly from Lindahl et al. (1991a) and Abildgaard et al. (1991).

When clotting is induced by TF, blocking of either AT or TFPI reduces the clotting times of heparinized plasma significantly (Lindahl et al. 1991). In normal plasma heparinized in vitro, AT dominates. In blood sampled after injection of heparin to cancer patients, the blocking of TFPI may reduce inhibition as much as blocking AT (Lindahl et al. 1991 b).

The release of fibrinopeptide A (FPA) by thrombin is a sensitive indicator of thrombin generation following addition of TF and $CaCl_2$ to citrated blood. Preincubation of heparinized blood with antibodies blocking AT results in a FPA release rate close to that in non-heparinized blood as shown in Fig. 3. This might suggest that AT was the sole mediator of the heparin mediated effect. But the use of antibodies blocking TFPI again showed the significance of this inhibitor. First, blocking only TFPI resulted in an appearance of FPA that was between that in heparinized and in non-heparinized blood. Blocking of both AT and TFPI with specific antibodies resulted in an appearance rate of thrombin of 30 % greater than in normal, non-heparinized blood (Abildgaard et al. 1991). These findings suggest that about one third of the total anticoagulant effect induced by therapeutic doses of heparin is in fact mediated via TFPI. Even when blood is heparinized in vitro, TFPI contributes significantly to the inhibition of coagulation triggered by tissue factor.

DOES TFPI CONTRIBUTE TO THE ANTITHROMBOTIC EFFECT OF HEPARIN?

Experiments comparing low molecular weight heparin and unfractionated heparin have shown that no single anticoagulant effect can explain the whole antithrombotic effect (Norrheim et al. 1991). To what extent the TFPI related effects contribute has not been examined. It should be kept in mind that the release of TFPI by heparin does not change the total amount of intravascular TFPI. The redistribution of TFPI might be advantageous in certain situations. It is possible that an increased TFPI activity in stagnant venous blood may have antithrombotic effects. In sepsis, circulating monocytes expose TF on the surface which triggers disseminated intravascular coagulation. Increased amounts of circulating TFPI could more effectively control this activation.

With purified reagents, heparin accelerates the action of TFPI, whereas it was abrogated in the presence of both heparin and AT (Broze et al. 1988). But the FPA release studies reported here show that even when heparin is added to blood in vitro, part of the anticoagulant response is mediated through TFPI. Thus blocking TFPI reduced the heparin effect considerably. When only AT was blocked, release rates were similar in the absence and presence of heparin. These findings imply that in whole blood, heparin enhanced TFPI only when AT was available. This indicates a synergism between TFPI and AT in their enhancement by heparin. It has been shown that endothelial glycosaminoglycans enhance the anticoagulant effect of AT (Rosenberg and Rosenberg 1984). It is tempting to suggest that both AT and TFPI are enhanced by glycosaminoglycans on the vascular surface. If this is correct, vascular GAGS play a greater role in the antithrombotic effect of vascular surfaces than previously assumed.

REFERENCES

Abildgaard, U., Lindahl, A. K., Sandset, P. M., 1991, Heparin requires both antithrombin and extrinsic pathway inhibitor for its anticoagulant effect in human blood, Haemostas.,in press.

Broze, G. J., Warren, L. A., Novotny W.F., Higuchi D. A., Girard T.J., 1988, Miletich J.P., The lipoprotein - associated coagulation inhibitor that inhibits the factor VII tissue factor complex also inhibits factor Xa: Insight into its possible mechanism of action, Blood, 71:335-343.

Hinz, U., Klaus, J. G-H., Müller-Berghaus, P., Müller-Berghaus, G., 1991, Mechanism of extrinsic pathway inhibitor stimulation by heparin in umbilical vein endothelial cells, Thromb Haemost., 65:952.

Jacobsen, P. B., Lindahl, A. K., Norrheim, L., Nordfang, O., Sandset, P. M., Abildgaard, U., 1990, Separation of plasma by heparin sepharose chromatography, Fibrinolysis, 4 (suppl.1): 24.

Lindahl, A. K., Sandset, P. M., Abildgaard, U., Andersson, T. R., Harbitz, T. B., 1989, Increased levels of extrinsic pathway inhibitor and decreased levels of other coagulation inhibitors in advanced cancer, Acta Chir Scand., 155:389-393.

Lindahl, A.K., Abildgaard, U., Sandset, P. M., Nordfang, O., 1990 a, Increased amounts of uncomplexed extrinsic pathway inhibitor in cancer patients' plasma. The effect of heparin injection, Arteriosclerosis, 10 (5):947-948.

Lindahl, A.K., Abildgaard, U., Stokke G., 1990 b, Extrinsic pathway inhibitor after heparin injection: Increased response in cancer patients, Thromb Res., 59:651-656.

Lindahl, A. K., Abildgaard, U., Larsen, M. L., Aamodt, L. M., Nordfang, O., Beck, T. C., 1991 a, Extrinsic pathway inhibitor (EPI) and the post-heparin anticoagulant effect in tissue thromboplastin induced coagulation, Thromb Res., in press.

Lindahl, A. K., Abildgaard, U., Staalesen, R., 1991 b, The anticoagulant effect of heparin resulting from interaction with extrinsic pathway inhibitor, Thromb Res., in press.

Michalski, R., Lane, D. A., Pepper, D. S., Kakkar, V. V., 1978, Neutralization of heparin in plasma by platelet factor 4 and protamine sulphate, Br J Haematol., 28: 561-571.

Nordøy, A., 1963, Difference in the heparin-neutralizing effect of protamine and polybrene as tested by thrombotest, Scand J Clin Lab Invest., 15 (3):205-210.

Norrheim, L., Abildgaard, U., Larsen, M. L., Lindahl, A. K., 1991, Involvement of the extrinsic pathway in the activities of low molecular weight heparins, Thromb Res., in press.

Rosenberg, R. D., Rosenberg, J. S., 1984, Natural anticoagulant mechanisms, J Clin Invest., 74:1-6.

Sandset, P. M., Abildgaard, U., Pettersen, M. A., 1987, A sensitive assay of extrinsic coagulation pathway inhibitor (EPI) in plasma and plasma fractions, Thromb Res., 35:389-400.

Sandset, P. M., Abildgaard, U., Larsen, M. L., 1988, Heparin induces release of extrinsic pathway inhibitor (EPI), Thromb Res., 50:803-313.

LMW HEPARIN: RELATIONSHIP BETWEEN ANTITHROMBOTIC AND ANTICOAGULANT EFFECTS

T. W. Barrowcliffe

National Institute for Biological Standards and Control, Potters Bar, UK

INTRODUCTION

LMW heparin is now becoming a well established alternative to unfractionated heparin (UFH), particularly for the prophylaxis of deep vein thrombosis, where the convenience of a single injection a day, instead of two or three as for UFH, is a decided advantage. Whilst the mechanisms of the anticoagulant actions of heparin and LMW heparin at the molecular level are now well understood, there is still some uncertainty and controversy over the mechanisms of their antithrombotic effect in vivo, and in particular about the relevance of the various in vitro anticoagulant assays to their in vivo actions.

This review, after briefly describing the similarities and differences between the various commercial LMW heparins and their methods of preparation, will focus mainly on the relationship between the antithrombotic activities of LMW heparin, as determined in animal models and clinical trials, and their anticoagulant activities measured in vitro.

METHODS OF PREPARATION OF LMW HEPARIN

The earliest LMW heparin to be investigated was prepared from UFH by gel permeation chromatography[1]. However, this is not an economic proposition for large scale manufacture because of low yield, and with the exception of the first preparation of CY216 (Institut Choay), which used alcohol fractionation, most manufacturers have used various methods of depolymerisation. The most common method uses nitrous acid, which breaks heparin chain glycosidic linkages between the 1-position of N-sulphated glucosamine and the 4-position of the adjacent uronate. The uronate remains intact at the non-reducing end of the fragment, but the glucosamine is converted to 2,5-anhydro-D-mannose, which is reduced by some manufacturers to anhydromannitol, though this is not essential for in vivo use. The two other major methods are enzymatic

Heparin and Related Polysaccharides
Edited by D.A. Lane *et al.*, Plenum Press, New York, 1992

Table 1. Methods of Preparation of Commercial LMW Heparin and their Mean Molecular Weights

Manufacturer	Code Numbers	Trade Names	Preparation Method	Mr (wt av)
Choay (Sanofi)	CY216*	Fraxiparine	Nitrous acid	4200
KabiVitrum	Kabi 2165	Fragmin	Nitrous acid	5700
Novo	LHN1	Logiparin	Heparinase	6300
Pharmuka (Rhone-Poulenc)	PK10169	Enoxaparine Clexane	ß-Elimination	3900
Sandoz		Sandoparin Mono-Embolex NM	Nitrous acid	5100

* CY216 may be available in two forms: one as indicated, and the other apparently a true fraction of heparin (without depolymerisation).

cleavage using heparinase, and B-elimination of esterified heparin by alkaline hydrolysis. Both of these methods yield fragments with the same end-groups; intact sulphated glucosamine at the reducing end and 4,5-unsaturated uronic acid at the non-reducing end. The methods used for the various commercial products which are currently available are summarised in Table 1.

SIMILARITIES AND DIFFERENCES

The fact that different processes are used for different manufacturers' products has led some authors to consider each product as a separate drug. There is a sense in which this view is true, particularly for regulatory purposes; each manufacturer has to demonstrate the consistency of his product and to provide details of toxicology and pharmacology in animals and humans, as well as evidence of clinical efficacy. However in many other respects the similarities between the various products outweigh the differences. There is no evidence that the different chemical end groups play any role in vivo, and this was emphasised in a study by Ostergaard et al[2], in which three different methods were used to prepare LMW heparins with a similar molecular weight distribution - all three products had a similar antithrombotic activity in experimental animals.

When compared with UFH, LMW heparins share many common characteristics :

1) anti-Xa activity higher than anti-IIa activity;

2) longer half-life and greater bioavailability;

3) less neutralisation by PF4;

4) less lipase release.

In addition it has been shown that for measurement of in vitro activities a single LMW heparin preparation can serve as a standard for the whole group[3,4].

The main aspect in which differences between products might be important is molecular weight distribution, which in turn largely determines the ratio between anti-Xa and anti-IIa activities. For the currently marketed products this ratio ranges from 1.8 to 3.5, reflecting mainly differences in the proportion of material at the higher end of the molecular weight range. However such differences as there are will be minimised after subcutaneous injection, because the higher M Wt components will be less readily absorbed, resulting in a "filtering" effect, and a closer similarity in molecular weight of the heparin chains actually circulating in the blood after injection of the different products. This is reflected in the fact that the dosage of the different products required for prophylaxis of DVT falls in a fairly narrow range, as discussed later.

MECHANISMS OF ANTICOAGULANT ACTIVITIES

The molecular mechanisms of the in vitro anticoagulant activities of heparin and LMW heparin are discussed in detail elsewhere in this symposium. However since it is relevant to the discussion of mechanisms of action in vivo a brief description will be given here. The specific pentasaccharide sequence is the minimum requirement for binding to antithrombin III (At III) and consequent conformational change which results in enhanced activity of the inhibitor. However an important characteristic is that the requirements for potentiation of inhibitory activity differ for the various coagulation enzymes. In particular, enhanced neutralisation of thrombin requires an additional polysaccharide sequence, adjacent to the pentasaccharide, of at least 13 saccharide units, so that both enzyme and inhibitor can be bound to the same heparin chain - the so-called "template" mechanism[5]. This is an absolute requirement, so that heparin molecules below 18 saccharides in length (Mr approx 5,400) have virtually no activity in anti-IIa assays, as well as in global anticoagulant assays such as the APTT, which depend on inhibition of thrombin. Potentiation of inhibition of factor Xa, however, requires only the minimum pentasaccharide sequence; there is no requirement for enzyme binding, and hence molecules below 18 saccharides, as well as the synthetic pentasaccharide have high anti-Xa activity, despite being completely unable to potentiate the inhibition of thrombin by At III.

This difference in mechanism is responsible for the fact

that commercial LMW heparins, which consist of various mixtures of molecules above and below 18 saccharides, have higher activities by anti-Xa than anti-IIa assays. Two further aspects of the anti-Xa activity of LMW heparin deserve further mention. First, the activity in most assay systems is measured using free enzymes in the absence of calcium or other components. Although the molar specific activity under these conditions is almost constant over a wide molecular weight range[6], this is not the case when anti-Xa activity is measured in the presence of calcium ions. The addition of calcium ions potentiates anti-Xa activity, the degree of potentiation increasing the longer the chain length[7], resulting in a marked increase in molar specific activity with molecular weight[8]. Secondly, the inhibitory action of At III on factor Xa, both in the presence and absence of heparin, is markedly diminished when factor Xa is incorporated into prothrombinase, and studies in our laboratory[9], and by Schoen et al[10] showed a decrease of anti-Xa activity when measured in this more physiological form. Taken together, these two aspects suggest that the high anti-Xa activities of the very low M Wt components of LMW heparin measured using free enzyme, may exaggerate the influence of these components in vivo.

Considering the other coagulation enzymes, factors IXa and XIa behave like thrombin in requiring enzyme binding to the same heparin chain as At III, whereas factor XIIa is similar to factor Xa in having no such requirement. However, inhibition of the contact system enzymes does not seem to be important for the overall anticoagulant action of heparin or LMW heparin, and studies by Griffith et al[11] suggest that inhibition of factor IXa plays only a minor role. In practice, therefore, discussion of the relevance of in vitro activities to in vivo actions of LMW heparin is centred mainly on their anti-Xa and anti-IIa activities.

Inhibition of Thrombin Generation

As described by Professor Hemker elsewhere in his symposium, studies with isolated enzymes, even when performed under physiological conditions, are not necessarily a good guide to the overall ability of heparin or LMW heparin to inhibit the generation of thrombin in clotting blood or plasma. One of the main reasons for this is the existence of the thrombin feedback loops, whereby the first traces of thrombin produced activate factor VIII and factor V, resulting in a rapid enhancement of the rate of subsequent thrombin generation. Studies by Ofosu et al[12] and Hemker and colleagues[13] have emphasised the importance of inhibition of these thrombin feedback loops by heparin and LMW heparin. Molecules below 18 saccharides, including the synthetic pentasaccharide, are relatively weak inhibitors of thrombin generation, in spite of their high anti-Xa activity, whereas heparin chains above 18 saccharides are much more potent inhibitors of thrombin generation[12,14]. The importance of thrombin inhibition is emphasised by the fact that synthetic inhibitors of thrombin, such as PPACK and recombinant hirudin, are effective inhibitors of thrombin generation in vitro[12,15].

INTERACTION WITH HEPARIN NEUTRALISING PROTEINS

Although binding to At III is the major determinant of the anticoagulant activities of heparin and LMW heparin, their overall action in vivo may be affected considerably by interaction with other plasma proteins, some of which may modulate anticoagulant activity. The most important protein in this category is platelet factor 4 (PF4), which is present in the a-granules of platelets and released upon platelet activation. PF4 binds strongly to heparin and in doing so neutralises its anticoagulant activity. There are two important aspects of the interaction of PF4 with heparin which are relevant for the action of LMW heparins in vivo. Firstly, PF4 interacts equally well with heparin of both low and high affinity to At III. Thus at a given concentration of PF4, the presence of low affinity heparin molecules will 'protect' the high affinity molecules from neutralisation, and such protection will be greater in heparin preparations with a higher proportion of low affinity molecules, as is the case with LMW heparins compared with UFH. Secondly, the interaction is dependent on M Wt, a chain length of at least 18 saccharides being required for optimum binding. This is the same requirement as for thrombin inhibition, so that molecules above 18 saccharides which have anti-IIa as well as anti-Xa activity, can be completely neutralised by PF4, whereas molecules below 18 saccharides, which retain anti-Xa activity, are more resistant to neutralisation by PF4. This accounts for the fact that, in the commercially developed LMW heparin preparations, all the anti-IIa and APTT activity can be neutralised by PF4, whereas a considerable portion of the anti-Xa activity remains non-neutralised. The same considerations apply to protamine, the heparin-neutralising agent used clinically for reversal of the anticoagulant effects.

PROTEINS RELEASED BY HEPARIN

Another aspect in which behaviour in vivo of heparin and LMW heparin differs from that in vitro is in release of components from the vessel wall after injection. The most well-known of these components are the lipase enzymes, lipoprotein lipase and hepatic triglyceride lipase (HTGL). Release of lipase enzymes has not been thought to play any role in the anticoagulant actions of heparin. However, there is evidence that injection of heparin, LMW heparin, or other charged polysaccharides results in appearance of "extra" anti-Xa activity, in addition to that of the injected drug[16], and studies in our laboratory suggested that HTGL might be responsible for this released anti-Xa activity[17]. More recently, it has been shown that the extrinsic pathway inhibitor, EPI, is also released by heparin and LMW heparin[18], and it now seems more likely that EPI release is responsible for this extra anti-Xa activity.

Whether this released activity is relevant to the antithrombotic action of heparin amd LMW heparin is not known. However, the fact that highly sulphated polyasaccharides such as SSHA and pentosan polysulphate, which have little anticoagulant activity in vitro but are potent releasing agents, are clinically effective antithrombotic agents is indirect evidence that this can be an important contributory mechanism.

ANTITHROMBOTIC ACTIONS OF LMW HEPARIN IN EXPERIMENTAL ANIMALS

Although data from animal models are inevitably only a substitute for clinical results, much useful information can be gained provided that the model itself is relevant to the pathogenic process of thrombosis in man, and that well standardised techniques are used. Most clinical uses of heparin and LMW heparin are in prevention and treatment of venous thrombosis. The pathogenesis of venous thrombosis is thought to involve mainly hypercoagulability and stasis, with platelet activation and vessel wall damage playing only a minor role[19]. One of the most widely used animal models for study of heparin and similar anticoagulants is the rabbit stasis model, orignally developed by Wessler[20]. In the original model, human serum, containing several activated clotting factors, was used as the source of hypercoagulability, and the combination of serum injection and local venous stasis gives a reproducible thrombus, which can be used as a basis for study of the effects of various anticoagulant drugs. A number of variations have been introduced to this method, the most important being the use of different stimuli, such as thromboplastin, as a cause of hypercoagulability - as discussed later, this can have an influence on the results.

ANTITHROMBOTIC ACTION AND ANTI-XA ACTIVITY

The hypothesis that prevention of thrombosis by heparin was primarily due to its anti-Xa activity was based on two lines of reasoning. Firstly, the clinical success of low-dose heparin in prevention of post-operative DVT was difficult to explain in terms of traditional anticoagulant effects, since at this low dosage there is virtually no prolongation of APTT. Secondly, because of the amplification nature of the coagulation cascade, with one molecule of factor Xa being able to promote the generation of over 1,000 molecules of thrombin, the inhibition of factor Xa could provide a more effective antithrombotic action than the inhibition of thrombin itself.

The availability of LMW heparin preparations with higher anti-Xa than anti-IIa activities, as well as individual oligosaccharides and the synthetic pentasaccharide, has enabled this hypothesis to be tested. Initial studies with LMW heparin preparations supported the view that anti-Xa activity was linked to antithrombotic effects. For instance, Thomas et al, using a Wessler rabbit stasis model with serum as the stimulus, found that a fraction with a mean M Wt of around 8,000 was equally as effective as UFH in terms of dry weight or anti-Xa units, despite having only half the activity of UFH in an APTT assay[21]. Similar results were found by Holmer et al[22] with a preparation of mean M Wt around 4,000 which had only 20% of the APTT activity of UFH, and by other groups using a variety of LMW heparin preparations with mean M Wt between 4,000 amd 6,000[23,24].

However all these LMW heparin preparations were heterogeneous and thus could not answer the question as to whether anti-Xa activity alone were primarily responsible for the antithrombotic effect.

In 1982 three papers were published which appeared to provide evidence against this hypothesis. Holmer et al[22], using fragments of M Wt 2,100, 3,300 and 4,000, found that the lower M Wt fragments, which had high anti-Xa activity and almost no effect on the APTT, were poor antithrombotic agents. Thomas et al[25] found that a homogeneous decasaccharide with high affinity to At III had almost no antithrombotic action, despite having an anti-Xa activity of over 1,000 iu/mg. A larger fragment of predominantly 16 saccharides was more effective, but was still only half as active as UFH, as was a 20-22 saccharides fragment. A similar discordance between anti-Xa activity and antithrombotic action was reported by Ockelford et al for a fragment of 3,000 M Wt[26].

The Role of Low-affinity Heparin

One reason for the relative ineffectiveness of the high affinity fragments studied by Thomas et al was subsequently shown to be the absence of heparin with low affinity to At III (LAH). Although LAH had no antithrombotic activity itself, its addition to the high affinity decasaccharide potentiated its antithrombotic action significantly, and a similar potentiation was found with the higher M Wt oligosaccharides[27]. This potentiating effect of LAH is probably due to its action in protecting the high affinity oligosaccharides from other heparin binding proteins such as PF4 as already discussed, and emphasises the importance of the low affinity heparin molecules for full antithrombotic action in vivo. However, even when LAH is present, the correlation between antithrombotic action and anti-Xa activity appears to break down below 18 saccharides, and indeed in this range, M Wt itself seems to be the most important parameter of antithrombotic action. This was emphasised in a study by Mattson et al[28] who tested equal dry weight doses of individual oligosaccharides between 8 and 20 units in a rabbit stasis model, and found a gradual decrease in antithrombotic potency with decreasing chain length.

Inhibition of Free and Complexed Factor Xa

Another reason for the poor correlation with anti-Xa activity at very low M Wt may be the method of measurement of anti-Xa activity, in which free factor Xa is used, in the absence of calcium ions or other components of the prothrombinase complex. Studies in our laboratory and elsewhere[7,9,10] have shown that the presence of these components has a profound influence on the anti-Xa activities of the lower M Wt oligosaccharides; when anti-Xa activity is measured in the presence of calcium ions, or in prothrombinase, there is a marked decrease in activity with decreasing M Wt. This can help to explain the gradual decrease in antithrombotic activity with decreasing M Wt, found by Mattson et al[28]. The relevance of measurement of anti-Xa activity in the presence of prothrombinase components is supported by studies of the synthetic pentasaccharide. Studies by Walenga et al[29] showed that, on a molar basis, 36 times more pentasaccharide than UFH was required to get an equivalent antithrombotic effect in a rabbit model. This compares very well with in vitro results showing that, when measured using prothrombinase, 40 times more molecules of pentasaccharide than UFH are required for an equivalent anti-Xa activity[9].

Nature of Thrombogenic Stimulus

Studies of the antithrombotic action of the synthetic pentasaccharide by different groups have given quite different results. For instance, whereas 150 ug/kg produced 80% antithrombotic activity in the studies of Walenga et al[29], the same dose gave only 20% activity in the study of Thomas et al[30]. Although both groups used the Wessler model with serum as the thrombogenic challenge, the method of preparation of the serum differed, that used by Walenga et al being collected after a short period of glass activation, compared with extensive kaolin activation, resulting in a higher concentration of activated factors, in the method of Thomas et al. This emphasises the difficulties of comparing absolute antithrombotic activities between different investigators and different animal models.

The importance of the thrombogenic stimulus was investigated further by Amar et al[31], who measured the antithrombotic activities of the pentasaccharide, two LMW heparins, and UFH using the Wessler model with either serum or thromboplastin as thrombogenic stimulus. Using thromboplastin, the pentasaccharide was largely ineffective as an antithrombotic agent, and the antithrombotic activites of the various compounds correlated best with their anti-IIa activities. However, when serum was used, the pentasaccharide was much more effective, and the anti-Xa activities of the LMW heparins then became more important. Furthermore, Amar et al[31] showed that, after a single subcutaneous injection of LMW heparin, the antithrombotic activity using thromboplastin disappeared after 9 hours, as did the anti-IIa activity, whereas the antithrombotic activity using serum persisted up to 18 hours, and was associated with detectable anti-Xa activity in the blood. Although the nature of the thrombogenic stimulus in human thrombosis is unknown, the fact that a single subcutaneous injection of LMW heparin gives protection against DVT for 24 hours in clinical practice suggests that the serum model may be more relevant than the thromboplastin model in the studies of Amar et al[31].

ANTITHROMBOTIC ACTION AND ANTI-IIA ACTIVITY

The importance of the thrombin feedback loops, and the relative ineffectiveness of oligosaccharides with only anti-Xa activity as inhibitors of thrombin generation, has emphasised that potentiation of thrombin inhibition by heparin and LMW heparin is the most relevant parameter for their overall ability to inhibit the generation of thrombin. Since oligosaccharides below 18 units, which have no anti-IIa activity, are also less effective antithrombotic agents than larger molecules, the question arises as to whether the anti-IIa activity of these longer chains is important for their antithrombotic action. This hypothesis is difficult to test with heparin and LMW heparin, because there are no chains which have anti-IIa activity that do not also have anti-Xa activity. However, Ofosu et al[32] showed that other sulphated polysaccharides, namely dermatan sulphate and pentosan polysulphate, which have no anti-Xa activity but catalyse the inhibition of thrombin via heparin co-factor II, were effective antithrombotic agents when given at a sufficiently high dose. Furthermore, the effective doses of these two agents, and of UFH and high-affinity heparin, produced a similar, almost two-fold increase in thrombin inhibition, as measured by gel analysis of thrombin-inhibitor complexes after addition of exogenous thrombin to ex vivo rabbit plasmas.

These studies demonstrate that agents which inhibit thrombin alone can be effective antithrombotic drugs, albeit at much higher doses than UFH, and that a relatively modest increase in the ability of the plasmas to inhibit thrombin, which may not be detected by conventional anticoagulant assays, may be sufficient to give an antithrombotic effect, at least in this particular animal model. Further support for the role of thrombin inhibition in antithrombotic actions in vivo comes from the studies of Amar et al[31], as already mentioned; using thromboplastin as the thrombogenic challenge, a good correlation was found between antithrombotic and anti-IIa activities of UFH and two LMW heparins.

CLINICAL RELEVANCE OF ANTI-XA AND ANTI-IIA ACTIVITIES

Clinical aspects of LMW heparin are dealt with in detail elsewhere in this symposium by Professor Hirsh. However, brief consideration can be given here to the relationship between clinical antithrombotic effects and in vitro anticoagulant activities. Considering first the prophylaxis of DVT, which is the main clinical indication for LMW heparin, Table 2 shows a summary of the dosages of the various LMW heparins which have been found effective in clinical trials. When expressed as anti-Xa activities, the dosages cover a fairly narrow range, from 2,000 to 3,500 iu, measured against the International Standard for LMW heparin. When expressed in terms of anti-IIa activities, however, the dosage range is much wider, almost 5-fold. In this clinical situation there appears to be a better correlation of the effective dosage with anti-Xa activity than with anti-IIa activity. The situation in prophylaxis is complicated by the fact that the drugs are given subcutaneously, and differences in absorption of the molecules with anti-Xa and anti-IIa activities could play a role. This does not apply to two other clinical situations, namely haemodialysis and treatment of established thrombosis.

In haemodialysis, the clinical effect can be assessed by

Table 2. Dosages of Prophylaxis of DVT

	mg	Anti-Xa iu	Anti-IIa iu
Clexane	~ 20	2,000	500
Fragmin	~ 18	2,500	1,100
Fraxiparine	~ 32	3,100	860
Logiparin	~ 45	3,500	2,400

prevention of clot formation or by prevention of thrombin generation using FpA assays. Using both assessment methods, Ireland et al[33] and Tew et al[34] found that the clinical effect of a LMW heparin, CY222, compared with UFH, correlated with their anti-Xa activities rather than with their anti-IIa activities. Similar results in haemodialysis have been found for other LMW heparins by other groups[35,36]. In the treatment of established venous thrombosis there have been few published studies so far of LMW heparin. Albada et al[37] compared UFH and a LMW heparin, Fragmin, in a double blind study. Doses were adjusted to maintain a constant anti-Xa activity with either drug, and there were no significant differences between UFH and LMW heparin in therapeutic effect. In this study, therefore there was again a good correlation between anti-Xa activity and clinical effects.

Overall, the clinical results do not support the hypothesis that the anti-IIa activity of LMW heparins is more important than their anti-Xa activity for antithrombotic action in vivo.

GENERAL DISCUSSION

The original rationale for the development of LMW heparin was that, by retaining the anti-Xa activity of UFH but with reduced anti-IIa activity they might display similiar antithrombotic action to UFH with reduced haemorrhagic side effects. The idea that anti-Xa activity might equate with antithrombotic action was based on the notion of prevention of thrombin generation; because of the amplification nature of the coagulation system, inhibition of one molecule of factor Xa could theoretically prevent the generation of over 1,000 molecules of thrombin. However, as discussed earlier, this is an over-simplification of the situation in clotting plasma, and does not take into account the importance of the thrombin feedback loops and the protection of factor Xa from inhibition when in prothrombinase. The elegant experiments of Hemker, Ofosu and their colleagues have shown that potentiation of inhibition of free factor Xa, such as occurs with oligosaccharides below 18 units and with the synthetic pentasaccharide, is an inefficient way to inhibit thrombin generation, whereas inhibition of thrombin feedback loops is much more effective.

The question may be asked, therefore, whether the anti-Xa activity of LMW heparin, measured by inhibition of free enzyme in plasma or At III, has any value at all. First of all it must be stressed that the arguments against the role of anti-Xa activity are based only on in vitro measurements. As already indicated, the results from in vivo studies, both in experimental animals and patients show considerable evidence for a correlation between anti-Xa activity and antithrombotic action. The arguments against the relationship between anti-Xa activity and antithrombotic action are mainly based on three papers published in 1982, which showed that ultra-LMW heparin fractions and fragments, possessing only anti-Xa activity, were much weaker antithrombotic agents than UFH[22,25,26]. Although these studies showed that anti-Xa activity per se could not account fully for the antithrombotic action of UFH, it should be emphasised that the fragments studied were not devoid of antithrombotic action, as indeed was found also for the synthetic pentasaccharide. Clearly these fragments could not be working by inhibition of thrombin, yet their anti-Xa activities were apparently too high when compared with their antithrombotic actions. This could be explained by the studies which indicated that anti-Xa activities of the very LMW saccharides, including the pentasaccharide, were much lower when factor Xa was incorporated into prothrombinase, and then correlated much better with their antithrombotic action.

The experiments reported by Mattson et al[28] who found a gradual increase in the antithrombotic activity of a series of oligosaccharides as their size increased from 8 to over 20 are critical for the anti-IIa hypothesis. If, as has been suggested from in vitro experiments, anti-IIa activity were the main determining factor in antithrombotic action, one would expect a marked increase in antithrombotic activity above 18 saccharides, when thrombin inhibitory activity first appears, or conversely a marked decrease in antithrombotic action below 18 saccharides. Neither of these predictions is borne out by the experimental data, and the gradual increase in antithrombotic activity with increasing Mr is best explained by inhibition of factor Xa, which shows a similar increase with increasing Mr when measured in the presence of calcium or in prothrombinase.

As already mentioned, clinical results suggest that anti-Xa activity may be more important in vivo than would be expected from the in vitro data. Ex vivo analyses also support a correlation between anti-Xa activity and protection against thrombosis; the half life of the anti-Xa activity of LMW heparin is much longer than its anti-IIa activity, and if the latter were the main determinant of antithrombotic action, it seems unlikely that one subcutaneous injection a day would be effective. In an analysis of ex vivo assay data Levine et al[38] proposed a "therapeutic window" based on anti-Xa levels between about 0.1 and 0.3 iu/ml after subcutaneous injection.

Of course in all these clinical studies the anti-Xa activity could simply be a marker for the real biological activity, such as inhibition of thrombin feedback loops. Certainly anti-Xa activity is probably the best indicator of the concentration of At III-binding sites, which are a prerequisite for antithrombotic activity. However if the antithrombotic action of LMW heparin were due entirely to the inhibition of thrombin, it seems difficult to understand why there is no correlation in any of the clinical studies between clinical efficacy and anti-IIa or APTT activity. One possibility, suggested by Professor Ofosu, is that the anti-IIa assays, measured normally in diluted plasma, are insufficiently sensitive, and Ofosu and colleagues found that inhibition of thrombin added to undiluted plasma did correlate with antithrombotic activity in animals[32]. On the other hand, the APTT also uses undiluted plasma, and is a measure of the delay in thrombin feedback activation of factors VIII and V, yet LMW heparin is effective at concentrations which give much less prolongation of the APTT than UFH, and in some studies no detectable changes in the APTT at all.

Another possible explanation for the poor correlation of anti-IIa assays with in vivo events, advocated by Professor Hemker, is that in vitro and ex vivo assays are all measured in the absence of platelets. Since PF4 neutralises LMW heparin less than UFH, their relative activities could be quite different in a situation where there is extensive platelet activation. Whilst there is some experimental evidence in favour of this hypothesis from in vitro studies, venous thrombi, unlike arterial thrombi, generally have a low platelet content, and the extent to which platelet activation and PF4 release actually occurs in, for instance, post-operative DVT is a matter for conjecture.

A third possibility is that that there could in fact be a real causal relationship between anti-Xa activity and antithrombotic action. It is important to distinguish "physiological" thrombin generation, which is required at the site of surgery, and "pathological" thrombin generation, which is due to a combination of hypercoagulability and stasis, and normally takes place at a distance from the initial site of triggering of the coagulation system. Clearly enhancement of Xa inhibition is largely ineffective in preventing "physiological" thrombin generation, because of the role of the thrombin feedback loops and the protective effect of the prothrombinase complex. However, once the prothrombinase complex has done its job, it is likely that factor Va will be rapidly inactivated by protein C, and factor Xa freed into the circulation. If this factor Xa were not inactivated, it would remain free to promote further thrombin generation at localised sites where cell activation, not only platelets but also monocytes and macrophages, has occurred. Thus it may be that an important part of the antithrombotic action of LMW heparin is to "scavenge" free factor Xa and thus prevent "pathological" thrombin generation occurring. If this hypothesis is correct, the release of EPI, which is an effective Xa scavenger, would also contribute to the antithrombotic action of LMW heparin, as well as their direct action via At III.

Whether or not there is a causal relationship, it seems

clear that measurement of anti-Xa activity, for dosage purposes and in ex vivo plasma, is a useful, though imperfect guide to the antithrombotic action of LMW heparins.

REFERENCES

1. E. A. Johnson, T. B. L. Kirkwood, Y. Stirling, J. L. Perez-Requejo, G. I. C. Ingram, D. R. Bangham, and M. Brozovic. Four heparin preparations: anti-Xa potentiating effect of heparin after subcutaneous injection. Thromb. Haemost. 35:586-591 (1976).

2. P. B. Ostergaard, B. Nilsson, D. Bergvist, U. Hedner, and P. C Pederson. The effect of low molecular weight heparin on experimental thrombosis and haemostasis - the influence of production method. Thromb. Res. 45:739-749 (1987).

3. T. W. Barrowcliffe, A. D. Curtis, T. P. Tomlinson, A. R. Hubbard, E. W. Johnson, and D. P. Thomas. Standardisation of low molecular heparins: a collaborative study. Thromb. Haemost. 54:675-679 (1985).

4. T. W. Barrowcliffe, A. D. Curtis, E. A. Johnson, and D. P. Thomas. An International Standard for low molecular weight heparin. Thromb. Haemost. 60:1-7 (1988).

5. M. W. Pomerantz, and W. G. Owen. A catalytic role for heparin. Evidence for a ternary complex of heparin cofactor thrombin and heparin. Biochim. Biophys. Acta 535:66-77 (1978).

6. D. A. Lane, J. Denton, A. M. Flynn, L. Thunberg, and U. Lindahl. Anticoagulant activities of heparin oligosaccharides and their neutralization by platelet factor 4. Biochem. J. 218:725-732 (1984).

7. T. W. Barrowcliffe, and Y. Le Shirley. The effect of calcium chloride on anti-Xa activity of heparin and its molecular weight fractions. Thromb. Haemost. 62:950-954 (1989).

8. V. Ellis, M. F. Scully, and V. V. Kakkar. The relative molecular mass dependence of the anti-factor Xa properties of heparin. Biochem. J. 238:329-333 (1986).

9. T. W. Barrowcliffe, S. J. Havercroft, G. Kemball-Cook, and U. Lindahl. The effect of Ca^{2+}, phospholipid, and Factor V on the anti-(Factor Xa) activity of heparin and its high-affinity oligosaccharides. Biochem. J. 243:31-37 (1987).

10. P. Schoen, T. Lindhout, G. Willems, and H. C. Hemker. Antithrombin III- dependent anti-prothrombinase activity of heparin and heparin fragments. J. Biol. Chem. 264:10002-10007 (1989).

11. M. J. Griffith. Kinetics of the heparin-enhanced antithrombin III/thrombin reaction. J. Biol. Chem. 257:7360-7365 (1982).

12. F. A. Ofosu, M. A. Blajchman, G. J. Modi, L. M. Smith, M. R. Buchanan, and J. Hirsh. The importance of thrombin inhibition for the expression of the anticoagulant activity of heparin, dermatan sulphate, low molecular weight heparin and pentosan polysulphate. Brit. J. Haem. 60:695-704 (1985).

13. H. C. Hemker. The mode of action of heparin in plasma, in: "Thrombosis and Haemostasis," M. Verstraete, J. Vermylen, R. Lijnen, J. Arnout, eds., Leuven University Press, Leuven (1987).

14. T. W. Barrowcliffe, and D. P. Thomas. Anticoagulant activities of heparin and fragments, in: "Heparin and Related Polysaccharides," F. A. Ofosu, I. Danishefsky, J. Hirsh, eds., Ann. New York Acad. Sci. 556:132-145 (1989).

15. E. Gray, J. Watton, S. Cesmeli, T. W. Barrowcliffe, and D. P. Thomas. Experimental studies on a recombinant hirudin, CGP 39393. Thromb. Haemost. 65:355-359 (1991).

16. D. P. Thomas, T. W. Barrowcliffe, R. E. Merton, J. Stocks, J. Dawes, and D. S. Pepper. In vivo release of anti-Xa clotting activity by a heparin analogue. Thromb. Res. 17:831-840 (1980).

17. E. Gray, G. Bengtsson-Olivecrona, T. Olivecrona, and T. W. Barrowcliffe. The anti-Xa activity of human hepatic triglyceride lipase. J. Lab. Clin. Med. 109:653-659 (1987).

18. M. Sandset, U. Abildgaard, and M. L. Larsen. Heparin induces release of extrinsic pathway inhibitor (EPI). Thromb. Res. 50:803-813 (1988).

19. D. P. Thomas. Pathogenesis of venous thrombosis, in: "Haemostasis and Thrombosis," A. L. Bloom, D. P. Thomas, eds., Churchill Livingstone, Edinburgh (1987).

20. S. Wessler. Thrombosis in the presence of vascular stasis. Amer. J. Med. 33:648-666 (1962).

21. D. P. Thomas, R. E. Merton, W. E. Lewis, and T. W. Barrowcliffe. Studies in man and experimental animals of a low molecular weight heparin fraction. Thromb. Haemost. 45:214-218 (1981).

22. E. Holmer, C. Mattson, and S. Nilsson. Anticoagulant and antithrombotic effects of heparin and low molecular weight heparin fragments in rabbits. Thromb. Res. 25:475-485 (1982).

23. D. Bergvist, B. Nilsson, U. Hedner, P. C. Petersen, and P. B. Ostergaard. The effect of heparin fragments of different molecular weights on experimental thrombosis and haemostasis. Thromb. Res. 38:589-601 (1985).

24. V. Diness, J. I. Nielsen, P. C. Perdersen, K. H. Wolffbrandt, and P. B. Ostergaard. A comparison of the antithrombotic and haemorrhagic effects of a low

molecular weight heparin (LHN-1) and conventional heparin. Thromb. Hasmost. 55:410-414 (1986).

25. D. P. Thomas, R. E. Merton, T. W. Barrowcliffe, L. Thunberg, and U. Lindahl. Effects of heparin oligosaccharides with high affinity for antithrombin III in experimental venous thrombosis. Thromb. Haemost. 47:244-248 (1982).

26. P. A. Ockelford, C. J. Carter, L. Mitchell, and J. Hirsh. Discordance between the anti-Xa activity and the antithrombotic activity of an ultra-low molecular weight heparin fraction. Thromb. Res. 28:401-409 (1982).

27. T. W. Barrowcliffe, R. E. Merton, S. J. Havercroft, L. Thunberg, U. Lindahl, and D. P. Thomas. Low-affinity heparin potentiaties the action of high-affinity heparin oligosaccharides. Thromb. Res. 34:125-133 (1984).

28. C. Mattson, M. Palm, K. Soderberg, and E. Holmer. Antithrombotic effect of heparin oligosaccharides. Ann. New York Acad. Sci. 556:323-332 (1989).

29. J. M. Walenga, M. Petitou, J. C. Lormeau, M. Samama, J. Fareed, and J. Choay. Antithrombotic activity of a synthetic heparin pentasaccharide in a rabbit stasis thrombosis model using different thrombogenic challenges. Thromb. Res. 46:187-198 (1987).

30. D. P. Thomas, R. E. Merton, E. Gray, and T. W. Barrowcliffe. The relative antithrombotic effectiveness of heparin, a low molecular weight heparin and a pentasaccharide fragment in an animal model. Thromb. Haemost. 61:204-207 (1989).

31. J. Amar, C. Caranobe, P. Sie, and B. Boneu. Antithrombotic potencies of heparins in relation to their antifactor Xa and antithrombin activities: an experimental study in two models of thrombosis in the rabbit. Brit. J. Haemat. 76:94-100 (1990).

32. F. A. Ofosu, F. Fernandez, N. Anvari, C. Caranobe, F. Dol, Y. Cadroy, M. Petitou, J. Mardiguain, P. Sie, and B. Boneu. Further studies on the mechanisms for the antithrombotic effects of sulphated polysaccharides in rabbits. Thromb. Haemost. 60:188-192 (1988).

33. H. Ireland, D. A. Lane, A. Flynn, A. C. Pegrum, and J. R. Curtis. Low molecular weight heparin in haemodialysis for chronic renal failure: dose finding study of CY222. Thromb. Haemost. 59:240-247 (1988).

34. C. J. Tew, D. A. Lane, E. Thompson, H. Ireland, and J. R. Curtis. Relationship between ex vivo anti-proteinase (factor Xa and thrombin) assays and in vivo anticoagulant effect of very low molecular weight heparin, CY222. Brit. J. Haemat. 70:335-340 (1988).

35. J. J. J. Borm, R. Krediet, A. Sturk, and J. W. ten Cate. Heparin versus low molecular weight heparin K2165 in chronic haemodialysis patients: a randomised cross-over study. Haemostasis 16:Suppl 2, 59-68 (1986).

36. B. Ljungberg, M. Blomback, H. Johnsson, and L. E. Lins. A single dose of low molecular weight heparin fragment for anticoagulation during haemodialysis. Clin. Nephrol. 27:31-35 (1987).

37. J. Albada, H. K. Nieuwenhuis, and J. J. Sixma. Treatment of acute venous thromboembolism with low molecular weight heparin (Fragmin). Results of a double-blind randomised study. Circulation 80:935-940 (1989).

38. M. N. Levine, A. Planes, J. Hirsh, M. Goodyear, N. Vochelle, and M. Gent. The relationship between anti-Factor Xa level and clinical outcome in patients receiving Enoxaparine low molecular weight heparin to prevent deep vein thrombosis after hip replacement. Thromb. Haemost. 62:940-944 (1989).

THE MODE OF ACTION OF HEPARINS IN VITRO AND IN VIVO

H.C. Hemker and S. Béguin

Department of Biochemistry, Cardiovascular
Research Institute and University of Limburg
Maastricht, the Netherlands

The antithrombotic action of a heparin is not necessarily confined to its effects on the clotting mechanism. Yet we restrict ourselves here to a discussion of the action of heparin on thrombin generation in platelet poor and platelet rich plasma for two reasons. In the first place it is more likely than not that inhibition of the clotting mechanism is at least one of the major working arms of heparin. All medication that inhibits blood clotting has an antithrombotic effect and all dis-orders that are known to impede clotting inhibition are accompanied by a thrombotic syndrome. Also the assumption that heparins act through their interference with blood coagulation is tacitly at the basis of virtually all studies of the pharmacokinetics and pharmacodynamics of heparin action, both clinical and experimental. Finally it is the subject that we studied and the only one that we dare to express ourselves about.

SPECIFIC AND COMPOSITE EFFECTS

The actions of heparin on haemostasis and thrombosis can be devided in specific biochemical effects, that are the effects on those biochemical processes that are at the basis of the action of heparin, and composite biological effects that are the consequence of the specific effects acting in some biological system, either in vivo or in vitro. The specific effects are: Increasing thrombin activation; increasing the activation of factor Xa; increasing the activation of factor IXa (1). There are good grounds to think that inactivation of other clotting factors is of no practical importance. All these activations take place through binding of heparin to antithrombin III (AT III). The action of heparin on thrombin via heparin cofactor II (HC II) occurs at such high concentrations of heparin as are seldomly obtained in clinical practice and will not be subject of discussion (2). Other negatively charged polyelectrolytes such as pentosan polysulfate (3) and lactobionic acid (4) do act via HC II, but they are not subject of this article.

The composite biological effects of heparin on the haemostatic-thrombotic apparatus are many and varied. The main ones are: The antithrombotic effect and the haemorrhagic effect. In fact these two are the only ones that matter if our aim is to find a good antithrombotic drug. They need extensive, laborious clinical experimentation to

Heparin and Related Polysaccharides
Edited by D.A. Lane *et al.*, Plenum Press, New York, 1992

be determined, so it is logical that we resource to model systems in order to find indications on these effects by laboratory shortcuts. They come in two kinds: determination of antithrombotic and haemorrhagic effects in animals and measurements of other composite effects on (human) blood or plasma.

The biological effects observed in the laboratory on isolated blood or plasma, either platelet poor (PPP) or platelet rich (PRP), again come in a great many varieties. Many of them are a variation on the clotting times, such as the activated partial thromboplastin time (APTT) or the Heptest. A clotting time is essentially the time that is necessary for a threshold amount of thrombin (around 15 nM) to be formed. This means that it is especially sensitive to influence on the lag phase of thrombin formation, where thrombin mediated feedback activation of factor VIII takes an especially important place. That is why tests like the APTT, where factor VIIIa formation is a rate limiting step in the lag phase, are sensitive to the action of heparins (fig.1). The prothrombin time (PT), where factor VIIIa does not play a role, is much less, if at all, affected (fig.2).

Our studies since 1985 focussed our attention on an essentially different type of biological effect. We determined the entire course of the thrombin generation curve under the influence of heparin (5,6). These curves yield a wealth of information on the

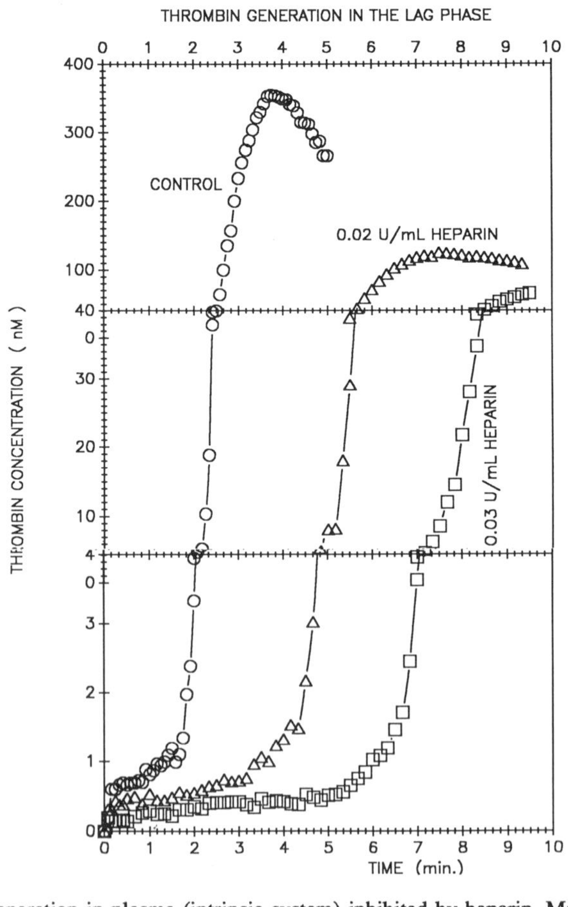

Figure 1. Thrombin generation in plasma (intrinsic system) inhibited by heparin. Methods as in ref. 6 with adaptation of the incubation times of the sample and the chromogenic substrate to the levels of thrombin.

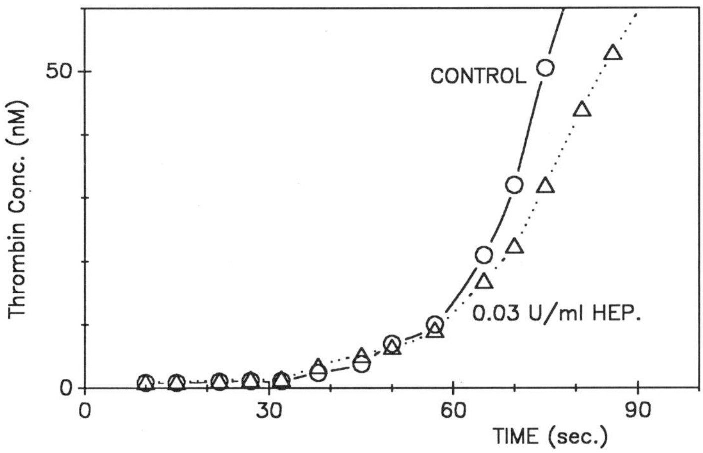

Figure 2. Thrombin generation in Plasma (extrinsic system) inhibited by heparin. Methods as in ref.6 with adaptation of the incubation times of the sample and the chromogenic substrate to the levels of thrombin.

mode of action of heparins, particularly because they allow to separate the effect of heparin on prothrombin conversion (thrombin generation without thrombin decay effects) from that on thrombin inactivation (7). For practical everyday use outside the research laboratory they are much too elaborate however. Reasoning that the action of the enzyme thrombin will be proportional to its concentration and to the time that it is allowed to act we determined the product of time and concentration, i.e. the area under the thrombin generation curve, i.e. the time-concentration integral of thrombin generated in clotting plasma, which we called the *thrombin potential.* We also found a way to determine the thrombin potential in a one-stage spectrophotometric procedure (8).

Any of the composite effects is a complicated function of the ensemble of the specific effects. The example of the APTT and the PT already shows that even in one class of composite effects the sensitivities for the specific effects may differ enormously. There is no *a priori* reason to assume that any one composite or specific effect will be a good yardstick for the antithrombotic- or antihaemostatic action of a heparin. Yet the observation that all anticoagulant drugs invariably are a) antithrombotic, in a dose dependent way, b) antihaemostatic when overdosed, and c) decrease the thrombin potential in a dose dependent way, whereas clotting times do not show such a systematic variation, made us think that the thrombin potential may be a good model for the antithrombotic potency of an anticoagulant drug.

HOW TO CHARACTERISE A HEPARIN

The specific effects are well defined biochemical properties whereas the biological effects are unknown functions of the specific effects. From this consideration it follows that a heparin should be characterised by its specific effects rather than by its biological ones. Unfortunately a biological effect, and a rather bizarre one viz. the prolongation of the prothrombin time of goat plasma, has been chosen to define the pharmacopea unit of heparin activity. Here we will propose a fundamentally different approach that allows to express heparin activity in standard-independent units of activity and to determine the levels of functional heparin in plasma.

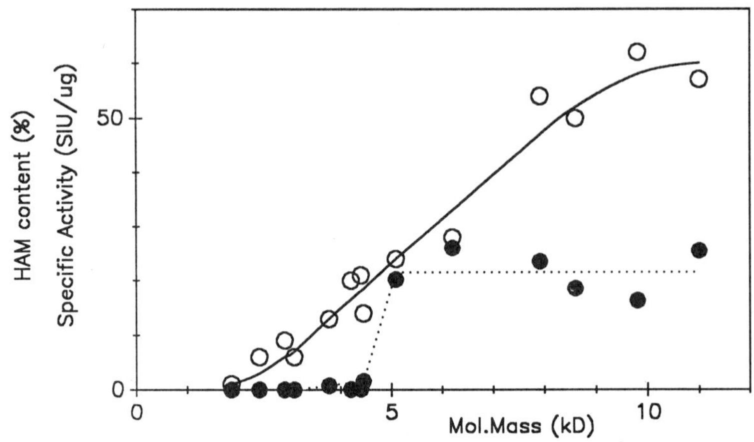

Figure 3. The HAM content and the HAM-specific activity of a series of LMWHs. o——o HAM content; ●·······● specific antithrombin activity.

THE FUNCTIONAL HETEROGENEITY OF HEPARIN

In fact the definition of heparin activity as an arbitrary composite activity did no harm as long as the proportion of the specific activities were similar in the different preparations that were to be compared. This was the case until the low molecular weight heparins (LMWHs) appeared on the scene. In the LMWHs however the antifactor Xa shifts independently from the antithrombin activity. At that moment one can imagine that an infinite variety of different combinations of the antifactor Xa activity and the antithrombin activity can result in the same composite effect on the prothrombin time in goat plasma, so that the unit of heparin activity becomes essentially senseless.

With decreasing molecular weight, glycosaminoglycans loose their capacity to catalyse the AT III dependent inhibition of thrombin (9-15). Barrowcliffe et al (loc.cit.), and Thomas et al (loc.cit.) have shown that heparin fragments with a chainlength of 10-18 monosaccharide units have a high anti-factor Xa activity, and that a length of 20-22 saccharides is necessary for an activity against thrombin. Lane and coworkers (loc.cit.) studied heparins of 8 to >18 monosaccharides and concluded that 18 units is the smallest chainlength that will allow to potentiate the inactivation of thrombin by AT III, whereas the activity against factor Xa was high in all fragments. From this it follows that all the high affinity material (HAM) in a heparin has anti factor Xa activity, whereas antithrombin activity is only expressed in HAM above the critical chainlength of 18 monosaccharides (ACLM)[1]. Below critical chainlength material (BCLM), in order to be active, has to bind to AT III, i.e. it has to contain minimally the AT III binding pentasaccharide. It thus comprises HAM with a length of between 5 and 18 sugar units i.e. a Mr of 1500 - 5400. Classical unfractionated heparin (UFH) hardly contains any BCLM. That explains why, before the advent of low molecular weight heparins, arbitrary units would suffice to indicate the potency of a heparin.

It has often been suggested that the smaller a heparin is, the more it exhibits an anti-factor Xa activity. Apart from the fact that this *per se* is not true (see below) this

[1] In order to prevent long and awkward abbreviations we use ACLM (resp. BCLM) to indicate Above (resp. Below) Critical chainLength *high affinity* Material.

can mean either of two things, given the usually highly disperse LMWH preparations . First, is it possible that in polydisperse heparin preparations the proportion of molecules with uniquely anti-factor Xa activity increases with decreasing mean molecular weight. Second the specific anti-factor Xa activity might increase with the chainlength.

In order to investigate this question, we determined the HAM and the ACLM content in a series of subfractions of LMWH, with the purpose to determine the specific activities per amount of active material.

The first thing that met the eye (16,17), was that the HAM content increases with molecular weight of the heparin (Fig. 3).

This is a natural consequence of the fact that upon scission of a high-affinity heparin chain the probability to hit the AT III binding pentasaccharide increases with decreasing molecular weight until it reaches 100% in a pentasaccharide. We suspect that this phenomenon may be at the basis of much of the reported variability of biological activity with chain length. We therefore expressed the observed activities per amount of HAM as far as they were related to factor Xa inhibition and per amount of ACLM for thrombin related phenomena. As a natural consequence the peak- and mean molecular weights of the active fraction in a LMWH will always be higher than that of the total heparin (fig.4).

THE SPECIFIC ACTIVITY OF A HEPARIN

The catalytic activities of a heparin in the interaction between AT III and thrombin (resp. factor Xa) can be used to quantitate its specific activity.

The reaction:

$$\text{Thrombin} + \text{AT III} \xrightarrow{\text{Heparin}} \text{Inactive product}$$

under conditions occuring in clotting plasma is pseudo first order in thrombin. It is

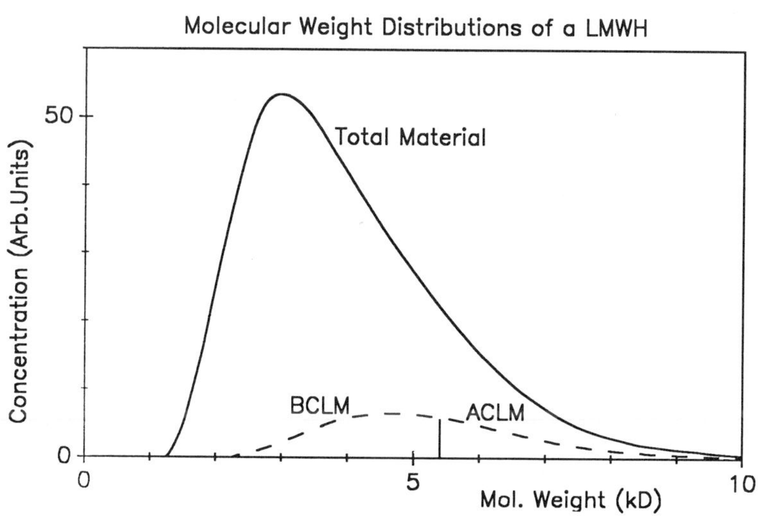

Figure 4. Molecular weight distribution of a LMWH. The area under the hatched line is the high affinity material.

characterised by the half life time of thrombin (resp. F.Xa), or, which is the same thing, by a decay constant, (k_{decay}) that is inversely proportional to the half life time. Upon addition of heparin to plasma, k_{decay} increases from its blank value (k_{blank}), proportional to the concentrations of AT III and heparin in the sample (Béguin et al. in press), hence:

$$k_{decay} = k_{spec}.[AT\ III].[heparin] + k_{blank}$$

Any well designed anti-thrombin or anti-factor Xa activity gives a result that is proportional to k_{decay}. This value thus can be obtained from current laboratory tests if adequate standards are available.

k_{spec} is the *specific activity* of the heparin, i.e. the increase of the decay constant that is brought about by 1µg/ml of heparin in the presence of 1 µM of AT III. It is influenced by a) Ca^{++} ions, b) plasma proteins, c) ionic strength, pH,temperature etc., so it should be determined under well defined, and specified conditions. The decrease of HAM content with Mr in a LMWH preparation is proportional to its molecular weight. This explains most of the variation of specific antifactor Xa activity with molecular weight. If antifactor Xa activity of a number of LMWHs is expressed per µg of ACLM is remarkably constant: (20.4 ± 4.1 SIU/µg, range 15.2 - 26.8, n=11)(fig.3). UFH and the 1st LMW standard are exceptions: 39 ± 0.5 SIU/µg. This may be due to a different affinity to other plasma proteins than that of most other LMWHs.

From the above it will be clear that the simplest possible rational approach to heparin standardisation is somewhat more complicated than accepted practice at this moment. This may be regretted but it may not be overlooked. One can not standardise away the fact that LMWHs have two different activities that cannot be measured with a common yardstick. A good example is the use of antifactor Xa units to compare the potency of UFH and pentasaccharide. One needs much more units of pentasaccharide to achieve the same antithrombotic effect than of UFH (18). This does not mean that pentasaccharide is a "bad" antithrombotic, but merely that it scores high on the antifactor Xa scale and not at all on the antithrombin scale, whereas the antithrombotic effect is a composite effect that responds to both antifactor Xa-and antithrombin specific activity; but in fact better to the latter.

THE USE OF SPECIFIC ACTIVITIES TO DETERMINE PLASMA LEVELS

When k_{spec} is known, the k_{decay} observed in a plasma sample, can be used to determine the heparin concentration in that sample. After adequate standardisation the results of anti-thrombin and anti-factor Xa tests then can be expressed in terms of ng/ml of functional heparin. Because all high affinity material (HAM) in a LMWH catalyses the inactivation of factor Xa, the anti-factor Xa activity will indicate the level of total HAM. The antithrombin activity will indicate the level of ACLM, i.e. the HAM with a $M_r > 5400$. BCLM, the below critical chainlength material that catalyses factor Xa inactivation only, can be obtained from the difference of HAM and ACLM. All LMWHs in practical use are a mixture of ACLM and BCLM. (Only the synthetic high affinity pentasaccharide is a pure BCLM (19,20). This makes it obligatory to characterise each LMWH via two specific activities: the antithrombin specific activity that is a property of ACLM and the anti-factor Xa specific activity that is a property of all HAM. Because of the proportionallity with AT III, the AT III content of the plasma has to be determined independently. If specific activities are to be used for the determination of the levels (in µg/ml) of ACLM and BCLM in plasma samples, different ACLM and BCLM standards should be available.

STANDARD INDEPENDANT UNIT OF HEPARIN ACTIVITY

If one wants to short-circuit the two steps of first determining the specific activity and then the heparin concentration, or when working with insufficiently defined materials, one can always express heparin activity in terms of an unambiguous ***Standard Independent Unit (SIU) of heparin activity***, that can be simply defined as that amount of heparin that, when added to 1 ml of normal plasma, containing 1μM of AT III, will raise the pseudo-first order decay constant of a clotting enzyme by one inverse minute. Standard materials can be defined in terms of SIU units. Of course there are two types of SI units, that based on factor Xa inactivation (Xa-SIU) and that based on the inactivation of thrombin (IIa-SIU). Their use is not restricted to heparins, because one can well define a thrombin based SI unit for dermatan sulfate, in that case reported on the heparin cofactor II content. It will be clear that pentasaccharide activities can be expressed in Xa-SI units and dermatan sulfates in IIa-SI units, but that heparins in general will need to be expressed in both. Because of the functional and pharmacological heterogeneity of heparins it can be expected that the course of heparin activity in vivo in terms of the two different units will not be identical (see below).

STANDARD UNITS AND STANDARD INDEPENDENT UNITS, THE EFFECT OF CA^{++}

A classical unit of standard heparin is an amount of heparin, e.g. 5 μg.
It has been automatically assumed to be a unit of antithrombin activity as well as a unit of antifactor Xa activity. In terms of activity one unit is far from being the same thing however. In fact one unit of the 4th international standad unit of heparin is a much more active antithrombin than antifactor Xa agent, as is immediately seen when we express its activity in standard independent units (table 1).

TABLE 1. EQUIVALENCES OF STANDARD UNITS AND STANDARD INDEPENDENT UNITS

STANDARD	AMOUNT	EQUIVALENT AMOUNT				
		μg	IU-IIa	IU-Xa	SIU-IIa	SIU-Xa
UFH	1 mg	1000	193.0	193.0	14.5	4.4
UFH	1 IU.	5.2	1	1	0.075	0.023
LMWH	1 mg	1000	67.0	168.0	7.0	1.7
LMWH	1 IU-IIa	14.9	1	0.4	0.105	0.026
LMWH	1 IU-Xa	6.0	2.5	1	0.042	0.010

There is still more confusion to be cleared. As a rule, in general laboratory practice, it is common to determine antithrombin and antifactor Xa actitvities in the absence of Ca^{++}ions. This is quite natural because the last thing one wants to have in these decay experiments is to complicate them by the simultaneous generation of endogenous factor Xa or thrombin. If one designs experiments that allow te determination of decay constants in the presence of physiological amounts of Ca^{++}, it appears that the inhibition of factor Xa by standard unfractionated heparin is about twice as efficient as in the absence of Ca^{++} (21). Low molecular weight heparins are hardly affected by these concentrations of Ca^{++}. So in the absence of Ca^{++}, i.e. under the usual laboratory conditions, the antithrombin activity of UFH is significantly underestimated. This makes that the current opinion about a high antifactor Xa over antithrombin ratio of low molecular weight heparins is based on a laboratory artifact.

A SUMMARY OF PREVIOUS IN VITRO RESULTS

At different occasions we have given an overview of our previous results obtained with PPP and PRP in vitro. The main conclusions are therefore only briefly summarised here:

a) The main inhibitory action of UFH in clotting plasma is on thrombin. At concentrations of UFH that inhibit almost completely the appearance of free thrombin, the (extrinsic) conversion of prothrombin into thrombin is inhibited only 20-30 %. This holds for the extrinsic system, where the secundary effects of inhibition of thrombin via feedback mechanisms (see below) is not rate limiting (6).

b) Despite their reputedly high anti-factor Xa action also LMWHs act mainly through inhibition of thrombin, unless they are so small as to have no antithrombin action to speak of, i.e. pentasaccharide and other P-type heparins. In the terms proposed in this article we must conclude that the P-type heparins that we recognised before actually are those heparins that consist almost entirely of BCL material (5,20).

c) The reason for a lack of a direct effect on prothrombin conversion despite a definite anti-factor Xa action is that in clotting plasma, when enough procoagulant phospholipid is added, factor Va is the rate limiting factor and factor Xa is present in excess. Factor Xa levels must be lowered to less than 10 % of their normal values before this diminution shows up as a lack of saturation of phospholipid-adsorbed factor Va, i.e. as an inhibition of prothrombinase (6,22).

d) In the intrinsic system factor VIIIa is rate limiting. Factor VIIIa generates as a result of activation by thrombin. Inhibition of thrombin therefore inhibits the activation of factor VIII and so, indirectly the generation of factor Xa. Therefore, in the intrinsic system, heparins inhibit prothrombin conversion via their antithrombin effect (6,23,24).

e) In platelet rich plasma without added procoagulant phospholipids, phospholipids are rate limiting. The burst of thrombin formation therefore occurs after traces of thrombin have activated the platelets. Heparin will delay the burst because of its antithrombin action. Activated platelets shed heparin-neutralising material (pf4). This neutralises UFH to concentrations of up to 0.4 U/ml. LMWH partly escape this inhibition (25).

AN APPROACH TO IN VIVO EFFECTS

In a preliminary experiment (26) we injected subcutaneously a dose of unfractionated heparin and of two kinds of LMWH (5000 U of UFH, 7500 Choay Units of LMWH-A, 40 mg of LMWH-B). The fact that we only have data on one volunteer per heparin precludes any quantitative conclusion. We present the results here only as a proposal for a rational approach to LMWH pharmacology.

In our volunteers we determined the course of the heparin activity in the blood in terms of SIUs and, via the specific activities, also the course of the blood levels of ACLM and BCLM. Then we determined the main composite biological activities in the samples, i.e. the thrombin potential, the inhibition of the peak of factor Xa generation, the inhibition of the peak of prothrombin converting activity and the thrombin potential in platelet rich plasma.

The main conclusions are

a) UFH causes no inhibition of prothrombinase and no inhibition in platelet rich plasma. Both LMWHs cause some inhibition of prothrombinase and retain significant activity in PRP.

b) ACLM from LMWHs has a longer half life time than that of UFH. BCLM has a longer half life time than ACLM from the same LMWH. This results in a "fractionation in vivo" of the injected LMWH. Late samples are enriched in BCLM.

c) ACLM in all instances is the most active material. The biological activities correlate much better with the ACLM content than with the BCLM content. Even the inhibition of the factor Xa peak is primarily caused by the ACLM present. Evidently by the antifactor Xa activity of the ACLM, because when expressed in SI-units, i.e. in terms of anti-factor Xa activity, then the inhibition of factor Xa shows a correlation with the Xa-SIU level. This indicates that in practice the anti-factor Xa level will always correlate with any biological effect, even if the anti-factor Xa action itself does not influence the effect. Interestingly the inhibition of overall thrombin generation in PRP, in comparison to that in PPP, is relatively dependent upon BCLM, which possibly reflects the fact that ACLM is more sensitive to neutralisation by activated platelets than BCLM is.

It should once more be stressed that these are preliminary results that may indicate a trend but that are to be substantiated by more data.

ACKNOWLEDGEMENTS

We thank all the members of our heparin team and especially A.V. Bendetowicz and S.Wielders for their many contributions to this work.

REFERENCES

1. Béguin, S., Dol, F., Hemker, H.C. Anti-factor IXa activity contributes to the heparin effect. Thromb Haemostas 66: 306-309 (1991).
2. Tollefsen D M, Activation of heparin cofactor II by heparin and dermatan sulfate. Nouv. Rev. Fr. Hematol. 26: 233-237 (1984).
3. Wagenvoord R, Hendrix H, Soria C,Hemker H C, Localisation of the inhibitory site(s) of pentosan polysulfate in blood coagulation. Thromb Haemostas 60:220-225 (1988).
4. Béguin S, Dol F, Hemker H C, Influence of lactobionic acid on the kinetics of thrombin in human plasma, Sem Thromb Hemostas 17: suppl.1, 126-128 (1991).
5. Hemker H C. The mode of action of heparin in plasma. XIth Congress Thrombosis Haemostasis, Brussels. Verstraete M, Vermylen J, Lijnen H R, Arnout J (eds). Leuven University Press, Leuven 1987; pp 17-36.
6. Béguin S, Lindhout T, Hemker H C. The mode of action of heparin in plasma. Thromb Haemostas 60: 457-62 (1988).
7. Hemker H C, Willems G M, Béguin S. A computer assisted method to obtain the prothrombin activation velocity in whole plasma independent of thrombin decay processes. Thromb Haemostas 56: 9-17 (1986).
8. Hemker HC, Wielders S, Béguin S. The thrombin potential, a parameter to assess the effect of antithrombotic drugs on thrombin generation. In Fraxiparine, H Bounameaux, M Samama and J W ten Cate edts. Schattauer, Stuttgart (1990), p89-101.
9. Thunberg L, Lindahl U. On the molecular-weight-dependence of the anticoagulant activity of heparin. Biochem J. 181: 241-243 (1979).
10. Andersson L O, Barrowcliffe T W, Holmer E, Johnson E A, Södenström G. Molecular weight

dependency of the heparin potentiated inhibition of thrombin and activated factor X. Effect of heparin neutralization in plasma. Thromb Res 15: 531-541 (1979).
11. Lane D A, Denton J, Flynn A M, Thunberg L, Lindahl U. Anticoagulant activities of heparin oligosaccharides and their neutralization by platelet factor 4. Biochem J. 1984; 218: 725-732.
12. Barrowcliffe T W, Merton R E, Havercroft S J, Thunberg L, Lindhal U, Thomas D P. Anticoagulant activities of heparin oligosaccharides. Thromb Res 34: 125-133 (1984).
13. Thomas D P, Merton R E, Barrowcliffe T W, Thunberg L, Lindhal U. Effects of heparin oligosaccharides with high affinity for antithrombin III in experimental venous thrombosis. Thrombos Haemostas. 47:244-248 (1982).
14. Thunberg L, Lindahl U. On the molecular-weight-dependence of the anticoagulant activity of heparin. Biochem J. 181: 241-243, (1979).
15. Ellis, V., Scully, M.F., Kakkar, V.V. The relative molecular mass dependence of the anti-factor Xa properties of heparin. Biochem. J. 238: 329-323, (1986).
16. Bendetowicz, A V ,Béguin S,Hemker H C, On the relationship between molecular mass and anticoagulant activity in a low molecular weight heparin. Thromb Haemostas 65, 1300 (1991)(abstr).
17. Bendetowicz A V, Pacaud E, Béguin S, Uzan A, Hemker H C, On the relationship between molecular mass and anticoagulant activity in a low molecular weight heparin (enoxaparin). Submitted.
18. Walenga, J.M., Bara, L., Petitou, M., Fareed, J., Samama, M., Choay, J. The inhibition of the generation of thrombin and the antihrombotic effect of a pentasaccharide with sole anti-factor Xa activity. Thromb Res 51: 23-33, (1988).
19. Choay J, Petitou M, Lormeau J C, Synai P, Casu B, Gatti G. Structure-activity relationship in heparin: A synthetic pentasaccharide with high affinity for antithrombin III and eliciting high anti-factor Xa activity. Biochem Biophys Res Commun. 116: 492-99 (1983).
20. Béguin S, Choay J, Hemker H C. The action of a synthetic pentasaccharide on thrombin generation in whole plasma. Thromb Haemostas 61: 397-401 (1989)
21. Schoen P, Lindhout T, Hemker H C, Ratios of anti-factor Xa to anti-thrombin activities of heparins determined in recalcified human plasma. J Clin Lab Invest. Accepted (1991).
22. Hemker H C, Béguin S, Mode of action of heparin and related drugs, Sem Thromb Hemost, 17: suppl.1, 29-34 (1991).
23. Pieters. J., Lindhout, T., Hemker, H.C. In situ generated thrombin is the only enzyme that effectively activates factor VIII and factor V in plasma. Blood 74: 1021-1024, (1989).
24. Pieters, J., Lindhout, T. The limited importance of factor Xa inhibition to the anticoagulant property of heparin in thromboplastin-activated plasma. Blood 72: 2048-2052, (1988).
25. Béguin, S., Lindhout, T., Hemker, H.C. The effect of trace amounts of tissue factor on thrombin generation in platelet rich plasma its inhibition by heparin. Thromb Haemostas. 61: 25-29, (1989).
26. H C Hemker, S Béguin, A V Bendetowicz, S Wielders. The determination of the levels of unfractionated and low molecular weight heparins in plasma. Their effect on thrombin mediated feedback reactions in vivo. Preliminary results on samples after subcutaneous injection. Haemostasis, Accepted for publication (1991).

PROPHYLACTICALLY EFFECTIVE DOSES OF ENOXAPARIN AND HEPARIN INHIBIT PROTHROMBIN ACTIVATION

Frederick A. Ofosu

Red Cross Society, Blood Services and
Department of Pathology, McMaster University
Hamilton, Ontario, Canada L8N 3Z5.

INTRODUCTION

Results of randomized clinical trials have demonstrated that unfractionated (UF) and low molecular weight (LMW) heparins reduce the frequency of post-operative deep vein thrombosis (DVT) and pulmonary embolism[1-6]. UF and LMW heparins are given by subcutaneous injection 2-12 hours pre-operatively, and the injections are then continued for up to 10 days. The prophylactic injections can also begin post-operatively. Prophylactic use of UF and LMW heparins can reduce the incidence of post-operative DVT by up to 70%[1-8]. UF and LMW heparins are also used to initiate the treatment of deep vein thrombi and pulmonary embolism[9,10]. The need to administer UF or LMW heparins for up to 10 days (to prevent post-operative DVT or to treat established DVT) suggests that the stimuli which contribute to the establishment and propagation of clinically significant thrombi persist for several days.

UF and LMW heparins catalyze the inhibition of activated clotting factors by antithrombin III[11]. One important consequence of their catalytic actions is that when present in high enough concentrations in plasma, UF and LMW heparins can markedly delay the onset of the generation of tenase and prothrombinase[12-14]. The principal mechanisms by which UF and LMW heparins inhibit the formation of the two enzyme systems is to delay factor V and factor VIII activation. The two amplification reactions are usually catalyzed by endogenously generated α-thrombin during coagulation[12-14]. Following the generation of tenase and prothrombinase in plasma containing an UF or a LMW heparin, UF and LMW heparins also catalyze the inhibition of factor Xa and thrombin produced. These anticoagulant actions of UF and LMW heparins can be demonstrated first by a delay of the onset of factor X and prothrombin activation, followed by increased inhibition of thrombin and factor Xa by antithrombin III[12-14]. I will provide evidence consistent with the view that when an UF heparin and a LMW heparin (Enoxaparin) were given after elective orthopedic surgery to prevent post-operative DVT, their principal action on in vivo coagulation was to inhibit the formation of prothrominase. Since they inhibited the formation of prothrombinase <u>in vivo</u>, it is highly likely that the UF heparin and Enoxaparin also inhibited tenase formation <u>in vivo.</u>

APPROACHES FOR MEASURING THE <u>IN VIVO</u> ACTIONS OF AN ANTITHROMBOTIC AGENT

Before presenting evidence that LMW and UF heparin inhibit prothrombinase formation <u>in vivo</u> after elective surgery, I should note that prothrombin activation <u>in vivo</u> is a tightly regulated process. Based on the concentrations of prothrombin, prothrombin fragment 1+2, and thrombin-antithrombin III in plasma, less than 0.1% of the total prothrombin in the circulation (~1.5μM) is usually converted to prothrombin fragment 1+2 and thrombin during the course of normal hemostasis[15]. The tight regulation of prothrombin activation is also maintained following major orthopedic surgery (see below). The direct way to investigate the effects of antithrombotic drugs on <u>in vivo</u> prothrombinase activity is to measure the degree to which anticoagulants affect the levels of prothrombin fragment 1+2 in plasma. Ideally, one would also measure the degree to which antithrombotic agents can inhibit the activation of factor VII, factor IX, and factor X. Activation of the last three clotting factors accelerate the generation of prothrombinase. One additional marker for demonstrating the effects of UF and LMW heparins on <u>in vivo</u> coagulation is to measure the concentration of endogenous thrombin-antithrombin III in plasma. If UF and LMW heparins inhibit the formation of prothrombinase following elective orthopaedic surgery, one would expect that the endogenous plasma levels of both prothrombin fragment 1+2 and thrombin-antithrombin III would decrease following injection of either drug. On the other hand, if the primary effect of UF and LMW heparins on <u>in vivo</u> coagulation is to accelerate the inhibition of thrombin instead of inhibiting prothrombinase formation, then the concentrations of endogenous thrombin-antithrombin III in plasmas of patients receiving an UF and a LMW heparin for prophylaxis of DVT would increase following injection of the drug.

ACTIONS OF UF HEPARIN AND ENOXAPARIN ON PROTHROMBIN ACTIVATION <u>IN VIVO</u>

On the basis of the above considerations, we have measured the concentrations of endogenous thrombin-antithrombin III in the plasmas of patients randomized to receive Enoxaparin or placebo following elective knee surgery. At the time this study was carried out, the commercial kits for measuring the concentrations of prothrombin fragment 1+2 and other markers of <u>in vivo</u> coagulation were not yet available. Nonetheless, one can anticipate that the concentrations of prothrombin fragment 1+2 in the plasmas of patients receiving placebo would be higher than those found in the plasmas of patients receiving Enoxaparin. Active treatment with Enoxaparin reduced the incidence of post-operative DVT by 70%[7]. As summarized in Figure 1, plasmas of patients receiving Enoxaparin contain significantly lower ($p <0.001$ to 0.03) concentrations of endogenous thrombin-antithrombin III than the levels in comparable plasmas of patients on placebo. When thrombin was added to plasmas receiving Enoxpararin and placebo, the post-

Enoxaparin plasmas inactivated more thrombin than post-placebo plasmas (not shown). This last observation also provides supportive evidence that Enoxaparin inhibits the activation of prothrombin in vivo.

The next question we have attempted to answer is whether UF heparin, like Enoxaparin, inhibits the activation of prothrombin when it is given prophylactically after elective hip surgery. Levine and associates have reported that 30 mg of Enoxaparin given twice daily, and 7500 I.U. of UF heparin given twice daily, reduce the frequency of post-operative DVT associated with elective hip surgery to ~20% [8]. We have evidence consistent with the view that prophylactically equivalent doses of Enoxaparin and UF heparin also have equivalent actions on in vivo coagulation. Figure 2 compares the levels of endogenous thrombin-antithrombin III in the plasmas of patients who developed post-operative DVT with plasmas of patients who remained DVT-negative. Pre- and post-operative plasmas of patients who remained DVT-negative had significantly lower concentrations of endogenous thrombin-antithrombin III levels than comparable plasma of patients who developed post-operative DVT. Since there were no differences in the concentrations of heparin or Enoxaparin in the plasmas of DVT-positive and DVT-negative patients (not shown), I would like to suggest that a higher pre-operative endogenous prothrombinase activity likely represents a high risk for developing post-operative DVT.

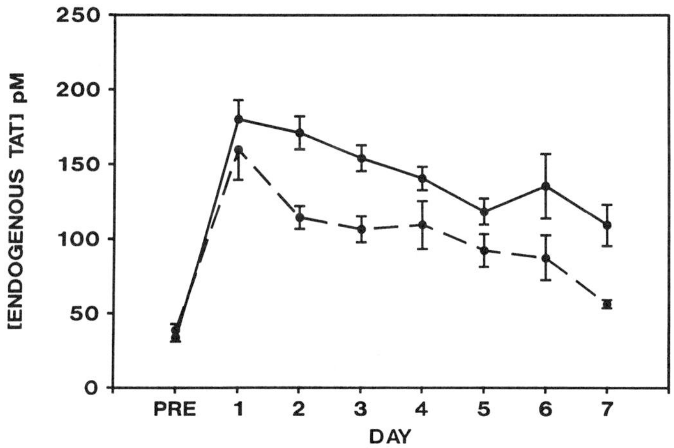

Figure 1. The endogenous prothrombinase activity (measured as endogenous thrombin-ATIII) in pre- and post-operative plasmas of patients randomized to receive placebo (solid lines) or Enoxaparin (broken-line) following elective knee surgery. The post-operative samples (days 1 to 7) were taken six hours after the first of the two daily subcutaneous injection of Enoxaparin. There were approximately 60 patients in each group.

ADDITIONAL METHODS CRUCIAL FOR DETERMINING HOW ANTI-THROMBOTIC DRUGS INHIBIT PROTHROMBINASE FORMATION IN VIVO

An important challenge for the future is to determine the mechanisms by which UF and LMW heparins inhibit prothrombinase formation (or activity) in vivo. Another important challenge is to determine the reasons why plasmas of patients who develop post-operative DVT have higher endogenous prothrombinase activity before the surgery, than the levels found in plasmas of patients in whom prophylaxis is successful. It is not unreasonable to expect that following major orthopedic surgery, increased tissue factor availability is an important stimulus

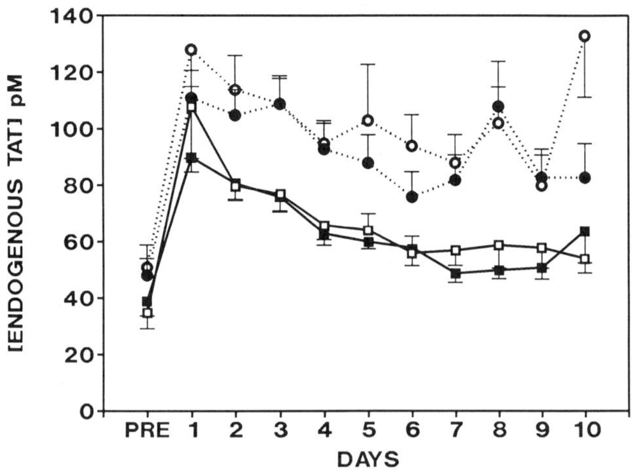

Figure 2. The endogenous prothrombinase activity (measured as endogenous thrombin-antithrombin III) in pre- and post-operative plasmas of patients randomized to receive Enoxaparin or heparin following elective hip surgery. Symbols: ■ DVT-negative Enoxaparin; ● DVT-positive Enoxaparin; □ DVT-negative heparin; ○ DVT-positive heparin.

for both post-operative hemostasis and thrombosis. As a direct consequence of increased tissue factor availability, there would be a coordinated activation of factor IX, factor X and prothrombin following surgery. I would like to propose that the prophylactic effectiveness of UF and LMW heparins after elective orthopedic surgery depends primarily on the degree to which the antithrombotic drugs can assure a coordinated suppression of the tissue-factor-dependent activation of coagulation factors in vivo.

We have recently observed that hirudin, a direct inhibitor of thrombin[16], can simultaneously suppress the activation of factor IX, factor X and prothrombin in vitro (Unpublished). A striking feature of our results was that if tissue factor diluted 1:10 was used to initiate coagulation, hirudin abrogated the activation of all three clotting factors. Based on the effects of hirudin on tissue factor-dependent coagulation, I propose that the injected UF or LMW heparin, vessel wall thrombomodulin, and vessel wall heparan sulphate all act in concert to simultaneously inhibit the tissue-factor-dependent activation of all three clotting factors following high risk orthopaedic surgery. If the hypothesis that heparin and Enoxaparin inhibit <u>in vivo</u> coagulation in a manner analogous to hirudin proves to be a realistic one, one should also see production of lower levels of factor IXa, factor IXa-antithrombin III, factor Xa, factor Xa-antithrombin III and prothrombin fragment 1+2 in the plasmas of patients who have received prophylactically effective doses of Enoxaparin or UFH, when compared with the concentrations seen in the plasmas of patients receiving no prophylaxis or inadequate prophylaxis.

Tools to directly measure the endogenous concentrations of factor VIIa, and subnanomolar levels of factor IXa-antithrombin III and factor IXa in the presence of significantly higher (>50nM) concentrations of factor IX will be crucial before one can accurately determine the degree to which UF and LMW heparins inhibit <u>in vivo</u> coagulation in DVT-negative and DVT-positive patients. Similar tools for measuring the subnanomolar concentrations of factor Xa and factor Xa-antithrombin III in the presence of >100nM factor X also need to be developed.

ACKNOWLEDGEMENTS

The experimental work described above was supported in part by a Grant-In-Aid from the Canadian Red Cross Society Research and Development Fund. I would like to acknowledge the support of Rhone-Poulenc Canada and Behringwerke in providing the Enzygnost-TAT kits. The active collaboration of Jacques Leclerc and Mark Levine in bringing this study to fruition is also gratefully acknowledged.

REFERENCES

1. S. Sagar, J. Massey, and J.M. Sanderson. Low-dose heparin prophylaxis against fatal pulmonary embolism. Br. Med. J. 4: 257 (1975).
2. V.V. Kakkar, B. Djazaeri, J. Fok, M. Fletcher, M.F. Scully, and J. Westick. Low molecular weight heparin and prevention of post-operative deep vein thrombosis. Br. Med. J. 284:375 (1982).
3. V.V. Kakkar, and M.J.G. Murray. Efficacy and safety of low molecular weight heparin (CY216) in preventing post-operative venous thromboembolism: A cooperative study. Br. J. Surg. 72:786 (1985).
4. T. Welin-Berger, S. Bygdeman, and C. Mebius. Deep vein thrombosis following hip surgery: Relationship to activated factor X activity: Effect of heparin and dextran. Acta Orthop. Scand. 53:937 (1982).
5. R. Leyvraz, J. Richard, F. Bachman, C. van Melle, J.M. Treyvand, J.J. Livio, and G. Candardjis. Adjusted versus fixed-dose subcutaneous heparin in the prevention of deep vein thrombosis after total hip replacement. N. Eng. J. Med. 309:954 (1983).

6. A.G.G. Turpie, M. Levine, J. Hirsh, C. Carter, R.M. Jay, P.J. Powers, M. Andrew, R. Gull, and M. Gent. A randomized controlled trial of a low molecular weight heparin (Enoxaparine) to prevent deep-vein thrombosis in patients undergoing elective hip surgery. N. Eng. J. Med. 315:925 (1986).
7. J.R. Leclerc, W. Geerts, L. Desjardins, F. Jobin, F. Delorme, and J. Bourgouin. A randomized trial of Enoxaparin for the prevention of deep vein thrombosis after major knee surgery. Thromb. Haemostas. 65:753 (1991).
8. M.N. Levine, J. Hirsh, M. Gent, A.G. Turpie, J. Leclerc, P. Powers, R.M. Jay, and J. Neemeth. Prevention of deep vein thrombosis in patients undergoing elective hip surgery: A randomized trial comparing Enoxaparin low molecular weight heparin with standard unfractionated heparin. Ann. Int. Med. 114:545 (1991).
9. R.D. Hull, G.E. Rascob, J. Hirsh, R.M. Jay, J.R. Leclerc, W.H. Geerts, D. Rosenbloom, C. Anderson and L. Harrison. Continuous intravenous heparin compared with intermittent subcutaneous heparin in the initial treatment of proximal-vein thrombosis. N. Engl. J. Med. 315: 1109 (1986).
10. R.D. Hull, G.E. Rascob, D. Rosenbloom, A.A. Panju, P. Brill-Edwards, J.S. Ginsberg, J. Hirsh, G.J. Martin, D. Green. Heparin for 5 days versus for 10 days in the initial treatment of proximal venous thrombosis. N. Engl. J. Med. 322: 1260 (1990).
11. R.E. Jordan, G.M. Oosta, W.T. Gardner, and R.D. Rosenberg. The kinetics of hemostatic enzyme-antithrombin interactions in the presence of low molecular weight heparin. J. Biol. Chem. 255:10081 (1980).
12. F.A. Ofosu, P. Sie, G.J. Modi, F. Fernandez, M.R. Buchanan, M.A. Blajchman, B. Boneu, and J. Hirsh. The inhibition of thrombin-dependent feedback reactions is critical to the expression of the anticoagulant effects of heparin. Biochem. J. 243:579 (1987).
13. F.A. Ofosu, J. Choay, N. Anvari, L.M. Smith, and M.A. Blajchman. Inhibition of factor X and factor V activation by dermatan sulfate and the pentasaccharide with high affinity to antithrombin III. Eur. J. Biochem. 193:485 (1990).
14. F.A. Ofosu, J. Hirsh, C.T. Esmon, J.W. Fenton II, and M.A. Blajchman. Unfractionated heparin inhibits thrombin-catalyzed amplification reactions more efficiently than those catalyzed by factor Xa. Biochem. J. 257:143 (1989).
15. J.M. Teitel, K.A. Bauer, H.K. Lau, and R.D. Rosenberg. Studies of the prothrombin activation pathway utilizing radioimmunoassays for the F2/F1+2 fragment and thrombin antithrombin complex. Blood 59: 1086 (1982).
16. P. Walsmann, F. Markward. Biochemical and pharmacological aspects of the thrombin inhibitor hirudin. Pharmazie 36: 653 (1981).

PHARMACOKINETICS OF HEPARIN AND OF DERMATAN SULFATE : CLINICAL IMPLICATIONS

B. Boneu, C. Caranobe, S. Saivin, F. Dol, P. Sié

Laboratoire d'Hémostase
Centre de Transfusion Sanguine
31052 Toulouse Cedex, France

INTRODUCTION

The pharmacokinetic data concerning heparin and dermatan sulfate are usually based upon the disappearance of the biological activities generated after parenteral administration, and not upon the direct determination of their chemical concentrations. At least for heparin, these biological activities are mainly related to the polysaccharide chain length and to the antithrombin III affinity, two factors which largely influence heparin clearance. A good understanding of the pharmacokinetic properties of these glycosaminoglycans needs several complementary approaches which may provide conflicting results reflecting the functional and structural heterogeneity of these compounds.

PHARMACOKINETICS OF UNFRACTIONATED HEPARIN (UH)

After parenteral administration of trace amounts of radiolabeled heparin to animals, the radioactivity is first concentrated in the liver, spleen, bone marrow and lungs. Desulfated and depolymerised material is then excreted in the urine (1). Histochemical methods allow UH to be identified in the endothelial and the reticulo-endothelial cells (2, 3). The blockade of the phagocytic capacities of the reticulo-endothelial system by dextran sulfate dramatically prolongs the half life of UH injected intravenously (4). UH also binds to endothelial cells in culture, a phenomenon independent of the antithrombin III affinity (5-7). Almost 30 % of bound UH is internalized and depolymerised by lysosomal enzymes (8). This cell compartment represents the saturable mechanism of clearance which is pre-eminent for the elimination of the doses usually prescribed in clinical practice. At higher doses, the clearance capacities of this mechanism reach a maximum and the excess of heparin which can not be metabolized is excreted as undegraded and still active material in the urine (1, 4, 9).

This model allows the major pharmacokinetic properties of unfractionated heparin to be understood. After bolus intravenous injection, the biological activities of heparin are cleared exponentially with a progressive prolongation of the half-life when the dose increases (10-12). Over 100 $IU.Kg^{-1}$ the disappearance of the biological activity follows a concave-convex pattern, which is typical of a saturable-non saturable mechanism of clearance (12, 13). In the rabbit (12), we have established that

Table 1. Half-Lives (min) of ^{125}I Heparin and of ^{125}I-CY216 After Bolus Intravenous Injection of Various Doses to Rabbits

Dose injected Anti Xa $IU.Kg^{-1}$	unfractionated heparin	CY 216
1.5	-	8.1
2.5	-	9.2
5	3.8	9.8
10	4.3	9.6
25	7.9	11.0
50	8.9	12.5
100	14.7	12.9
250	32.0	-
375	24.0	-
500	30.0	-

These data are taken from ref. 16. Trace amounts of radiolabeled UH and CY 216 were injected to animals with increasing doses of unlabeled material. Blood was taken at time intervals to determine the circulating radioactivity. The curves obtained were decomposed into 3 exponentials (alpha, beta, gamma). The beta phase half-life was found to be dose-dependent for UH and almost constant for CY 216.

there is almost 10 fold difference between the half-life and the clearance calculated after injection of 5 $IU.kg^{-1}$ and 250 $IU.kg^{-1}$ (Table 1). Over this dose, *i.e.* after saturation of the cellular mechanism of clearance, there is no further change in the pharmacokinetic parameters. Moreover, we have shown that after bolus intravenous injection the instantaneous half-life continuously shortens as the plasma concentration decreases; this accounts for the convexity of the last part of the curve (Fig 1). The non-linearity of the pharmacokinetic parameters also explains the apparent low bioavailability of UH, roughly 30 %, when delivered subcutaneously at low doses to prevent deep vein thrombosis. We hypothesized that, at any dose, the bioavailability of heparin was in fact close to 100 % (14). The bioavailability of heparin injected by subcutaneous route reflects the amount which enters the plasma compartment from the subcutaneous depot and the velocity at which it disappears from the plasma. Since the plasma concentrations generated are lower following subcutaneous compared with intravenous injection, the amounts of UH delivered by the subcutaneous route are cleared more rapidly by the saturable mechanism which continuously shortens the half-life as the plasma concentration declines (12). The relative importance of this dose-dependent effect decreases as the dose delivered by the subcutaneous route increases, *i.e.* when plasma UH concentrations generated are higher and when the cellular mechanism of clearance is saturated (Fig 2).

FACTORS INFLUENCING THE PHARMACOKINETIC BEHAVIOR OF HEPARIN

The pharmacokinetic behavior of the polysaccharide chains of heparin may be altered by the molecular weight, the affinity to antithrombin III and the non-specific binding to plasma proteins. Clinical experience indicates that other unidentified factors may significantly modulate the pharmacokinetic properties of heparin.

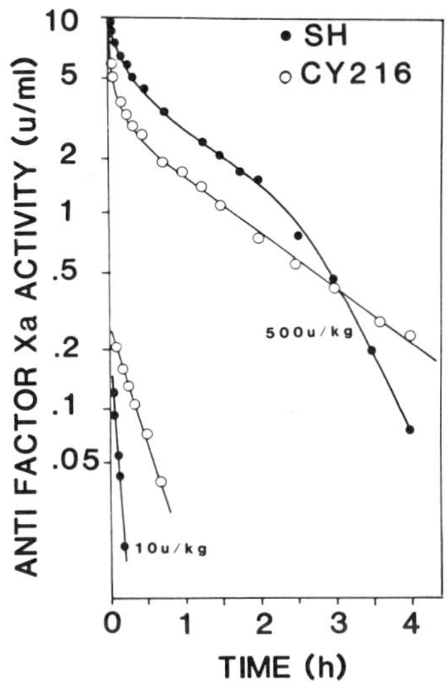

Fig. 1 - Disappearance of the anti-factor Xa activity after bolus intravenous injection of 10 or 500 anti-Xa U.kg^{-1} of standard heparin or of CY 216 to rabbits. The low dose of CY 216 is cleared at a slower rate than that of UH. The high dose of standard heparin is cleared according to a typical concave-convex pattern which is characteristic of a saturable / non-saturable mechanism of elimination. This figure also shows that the very high plasma concentrations of CY 216 are cleared at a faster rate than those of standard heparin.

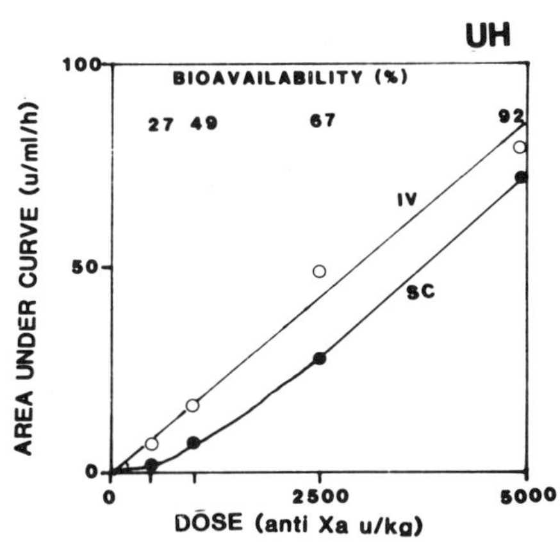

Fig. 2 - Evolution of the area under the curves after bolus intravenous or subcutaneous injection of increasing doses of unfractionated heparin to rabbits. For the IV route, the AUC increase linearly with dose. Note that at the lowest dose, the saturable mechanism of clearance is already saturated (Table 1). For the SC route, the low dose generates low plasma concentrations which are cleared at a faster rate (Table 1). Therefore, the AUC is underestimated. The relative importance of this phenomenon decreases as the dose delivered increases, thus the bioavailability tends towards 100%.

Table 2. **Pharmacokinetic Parameters Calculated from the Anti-factor Xa and Anti-Thrombin Activities Generated by a Single Subcutanueous Injection of 1 500 Antifactor U.Kg^{-1} of CY 216 in Rabbits**

	Dose IU.Kg^{-1}	C max IU.ml^{-1}.h^{-1}	AUC IU.ml^{-1}.h^{-1}	Clearance ml.Kg-1.h^{-1}
anti-Xa	1 500	3.9	38	40
anti-thrombin	417	0.6	6	74
ratio	3.6	6.5	6.3	0.54

While the *in vitro* anti-factor Xa / anti-thrombin ratio is 3.6, the *ex vivo* ratio for C max, and AUC is twice as high. This results from the higher clearance of the anti-thrombin activity in comparison with that of the anti-factor Xa activity.

Molecular weight : low molecular weight heparin (LMWH)

The affinity of LMWHs for endothelial cells in culture is at least 100 times lower than that of UH (7). After injection of radiolabeled heparins, more than 60 % of UH is recovered in the liver and the kidneys but less than 10 % for LMWH (4). The blockade of the functional properties of the reticulo-endothelial system does not alter the pharmacokinetics of LMWH, while it dramatically prolongs the half life of UH (4). Binephrectomy moderately alters the half life of UH while it impairs LMWH disposition even after injection of low doses (4, 15). These data indicate that the relative contributions of the saturable and cellular mechanism to the clearance of LMWH is negligible and that a major part of this material is excreted in the urine. This will have a number of consequences for the pharmacokinetic properties of LMWH.

The half-life and the clearance of a LMWH are independent of the dose injected (16, 17). Table 1 indicates that, below 50 anti-factor Xa IU.Kg^{-1}, the half-lives of UH are shorter than those of LMWH, but that, above this dose, they become longer. These results explain the behavior of UH and of CY 216 after bolus injection of a low and a high dose (Fig. 1) and they have been confirmed by experiments where both compounds were delivered by constant infusion : the plasma activities generated by LMWH are higher than those generated by UH only at low dose regimens (16). Fortunately, the doses prescribed in man for clinical purposes are relatively low and at these doses LMWHs possess a longer half life and a lower clearance than UH. Therefore it is possible to prevent deep vein thrombosis with one single daily subcutaneous injection of a LMWH. Similarly, to treat established deep vein thrombosis, the doses required are 2 to 3 times lower than those of UH. Another practical consequence of the non dose-dependency of LMWH half-life and clearance is the bioavailability close to 100 % after subcutaneous administration of any dose (14, 17, 18). This is not the case for UH, as discussed in the previous section. Finally, one can expect a half-life prolongation and a lower clearance in case of renal failure. Indeed, at least for CY 216 (Fraxiparine^{R}), PK 10169 (Lovenox^{R}, Clexane^{R}) and CY 222 we have found an average of 2-fold half-life prolongation and clearance reduction in patients affected by chronic renal insufficiency, in comparison with healthy volunteers (19-21).

The residual anti-thrombin activity of a LMWH preparation is linked to its content in polysaccharidic chains having more than 18 sugar units. Since the affinity of heparin to endothelial cells is largely modulated by the molecular weight, one can expect a faster clearance of the longer chains. We and others (14, 17, 22) have demonstrated that after parenteral injection of a

Table 3. Evaluation of the Clearance of Unfractionnated Heparin, CY 216 and of their Butyryl Derivatives ($ml.kg^{-1}.h^{-1}$) (C4-UH and C4-CY216) After Bolus Intravenous Injection of Increasing Doses to Rabbits

Dose $mg.kg^{-1}$	UH	C4-UH	CY216	C4-CY216
0.15	423±81	437±141	-	-
0.30	116±22	254±145	76±31	146±59
0.60	112±11	137± 45	49±14	63±42
1.2	-	-	49± 6	41±12
1.5	79± 9	47± 12	-	-
3	52±11	46± 4	38± 6	32± 9
6	38± 8	22± 5	45± 4	20± 6
12	-	-	31± 8	20± 2
15	54±12	17± 3	-	-
30	50± 5	12± 2	38± 7	13± 2

The above results represent the clearances calculated from the plasma anti-factor Xa activities. The results calculated from the anti-thrombin activities were comparable. These data are taken from Ref. 30.

given dose of LMWH, the clearance of the anti-thrombin activity was 1.5 to 2 times higher than that of the anti-factor Xa activity (Table 2). Therefore the *ex vivo* anti-factor Xa/anti-thrombin ratio of a LMWH is higher than the *in vitro* ratio and continuously increases after the injection. At a given time after the injection, there is no longer detectable anti-thrombin activity in the plasma, while the anti-factor Xa activity still remains high. This property raises unanswered questions upon the mechanisms of the antithrombotic action of heparin.

Antithrombin III affinity

The affinity of the polysaccharidic chains of heparin to antithrombin III constitutes another factor which influences its disposition from the plasma independently of the molecular weight. We injected rabbits with radiolabeled heparin fractions (mean MW : 14 000 D) having high affinity and low affinity to antithrombin III and observed that the low-affinity fraction was cleared faster than the high-affinity fraction (23). These results were confirmed independently for Fragmin : the fraction with low affinity to antithrombin III has a 2-fold higher clearance than the fraction with high affinity (24). Using synthetic pentasacharides of different structure and having different affinities to antithrombin III, Vogel et al (25) reported highly significant correlations ($r > 0.91$) between the half-life of these compounds and the Kd to antithrombin III. These observations explain why the pharmacokinetic parameters calculated from radiolabeled heparin and from the biological activities generated by these heparins are not superimposable (23). The experiments performed with radiolabeled material reflect the behavior of all the polysaccharidic chains, independently of their affinity to antithrombin III while the data obtained from the disappearance of the biological activity reflect the behavior of the moiety which has a high affinity to antithrombin III. Moreover, these observations and those of the previous section indicate that the chemical composition of heparin continuously changes after its parenteral administration.

Covalent heparin-antithrombin III complexes have been prepared. These complexes retain the anticoagulant and antithrombotic properties of the parent compounds. The half-life of disappearance of such complexes is prolonged and tends towards that of antithrombin III (26, 27).

Non-specific binding to plasma proteins

In collaboration with Sanofi-Choay we have recently reported (28-30) on heparins having long lasting effects. These heparins are O-acylated and contain an average of 1.7 butyryl chains per disaccharide unit. In a purified system the catalytic activity of these heparins upon thrombin/factor Xa inhibition by antithrombin III is unaffected but the addition of albumin to the reagents dramatically reduces the catalytic activities. In plasma, the anticoagulant and the specific activities are reduced roughly 2-fold. The antithrombotic activity and the prohemorrhagic potential is comparable to the parent compounds. Table 3 summarizes the evolution of the clearance after bolus injection of increasing doses. The strong reduction of the clearance of UH at the higher doses is confirmed, while that of CY 216 is almost constant. Over the dose of 3 $mg.Kg^{-1}$, both butyryl derivatives have a 2 to 3 times lower clearance. The mechanism responsible for this delayed elimination is not fully understood. The non-specific binding of the butyryl derivative to plasma albumin could reduce and delay the specific binding to antithrombin III. If such a pattern was confirmed in human volunteers, such compounds could be of interest for the treatment of established deep vein thrombosis with one single subcutaneous injection a day, as has been reported recently.

Unidentified factors affecting heparin pharmacokinetics

A number of unidentified factors may alter the pharmacodynamic behaviour of UH and of LMWH in patients affected by thrombotic disorders and in normal volunteers. Variations as high as 400 % have been found in the pharmacokinetic parameters (31-34). Some retrospective studies suggest that the level of UH required to achieve proper anticoagulation is higher in patients affected by thromboembolic disorders (35-36). However, other prospective studies have not confirmed the heparin hyperconsumption phenomenon, except in cases of acute pulmonary embolism (33, 34, 37, 38).

Circadian changes in the anticoagulant effect of heparin infused at a constant rate have also been reported (39). These observations could result from circadian changes in the sensitivity of the plasma to a given dose of heparin or to circadian alterations in the pharmacokinetic parameters. However a more recent report failed to confirm these observations (40).

PHARMACOKINETICS OF DERMATAN SULFATE (DS)

Since DS is not yet widely used for therapeutic purposes, the information concerning its pharmacokinetic properties is relatively scarce. In the competitive binding experiments performed with cultured endothelial cells and heparin, it was noticed that DS did not displace heparin from cell surface (5, 7). This suggests that DS does not bind to endothelial cells. After injection of ^{125}I-DS to human volunteers, unmetabolized compound is recovered very soon in the urine (41). These observations fit with report of Dol et al who found that renal excretion was critical for the disposition of dermatan sulfate after intravenous injection to rabbits (42).

Therefore, for the same reasons as those discussed above for LMWH, the pharmacokinetic parameters of DS have been found constant or moderately affected by the doses delivered by intravenous route (41-43). Fig. 3 allows a direct comparison of the half-lives, and the clearance of UH, LMWH and DS in man. The faster disappearance of DS may be explained as follows. The pharmacokinetic data of heparins were calculated from the circulating biological activities. Thus, they reflect the behavior of the moiety which has a high affinity to antithrombin III. We have indicated in the previous section that this moiety was cleared more slowly than the low-affinity moiety. Such functional heterogeneity towards heparin cofactor II has not been reported for dermatan sulfate.

Table 4. Influence of Molecular Weight Distribution upon the Clearance (IV route) and the Bioavailability (SC route) of Natural Dermatan Sulfate and of its Low Molecular Weight Fractions

	MW Distribution (Kd)	Clearance $ml.kg^{-1}.h^{-1}$	Bioavailability %
Natural DS	12-45	170±22	29
LMW - DS	14-16	190±35	74
LMW - DS	7.9-10	218±64	92
LMW - DS	1.6-8	323±49	120

Dose : 2 $mg.kg^{-1}$.

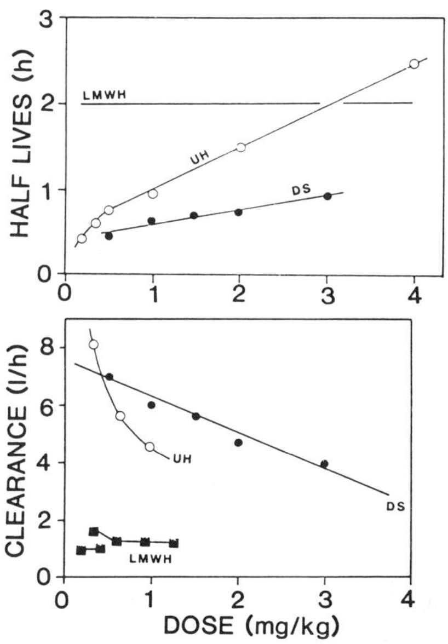

Fig. 3 - This figure summarizes several results taken from the literature and obtained in man. The doses delivered are expressed in weight. The clearances are calculated from the anti-factor Xa activity for LMWH (17) anti-factor Xa or anti-thrombin activities for UH (10, 11), and from the anti-thrombin activity for DS (43).

After subcutaneous or intramuscular administration, the peak of DS activity is obtained at 3 h and this activity disappears with a very long half-life (8 h). However, in contrast to heparin, the bioavailability of DS after subcutaneous or intramuscular administration is low (25 % and 13 % respectively) and not improved when the dose injected is increased, as is the case for UH (41, 43). This indicates that a significant fraction of the amount injected remains at the site of the injection, from where it is resorbed very slowly. This explains why, when the drug is delivered for several consecutive days by intramuscular route, there is a progressive increase in the plasma concentration. At least 10 days are required to obtain a steady state plasma concentration.

Thus, in comparison with heparins, the 2 major pharmacokinetic drawbacks of DS are a relatively short half-life and a low bioavailability after subcutaneous and intramuscular administration.

ATTEMPTS TO IMPROVE THE PHARMACOKINETIC PROPERTIES OF DERMATAN SULFATE

The pharmacokinetic properties of DS may be modulated using the same strategies as those used for heparins.

We have investigated the pharmacokinetic behavior of 2 oversulfated DS containing 2 and 3.7 sulfate groups per disaccharide unit *i.e.* a 2 fold and 3.7 fold increase compared to the natural compound (44). As for UH, there was a prolongation of the half-life of disappearance when the dose was increased and, at the higher dose, the curve of disappearance followed a concave-convex pattern. This suggests that these oversulfated compounds are cleared according the same complex mechanism as UH. Indeed, oversulfated DS binds to endothelial cells with high affinity (45), while the natural compound does not (5, 7). Unfortunately these oversulfated DS are not more efficient in preventing thrombosis (44) and their hemorrhagic potential is increased (46).

DS-oligosaccharides with high affinity to heparin cofactor II have been prepared. These oligosaccharides have enhanced catalytic activities upon thrombin inhibition by heparin cofactor II as well as enhanced antithrombotic activities (47). Whether or not these oligosaccharides have a longer half-life remains to be established.

Another possibility to improve the pharmacokinetic behaviour of DS appears to be depolymerisation yielding low molecular weight compounds (LMWH-DS). We have reported that such a compound has a longer half-life and a bioavailability close to 100 % after subcutaneous administration (48). Unfortunately, due to a higher distribution volume, the total clearance is doubled and is inversely related to the molecular weight distribution (Table 4). Further studies are required to find the best compromise between molecular weight, clearance and bioavailability.

ACKNOWLEDGEMENTS

The authors are indebted to Professor G. Houin (Pharmaceutic Science Faculty, Toulouse, France), M.R. Buchanan (McMaster University, Hamilton, Ontario), J. Choay, M. Petitou, J.C. Lormeau, H. Toulemonde, B. Bayrou (Sanofi-Choay, Paris, France) and F. Gianese (Mediolanum Farmaceutici, Milan, Italy) for stimulating discussions. The excellent technical assistance of A.M. Gabaig is gratefully acknowledged.

REFERENCES

1. J. Dawes, and D. S. Pepper. Catabolism of low-dose heparin in man. Thromb Res, 14:845 (1979).
2. T. H. Oh, S. S. Naidoo, and L. B. Jaques. The uptake and disposition of ^{35}S-heparin by macrophages in vitro. J Reticuloendothel Soc, 13:134 (1973).
3. J. Mahadoo, L. Hiebert, and L. B. Jaques. Vascular sequestration of heparin. Thromb Res, 12:79 (1977).
4. M. Palm, and C. Matsson. Pharmacokinetics of heparin and low molecular weight heparin fragment (Fragmin^{r}) in rabbits with impaired renal or metabolic clearance. Thromb Haemost, 58:932 (1987a).
5. B. Glimelius, C. Busch, and M. Hook. Binding of heparin on the surface of cultured human endothelial cells. Thromb Res, 12:773 (1978).
6. T. Barzu, P. Molho, G. Tobelem, M. Petitou, and J. Caen. Binding and endocytosis of heparin by human endothelial cells in culture. Biochim Biophys Acta, 945:196 (1985).

7. T. Barzu, J. L. M. L. Van Rijn, M. Petitou, P. Molho, G. Tobelem, and J.P. Caen. Endothelial binding sites for heparin ; specificity and role in heparin neutralization. Biochem J, 238:847 (1986).
8. T. Barzu, J. L. M. L. Van Rijn, M. Petitou, and J.P. Caen. Heparin degradation in the endothelial cell. Thromb Res, 47:601 (1987).
9. J. Pipper. The fate of heparin in rabbits after intravenous injection. Filtration and tubular secretion in the kidneys. Acta Pharmacol, 3:373 (1947).
10. P. Olsson, H. Lagergren, and S. Ek. The elimination from plasma of intravenous heparin. An experimental study on dogs and humans. Acta Med Scand, 173:619 (1963).
11. T. D. Bjornsson, K. M. Wolfram, and B. Kitchell. Heparin kinetics determined by three assay methods. Clin Pharmacol Ther, 31:104 (1982).
12. B. Boneu, M. R. Buchanan, C. Caranobe, A. M. Gabaig, D. Dupouy, P. Sié, and J. Hirsh. Evidence for a saturable mechanism of disappearance of standard heparin in rabbits. Thromb Res, 46:835 (1987).
13. C. A. M. De Swart, B. Nijmeyer, J. M. M. Roelofs, and J. J. Sixma. Kinetics of intravenously administered heparin in normal humans. Blood, 60:1251 (1982).
14. L. Briant, C. Caranobe, S. Saivin, P. Sié, B. Bayrou, G. Houin, and B. Boneu. Unfractionated heparin and CY 216 : pharmacokinetics and bioavailabilities of the antifactor Xa and IIa effects after intravenous and subcutaneous injection in the rabbit. Thromb Haemost, 61:348 (1989).
15. C. Caranobe, A. Barret, A. M. Gabaig, D. Dupouy, P. Sié, and B. Boneu. Disappearance of circulating anti-Xa activity after intravenous injection of standard heparin and of a low molecular weight heparin (CY 216) in normal and nephrectomized rabbits. Thromb Res, 40:129 (1985).
16. B. Boneu, M. R. Buchanan, C. Caranobe, A. M. Gabaig, D. Dupouy, P. Sié, and J. Hirsh. The disappearance of a low molecular weight heparin fraction (CY 216) differs from standard heparin in rabbits. Thromb Res, 46:845 (1987).
17. A. M. Frydman, L. Bara, Y. Le Roux, M. Woler, F. Chauliac, and M. Samama. The antithrombotic activity and pharmacokinetics of enoxaparine, a low molecular weight heparin, in humans given single subcutaneous doses of 20 to 80 mg. J Clin Pharmacol, 28:609 (1988).
18. J. Dawes, L. Bara, E. Billaud, and M. Samama. Relationship between biological activity and concentration of a low molecular weight heparin (PK 10169) and unfractionated heparin after intravenous and subcutaneous administration. Haemostas, 16:116 (1986).
19. C. Goudable, H. Ton That, A. Damani, D. Durand, C. Caranobe, P. Sié, and B. Boneu. Low molecular weight heparin half life is prolonged in hemodialysed patients. Thromb Res, 43:1 (1986).
20. Y. Cadroy, J. Pourrat, M.F. Baladre, S. Saivin, G. Houin, J.L. Montastruc, I. Vernier, and B. Boneu. Delayed elimination of enoxaparine in patients with chronic renal insufficiency. Thromb Res, 63:385 (1991).
21. C. Goudable, S. Saivin, G. Houin, P. Sié, B. Boneu, H. Tonthat, and J.M. Suc. Pharmacokinetics of a low molecular weight heparin (fraxiparine[r]) in various stages of chronic renal failure. Nephron, 1991 (In press).
22. T. Matzsch, D. Bergqvist, U. Hedner, and P. Ostergaard. Effects of an enzymatically depolymerized heparin as compared with conventional heparin in healthy volunteers. Thromb Haemost, 57:97 (1987).
23. C. Caranobe, M. Petitou, D. Dupouy, A.M. Gabaig, P. Sié, M. R. Buchanan, and B. Boneu. Heparin fractions with high and low affinities to antithrombin III are cleared at different rates. Thromb Res, 43:635 (1986).
24. M. Palm, and C. Matsson. Pharmacokinetics of fragmin. A comparative study in the rabbit of its high and low affinity to antithrombin. Thromb Res, 48:51 (1987b).

25. G. Vogel, T. Von Dinther, A. Visser, M. T. Buiting, R. Von Amsterdan, and D. Meuleman. Significance of the AT III binding affinity of some pentasaccharide analogues for their anti-Xa activities and elimination rates in rats. Thromb Haemost, 65:930 (1991).
26. M. Hoylaerts, E. Holmer, M. De Mol, and D. Collen. Covalent complexes between low molecular weight heparin fragments and antithrombin III. Inhibition kinetics and turnover parameters. Thromb Haemost, 49:109 (1983).
27. C. Mattson, M. Hoylaerts, E. Holmer, T. Uthne, and D. Collen. Antithrombotic properties in rabbits of heparin and heparin fragments covalently coupled to human antithrombin III. J Clin Invest, 75:1169 (1985).
28. S. Saivin, M. Petitou, J.C. Lormeau, D. Dupouy, P. Sié, C. Caranobe, G. Houin, and B. Boneu. Pharmacological properties of unfractionated heparin derivative with long lasting effects. J Lab clin Med, 1991 (In press).
29. S. Saivin, M. Petitou, J.C. Lormeau, D. Dupouy, P. Sié, C. Caranobe, G. Houin, and B. Boneu. Pharmacological properties of a low molecular weight heparin derivative with long lasting effects. Thromb Haemost, 1991 (In press).
30. S. Saivin, C. Caranobe, M. Petitou, J.C. Lormeau, G. Houin, and B. Boneu. Pharmacodynamic properties of long lasting butyryl heparin derivatives in the rabbits. Thromb Haemost, 1991 (submitted).
31. R. J. Cipolle, R. D. Siefert, B. A. Neilan, D. E. Zaske, and E. Haus. Heparin kinetics : variables related to disposition and dosage. Clin Pharmacol Ther, 29:387 (1981).
32. J. W. Estes. The heterogeneity of the anticoagulant response to heparin. J Clin Path, 25:45 (1972).
33. J. Hirsh, W. G. Van Aken, A. S. Gallus, C. T. Dollery, J. F. Cade, and W. L. Yung. Heparin kinetics in venous thrombosis and pulmonary embolism. Circulation, 53:691 (1976).
34. S. Benchekroun, B. Eychenne, O. Mericq, A. Colombani, P. Douste-Blazy, A. Barret, P. Sié, and B. Boneu. Heparin half-life and sensitivity in normal subjects and in patients affected by deep vein thrombosis. Eur J Clin Invest, 16:536 (1986).
35. T. M. White, J. L. Bernene, and A. M. Marino. Continuous heparin infusion requirements. Diagnostic and therapeutic implication. JAMA, 241:2717 (1979).
36. B. L. Beavers, D. Young, and B. Satiani. Prediction of heparin requirements in acute thromboplastic venous disease. Arch Surg, 120:436 (1985).
37. T. L. Simon, T. M. Hyers, J. P. Gaston, and L. A. Harker. Heparin pharmacokinetics : increased requirements in pulmonary embolism. Br J Haematol, 39:111 (1978).
38. C. G. Elliott, R. J. Michocki, R. Brown, and O. E. Ottesen. Heparin requirements in pulmonary embolism and venous thrombosis : a prospective study. J Clin Pharmacol, 22:102 (1982).
39. H. A. Decousus, M. Croze, F. A. Levi, J. G. Jaubert, B. M. Perpoint, J. F. De Bonadona, A. Reinberg, and P. M. Queneou. Circadian changes in anticoagulant effect of heparin infused at a constant rate. Brit Med J, 290:341 (1985).
40. P. Toulon, J. F. Vitoux, C Leroy, T. Lecomte, M. Roncato, Y. Motobashi, M. Aiach, and J. N. Fiessinger. Circulating activities during constant infusion of heparin or a low molecular weight derivative (Enoxaparine) : failure to demonstrateany circadian variations. Thromb Haemost, 58:1068 (1987).
41. J. Dawes, B. A. Hodson, and D. S. Pepper. The absorption, clearance and metabolic fate of dermatan sulphate administered to man-studies using a radioiodinated derivative. Thromb Haemost, 62:945 (1989).

42. F. Dol, G. Houin, D. Dupouy, Y. Cadroy, C. Caranobe, A.M. Gabaig, J. Mardiguian, P. Sié, and B. Boneu. Pharmacokinetics of dermatan sulfate in the rabbit after intravenous injection. Thromb Haemost, 59:255 (1988).
43. F. Dol, G. Houin, M. Rostin, J.L. Montastruc, D. Dupouy, F. Gianese, P. Sié, and B. Boneu. Pharmacodynamics and pharmacokinetics of dermatan sulfate in humans. Blood, 74:1577 (1989).
44. F. Dol, C. Caranobe, D. Dupouy, M. Petitou, J.C. Lormeau, J. Choay, P. Sié, and B. Boneu. Effects of increased sulfation of dermatan sulfate on its in vitro and in vivo pharmacological properties. Thromb Res, 52:153 (1988).
45. J. L. M. L. Van Rijn, M. Trillou, J. Mardiguian, G. Tobelem, and J. Caen. Selective binding of heparins to human endothelial cells. Implications for pharmacokinetics. Thromb Res, 45:211 (1987).
46. J. Van Ryn-McKenna, F. A. Ofosu, J. Hirsh, and M. R. Buchanan. The antithrombotic and bleeding effects of glycosaminoglycans with different degrees of sulfation. Br J Haematol, 71:265 (1989).
47. P. Sié, D. Dupouy, C. Caranobe, B. Boneu, and M. Petitou. Antithrombotic properties of a dermatan sulfate hexadecasaccharide fractionated by affinity for heparin cofactor II. 1991 (submitted).
48. F. Dol, M. Petitou, J.C. Lormeau, J. Choay, C. Caranobe, P. Sié, S. Saivin, G. Houin, and B. Boneu. Pharmacologic properties of a low molecular weight dermatan sulfate : comparison with unfractionated dermatan sulfate. J Lab Clin Med, 115:43 (1990).

HEPARIN IN THE PREVENTION AND TREATMENT OF ARTERIAL THROMBOEMBOLISM

Marc Verstraete

Center for Thrombosis and Vascular Research
University of Leuven, Campus Gasthuisberg, O & N
Herestraat 49, B-3000 Leuven

INTRODUCTION

The aim of treatment of thromboembolic occlusion is prompt restoration of blood flow. Ideal treatment of thrombosis is therefore to remove the obstruction. If this is not possible, treatment should ensure that the thrombus does not extend or generate an embolus.

Prevention of thromboembolism is at present easier to achieve in most situations than treatment of pre-formed thrombi. In treating established thrombosis, antithrombotic drugs can at best only prevent extension of thrombi and emboli formation, and these limitations sould be borne in mind. At present the anticoagulants in clinical use are heparin (standard unfractionated heparin and low molecular weight heparin) and oral anticoagulants (coumarin and indanedione derivatives).

ATRIAL FIBRILLATION

Atrial fibrillation unrelated to valvular heart disease starts at a mean of 64 years, affects 2 to 5 percent of the general population over the age of 60, and is associated with a 5 to 7 percent yearly risk of thromboembolism (Chesebro et al., 1992). In patients with valvular heart disease, atrial fibrillation is nearly always associated with mitral disease. Patients with paroxysmal atrial fibrillation are at equal risk of thromboembolism as those with chronic atrial fibrillation, even in patients without valvular heart disease (Stroke Prevention in Atrial Fibrillation Study Group Investigators, 1990). These patients should be on long-term oral anticoagulation at a low international normalized ratio (1.5 to 4.5) or on aspirin in older patients.

VALVULAR HEART DISEASE AND PROSTHETIC HEART VALVES

Not all patients with valvular heart disease require anticoagulation; as the incidence of thromboembolism depends on the valve involved (more frequent in mitral than in aortic valve disease), the degree of left ventricular dysfunction, left atrial size, atrial fibrillation and history of previous thromboembolism (Chesebro et al., 1992). If anticoagulation is required it should be long-term and thus oral anticoagulation.

Heparin and Related Polysaccharides
Edited by D.A. Lane *et al.*, Plenum Press, New York, 1992

In the first decade of valve replacement most patients had atrial fibrillation before but also after valve replacement because surgery was postponed untill the later stages of the disease (Friedli et al., 1971; Cleland et al., 1973; Fuster et al., 1982). This situation has changed and up to one-third of patients with atrial fibrillation before mitral valve replacement may convert to sinus rhythm after surgery (Cohn et al., 1985). However, atrial fibrillation increases with age and is present in many patients with mitral or aortic valve replacement when they reach the age of 65 or 70 years (Douglas et al., 1984). Despite treatment with oral anti-coagulants, the incidence of thromboembolic complications with the best mechanical prostheses is still about 0.2 (fatal) to 2 (non fatal) per 100 patient-years.

Antithrombotic prevention in patients with mechanical heart valve replacement

The present recommendation is to start with intravenous heparin four to six hours after operation with a dose maintaining the partial thromboplastin time at the upper limit of normal (circa 600 U/h). This infusion is continued until chest tubes are removed and then substituted by subcutaneous unfractionated heparin (12,500 U twice daily), this dose being adjusted to increase the partial thromboplastin time 1.5 to 2 times the normal level. After a few days oral anticoagulation is started and maintained at a prothrombin time corresponding with an International Normalized Ratio (INR) between 3.0-4.5.

No trial was conducted in patients with mechanical heart valves to determine whether a lower level of anticoagulation (e.g. INR 2.5-3.5) would be as effective but reduce the bleeding incidence observed with the deeper anticoagulation corresponding to the presently recommended INR range of 3.0-4.5. A step in the right direction comes from Argentina where patients with mechanical prosthetic valves were randomised to an oral anticoagulation aiming at an INR of 2.0-3.0 or 3.0-4.5. The protection from embolism was equal in the two treatment groups but at the cost of more bleeding in the higher dose range (Altman et al., 1991). However, it is important to note that both groups of patients also received a combination of aspirin and dipyridamole in addition to the oral anticoagulant.

Antithrombotic prevention in patients with bioprosthetic valve replacement

To overcome the risk of thromboembolism that is inherent to all prosthetic valves and the attendant hazards and inconveniences of life-long anticoagulation, considerable research has been directed to the development of non-thrombogenic tissue valves. The first of these were chemically sterilised homografts which were soon abandoned because the formalin used dissolves the collagen cross linkages in the valve cusps. In a further development porcine heterografts were sterilized in a dilute solution of gluraldehyde which renders the valve essentially inert collagen shells with little if any antigenicity. These valves are then mounted on semiflexible frames.

The present recommendation is that patients with bioprosthetic valves in a mitral position be treated for the first three months after valve insertion with oral anticoagulants at an INR of 2.0-3.0 (ACC/NHLBI, 1986). The lower level of anticoagulation is based on a randomized study of patients with a bioprosthetic valve in the mitral position which showed that a less intensive oral anticoagulation (INR 2.0-3.0) gave the same protection from thromboembolism (1.2%) as deeper anticoagulation (INR 3.0-4.5) but with significantly less bleeding (5.7% vs 20.6%) (Turpie et al., 1988).

However, after three months these patients can discontinue coumarin drugs and switch to aspirin (80 mg/day) but only in the absence of atrial fibrillation, a large left atrium or a history of previous systemic thromboembolism.

UNSTABLE ANGINA

The use of heparin appears to be of value in patients hospitalized with unstable angina (Telford et al., 1981; Williams et al., 1986; Neri Serneri et al., 1988,1990; Théroux et al., 1988). In the acute phase of unstable angina, initial bolus of 5,000 heparin, followed by a continuous infusion of 1,000 U heparin per hour with maintenance of the partial thromboplastin time in the 1.5 to 2 times normal range may be even more effective than aspirin alone in the prevention of recurrent ischemic episodes and early events. Combination therapy with aspirin and heparin during the early hospital phase may confer additional benefit over either agent used alone (The RISC Study Group). Aspirin should be initiated immediately, and maintained at a dose of 160 to 325 milligram/day. A lower daily dose (75 milligrams) may be sufficient, and in conjunction with heparin may be safer, although the evidence in support of such a low dose is relatively small by comparison to that for the higher doses. The heparin should be sustained for at least seven days in sufficient dose to maintain the partial thromboplastin time in the 1.5 to 2 times normal range.

MURAL THROMBOSIS, ARTERIAL EMBOLISM AND EARLY MORTALITY IN ACUTE MYOCARDIAL INFARCTION

Without antithrombotic therapy, left ventricular mural thrombi develop in 20 percent of cases of acute myocardial infarction, 40 percent of anterior infarcts, and 60 percent of those large anterior infarcts which involve the ventricular apex and produce peak serum creatine kinase concentrations greater than 2,000 units/liter in the absence of thrombolytic reperfusion. Autopsy studies after fatal myocardial infarction, representing the largest infarcts, have found the incidence of mural thrombi in the range of 40 to 70 percent when anticoagulant therapy was not given (Kothari et al., 1984; Funke Kupper et al., 1989; Veterans Administration Cooperative Study Investigators, 1973; Hilden et al., 1961; Penny et al., 1988). While more than one third of patients with acute anterior myocardial infarction develop left ventricular thrombi, this complication occurs in less than 5 percent of those with inferior infarction.

Within the past decade, six randomized studies involving patients with acute myocardial infarction have addressed the relationship of echocardiographically detected left ventricular thrombi and cerebral embolism (Table 1) (Working Party on Anticoagulant Therapy in Coronary Thrombosis, 1969; Drapkin et al., 1972; Nordrehaug et al., 1985; Davis et al., 1986; Gueret et al., 1986; Arvan et al., 1987; Turpie et al., 1989, the SCATI Group, 1989). In aggregate, thrombus formation was reduced by more than 50 percent with anticoagulation: individually, however, each trial had insufficient sample size to detect significant differences in embolism.

Regarding the intensity of anticoagulation for prevention of left ventricular mural thrombus formation in patients with acute myocardial infarcton, Turpie et al. (1989) described a randomized prospective trial involving 221 patients with acute anterior myocardial infarction who did not receive thrombolytic agents in the acute phase, where subcutaneous administration of calcium heparin in a dose of 12,500 units every 12 hours offered greater benefit for prevention of left ventricular mural thrombi during the first ten days than the lower dosage of 5,000 units every 12 hours conven-

Table 1. Randomized trials of anticoagulant therapy in acute anterior myocardial infarction for prevention of left ventricular thrombosis

	Number of patients	Antithrombotic regimen	Patients with left ventricular thrombi (%)		Results
			Treatment	Control	
Nodrehaug	53	IV heparin (400 units/kilogram/day) followed by warfarin versus placebo	0	26	Positive ($p < 0.01$)
Davis	52	IV heparin (1,000 units/hour) followed by warfarin versus low-dose A/C	56	56	Negative
Gueret	46	IV heparin (1,000 units/hour) followed by SQ heparin versus no A/C	38	52	No significant difference
Arvan	30	IV heparin (1,000-3,000 units/hour) followed by warfarin versus no A/C	31	35	Negative
Turpie	183	High-dose SQ heparin (12,500 units) versus low-dose (5,000 units) every 12 hours	11	32	Positive ($p= 0.0004$)
SCATI	200	SQ heparin (12,500 units) every 12 hours versus no A/C	18	36	Positive ($p < 0.01$)*

*Some patients admitted within six hours of onset of infarction received intravenous streptokinase.
IV= intravenous; SQ= subcutaneous; A/C= anticoagulation

Table 2. Frequency of systemic arterial and pulmonary embolism by short-term anticoagulation.

	Number of patients	Systemic arterial embolism			Pulmonary embolism		
		AC	Control	Reduction	AC	Control	Reduction
MRC	1,427	1.3	3.4	62*	2.2	5.6	61*
BMH	1,136	1.7	2.3	26	3.8	6.1	38
VACT	999	0.8	5.4	85*	2.0	4.8	58*

*Denotes a statistically significant difference ($p < 0.01$).
AC= anticoagulant group; MRC= Medical Research Council Trial; BMH= Bronx Municipal Hospital Study; VACT= Veterans Administration Cooperative Trial.

tionally given for prevention of venous thrombosis. Thrombi were detected echocardiographically in 11 percent of the group given the higher heparin dose, compared with 32 percent of those receiving the lower dose ($p = 0.0004$). This beneficial effect correlated strongly with plasma heparin activity. The overall incidence of nonhemorrhagic stroke in the early period following infarction was just under 1 percent in those initially treated with high-dose heparin, and close to 4 percent in the low-dose group ($p = 0.17$), but insufficient event rates may have been the reason why there was no significant differnce in embolic complications. Hemorrhagic complications were equally frequent in both treatment groups in this study.

In the SCATI (The SCATI Group, 1989) randomized trial of 771 patients with acute myocardial infarction, 433 were given the thrombolytic agent streptokinase (1.5 million units over one hour). Half received heparin (12,500 units every 12 hours) and the other no heparin. Of the 235 cases of initial anterior infarction, satisfactory two-dimensional echocardiographic examinations were obtained both early after admission and near the time of discharge in 200. The predischarge echocardiograms showed a significantly lower incidence of left ventricular thrombi in those receiving heparin (18 percent) than in the control group (37 percent: $p < 0.01$). The earlier echocardiogram was obtained within 24 hours of admission in 86 percent of patients, at which time 20 (18 percent) of those assigned to receive heparin and 13 (14 percent) of those in the control groups showed evidence of thrombi, not a significant difference. Among those free of early echocardiographically manifest thrombi, 6 percent of those treated with heparin developed thrombi by the time of the predischarge examination, compared with 26 percent of the controls (relative risk 0.17; 95 percent conficence interval 0.06 to 0.47; $p < 0.0002$). It is not clear, however, what proportion of these patients were given thrombolytic therapy and whether this intervention affected the rate of development of ventricular thrombi. Stroke occurred in two patients in the control group, and both had echocardiographic evidence of mural thrombi; no strokes were detected in the treated patients. One fatal cerebral hemorrhage occurred in a patient in the heparin group.

Over the past 20 years, three large trials involving patients with acute inferior and anterior myocardial infarctions concluded that initial treatment with heparin followed by administration of an oral anticoagulant reduced the occurrence of cerebral embolism from 3 percent to 1 percent when compared with no anticoagulation (Table 2). Differences were statistically significant in two of the studies, with a concordant trend in the third (Veterans Administration Cooperative Study Investigators, 1973; Working Party on Anticoagulant Therapy in Coronary Thrombosis, 1969; Drapkin et al., 1972).

Both the GISSI-2 (Gruppo Italiano per lo Studio della Supravvivenza nell Infarto Myocardico, 1990; The International Study Group, 1990) and ISIS-3 (Sleight, 1991) trials examined the benefit of subcutaneous heparin (after 12 or 4 hours, respectively) added to a regimen of thrombolysis and aspirin therapy. The results of the two trials were similar. Considering the number of events per 1,000 patients, for early mortality, the reductions associated with heparin were 5 deaths per 1,000 in ISIS-3 and 4 deaths per 1,000 in GISSI-2. Reinfarction was reduced by heparin by 3 events per 1,000 patients in ISIS-3 and by 1 event per 1,000 patients in GISSI-2.

CHRONIC PHASE OF MYOCARDIAL INFARCTION

There is very limited experience with long-term subcutaneous administration of heparin after myocardial infarction. An open Italian trial recruited 728 patients 6 to 18 months after myocardial infarction; they were randomized to a daily subcutaneous injection of 12,500 U standard heparin during in average 23 months or no heparin (Neri Serneri et al., 1987). Heparin reduced the cumulative mortality rate by 34 percent on intention-to-treat basis (not significant) and by 48 percent on on-treatment basis ($p < 0.05$). The reinfarction rate was 63 percent lower in the heparin group in the two types of analysis. It should be noted that 15 percent of the patients were on beta-blockers and more than 30 percent of those assigned to heparin and more than 40 percent of those assigned to the control group took antiplatelet agents. Nevertheless, this study provides evidence of a benefit of subcutaneous heparin after myocardial infarction.

HEPARIN AS ADJUNCTIVE ANTITHROMBOTIC THERAPY TO THROMBOLYSIS IN ACUTE MYOCARDIAL INFARCTION

In the early eighties streptokinase and alteplase have been infused directly in the coronary artery and a bolus heparin routinely injected immediately after arterial entry, most often followed by a maintenance infusion started immediately or after recanalization was obtained. When it became apparent that the intravenous route of administration of a thrombolytic drug is not markedly less effective than the intracoronary route, co-administration of heparin was maintained in only a few trials using streptokinase and in all trials with alteplase. The reasoning was that with the high dose of streptokinase required for its intravenous use, the severe depletion of fibrinogen could compensate during the first 1 or 2 days for heparin anticoagulation while the short plasma half-life and moderate fibrinogen decrease during intravenous treatment with alteplase may be associated with a high risk of reocclusion.

In a recent trial in patients receiving a 3 hour infusion of alteplase, immediate intravenous heparin did not appear to improve the patency rate assessed at 90 minutes after initiation of thrombolytic treatment (Topol et al. 1989). The dosage of heparin used may have been too low and too few patients were involved in this trial to judge effectiveness. Three other trials revealed a significantly higher patency rate at 18 hours (Hsia et al. 1990), 60 hours (Bleich et al. 1990) and 81 hours (de Bono et al. 1991) in patients given alteplase plus immediate intravenous heparin compared to alteplase without heparin. The latter three trials are concordant in their conclusion that a higher patency rate after alteplase is maintained in the presence than in absence of coadministered intravenous heparin.

The need for intravenous concomitant heparin is also evident when saruplase is used as thrombolytic agent in patients with acute myocardial infarction (Tebbe et al., 1990). The time from treatment from onset to patency-determining angiogram was 6 to 12 hours. Eighty-one percent of those receiving saruplase and intravenous heparin had a patent infarct-related artery, compared with only 60 percent in those in whom heparin was omitted (Tebbe et al., 1990). It thus seems from these multiple conformatory studies that short-acting, relatively fibrin-specific thrombolytic agents as alteplase and saruplase do require adequate and concomitant intravenous heparin anticoagulation to achieve and sustain therapeutic prolongations of the partial thromboplastin time.

The important role of intravenous heparin is also evident from the high levels of fibrinopeptide A and a 16-amino-acid peptide sensitive marker of circulating thrombin activity, which are high during treamtent with alteplase in the absence of heparin and fall as soon as heparin is added (Eisenberg et al., 1986; Rapold et al., 1989; Owen et al., 1990; Rapold et al., 1991).

It should be noted that the most recent studies performed by the ISIS and GISSI collaborators had heparin administered subcutaneously and only 4 hours (ISIS-2) or 12 hours (GISSI-2, 1990; The International Group, 1990) after thrombolytic therpay with respectively duteplase or alteplase. As alteplase has a relatively sparing effect on fibrinogen the additional influence of the moderate anticoagulant effecct of the relatively low dose of heparin which, moreover, was administered late, may have been insufficient to prevent early reocclusion in these two megatrials.

REFERENCES

ACC/NHLBI, 1986, National conference on antithrombotic therapy, Chest, 89 (Suppl): S1.

Altman, R., Rouvier, J., Gurfinkel, E., D'Ortencio, O., Manzanel, R., de la Fuente, L., and Favaloro, R. G., 1991, Comparison of two levels of anticoagulant therapy in patients with substitute heart valve, J. Thorac. Cardiovasc. Surg., 101: 427.

Arvan, S., and Boscha, K., 1987, Prophylactic anticoagulation for left ventricular thrombi after acute myocardial infarction: a prospective randomised trial, Am. Heart J., 113: 688.

Bleich, S. D., Nichols, T., Schumacher, R., Cooke, D., Tate, D., and Brinkman, D., 1990, The role of heparin following coronary thrombolysis with tissue plasminogen activator (t-PA), Am. J. Cardiol., 66: 1412.

Cleland, J., and Molloy, P. J., 1973, Thromboembolic complications of the cloth-covered Starr-Edwards prostheses no. 2300 aortic and no. 6300 mitral, Thorax, 28: 41.

Chesebro, J. H., and Fuster, V., 1991, Valvular heart diseases and prosthetic heart valves, in: "Thrombosis in Cardiovascular Disorders", V. Fuster, and M. Verstraete, eds., W.B. Saunders, in press.

Cohn, L. H., Allred, E. N., Cohn L. A., Austin, J. C., Sabik, J., DiSesa, V. J., Shemin, R. J., and Collins Jr., J. J., 1985, Early and late risk of mitral valve replacement: a 12 year concomitant comparison of the porcine bioprosthetic and prosthetic disc mitral valves, J. Thorac. Cardiovasc. Surg., 90: 872.

de Bono, D. P., Simoons, M. L., Tijssen, J., Arnold, A. E. R., Betriu, A., Burgersdijk, C., Bescos, L. L., Mueller, E., Pfisterer, M., Van De Werf, F., Zijlstra, F., and Verstraete, M., for the ECSG, 1991, Early intravenous heparin enhances coronary patency after alteplase thrombolysis: results of a randomized double blind European Cooperative Study Group Trial, Br. Heart J., in press.

Eisenberg, P. R., Sherman, L., Rich, M., Schwartz, D., Schectman, K., Geltman, E. M., Sobel, B. E., and Jaffe, A. S., 1986, Importance of continued activation of thrombin reflected by fibrinopeptide A to the efficacy of thrombolysis, J. Am. Coll. Cardiol., 7: 1255.

Davis, M. J. E., and Ireland, M., 1986, Effects of early anticoagulation on the frequency of left ventricular thrombi after anterior wall acute myocardial infarction, Am. J. Cardiol., 57: 1244.

Douglas, P., Hirshfeld, J. W., and Edmunds, L. H., 1984, Clinical correlates of post-operative atrial fibrillation (abstr), Circulation, 70 (Suppl II): II-165.

Drapkin, A., and Merskey, C., 1972, Anticoagulant therapy after acute myocardial infarction: relation of therapeutic benefit to patient's age and severity of infarction, J. Am. Med. Assoc., 222: 541.

Friedli, B., Aerichide, N., Grondin, P., and Campeau, L., 1971, Thromboembolic complications of heart valve replacement, Am. Heart J., 81: 702.

Funke Kupper, A. J., Verheugt, F. W. A., and Peels, C. H., 1989, Left ventricular thrombus incidence and behavior studied by serial two-dimensional echocardiography in acute anterior myocardial infarction: left ventricular wall motion, systemic embolism and oral anticoagulation, J. Am. Coll. Cardiol., 13: 1514.
Fuster, V., Pumphrey, C. W., McGoon, M. D., Chesebro, J. H., Pluth, J. R., and McGoon, D. C., 1982, Systemic thromboembolism in mitral and aortic Starr-Edwards prostheses: A 10-19 year follow-up, Circulation, 66 (Suppl I): 157.
Gruppo Italiano per lo Studio della sopravvivenza nell Infarto Miocardioco, 1990, GISSI-2: a factorial randomised trial of alteplase versus streptokinase and heparin versus no heparin among 12?490 patients with acute myocardial infarction, Lancet, 336: 65.
Gueret, P., Dubourg, O., Ferrier, A., Farcot, J. R., Rigaud, M., and Bourdanas, J. P., 1986, Effects of full dose heparin anticoagulation on the development of left ventricular thrombosis in acute transmural mycoardial infarction, J. Am. Coll. Cardiol., 8: 416.
Hilden, T., Iversen, K., Raaschou, F., et al., 1961, Anticoagulation in acute myocardial infarction, Lancet, 2: 327.
Hsia, J., Hamilton, W. P., Kleiman, N., Roberts, R., Chaitman, B. R., and Ross, A. M., 1990), A randomized trial of heparin versus aspirin adjunctive to tissue plasminogen activator-induced thrombolysis for acute myocardial infarction, N. Engl. J. Med., 323, 1433.
Kothari, A. J., Paczkowski, K., Baker, K. M., et al., 1984, Ventricular thrombi in acute myocardial infarction: incidence, complications and effects of anticoagulation, J. Am. Coll. Cardiol., 3 (Suppl.): 601.
Nordrehaug, J. E., Johannessen, K.A., and von der Lippe, G., 1985, Usefullness of high-dose anticoagulants in preventing left ventricular thrombus in acute myocardial infarction, Am. J. Cardiol., 55: 1491.
Owen, J., Friedman, K. D., Grossmann, B. A., Wilkins, C., Berke, A. D., and Powers, E. R., 1988, Thrombolytic therapy with tissue-type plasminogen activator or streptokinase induces transient thrombus activity. Blood, 72: 616.
Serneri, G. G. N., Abbate, R., Prisco, D., Carnovali, M., Fazi, A., Casolo, G. C., Bonechi, F., Rogasi, P. G., and Gensini, G. F., 1988, Decrease in frequency of anginal episodes by control of thrombin generation with low-dose heparin: a controlled cross-over randomized trial, Am. Heart J., 115: 60.
Serneri, G. G. N., Gensini, G. F., Carnovalli, M., Rovelli, F., Pirelli, S., and Fortinin, A., 1987, Effectiveness of low-dose heparin in prevention of myocardial infarction, Lancet, i: 937.
Serneri, G. G. N., Gensini, G. F., Poggesi, L., Trotta, F., Modesti, P. A., Boddi, M., Teri, A., Margheri, M., Casolo, G. C., Bini, M., et al., 1990, Effect of heparin, aspirin, or alteplase in reduction of myocardial ischaemia in refractory unstable angina, Lancet, 335: 615.
Sleight, P., 1991, Preliminary outcome data. ISIS-3 presented at the American College of Cardiology Annual Scientific Meeting, March 1991, Atlanta, GA.
Penny, W. J., Chesebro, J. H., Heras, M., and Fuster, V., 1988, Antithrombotic therapy for patients with cardiac disease, Curr. Probl. Cardiol., 13, 644.
Rapold, H. J., Kuemmerli, H., Weiss, M., Baur, H., and Haeberli, A., 1989, Monitoring of fibrin generation during thrombolytic therapy of acute myocardial infarction with recombinant tissue-type plasminogen activator, Circulation, 79: 980.
Rapold, H. J., de Bono, D., Arnold, A. E. R., Tijssen, J., De Cock, F., Collen, D., and Verstraete, M., for the European Cooperative Study Group, 1991, Fibrinopeptide A plasma levels and the significance of adequate anticoagulation for coronary patency, recurrent ischemia and left ventricular thrombosis in patients with acute myocardial infarction treated with alteplase, Circulation, in press.

Stroke Prevention in Atrial Fibrillation Study Group Investigators, 1990, Preliminary report of the SPAF Study, N. Engl. J. Med., 322: 863.

Tebbe, U., 1990, A randomized trial of heparin versus placebo after pro-urokinase induced thrombolysis. Presented at the George Washington University 6th International Workshop on Thrombolysis and Interventional Therapy in Acute Myocardial Infarction, Dallas, Texas, November 1990.

Telford, A. M., and Wilson, C., 1981, Trial of heparin versus atenolol in prevention of myocardial infarction in intermediate coronary syndrome, Lancet, 1: 1225.

The International Study Group, 1990, In-hospital mortality and clinical course of 20,891 patients with suspected acute myocardial infarction randomised between alteplase and streptokinase with or without heparin, Lancet, 336: 71.

The RISK Group, 1990, Risk of myocardial infarction and death during treatment with low-dose aspirin and intravenous heparin in men with unstable coronary artery disease, Lancet, 336: 827.

Théroux, P., Quimet, H., and McCans, J., 1988, Aspirin, heparin, or both to treat acute unstable angina ? N. Engl. J. Med., 1988, 319: 1105.

The SCATI (Studio della Calciparina nell'angina e nella Trombosi Ventricolare nell'infarto) group, 1989, Randomised controlled trial of subcutaneous calcium-heparin in acute myocardial infarction, Lancet, ii: 182.

Topol, E. J., George, B. S., Kereiakes, D. J., Stump, D. C., Candela, R. J., Abbottsmith, C. W., Aronson, L., Pickel, A., Boswick, J. M., Lee, K. L., Ellis, S. G., Califf, R. M., and the TAMI Study Group, 1989, A randomized controlled trial of intravenous tissue plasminogen activator and early intravenous heparin in acute myocardial infarction. Circulation, 79: 281.

Turpie, A. G. G., Gunstensen, J,. Hirsh, J., Nelson, H., and Gent, M., 1988, Randomised comparison of two intensities of oral anticoagulant therapy after tissue heart valve replacement, Lancet, 1: 1242.

Turpie, A. G. G., Robinson, J. G., Doyle, D. J., Mulji, A. S., Mishkel, G. J., Sealey, B. J., Cairns, J. A., Skingley, L., Hirsh, J., and Gent, M., 1989, Comparison of high-dose with low-dose subcutaneous heparin to prevent left ventricular mural thrombosis in patients with acute transmural anterior myocardial infarction, N. Engl. J. Med., 320: 352.

Veterans Administration Study Investigators, 1973, Anticoagulants in acute myocardial infarction: results of a cooperative clinical trial, J. Am. Med. Assoc., 225: 724.

Williams, D. O., Kirby, M. G., McPherson, K., and Phear, D. N., 1896, Anticoagulant treatment in unstable angina, Br. J. Clin. Pract., 40: 114.

Working Party on Anticoagulant Therapy in Coronary Thrombosis, 1969, Assessment of short-anticoagulant administration after cardiac infarction, Br. Med. J., i: 335.

GLYCOSAMINOGLYCANS IN PROPHYLAXIS AGAINST VENOUS THROMBOEMBOLISM

David Bergqvist, Bengt Lindblad, Thomas Mätzsch

Department of Surgery, Lund University
Malmö General Hospital, Malmö, Sweden

Ever since the 1930's when Jorpes and Crafoord in Stockholm started using heparin to prevent postoperative venous thromboembolism, a tremendous amount of basic as well as clinical research has been performed on heparin and related compounds to create effective means of prophylaxis, which at the same time is safe. This research is continuously progressing, several new substances being evaluated in animal experiments but so far without clinical documentation. Today it is a rather common view that it is possible to reduce the risk of developing deep vein thrombosis (DVT) and its acute consequence in the form of fatal pulmonary embolism (FPE) by various prophylactic modalities. Which risk factor definitions to use to determine the volume of patients needing prophylaxis is, however, still a matter for discussion. Heparin is world wide the dominating prophylactic method (Bergqvist, 1990) but in recent years low molecular weight heparins have gained a widespread interest and several of them are approved for clinical use by European health authorities. This chapter will review heparin and other glycosaminoglycans as pharmacological agents to diminish the frequency of postoperative venous thromboembolism. The discussion is mostly based on overviews, compilations and metaanalyses as the number of individual studies by now is very large, especially concerning low dose heparin and low molecular weight heparin and there is little meaning in this context to repeat the results of every single study. The original studies are all randomized and controlled with objective diagnostic methods for the detection of DVT. Relative risk reduction has been calculated according to Gent and Roberts (1986).

Low dose heparin

Heparin was initially used prophylactically with what is now known as therapeutic doses, which led to serious and unacceptable bleedings. However, already in 1942 JP Strömbäck, professor of surgery at Lund, Sweden, discussed the possibility to prevent DVT with lower heparin doses than what was needed to treat patients with established DVT. This idea was taken up by Sharnoff et al. (1962) but the concept of subcutaneous low dose heparin (5000 IU twice or three times daily) with preoperative start of prophylaxis was first published by Williams (1971). The beneficial effect was established in the International Multicentre Trial conducted by Kakkar (1975), where, beside the thromboprophylactic effect, a significant decrease in the frequency of FPE was seen as compared with placebo.

Low dose heparin was introduced for thromboprophylaxis in surgical patients at about the same time as the ^{125}I–fibrinogen uptake test (FUT) was developed as a diagnostic method for DVT. The necessity to base diagnosis of DVT on objective criteria had become obvious, and FUT was simple and reliable in the postoperative situation and made it possible to survey fairly large number of patients simultaneously. FUT had a great impact on our understanding of the natural history of postoperative DVT as well as of thromboprophylaxis. This parallel development of low dose heparin and FUT made it possible to perform a large number of studies on low dose heparin prophylaxis and most of them are thoroughly evaluated in the metaanalysis by Collins et al. (1988). In Tables 1, 2 and 3 (from Lindblad 1988) are summarized the thromboprophylactic effect compared to no prophylaxis in nonorthopaedic surgery (39 studies), total hip replacement (15 studies) and hip fracture surgery (10 studies). The relative risk reduction is 61 %, 59 % and 46 % respectively.

As pointed out by Collins et al. (1988) the risk reduction is similar between these three groups of patients, the problem, however, being that the orthopaedic patients have a much higher frequency in the untreated group, which means that a considerable number of patients still develop DVT, even with low dose heparin prophylaxis. Patients operated on for a colorectal cancer also belong to a group fairly resistent to prophylaxis (Wille–Jørgensen et al., 1988).

Table 1. Frequency of DVT in non–orthopaedic studies comparing low dose heparin (LDH) with no prophylaxis.

	LDH	Control
No of patients	3042	2951
No of DVT	289	721
Frequency of DVT (%)	9.5	24.4
Relative risk reduction	61 %	

Table 2. Frequency of DVT in patients undergoing total hip replacement comparing low dose heparin (LDH) with no prophylaxis.

	LDH	Control
No of patients	546	543
No of DVT	111	269
Frequency of DVT (%)	20.3	49.5
Relative risk reduction	59 %	

Table 3. Frequency of DVT in patients undergoing surgery for hip fracture comparing low dose heparin (LDH) with no prophylaxis.

	LDH	Control
No of patients	342	327
No of DVT	93	166
Frequency of DVT (%)	27.2	50.8
Relative risk reduction	46 %	

Table 4. Frequency of FPE in controlled studies comparing low dose heparin (LDH) with no prophylaxis.

	LDH	Control
No of patients	6705	6635
No of FPE	14	52
Frequency of FPE	0.21	0.78
Relative risk reduction	73 %	

The reduction by low dose heparin in FPE is shown in Table 4 and a significant risk reduction (73 %) is without any doubt (Bergqvist, 1983; Clagett & Reish, 1988; Collins et al., 1988; Lindblad, 1988). Only one single study is large enough to show this effect by itself (International Multicentre Trial, 1975). In the metaanalysis by Collins et al. (1988) this reduction also favourably influences the total mortality.

Low dose heparin is given twice or thrice daily. There does not seem to be a difference in prophylactic effect but the more frequent administrations lead to more bleeding complications (Seglias & Gruber, 1979; Bergqvist, 1979). Also with administration 12 hourly there is an increased but low risk for bleeding complications, although hardly of clinical significance except in patients where even very small bleedings may jeopardize surgical results such as in neurosurgery or eye surgery.

As a modification of low dose heparin prophylaxis, heparin given with adjusted doses has been popularized by Leyvraz et al. (1983, 1988). The first dose of 3500 IU was given 2 hours before elective hip replacement and thereafter subcutaneous injections were given 8 hourly and the dose adjusted in steps of 500–1000 IU to reach an APTT in the upper normal level 6 hours postinjection. It was possible to reduce the frequency of DVT from 39 to 13 % compared with conventional low dose heparin without increasing the bleeding risk. This model has later been verified (Green et al., 1988; Taberner et al., 1989).

The effect of low dose heparin may also be augmented by combining it with graded elastic compression stockings (Törngren, 1980; Borow & Goldson, 1983; Wille–Jørgensen et al., 1985).

Low dose heparin has a documented effect in reducing postoperative thromboembolic complications. However, it gives a small bleeding risk and the effect is not optimal in orthopaedic and colorectal cancer surgery. A practical disadvantage is the need for injections at least twice daily. Adjusted dose heparin needs laboratory controls which is also a disadvantage.

Low dose heparin combined with dihydroergotamine (DHE)

A combination prophylaxis with DHE is used to increase venous tonus, thereby increasing the flow rate and counteracting stasis. The experience with this combination has been summarized by Kuster & Gruber (1984); Gent & Roberts (1986) and Lindblad (1988). It has been possible to reduce the frequency of DVT compared with low dose heparin alone in patients undergoing elective surgery (both general and hip) but not in hip fracture surgery (although there is only one study with this last comparison, Lahnborg 1980). In Table 5 the results are summarized. It is possible to reduce the heparin dose to 2500 IU and maintain the prophylactic effect when combining it with DHE (Table 6).

Table 5. Compilation of data from controlled studies comparing the combination of low dose heparin (LDH) and dihydroergotamine (DHE) versus low dose heparin alone.

	LDH + DHE			LDH		
	No of patients	No of DVT	Frequency of DVT (%)	No of patients	No of DVT	Frequency of DVT (%)
General	1892	170	9.0	1599	277	17.3
Total hip replacement	505	72	14.3	343	123	35.9
Hip fracture	70	12	17.1	71	15	21.1

Relative risk reduction in favour of LDH + DHE:	general	48 %
	total hip replacement	60 %
	hip fracture	19 %

Table 6. The effect of lowering the heparin dose when combined with dihydroergotamine (DHE) versus standard low dose heparin. Non–orthopaedic patients.

	DHE + 2500 IU heparin	LDH 5000 IU heparin
No of patients*	962	965
No of DVT	105	103
Frequency of DVT (%)	10.9	10.7

* Nine studies. In seven the frequency of bleeding complications was registered and in five of those the frequency was significantly lower in the combination group.

Table 7. The effect on FPE in controlled studies comparing low dose heparin – dihydroergotamine (LDH + DHE) with low dose heparin alone.

	LDH + DHE	LDH
No of patients	2338	1719
No of FPE	4	5
Frequency of FPE	0.17	0.29
Relative risk reduction	41 %	

This is obtained with a significant reduction in bleeding complications. Concerning FPE these data are given in Table 7, the frequency being of a similar magnitude as with low dose heparin. Gruber (1982) showed the combination to be equally effective as dextran in reducing FPE in patients undergoing orthopaedic surgery.

Low dose heparin and DHE have been used particularly among orthopaedic surgeons. The main drawback is that DHE also may induce contraction of arteries (Lindblad et al., 1984). This may be of importance in patients with a preoperative compromized circulation, either elderly with atherosclerosis or patients with multitrauma and hypovolemia or shock. Although extremely rare, there has been a number of extremity gangrenes as a complication to low dose heparin–DHE prophylaxis, in a high frequency with a poor prognosis concerning the extremity (Mattsson et al., 1991). In an Austrian study 142 patients out of 61092 receiving low dose heparin in combination with DHE developed gangrene, seven of whom had to be amputated. Parallel with the development of alternative prophylactic methods DHE as an adjunctive substance seems to be on the way out from the prophylactic arsenal.

Low molecular weight heparin

The biochemical background of low molecular weight heparins is extensively reported in other chapters in this volume. Suffice it to state that their potential advantages from a clinical point of view are better bioavailability and longer biological half life after subcutaneous administration. Furthermore less platelet interaction and less influence on lipolysis are achieved. Today there are several types of low molecular weight heparins with differences in production methodology (Table 8) and thereby in

Table 8. The production methodology of various commercial low molecular weight heparins and their molecular weights.

Heparin fragment	Production methodology	Average molecular weight (Dalton)
CY 216	Ethanol precipitation, nitrous acid depolymerization	4500
CY 222	Exhaustive nitrous acid depolymerization	2500
Enoxaparin	Benzylation and alkaline depolymerization	3500 – 5500
Kabi–2165	Nitrous acid depolymerization, gel filtration	4000 – 6000
Novo LHN–1	Enzymatic depolymerization with heparinase from Flavobacterium heparinum	4900
OP 2123	Peroxidative depolymerization, ethanol extraction	3500 – 5000
Sandoz CH–8140	Isoamylnitrite depolymerization	4500 – 8000
Reveparin	Nitrous acid depolymerization	3500 – 4500

biochemical and biophysical characteristics as well as in pharmacological action (Cade et al., 1984, Fareed et al., 1989). However, the method of preparation per se seems to be of minor importance, at least if the molecular weight distributions of the products are similar (θstergaard et al., 1987). In this overview, however, the various low molecular weight heparins are discussed together since it has been difficult to show any clinically relevant differences between them and so far there are no comparative studies.

There are several reviews on the effect of low molecular weight heparins in preventing postoperative DVT (Samama et al., 1989; Levine & Hirsh, 1988; Verstraete, 1990; Bergqvist, 1991). Altogether today there are some 50 studies, most of them comparing low molecular weight heparins with standard low dose heparin. A few have made the comparisons with dextran and only a very few have used untreated controls or placebo (Ockelford et al., 1989; Turpie et al., 1986; Lassen et al., 1988; 1989; Törholm et al., 1991). In the studies where low molecular weight heparins were compared with no prophylaxis the decrease in frequency of DVT was highly significant.

In Table 9 are summarized the 22 studies in non-orthopaedic surgery where low molecular weight heparins are compared with standard low dose heparin. Ten of them were open, twelve double blind. In four of the studies dihydroergotamine was used in both groups (Sasahara et al., 1986; Welzel et al., 1988; Baumgartner et al., 1989; Kakkar et al., 1989), in one elastic compression stockings (Eur. Fraxiparine Study 1988). In most individual studies there has been no significant difference between the two prophylactic methods but overall on the compiled data there is a relative risk reduction of DVT of 21 % with low molecular weight heparin.

Table 9. The frequency of DVT in controlled non-orthopaedic studies comparing low molecular weight heparin fragments (LMWH) with low dose heparin (LDH).

	LMWH	LDH
No of patients	4919	4395
No of DVT	249	286
Frequency of DVT (%)	5.1	6.5
Relative risk reduction	21 %	

Table 10 shows the corresponding data for patients undergoing hip surgery. In eight trials total hip replacement was studied (Barre et al., 1987; Dechavanne et al., 1989; Eriksson et al., 1991; Haas et al., 1987a,b; Lassen et al., 1988; Levine et al., 1991; Planes et al., 1988) and in three hip fractures (Lassen et al., 1989; Monreal et al., 1989; Pini et al., 1989), the relative effect on DVT being similar, however, with a reduction in favour of low molecular heparins of 17 %. The diagnostic methodology when dealing with hip surgery should optimally be bilateral phlebography, used in five of the studies. The others used isotope tests, often with venographic confirmation.

Table 10. The frequency of DVT in controlled studies on hip surgery comparing low molecular weight heparin fragments (LMWH) with low dose heparin (LDH).

	LMWH	LDH
No of patients	948	904
No of DVT	185	212
Frequency of DVT (%)	19.5	23.5
Relative risk reduction	17 %	

Table 11. Compiled data from three studies comparing low molecular weight heparin (LMWH) with dextran in patients undergoing elective hip surgery.

	LMWH	Dextran
No of patients	216	221
No of DVT	38	71
Frequency of DVT (%)	17.6	32.1
Relative risk reduction	45 %	

In a recent study on elective hip arthroplasty there was a significant reduction of thigh thrombi and scintigraphic emboli with low molecular weight heparin compared with standard low dose heparin (Eriksson et al., 1991).

In three studies (Eriksson et al., 1988; Mätzsch et al., 1989; 1991a) low molecular weight heparin has been compared with dextran (Table 11), which in Scandinavia for a long time has been the method of choice in orthopaedic surgery (Bergqvist, 1990). The relative reduction of DVT strongly favours low molecular weight heparin (45 %).

Concerning fatal pulmonary embolism available data are summarized in Tables 12 and 13. In the 38 randomized studies comparing low molecular weight heparin and low dose heparin there is a low frequency of fatal pulmonary embolism and the beneficial effect is further stressed in Table 13. Another way of analyzing the results is to regard each study as a separate test that can lead to three outcomes regarding mortality: fewer deaths in the low molecular weight heparin group, fewer deaths in the low dose heparin group or an equal number of deaths in the two (Ljungström, 1988). FPE and total mortality are both significantly lower in the low molecular weight heparin groups.

Table 12. The frequency of FPE in studies comparing low molecular weight heparin (LMWH) with low dose heparin (LDH).

	LMWH	LDH
No of patients	6366	5813
No of FPE	1	7
Frequency of FPE (%)	0.02	0.12
No of deaths	79	76
Mortality (%)	1.2	1.3
Relative risk reduction in FPE	85 %	

Table 13. Frequency of FPE in studies on low molecular weight heparins.

	No of patients	No of FPE	Frequency of FPE (%)
From Table 12	6366	1	0.02
Pezzuoli et al.	2247	2	0.09
Haas & Flosbach	8738	3	0.03
Haas & Zielke	1060	0	0
Wolf et al.	33421	12	0.04
Other studies	3713	2	0.06
	55545	20	0.04

When low molecular weight heparin was introduced clinically there were in vitro as well as animal experimental evidence that a good thromboprophylactic effect was combined with a low bleeding potential. However, some of the early studies, which based their dosing of low molecular weight heparins on experimental data, showed an increased risk for haemorrhage (Schmitz–Hübner et al., 1984; Bergqvist et al., 1986; Koller et al., 1986; Borstad et al., 1988). What anti–Xa–levels really mean in terms of antithrombotic effect and bleeding risk is not yet fully understood but without doubt they do reflect variables of clinical importance (Levine & Hirsh, 1988; Levine et al., 1989). It seems clear that it is possible to define effective doses with minimal haemorrhagic risk for the various low molecular weight heparin fragments. When compiling data concerning bleeding complications it is very important to take the historical development into consideration, recent studies being without increased bleeding risk. One question, which has raised some concern, is whether or not it is safe to combine low molecular weight heparin and epidural/spinal anaesthesia. The frequency of epidural/spinal haematoma with neurological deficit in conjunction with low molecular weight heparin is extremely low and confined to one case report and there are no data suggesting it to be a clinically relevant problem in patients given prophylactic low molecular weight heparins (Bergqvist et al., 1991b). There is no indication that the time interval between the administration of low molecular weight heparin and epidural/ spinal puncture is critical and clinically relevant.

The optimal administration routines for low molecular weight heparins have to be established. In the majority of studies the first dose has been given around two hours before surgery and thereafter one dose every 24 hours. In one study the exact time for this first injection was registered and turned out to be in mean 2.3 hours before surgery with a variation of (0.2 – 9.3 h) (Bergqvist et al., 1988). In a couple of studies it has been shown that the first dose given in the evening before surgery with repetition every evening is highly effective (Bergqvist et al., 1988; Planes et al., 1988). This is of practical advantage, and in our study it also turned out to be more effective than standard low dose heparin. Moreover, there are data favouring such an approach from the point of view of circadian rhythm (Mätzsch et al., 1991b). Another new and practically interesting approach, which needs confirmation, is to start prophylaxis postoperatively as has been proposed and shown effective by the McMaster group (Turpie et al., 1986; Levine et al., 1991).

Semisynthetic heparin analogue (SSHA)

As indicated by the name this is a semisynthetic substance, a galactosaminoglycan polysulfate from the respiratory tract of horned cattle. It is composed of disaccharide units with almost equimolar amounts of hexuronic acid and hexosamine and has a

Table 14. Controlled studies on the prophylactic effect of SSHA in comparison with low dose heparin (LDH).

	SSHA		LDH	
	No of patients	No of DVT	No of patients	No of DVT
Kakkar et al., 1977	50	2	50	6
Kakkar et al., 1978	94	6	96	12
Törngren et al., 1984	279	37	148	21
Total no of patients	423	45	294	39

mean molecular weight of 7000 dalton (range 5000 – 15000). There are no sulfamino groups and the molecule contains insignificant amount of heparin. It inhibits factor Xa in vitro and is considered antithrombotic without significantly influencing overall clotting (Thomas et al., 1977). It seems independent of antithrombin and may involve the extrinsic coagulation pathway (Törngren et al., 1990).

In Table 14 are summarized the prophylactic trials on SSHA. All three studies were performed on general surgical patients with standard low dose heparin for comparison. The FUT was used for diagnosis. In the study by Törngren et al. (1984) there was comparison of two doses of SSHA (37.5 and 50 mg twice daily respectively) but no difference was found between the two why they are added together in the table. There does not seem to be an increased bleeding risk with SSHA in the doses used, compared to low dose heparin prophylaxis.

Sodium pentosan polysulfate (SP 54, PZ 68)

This semisynthetic sulphated polysaccharide is made from plants, the average molecular weight being about 4000 dalton with a narrow range of 1500 – 5000. It is composed of pentose units, contrary to the hexose units of heparin. After subcutaneous injection the substance has a stronger influence on the APT time and about the same influence on factor Xa inhibition as heparin in low doses (Bergqvist & Nilsson, 1981). Ryde et al. (1981) showed it to markedly inhibit factor Xa but not thrombin. It also stimulates fibrinolysis (Cocheri et al., 1978).

Table 15 shows the thromboprophylactic trials made. All studies are fairly small and the control groups vary but it seems reasonable to conclude that sodium pentosan polysulfate is better than no prophylaxis and better than dextran both in general and hip surgery and seems to reduce the frequency of DVT to a level seen with low dose heparin. This is obtained without increase in bleeding complications compared to other prophylactic modalities.

Table 15. Controlled studies on the prophylactic effect of pentosan polysulfate (SP 54, PZ 68)

Authors	Type of surgery	Control No of patients	No of DVT	SP 54 No of patients	No of DVT	LDH No of patients	No of DVT	Dextran No of patients	No of DVT
Bergqvist et al., 1980*	General			52	0	54	2		
Bergqvist et al., 1980*	Tot hip rep			53	5	59	6	56	22
Bergqvist et al., 1980*	Hip fract			29	6	25	5	29	6
Bergqvist, Ljungnér,1981*	General			34	1			52	10
Fredin et al., 1982	Hip fract			28	6			41	8
Joffe 1976	General	101	52	48	7				
Morin 1983	Tot hip rep	23	10	23	1	54	5		
Munich studies**	General			93	1			49	9

* LDH combined with DHE (Bergqvist et al., 1980)
** Theses quoted by Gruber & Döbeli, 1983.

Org 10172

Org 10172 (Lomoparan) is a mixture of sulfated glycosaminoglycuronans which are extracted from the porcine intestinal mucosa with a mean molecular weight of 6500 dalton (range 5000 – 10000). It consists mainly of heparan sulfate and dermatan sulfate and small amounts of chondroitin sulfate. It has little anti–IIa–activity but a well preserved anti–Xa–activity. It has a small effect on APT time. Experimentally it has a beneficial antithrombotic/bleeding profile (Meuleman et al., 1982; Ockelford et al., 1985).

Table 16. Controlled studies on the prophylactic effect of Org 10172

Type of surgery	Org 10172 No of patients	No of DVT	LDH No of patients	No of DVT	Dextran No of patients	No of DVT	Warfarin No of patients	No of DVT	Control No of patients	No of DVT
Total hip rep*	97	15							98	56
Total hip rep*	145	25	139	44						
Hip fracture**	107	14			115	40				
Hip fracture***	93	6					98	15		
General surg.*	204	20	211	33						

* Organon files
** Bergqvist et al., 1991a
*** Gerhart et al., 1991

In Table 16 the studies on Org 10172 are summarized. Just as with sodium pentosan polysulfate the overall material is heterogenous concerning type of surgery and group for comparison. The dose of Org 10172 has been 750 units twice daily subcutaneously. Increasing the dose seems to increase the risk for bleeding, at least in patients undergoing transurethral resection of the prostate (ten Cate et al., 1987). In the studies listed in the table there has, however, been no increase in haemorrhagic complications compared to other prophylactic modalities. In a double blind dose finding study by Cade et al. (1987) in general surgical patients, nine of 14 placebo patients developed DVT, 4 of 11 on Org 10172 500 units twice daily but none of ten patients each receiving 750 or 1000 units twice daily. The effectiveness has moreover been shown in a study on 75 patients with thrombotic stroke where the DVT frequency in the placebo group was 28 % versus 4 % in the Org 10172 group (Turpie et al., 1987).

To conclude, Org 10172 is better than no treatment and better than dextran and as least as effective as low dose heparin or warfarin. The effect has been shown for general and hip surgery and without increased risk of bleeding complications.

Dermatan sulfate

Dermatan sulfate is a natural glycosaminoglycan which potentiates only heparin cofactor II, which means a selective antithrombin activity. An antithrombotic effect with a low bleeding potential has been demonstrated in animal models (Maggi et al., 1987; Van Ryn-McKenna et al., 1989a,b). So far there is only one clinical trial where dermatan sulfate has been studied with a randomized, double-blind placebo controlled design in hip fracture patients (Agnelli et al. 1991). Bilateral venography showed a frequency of DVT of 64 % (23/36) in the placebo group and 38 % (28/74) in the group given 300 mg dermatan sulfate intramuscularly twice daily. There were no signs of an increased bleeding tendency. Obviously many more studies need to be done in various types of patients before more general conclusions on dermatan sulfate can be drawn, but this hip fracture study does not point to a remarkably good effect.

Concluding remarks

The use of glycosaminoglycans, starting with low dose heparin, has been a real step forward in the pharmacoprophylaxis against venous thromboembolism. The efficacy has been shown in numerous studies. There seems to be no doubt that the new low molecular weight heparins are as least as effective as standard heparin and with some theoretical as well as practical advantages. Other glycosaminoglycans have been much less extensively studied but they all seem to be as effective as low dose heparin. There are several practical clinical problems which are of importance to solve. Are there any clinically relevant differences between the various substances, especially the low molecular weight heparins? When is the time for optimal start of prophylaxis in relation to surgery, taking both safety and efficacy into consideration? Are there groups of patients where a prolonged prophylaxis may be indicated? Is it possible to increase the efficacy by combining glycosaminoglycans with other prophylactic methods such as mechanical or dextran?

REFERENCES

Agnelli, G., Cosmi, B., Di Filippo, P., Ranucci, V., Veschi, F., Longetti, M., Renga, C., Barzi, F., Gianese, F., Lupattelli, L., Rinonapoli, E., Nenci, G., 1991, A randomized, double-blind, placebo-controlled trial of dermatan sulphate for prevention of deep vein thrombosis in hip fracture. Thromb. Haemost., 65:704.

Baumgartner, A., Jacot, N., Moser, G., Krähenbühl, B., 1989, Prevention of postoperative deep vein thrombosis by one daily injection of low molecular weight heparin and dihydroergotamine. VASA, 18:152.

Barre, J., Phister, G., Potron, G., Droulle, C., Baudrillard, J.C., Barbier, P., Kher, A., 1987, Efficacité et tolerance comparée du Kabi 2165 et de l'héparine standards dans la prévention des thromboses veineuses profundes au cours des prothèses totales de hanche. J. Malad. Vasc., 12:90.

Bergqvist, D., 1979, Prophylaxis of postoperative thromboembolic complications with low-dose heparin. An analysis of different administration intervals. Acta Chir. Scand., 145:7.

Bergqvist, D., 1983, "Postoperative thromboembolism. Frequency, Etiology, Prophylaxis," Springer-Verlag, Berlin Heidelberg New York.

Bergqvist, D., 1990, Prophylaxis against postoperative venous thromboembolism - a survey of surveys. Thromb Haemorrh Disorders, 2:69.

Bergqvist, D. Overview of clinical trials of low molecular weight heparins, 1991. Manuscript.

Bergqvist, D., Ljungnér, H., 1981, A comparative study of dextran 70 and a sulphated polysaccharide in the prevention of postoperative thromboembolic complications in patients undergoing abdominal surgery. Br. J. Surg., 68:449.

Bergqvist, D., Nilsson, I.M., 1981, A sulphated polysaccharide - the effects in vivo and in vitro on the haemostatic system i healthy volunteers. Thromb. Res., 23:309.

Bergqvist, D., Efsing, H.O., Hallböök, T., Lindblad, B., 1980, Prevention of postoperative thromboembolic complications. A prospective comparison between dextran 70, dihydroergotamin heparin and a sulphated polysaccharide. Acta Chir. Scand., 146:559

Bergqvist, D., Burmark, U.S., Frisell, J., Hallböök, T., Lindblad, B., Risberg, B., Törngren, S., Wallin, G., 1986, Low molecular weight heparin once daily compared with conventional low-dose heparin twice daily. A prospective double-blind multicentre trial on prevention of postoperative thrombosis. Br. J. Surg., 73:204.

Bergqvist, D., Mätzsch, T., Burmark, U.S., Frisell, J., Guilbaud, O., Hallböök, T., Horn, A., Lindhagen, A., Ljungnér, H., Ljungström, K-G., Onarheim, H., Risberg, B., Törngren, S., Örtenwall, P., 1988, Low molecular weight heparin given the evening before surgery compared with conventional low dose heparin in the prevention of thrombosis after elective general abdominal surgery. Br. J. Surg., 75:888

Bergqvist, D., Kettunen, K., Fredin, H., Faunø, P., Suomalainen, O., Soimakallio, S., Karjalainen, P., Cederholm, C., Jensen, L.J., Justesen, T., Stiekema, J.C.J., 1991a, Thromboprophylaxis in patients with hip fractures. A prospective, randomized comparative study between Org 10172 and dextran 70. Surgery, 109:617.

Bergqvist, D., Lindblad, B., Mätzsch, T., 1991b, Low molecular weight heparin for thromboprophylaxis and epidural/spinal anaesthesia - is there a risk?. Review article Submitted.

Borow, M., Goldson, H.J., 1983, Prevention of postoperative venous thrombosis and pulmonary emboli with combined modalities. Am. Surg., 49:599.

Borstad, E., Urdal, K., Handeland, G., Abildgaard, U., 1988, Comparison of low molecular weight heparin vs. unfractionated heparin in gynecological surgery. Acta Obstet. Gynecol. Scand., 67:99.

Cade, J.F., Buchanan, M.R., Boneu, B., Ockelford, P., Carter, C.J., Cerskus, A.L., Hirsh, J., 1984, A comparison of the antithrombotic and hemorrhagic effects of low molecular weight heparin fractions: the influence of the method of preparation. Thromb. Res., 35:613.

Cade, J.F., Wood, M., Magnani, H.N., Westlake, G.W., 1987, Early clinical experience of a new heparinoid, Org 10172, in prevention of deep venous thrombosis. Thromb. Res., 45:497.

Clagett, G.P., Reisch, J.S., 1988, Prevention of venous thromboembolism in general surgical patients. Ann. Surg., 208.

Coccheri, S., de Rosa, V., Cavallaroni, K., Poggi, M., 1978, Activation of fibrinolysis by means of sulfated polysaccharides: present status and perspectives, in: "Progress in chemical fibrinolysis and thrombolysis," J.F. Davidson, R.M. Rowan, M.M. Samama, P.C. Desnoners, eds., Rawen Press, New York.

Collins, R., Scrimgeour, A., Yusef, S., Peto, R., 1988, Reduction in fatal pulmonary embolism and venous thrombosis by perioperative administration of subcutaneous heparin. Overview of results of randomized trials in general, orthopedic, and urologic surgery. N. Engl. J. Med., 318:1162.

Dechavanne, M., Ville, D., Berruyer, M., Trepo, F., Dalery, F., Clermont, N., Lerat, J.L., Moyen, B., Fischer, L.P., Kher, A., Barbier, P., 1989, Randomized trial of a low-molecular-weight heparin (Kabi 2165) versus adjusted-dose subcutaneous standard heparin in the prophylaxis of deep-vein thrombosis after elective hip surgery. Haemostasis, 1:5.

Eriksson, B.I., Zachrisson, B.E., Teger-Nilsson, A.-C., Risberg, B., 1988, Thrombosis prophylaxis with low molecular weight heparin in total hip replacement. Br. J. Surg., 75:1053.

Eriksson, B.I., Kälebo, P., Anthmyr, B.A., Wadenvik, H., Tengborn, L., Risberg, B., 1991, Prevention of deep-vein thrombosis and pulmonary embolism after total hip replacement. J. Bone Joint. Surg., 73-A:484.

European Fraxiparin Study Group, 1988, Comparison of a low molecular weight heparin and unfractionated heparin in the prevention of deep venous thrombosis in patients undergoing abdominal surgery. Br. J. Surg., 75:1058.

Fareed, J., Walenga, J.M., Hoppensteadt, D., Racanelli, A., Coyne, E., 1989, Chemical and biological heterogenecity in low molecular weight heparins: implications for clinical use and standardization. Semin. Thromb. Hemost., 15:440.

Fredin, H.O., Nillius, S.A., Bergqvist, D., 1982, Prophylaxis of deep vein thrombosis in patients with fractures of the femoral neck. A prospective comparison between dextran and a poly-sulphated saccharide. Acta Orthop. Scand., 53:413.

Gent, M., Roberts, R.S., 1986, A meta-analysis of the studies of dihydroergotamine plus heparin in the prophylaxis of deep vein thrombosis. Chest, 89:396S.

Gerhardt, T.N., Yett, H.S., Robertson, L.K., Lee, M.A., Smith, M., Salzman, E.W., 1991, Low-molecular-weight heparinoid compared with warfarin for prophylaxis of deep-vein thrombosis in patients who are operated on for fracture of the hip. J. Bone Joint. Surg., 73-A:494.

Green, D., Lee, M.Y., Ito, V.Y. Cohn, T., Press, J., Filbrandt, P.R., Vandenberg, W.C., Yarkony, G.M., Meyer, P.R. Fixed- vs adjusted-dose heparin in the prophylaxis of thromboembolism in spinal cord injury. JAMA, 260:1255.

Gruber, U.F., 1982, Prevention of fatal postoperative pulmonary embolism by heparin dihydroergotamine or Dextran 70. Br. J. Surg., 69:54.

Gruber, U.F., Döbeli, P.M., 1983, Wert der Heparinoide zur Prophylaxe thromboembolischer Komplikationen in der Chirurgie. Helv. chir. Acta, 50:463.

Haas, S.,Flosbach, C.W., 1990, Prevention of post-operative thromboembolism in general surgery with enoxaparin: preliminary findings. Acta Chir Scand Suppl., 556:96.

Haas, S., Zielke, E., 1987, Postoperative Thromboseprophylaxe mit 2500 anti Xa-E niedermolekularem Heparin Kabi in der Abdominalchirurgie. medwelt, 38:1658.

Haas, S., Stemberger, A., Fritsche, H-M., 1987a, Prophylaxis of deep vein thrombosis in high risk patients undergoing total hip replacement with low molecular weight heparin plus dihydroergotamine. Drug Res., 37:839.

Haas, S., Stemberger, A., Fritsche, H.-M., Lechner, F., 1987b, Untersuchungen zur antithrombotischen Wirksamkeit von niedermolekularen Heparin Kabi. Med. Welt., 38:714.

An International Multicentre Trial, 1975, Prevention of fatal postoperative pulmonary embolism by low doses of heparin. Lancet, II:45.

Joffe, S., 1976, Drug prevention of postoperative deep vein thrombosis. Arch. Surg., 111:37.

Kakkar, V.V., Lawrence, J.D., Stamatakis, J.D., Bentley, P.G., Ward, V.P., 1977, A comparative study of heparin and a semi–synthetic heparin analogue in the prevention of postoperative deep vein thrombosis. Br. J. Surg., 64:827.

Kakkar, V.V., Lawrence, D., Bentley, P.G., de Haas, H.A., Ward, V.P., Scully, M.F., 1978, A comparative study of low doses of heparin and a heparin analogue in the prevention of postoperative deep vein thrombosis. Thromb. Res., 13:111.

Kakkar, V.V., Stringer, M.D., Hedges, A.R., Parker, C.J., Welzel, D., Ward, V.P., Sanderson, R.M., Cooper, D., Kakkar, S., 1989, Fixed combinations of lowmolecular weight or unfractionated heparin plus dihydroergotamine in the prevention of postoperative deep vein thrombosis. Am. J. Surg., 157:413.

Koller, M., Schoch, U., Buchmann, P., Largiadèr, F., von Felten, A., Frick, P.G., 1986, Low molecular weight heparin (Kabi 2165) as thromboprophylaxis in elective visceral surgery. A randomized, double–blind study versus unfractionated heparin. Thromb. Haemost., 56:243.

Kuster, B., Gruber, U.F., 1984, Wert von Heparin–Dihydergot zur Prophylaxe thromboembolischer Komplikationen. Schweiz. Med. Wschr., 114:322.

Lahnborg, G., 1980, Effect of low–dose heparin and dihydroergotamine prophylaxis on frequency of postoperative deep–vein thrombosis in patients undergoing posttraumatic hip surgery. Acta Chir. Scand., 146:319.

Lassen, M.R., Borris, L.C., Christiansen, H.M., Møller–Larsen, F., Knudsen, V.E., Boris, P. Nehen, A.–M., Jurik, A.–G., de Carvalho, A., Nielsen, N.W., Lucht, U., 1988, Heparin/dihydroergotamine for venous thrombosis prophylaxis comparison of low–dose heparin and low molecular weight heparin in hip surgery. Br. J. Surg., 75:686.

Lassen, M.R., Borris, L.C., Christiansen, H.M., Møller–Larsen, F., Knudsen, V.E., Boris, P., Nehen, A.–M., Jurik, A.–G., de Carvallo, A., Nielsen, N.W., Lucht, U., 1989, Prevention of thromboembolism in hip–fracture patients. Comparison of low–dose heparin and low– molecular–weight heparin combined with dihydroergotamine. Arch. Ortop. Trauma Surg., 108:10.

Levine, M.N., Hirsh, J., 1988, An overview of clinical trials of low molecular weight heparin fractions. Acta Chir. Scand., Suppl 543:73.

Levine, M.N., Planes, A., Hirsh, J., Goodyear, M., Vochelle, N., Gent, M., 1989, The relationship between anti–factor Xa level and clinical outcome in patients receiving Enoxaparine low molecular weight heparin to prevent deep vein thrombosis after hip replacement. Thromb. Haemost., 62:940.

Levine, M.N., Hirsh, J., Gent, M., Turpie, A.G., Leclerc, J., Powers, P.J., Jay, R.M., Neehmeh, J., 1991, Prevention of deep vein thrombosis after elective hip surgery. A randomized trial comparing low molecular weight heparin with standard unfractionated heparin. Ann. Intern. Med., 114:545.

Leyvraz, F.P., Richard, J., Bachmann, F., van Melle, G., Treyvaud, J.M., Livio, J.–J., Candardjis, G., 1983, Adjusted versus fixed subcutaneous heparin in the prevention of deep vein thrombosis after total hip replacement. N. Engl. J. Med., 309:954.

Lindblad, B., 1988, Prophylaxis of postoperative thromboembolism with low dose heparin alone or in combination with dihydroergotamine. Acta Chir. Scand., Suppl. 543:31.

Lindblad, B., Bergqvist, D., Efsing, H.O., Hallböök, T., Lindell, S–E., 1984, Changes in peripheral haemodynamics induced by dextran 70, dihydroergotamine, and their combination. A study using ultrasonic, serial phlebography, pletysmography and isotope clearance techniques. VASA, 13:165.

Ljungström, K.G., 1988, The antithrombotic efficacy of dextran. Acta Chir. Scand., Suppl 543:26.

Maggi, A., Abbadini, N., Pagella, P.G., Borowska, A., Pangrazzi, J., Donati, M.B., 1987, Antithrombotic properties of dermatan sulphate in a rat venous thrombosis model. Haemostasis, 17:329.

Mattsson, E., Ohlin, A., Fredin, H., Nilsson, P., Bergqvist, D., 1991, Lower-limb vasospasm and renal failure during postoperative thromboprophylaxis. Eur. J. Surg., 157:289.

Meuleman, D.G., Hobbelen, P.M.J., van Dedem, G., Moelker, H.C.T., 1982, A novel antithrombotic heparinoid (Org 10172) devoid of bleeding-inducing capacity: a survey of its pharmacological properties in experimental animal models. Thromb. Res., 27:353.

Monreal, M., Lafoz, E., Navarro, A., Granero, X., Caja, V., Caceres, E., Salvador, R., Ruiz, J., 1989, A prospective double-blind trial of a low molecular weight heparin once daily compared with conventional low-dose heparin three times daily to prevent pulmonary embolism and venous thrombosis in patients with hip fracture. J Trauma, 29:873.

Morin, J.P., 1983, Note sur l'utilisation du polyester sulfurique de pentosane dans la prévention des thromboses veineuses profondes post-opératoires des membres inférieurs. Agressologie, 24:597.

Mätzsch, T., Bergqvist, D., Fredin, H., Hedner, U. Lindhagen, A., Nistor, L., 1989, Thromboprophylactic effect of a low molecular weight heparin as compared to dextran in elective hip surgery. Thromb. Haemost., 62:520.

Mätzsch, T., Bergqvist, D., Fredin, H., Hedner, U., Lindhagen, A., Nistor, L., 1991a, Comparison of the thromboprophylactic effect of a low molecular weight heparin versus dextran in total hip replacement. Thromb. Haemorrh. Disorders, 1:25.

Mätzsch, T., Bergqvist, D., Burmark, U.S., Holmer, E., Söderberg, K., 1991b, The influence of surgical trauma on factor XaI and IIaI activity and heparin concentration after single dose of low molecular weight heparin. Blood Coag. Fibrinolysis, In press.

Ockelford, P.A., Carter, C.J., Hirsh, J., 1985, In vivo activity of a new heparinoid. Pathology, 17:78.

Pezzuoli, G., Serneri, N., Settembrini, P., Coggi, G., Olivari, N., Buzzetti, G., Chierichetti, S., Scotti, A., Scatigna, M., Carnovali, M., and the STEP- Study Group, 1989, Int. Surg., 74:205.

Pini, M., Tagliaferri, A., Manotti, C., Lasagni, F., Rinaldi, E., Dettori, A.G., 1989, Low molecular weight heparin (Alfa LHWH) compared with unfractionated heparin in prevention of deep-vein thrombosis after hip fractures. Int. Angio., 8:134.

Planes, A., Vochelle, N., Mazas, F., Manzat, C., Zucman, J., Landais, A., Pascariello, J.C., Weill, D., Butel, J., 1988, Prevention of postoperative venous thrombosis: a randomized trial comparing unfractionated heparin with low molecular weight heparin in patients undergoing total hip replacement. Thromb. Haemost., 60:407.

Ryde, M., Eriksson, H., Tangen, O., 1981, Studies on the different mechanisms by which heparin and polysulphated xylan (PZ68) inhibit blood coagulation in man. Thromb. Res., 23:435.

Samama, M., Boissel J.P., Combe-Tamzali, S., Leizorovicz, 1989, Clinical studies with low molecular weight heparins in the prevention and treatment of venous thromboembolism. Ann. N.Y. Acad. Sci., 556:386.

Sasahara, A.A., Koppenhagen, K., Häring, R., Welzel, D., Wolf, H., 1986, Low molecular weight heparin plus dihydroergotamine for prophylaxis of postoperative deep vein thrombosis. Br. J. Surg., 73:697.

Schmitz-Hübner, U., Bünte, H., Freise, G., 1984, Clinical efficacy of low molecular weight heparin in postoperative thrombosis prophylaxis. Klin. Wochenschr., 62:249.

Seglias, J., Gruber, U.F., 1979, Dosage in low-dose heparin prophylaxis. Haemostasis, 8:361.

Sharnoff, J.G., Kass, H.H., Mistica, B.A., 1962, A plan of heparinization of the surgical patient to prevent postoperative thromboembolism. Surg. Gynaecol. Obstet., 115:75.

Taberner, D.A., Poller, L., Thomson, J.M., Lemon, G., Weighill, F.J., 1989, Randomized study of adjusted versus fixed low dose heparin prophylaxis of deep vein thrombosis in hip surgery. Br. J. Surg., 76:933.

ten Cate, H., Henny, Ch.P., ten Cate, J.W., Büller, H.R., Dabhiowala, N.F., 1987, Randomized double-blind, placebo controlled safety study of a low molecular weight heparinoid in patients undergoing transurethral resection of the prostate. Thromb. Haemost., 57:92.

Thomas, D.P., Lane, D.A., Michalski, R., Johnsson, E.A., Kakkar, V.V., 1977, A heparin analogue with specific action on antithrombin III. Lancet, I:120.

Turpie, A.G.G., Levine, M.N., Hirsh, J., Carter, C.J., Ray, R.M., Powers, P.J., Andrew, M., Hull, R.D., Gent, M., 1986, A randomized controlled trial of a low-molecular-weight heparin (Enoxaparine) to prevent deep vein thrombosis in patients undergoing elective hip surgery. N. Engl. J. Med., 315:925.

Turpie, A.G.G., Levine, M.N., Hirsch, J., Carter, C.J., Richard, M.J., Powers, P.J., Andrew, M., Magnani, H.N., Hull, R.D., Gent, M., 1987, Double-blind randomized trial of Org 10172 low-molecular-weight heparinoid in prevention of deep-vein thrombosis in thrombotic stroke. Lancet, 1:523.

Tørholm, C., Broeng, L., Seest Jørgensen, P., Bjerregaard, P., Josephsen, L., Korsholm Jørgensen, P., Hagen, K., Bjerre Knudsen, J., 1991, Thromboprophylaxis by low-molecular-weight heparin in elective hip surgery. A placebo controlled study. J.Bone Joint. Surg (Br.), 73-B:434.

Törngren, S., 1980, Low dose heparin and compression stockings in the prevention of postoperative deep venous thrombosis. Br. J. Surg., 67:482.

Törngren, S., Kettunen, K., Lahtinen, J., Koppenhagen, K., Brücke, P., Hartl, P., Hutter, O., Haller, H., Lahnborg, G., Forsskåhl, B., 1984, A randomized study of a semi-synthetic heparin analogue and heparin in prophylaxis of deep vein thrombosis. Br. J. Surg., 71:817.

Törngren, S, Blombäck, M., Almgård, L.-E., Norming, U., Nyman, C.R., 1990, Conventional heparin and semisynthetic heparin analogue (SSHA) alteration of blood coagulation after embolic occlusion of human renal circulation. Thromb. Res., 59:237.

Van Ryn-McKenna, J., Gray, E., Weber, E., Ofosu, F.A., Buchanan, M.R., 1989a, Effects of sulfated polysaccharides on inhibition of thrombus formation initiated by different stimuli. Thromb. Haemost., 61:7.

Van Ryn-McKenna, J., Ofosu, F.A., Gray, E., Hirsh, J., Buchanan, M.R., 1989b, Effects of dermatan sulfate and heparin on inhibition of thrombus growth "in vivo". Ann. N.Y. Acad. Sci. 556:304.

Verstraete, M., 1990, Pharmacotherapeutic aspects of unfractionated and low molecular weight heparins. Drugs, 40:498.

Welzel, D., Wolf, H., Koppenhagen, K., 1988, Antithrombotic defense during the postoperative period. Clinical documentation of low molecular weight heparin. Drug Res., 38:120.

Wille-Jørgensen, P., Thorup, J., Fischer, A., Holst-Christensen, J., Flamsholt, R., 1985, Heparin with and without graded compression stockings in the prevention of thromboembolic complications of major abdominal surgery: a randomized trial. Br.J. Surg., 72:579.

Wille-Jørgensen, P., Kjærsgaard, J., Jørgensen, T., Larsen, T.K., 1988, Failure of prophylactic management of thromboembolic disease of colorectal surgery. Dis. Colon Rectum, 31:384.

Williams, H.T., 1971, Prevention of postoperative deep-vein thrombosis with perioperative subcutaneous heparin. Lancet, II:950.

Wolf, H., Welzel, D., Kaiser, H., Majer, M., Schäfer, D., Husfeldt, K.-J., Voigt, J., Sunder-Plassmann, L., 1988, Bewertung der perioperativen Thromboembolie-Prophylaxe mit niedermolekularem Heparin und Dihydroergotamin. Eine Untersuchung zur Frage der Inzidenz tödlicher Lungenembolien sowie unerwünschter Begleiterscheinungen, inbesondere des Vasospasmus- und Herzinfarktrisikos. Arzneim.-Forsch./Drug Res., 38:1516.

Østergaard, P.B., Nilsson, B., Bergqvist, D., Hedner, U., Pedersen, P.C., 1987, The effect of low molecular weight heparin on experimental thrombosis and haemostasis - the influence of production method. Thromb. Res., 45:739.

TREATMENT OF DEEP VEIN THROMBOSIS (DVT) WITH LOW MOLECULAR WEIGHT HEPARINS (LMWH)

Michel-Meyer Samama

Laboratoire Central d'Hématologie
Hôtel-Dieu
75181 Paris cedex 04

INTRODUCTION

The medical treatment of established deep venous thrombosis (DVT) includes thrombolytic agents and/or anticoagulants (heparins followed by oral antivitamin K agents).

The use of thrombolytic therapy is debatable and reserved to a limited number of patients but there is a new interest since the advent of tissue plasminogen activator (tPA) (Rogers and Lutcher, 1990; Turpie 1989).

The use of heparin has been the gold standard for many years since the initial therapeutic trial of Barritt and Jordan (1960) who demonstrated a therapeutic efficacy on the prevention of pulmonary embolism (PE) in patients with DVT.

SUBCUTANEOUS VERSUS INTRAVENOUS ROUTE

More recently, there has been another developing debate over the use of subcutaneous (SC) versus intravenous (IV) administration of heparin. This debate had led to a further discourse of the relative merits of low molecular weight heparin (LMWH) as compared to unfractionated heparin (UH). Regarding the choice of route of administration of UH, there has been seven trials since 1979 with some variation of results, the most recent and largest series was by Pini et al (1990). This study failed to demonstrate any significant difference in these approaches, on 133 and 178 patients treated by IV or SC route, respectively. Pooling all these studies (Pini et al. 1990) seems to lead to the same conclusion (table 1).

However, the results of a meta-analysis of essentially the same studies showed a suggestion of better efficacy with SC route but was less conclusive regarding hemorrhages (Hommes et al. 1991).

It should be noted that using the SC approach rather than IV infusion may cause a delay in establishing the proper level of dosage to attain desired therapeutic range. This was clearly shown in the study of Hull et al. (1986).

Heparin and Related Polysaccharides
Edited by D.A. Lane *et al.*, Plenum Press, New York, 1992

Table 1. Main results of the randomized trials comparing IV versus SC UH treatment in DVT.
(From Pini et al. 1990)

	N of patients		Pulmonary Embolism		Major Bleeding	
	IV	SC	IV	SC	IV	SC
Pini et al (1990)	133	138	2 (1 fatal)	4 (1 fatal)	9	5
Pooling of results of the 6 previous studies	304	303	6	8	12	12
TOTAL	437	441	8	12	21	17

THERAPEUTIC TRIALS WITH LMWH

Let us now turn to the problem at hand which is the comparison of LMWH to UH. It should be recognized that, initially, it was believed that LMWH would be only useful in prophylaxis and not in therapy. This was believed because the main action of LMWH was attributed to anti-Xa activity and not to anti-thrombin activity. However, now it is becoming apparent that this hypothesis is not entirely correct (for review see Samama and Desnoyers, 1991). This has been shown by several therapeutic trials that will be discussed in this paper. To begin with, it should be noted that one or more studies in laboratory animals have demonstrated that LMWH does inhibit thrombus propagation and induce less bleeding at equipotent antithrombotic activity levels compared to UH (Carter et al. 1982). In earlier studies, there was a selection of patients with a greater risk of bleeding because it was felt there was a greater safety margin (Huet et al. 1987). Another evidence of usefulness of LMWH in treating DVT is its considerable efficacy in preventing clotting in hemodialysis circuits (Lane et al. 1986). There are several potential advantages of LMWH compared to UH: these include a better benefit to risk ratio (efficacy and safety), reduction of frequency required of doses per day, reduced necessity of laboratory monitoring. And since all the above appear to simplify treatment, then home treatment could be considered with these new therapeutic agents. To investigate further the relative efficacy and safety of these agents, studies had to be designed to distinguish clear end points: these could include clinical signs (swelling, pain, etc) serial phlebography with a scoring system, proven recurrence of venous thrombosis, recognizable pulmonary embolism, and development of post phlebitic syndrome.

The use of clinical signs is helpful but of questionable validity since it is not objective. The use of serial phlebography remains highly regarded, but is invasive. It requires subjective visualization which explains why scoring methods have been developed, and independent observations by different radiologists appear necessary. The objective of these scoring methods is to quantify the evolution of the size of the clot. It has to be noted that an apparent reduction of size of a thrombus in a leg vein could really result from embolization to the lung. This would be most difficult to evaluate where proof of clots in the lung are not sought. In the attempt to prove presence of thrombi in the lung, serial lung scans can be helpful and have been done in some studies. Naturally, evidence of pulmonary emboli or proven recurrence of venous thrombosis in the leg are excellent evidences of lack of efficacy of treatment. However, they are uncommon and therefore a large number of patients has to be included to make the study statistically valid.

Finally, the development of post-phlebitic syndrome is an excellent proof of lack of efficacy, but takes much too long to develop to be of much use in any study.

Thus far, the reference treatment has been UH as mentioned earlier, but the route of administration SC or IV is still in debate. That explains the reason that different studies have used one or the other of these means. It is significant to note that in these trials, in the interval between suspicion and proof of DVT, IV heparin has been started in all these studies. Only on establishment of proof have the actual therapeutic trials started.

The problem of having some studies using coumarins from the very onset of heparin therapy as compared to other studies where the oral anticoagulants were added later, produces a possibility of heterogeneity of results. This has also to be carefully considered in final analysis of the results of different trials. In addition, there is the situation that some studies are randomized open studies whereas others are double-blind randomized studies.

The first two pilot studies by Bratt et al. (1985, 1990) were conducted using a rather small number of patients, and compared IV LMWH versus IV UH. The advantage of these pilot studies was that they clearly demonstrated that the same anti-Xa unitage could not be used for LMWH and UH because of development of bleeding in the former group which caused the premature discontinuation of the trial. It was found that a more appropriate dose of LMWH was about 50 % of that originally attempted. It was also clearly demonstrated that APTT is a totally inappropriate laboratory test for monitoring patients on LMWH. A better test, anti-Xa plasma level was used in the following studies and found to be much more appropriate. Interestingly, in their recent publication, Bratt et al. (1990) found in a prospective, randomized open study that after a median follow-up of 23 months, the frequency of rethrombosis was similar in the standard heparin group and in the Fragmin group : 6 out of 43 and 4 out of 41 reexamined patients, respectively.

In 1986, Holm et al. conducted a double blind randomized study of 56 patients using LMWH or UH given subcutaneously twice daily. The dose of LMWH was half that of U.H. The results by Marder score did not show any significant difference in the two groups of patients neither was there any significant incidence of bleeding episodes.

Very recently 194 patients with DVT and/or pulmonary embolism were included in a double-blind randomized study comparing LMWH IV versus IV UH therapy (Albada et al. 1989). Doses were adjusted in both groups of patients in order to maintain anti-Xa plasma levels (amidolytic method) between 0.4 to 0.9 IU/mL or 0.3 to 0.6 IU/mL when there was high bleeding risk, blood being collected about 3-4 hours after injection. This study was of great importance because safety was being very carefully evaluated. The LMWH was found to produce fewer major and minor bleeding episodes than UH. However, this was not considered statistically significant. Discontinuation of treatment was considered necessary in about 2 % of each group. In regard to efficacy of each type of therapy, judged by death from P.E., worsening of thromboembolism, development of new defects on lung scan, changes in serial plethysmography, as well as clinical findings (disappearance of pain and decreased swelling, etc...) neither method was found to be statistically superior to the other.

Two multicentre randomized open studies can be considered together since they are of very similar design although using two different LMWH, Fraxiparine® and Fragmin®, respectively. The results were very similar when using Marder score and Arnesen score as criteria of efficacy with either type of therapy (Anonyme 1989) (Duroux 1991). In the French study which included 60 patients the difference of bleeding incidence seems to clearly favour LMWH, while in the European study of 176 patients this appears less clear in that regard (Duroux 1991).

In review, it should be noted that all the studies already mentioned did not show significant difference in efficacy between LMWH and UH, in the scores designed by Marder or Arnesen with the exception of the European Study.

There are four more published in part, or unpublished studies, which are to be considered:

In the Prandoni trial including two groups of approximately 80 patients and using Fraxiparin subcutaneously administered twice daily at adjusted dosage according to body weight, versus UH administered intravenously continuously achieving an APTT of 1.5 to 2.5 times normal, it was found that there were more thromboembolic episodes and DVT recurrences in the UH group, but the difference was not statistically significant. Interestingly, mortality was also improved, but, again, this result could not be proven by statistical analysis ($p = 0.2$) (Prandoni et al. 1990) (Prandoni 1991).

The recent study of Hull et al. (1991) using Logiparin®, a double-blind randomized study with a large number of participants, is a particularly important work. This was preceded by a dose ranging study (Siegbahn et al. 1989) which showed the logic of the dose used in Hull's study. Hull et al. used only one injection per day of the LMWH (175 IU/kg by day) and compared this against continuous IV UH, producing a level of 1.5 to 2.5 times normal APTT. His results are quite impressive in this series of 432 patients. DVT recurrences were reduced in the Logiparin® group from 5.9 to 3.3 % at a 3 months point. Three PE occured with UH and none with Logiparin®. Major bleeding occured in 10 patients with UH and 1 with Logiparin®, which has a p value of < 0.02. Minor bleeding was the same in the two groups. Mortality at 3 months was 21 (10 %) with UH and 9 (4.4 %) with Logiparin®, which has a p value of 0.03. If this result can be verified in other studies, then the one will have to consider its impact on future therapeutic choices.

Lopaciuk (1991), has just completed an open stratified randomized study using CY 216 at a dose of 225 Institut Choay Units (ICU)* kg b.w. subcutaneously twice daily for 10 days, compared to calcium heparin given also subcutaneously twice daily aiming at an APTT patient/control ratio ranging from 1.5 to 2.5, 3 to 4 hours after injection. Acenocoumarin was started for both groups on the 7th day. He studied 146 patients and there was no difference in either group in DVT, Arnesen score, PE incidence or bleeding (major or minor).

Finally, a randomized trial has used Enoxaparin, 1 mg/kg body weight, given subcutaneously twice a day versus UH aiming an APTT patient/control ratio close to 2. Two groups of 67 patients were included and results were analyzed in 117 patients. They were in favour of Enoxaparin regarding efficacy and safety (Simonneau 1991).

A very small number of trials (Janvier et al. 1987; Faivre et al. 1988) have used an ultralow molecular weight heparin CY 222 with a mean MW of about 50 % of the usual LMWH. The anti-Xa/Anti-IIa ratio was higher (anti-Xa 250 ICU/mg, Anti-IIa 25 IU/mg).

* 1 anti-Xa I.C.U. = 0.41 I.U.

In a preliminary dose ranging studies (Elias et al. 1986) the optimal dose was found to be 750 ICU/kg b.w./24 hrs corresponding to about 300 anti-Xa I.U. per day and which is about 50 % higher in unitage compared to the usual LMWH Fraxiparine (mean MW 5000, from the same company). This is an interesting observation suggesting that the reduction in molecular size increases the optimal dose required in anti-Xa units.

Efficacy of CY 222 in the treatment of DVT of lower limbs was evaluated in a prospective clinical trial including 95 patients who received 750 anti-Xa ICU/kg b.w. daily as 3 divided doses over a minimum of 10 days (Janvier et al. 1987). Clinical symptomatology regressed in 88 % of patients. Phlebography before and after treatment was available in a small number of patients (n = 12) and showed an improvement of Arnesen's score in 10 of them. There was no clinically significant bleeding except in patients who had an interruption of the vena cava before the initiation of CY 222 injections.

In another study with the same dose of CY 222 by Faivre et al. in 1988, the comparison was made against UH given in two S.C. injections aiming at an APTT patient/control ratio of 2, 3 to 4 hours after injection. The total number of patients included was 68. The Marder score showed similar efficacy. However there appeared to be less bleeding with CY 222.

LABORATORY MONITORING

It has been, for many years, generally accepted that UH must be monitored carefully by laboratory control for efficacy and safety factors.

For LMW Heparin there is a possibility that laboratory control may be less stringent in its requirements (Handeland et al. 1990). However, we probably need a larger number of studies to prove this point conclusively.

Although for U.H., the APTT test is the conventional and logical choice, for LMWH the logical choice as found thus far cannot be APTT but appears to be anti-Xa measurement in plasma.

In the studies mentioned in this review, the therapeutic range at peak level (3 to 4 hours after administration) was 0.5 to 1.0 IU per mL of plasma, using the new International LMWH standard.

The therapeutic usefulness of laboratory monitoring is very debatable, the appropriate test is also debatable (Abildgaard et al. 1990) since anti-IIa may be more closely related to antithrombotic activity, but the concentration is much lower than anti-Xa, and therefore more difficult to monitor. A clinical trial has been recently conducted in order to prove or disprove the need for laboratory monitoring in 109 patients treated with Fragmin®. The dosage was not modified during 10 days in 49 patients. In contrast, in the remaining 60 patients, the dosage was adjusted to reach the therapeutic range 0.5 to 1.0 IU/mL, 3 to 4 hours after the morning injection on Day 2 and Day 6. The results suggest that a laboratory control at Day 2 would optimize the therapeutic benefit (Aiach 1991). Finally, even if laboratory control is really necessary and the appropriate test well selected, the therapeutic range has still to be precisely determined.

CONCLUSION

Although the experience regarding the use of LMWH in established DVT is limited, the available results of clinical trials are very encouraging. Different LMWH have been used and it is clear that dose finding studies are essential.

There is some suggestion that LMWH are at least as efficacious as UH in the treatment of DVT without increase in the bleeding risk. Laboratory monitoring seems to be less stringent with LMWH than with UH.

Finally, the use of LMWH could allow a simplification of treatment, facilitating home treatment.

REFERENCES

Abildgaard, U., Norrheim, L., Larsen A.E., Nesvold, A., Sandset, P.M., Odegaard, O.R., 1990. Monitoring therapy with LMW Heparin; a comparison of three chromogenic substrate assays and the Heptest clotting assay. Haemostasis 20:193-203

Aiach, M. 1991. Should deep venous thrombosis treatment with low molecular weight heparin (Fragmin®) be monitored or not? Results of a randomized comparative multicentric french study. Thromb. Haemostas. 65:754, abst. 307

Albada, J., Nieuwenhuis, H.K. and Sixma, J.J., 1989. Treatment of acute venous thromboembolism with low molecular weight heparin (Fragmin®) : results of a double blind randomized study. Circulation, 80:935-940

Anonyme, 1989. Traitement des thromboses veineuses profondes constituées. Etude comparative d'un fragment d'héparine de bas poids moléculaire (Fragmine®) administrée par voie sous-cutanée, et de l'héparine standard administrée par voie intraveineuse continue. Etude multicentrique. Rev. Med. Interne, 10:375-381

Barritt, D.W., and Jordan, S.G., 1960. Anticoagulant drugs in the treatment of pulmonary embolism. Lancet 1:1309

Bratt, G., Tornebohm, E., Granqvist, S., Aberg, W., Lockner, B., 1985. A comparison between low molecular weight heparin (Kabi 2165) and standard heparin in the intravenous treatment of deep venous thrombosis. Thromb. Haemostas. 54:813-817

Bratt, G., Aberg, W., Johansson, M., Törnebohm, E., Granqvist, S., Lockner, D., 1990. Two daily subcutaneous injections of Fragmin as compared with intravenous standard heparin in the treatment of deep vein thrombosis (DVT). Thromb. Haemostas. 64:506-510

Carter, C.J., Kelton, J.G., Hirsh, J., Cerskus, A., Santos, A.V., Gent, M., 1982. The relationship between the haemorrhagic properties of low molecular weight heparin in rabbits. Blood 59:1239-1245

Duroux, P., 1991. A randomised trial of subcutaneous low molecular weight heparin (CY 216) compared with intravenous unfractionated heparin in the treatment of deep vein thrombosis. A collaborative European Multicentre Study. Thromb. Haemostas. 65:251-256

Elias, A., Lecorff, G., Bouvier, J.L., Aillaud, M.F., Juhan-Vague, I., Toulemonde, F., Serradimigni, A., 1986. Treatment of deep vein thrombosis by a very low molecular weight heparin fragment (CY 222): a dose range study. Thromb. Res. suppl. VI, 82, abst. 162

Faivre, R., Neuhart, Y., Kieffer, Y., Apfel, F., Magnin, D., Didier, D., Toulemonde, F., Bassand, J.P., Maurat, J.P., 1988. Un nouveau traitement des thromboses veineuses profondes; les fractions d'héparine de bas poids moléculaire. Etude randomisée. Presse Med. 17:197-200

Handeland, G.F., Abildgaard, U., Holm, H.A., Arnesen, K.E., 1990. Dose adjusted heparin treatment of deep venous thrombosis; a comparison of unfractionated and low molecular weight heparin. Eur. J. Clin. Pharmacol. 39:107-112

Hommes, D.W., Bura, A., Mazzolat, L., Büller, H.R., ten Cate, J.W., 1991. Subcutaneous heparin compared to continuous intravenous heparin administration in the initial treatment of deep vein thrombosis: A systemic overview and meta-analysis. Thromb. Haemostas. 65:753, abst. 305

Huet, Y., Gouault-Heilmann, M., Contant, G., Brun-Buisson, C., 1987. Treatment of acute pulmonary embolism by a low molecular weight heparin fraction. A preliminary study. Intensive Care Med. 13:126-130

Hull, R.D., Raskob, G.E., Hirsh, J., Jay, R.M., Leclerc, J.R., Geerts, W.H., Rosenblum, D., Sackett, D.L., Anderson, C., Harrison, L., Gent, M., 1986. Continuous intravenous heparin compared with intermittent subcutaneous heparin in the initial treatment of proximal-vein thrombosis. N. Engl. J. Med. 315:1109-1114

Hull, R.D., Raskob, G.E., Pineo, G.F., Green D., Trowbridge, A.A., Elliott, C.G. et al. 1991. A randomized double-blind trial of low-molecular weight heparin in the initial treatment of proximal-vein thrombosis. Thromb. Haemostas. 65:872. abst. 620

Janvier, G., Dugrais, G., Winnock, S., Vergnes, C., Boisseau, M., Broussin, J., Serise, J.M., Boissieras, P., Videau, J., Toulemonde, F. 1987. Le CY 222, héparine de très bas poids moléculaire dans le traitement curatif des thromboses veineuses profondes. A propos de 95 observations. Résultats cliniques et phlébographiques. J. Mal. Vascul. 12:141-144

Lane, D.A., Flynn, A., Ireland, H., Anastassiades, E., Curtis, J.R., 1986. On the evaluation of heparin and low molecular weight heparin in haemodialysis for chronic renal failure. Haemostasis 16 (suppl. 2):38-47

Lopaciuk, S., 1991. Activity of Fraxiparine® in the treatment of DVT. ISH - XIth Meeting, Basel, August 31, Fraxiparine in the nineties".

Pini, M., Pattacini, C., Quintavalla, R., Poli, T., Megha, A., Tagliaferri, A., Manotti, C., Dettori, A.G., 1990. Subcutaneous vs intravenous heparin in the treatment of deep venous thrombosis. A randomized Clinical Trial. Thromb. Haemostas. 64:222-226

Prandoni, P., Vigo, M., Cattelan, A.M., Ruol, A. 1990. Treatment of deep venous thrombosis by fixed doses of a low-molecular weight heparin (CY 216). Haemostasis 20 (suppl. 1):220-223

Prandoni, P., 1991. Fixed dose LMW Heparin (CY 216) as compared with adjusted dose intravenous heparin in the initial treatment of symptomatic proximal venous thrombosis. Thromb. Haemostas. 65:872 , abst. 618

Rogers, L.Q., Lutcher, C.L., 1990. Streptokinase therapy for deep vein thrombosis: a comprehensive review of the English literature. Am. J. Med. 88:389-395

Samama, M, Desnoyers, P., 1991. Low-molecular-weight heparins and related glycosaminoglycans in the prophylaxis and treatment of venous thromboembolism, in: "Recent Advances in Blood Coagulation", vol. 5. L. Poller, ed. Churchill Livingstone, Edinburgh. pp 177-222

Siegbahn, A., Shams, Y-Hassan, Roberg, J., Bylund, H., Neerstrand, H.S., Ostergaard, P., Hedner, U., 1989. Subcutaneous treatment of deep venous thrombosis with low molecular weight heparin. A dose finding study with LMWH-NOVO. Thromb. Res. 55:767-778

Simonneau, G. 1991. Subcutaneous fixed dose of Enoxaparine (E) versus intravenous adjusted dose of unfractionated heparin (UH) in the treatment of deep venous thrombosis (DVT). TVPENOX Group Study.. Thromb. Haemostas. 65: 754, abst. 306

Turpie, A.G.G., Thrombolysis in deep vein thrombosis, in: "Thrombolysis in cardiovascular disease"Julian, D.G., Kübler, W., Norris, R.M., Swan, H.J., Collen, D., Verstraete, M. eds., Marcel Dekker, Inc. New-York-Basel (1989) p.397-408

RELATIONSHIP BETWEEN DOSE, ANTICOAGULANT EFFECT AND THE CLINICAL EFFICACY AND SAFETY OF HEPARIN

Jack Hirsh

Director, Hamilton Civic Hospitals Research Centre
Professor of Medicine, McMaster University
Hamilton, Ontario L8V 1C3

Heparin is poorly absorbed from the gastrointestinal tract and is administrated by intravenous or subcutaneous injection. Heparin binds to many plasma proteins including vitronectin[1], histidine-rich glycoprotein[2] and fibronectin, which limits its access to antithrombin III and results in a variable loss of recovered anticoagulant activity.

After an intravenous injection of heparin into healthy volunteers, there is a rapid phase of elimination due to equilibration which is followed by a more gradual disappearance which can best be explained by a combination of a saturable and a first order mechanism of clearance[3,4]. The saturable phase of clearance is much more rapid than the non-saturable phase which follows first order kinetics. Because of these kinetics, the anticoagulant response to heparin at therapeutic doses is not linear but the anticoagulant effect increases disproportionally both in its intensity and duration with increasing dose[3,4]. Thus, the apparent biological half-life of heparin increases from approximately 30 minutes with an intravenous bolus of 25 U/kg to 60 minutes with an intravenous bolus of 100 U/kg to 150 minutes with a bolus of 400 U/kg[5,6] (Table 1).

The saturable phase of heparin clearance is thought to be the result of heparin binding to receptors on endothelial cells and macrophages where it is internalized,

TABLE 1

EFFECT OF HEPARIN DOSAGE ON APPARENT BIOLOGICAL HALF-LIFE

Dose of Heparin U/kg (70 kg)	Half-Life Min	Author
25 (1,750)	30	Bjornsson (6)
75 (5,250)	60	Bjornsson (6)
100 (7,000)	56	Olsson (5)
400 (28,000)	152	Olsson (5)

depolymerized and metabolized into smaller and less sulfated forms[3]. Results of animal experiments indicate that the capacity of the cellular mechanism to clear heparin is restored rapidly after heparin disappears from the blood[3].

A plausible explanation for these findings is that heparin bound to cell surface receptors is internalized, thereby, freeing up cell surface receptors to bind more circulating heparin and so providing a renewable site for rapid removal of sub–saturating concentrations of heparin from the plasma. When heparin is administered in progressively increasing doses which exceed the saturating concentration, an increased proportion is cleared by the slower non–saturable (possibly renal) mechanism. This proposed mechanism of heparin clearance explains why at a steady state, low doses of heparin do not produce measurable heparin levels[4] and why with increasing doses over a dose range of 0–30,000 U of heparin/24 hrs by continuous infusion, the recovered anticoagulant effect of heparin does not increase in a linear manner with dose but rather increases disproportionally with increasing dose. Studies in patients with venous thrombosis have shown that at doses of 30,000 U/24 hrs (after 5,000 intravenous bolus) the recovered anticoagulant effect of heparin administered by subcutaneous injection (in two divided doses) is significantly less than when administered by the continuous intravenous route[7]. Even when administered in a dose of 35,000 U/24 hrs, the plasma recovery of heparin is significantly less when administered by the subcutaneous route (12 hourly) than when administered intravenously[8].

This model of pharmacokinetics of heparin provides an explanation for some of the puzzling observations associated with heparin usage. Thus, the marked variability in dosage requirements between patients treated with heparin can be explained in part by differences in the concentrations of heparin binding proteins and, in part, by the variability in heparin clearance between patients. Since some heparin binding proteins are acute phase reactants, heparin binding to these proteins might explain the resistance to heparin seen in sick patients with inflammatory and malignant disorders[9]. The poor bioavailability of low doses of heparin is also contributed to by its rapid clearance by a renewable cellular clearance mechanism which is not fully saturated even at doses of 10,000 to 15,000 U/24 hrs administered for 6 hrs by continuous infusion[4]. The bioavailability of heparin administered by the subcutaneous route is reduced (compared to intravenous administration) because the gradual entry of heparin into the circulation does not keep pace with the clearance of heparin from the plasma by the saturable clearance mechanism. This also explains the delay in reaching steady state heparin levels when heparin is administered by subcutaneous injection 12 hourly even at a dose of 30,000 to 35,000 U/24 hrs[7,8] and why, the dosage requirement for subcutaneous heparin to achieve therapeutic heparin concentrations (0.2–0.4 U/ml by protamine titration) is approximately 10–15% higher than when heparin is administered by intravenous infusion[8].

The elimination of heparin from plasma is accelerated in clinical and experimental acute pulmonary embolism[10,11] by a mechanism which is poorly understood. There are reports that intravenous nitroglycerine may increase heparin dosage requirements acutely[12,13], but interaction was not seen in a randomized cross–over study using relatively low doses of nitroglycerin[14].

The anticoagulant effect of heparin is modified by platelets, fibrin, vascular surfaces and plasma proteins. Platelets inhibit the anticoagulant effect of heparin by binding Factor Xa and protecting it from inactivation by the heparin/AT III complex[15,16] and by secreting the heparin–neutralizing protein, platelet factor 4[17]. Fibrin binds thrombin and protects it from inactivation by heparin/AT III complex[18,19]. In plasma, much higher concentrations of heparin are needed to inactivate fibrin–bound thrombin than free thrombin[18]. In contrast, fibrin–bound thrombin is not protected from inactivation by AT III–independent thrombin inhibitors such as hirudin or hirudin fragments[18]; an observation which might explain why heparin is less effective than hirudin in preventing the formation of experimental arterial thrombosis[20] and in preventing the extension of experimental venous thrombosis[21]. The relative resistance of fibrin–bound thrombin to inhibition by heparin could also explain why higher concentrations of heparin are required to prevent extension of venous thrombosis than to prevent its formation[22] and why heparin fails to inhibit thrombin activity in some patients after successful coronary thrombolysis[23-25]. Thrombin bound to subendothelial surfaces is also protected from inactivation by heparin[26], possibly through mechanisms which are similar to those that protect fibrin–bound thrombin.

MONITORING HEPARIN TREATMENT

Heparin therapy is usually monitored to maintain its anticoagulant effect within a defined range in order to limit recurrence and extension of thrombosis while minimizing the risk of bleeding. This range is commonly referred to as the therapeutic range. The concept of a therapeutic range for the control of heparin therapy has been validated by studies in animals and by results from clinical trials in patients with venous thrombosis[7,27] and acute myocardial infarction[28-30]. The need for monitoring heparin therapy adds to the expense and complexity of heparin treatment because the dosage requirements differ between patients and can differ within patients during their course of treatment. The causes for this variability are unknown but are likely to be contributed to variations in levels of heparin binding proteins (which compete with AT III for heparin binding), the proportion of the administered dose cleared by the more rapid saturable phase of clearance and the concentration of coagulation factors which affect the relationship between heparin levels and anticoagulant response (see below). Heparin monitoring was performed initially by the whole blood clotting time. A whole blood clotting time (WBCT) of twice control was selected as the therapeutic range based on studies in the 1950's by Wessler examining the effect of heparin on experimental jugular vein thrombi in rabbits[31]. Because of its imprecision and inconvenience, the WBCT was replaced by activated partial thromboplastin time (APTT)[32-35]. In comparative in–vitro studies on heparinized blood, a WBCT of twice to three times control was found to be equivalent to an APTT ratio (patients APTT/mean laboratory APTT) of 1.5 to 2.5[36]. Unfortunately, the APTT in its present form is far from ideal as a test for monitoring heparin therapy because its responsiveness (sensitivity) to heparin varies considerably among different APTT reagents[36] and even among different batches of the same brand of APTT reagent[37]. Various technical factors influence the responsiveness of APTT reagents to heparin including: a) the type of contact activator (reagents using ellagic acid are less sensitive than kaolin–derived reagents)[38]; b) the buffering capacity and the phospholipid composition of the reagent [39]; and c) the method of clot detection including the use of automated procedures[40]. As a consequence of this variability, it is inappropriate to compare the results obtained when heparin is monitored by different APTT reagents[37]. The situation is complicated further because the relationship between the APTT result and heparin levels obtained on control plasma when increasing doses of heparin is added in–vitro may not reflect the relationship obtained when heparin is added in vitro to plasma of patients with thrombosis. Furthermore, the relationship between the APTT and heparin levels obtained in vitro does not correspond to the relationship obtained on ex–vivo samples when the APTT and heparin assays are performed on plasma obtained from patients who are being treated with heparin[41,42]. Standardization can be obtained within laboratories by establishing a relationship between the APTT ratio and heparin levels on plasma obtained from heparinized patients over a wide range of values and by comparing each new batch of reagent with the previous batch. If the curves describing the relationship between the heparin level and the APTT results obtained with the new batch and previous batch are parallel over the therapeutic range, a correction factor can be applied to avoid the confusion of changing the therapeutic range of the APTT ratio every time the batch of APTT reagent is changed. At the Hamilton District Laboratory Program which serves the hospitals affiliated with McMaster University we purchase a new batch of reagent from the same manufacturer every two years and have had to standardize each new reagent against the expiring batch to avoid confusing the clinicians by changing the recommended therapeutic range every time a new batch is obtained.

DOSING CONSIDERATIONS

The mean dose of heparin required to produce an APTT of 1.5 to 2.5 times control (heparin level by protamine titration of 0.2 to 0.4 U/ml) and by anti-factor Xa assay of 0.35 to 0.7 U/ml is approximately 32,000 U/24 hrs when heparin is administered by continuous intravenous infusion and 35,000 U/24 hrs when it is administered by subcutaneous injection 12 hourly. For continuous intravenous infusion, the heparin effect should be measured 6 hours after the bolus injection and then 6 hourly until a stable dose response is obtained[43]. For subcutaneous heparin, the heparin effect should be measured daily 6 hours after injection. Approximately 30% of patients with thrombosis require higher doses of heparin to achieve an APTT result in the therapeutic range. The mechanism of heparin resistance in these patients is not fully understood. In some (approximately a third of heparin resistant patients), the circulating heparin level is in the therapeutic range but the APTT is below the therapeutic range. These patients have circulating procoagulants which shorten the pre-treatment of the APTT and inhibit its prolongation by heparin. Preliminary results of a randomized study suggest that it is safe to monitor these patients by heparin assay. The remaining two-thirds of patients are truly heparin resistant. Some have high concentrations of heparin binding proteins which compete with AT III for binding while others may have an increased rate of heparin clearance.

Mechanism of Heparin-induced Bleeding

Heparin has the potential to induce bleeding by inhibiting blood coagulation, by impairing platelet function and by increasing capillary permeability. The major anticoagulant action of heparin when administered at therapeutic concentrations is through its ability to augment the inhibition of a number of activated coagulation factors by antithrombin III (ATIII)[44-46]. Heparin also binds to platelets and inhibits platelet aggregation and adhesion[47]. Heparin also increases vessel permeability to albumin and erythrocytes in a dose-related manner[48]. Heparin can also produce thrombocytopenia, but this is rarely an important cause of bleeding.

Relationship between Risk of Bleeding and Heparin Dose/Response

The heparin response (measured by a test of blood coagulation, i.e. APTT) is influenced by the heparin dose and it is not possible from reported studies to separate the effects of these two dependent variables (dose and laboratory response) on hemorrhagic rates. There have been no randomized trials in patients with established venous thromboembolism directly comparing different doses of heparin, but if hemorrhagic rates are correlated with the 24 hour dose of heparin (from the studies in Table 2), a significant relationship is observed between heparin dose and bleeding (correlation coefficient is 0.56)[49]. In a study evaluating prophylaxis in patients with recent onset traumatic spinal cord injuries, the incidence of bleeding was significantly greater in patients randomized to receive heparin adjusted to maintain the APTT at 1.5 times control than compared to heparin 5000 units twice daily[50]. The mean dose of heparin for the adjusted–dose regimen was 13,200 units. Bleeding occurred in seven adjusted–dose patients compared to none in the fixed–dose group.

Since there is experimental evidence that inhibition of platelet function and microvascular bleeding occur with fractions of heparin with low–affinity to antithrombin III, heparin–induced bleeding may not always be predicted by the anticoagulant response in vivo. Subgroup analysis of randomized trials and prospective cohort studies provide only suggestive evidence for an association between the incidence of bleeding and the anticoagulant response.

In the Urokinase Pulmonary Embolism Study[51], bleeding occurred in 20% of the 30 patients whose whole blood clotting time was greater than 60 minutes, but in only five percent of the 19 patients whose whole blood clotting was less than 60 minutes (relative risk 4.0). Norman and coworkers[52] reported five major bleeding episodes in ten patients whose APTT was prolonged to more than twice the upper limit of their therapeutic range for at least 50% of their assays but in only one of 40 patients whose APTT remained in the therapeutic range (relative risk 20.0). Wilson and Lampman described 80 non–surgical patients receiving heparin monitored by the whole blood clotting time[53]. Ten of eighteen patients (56%) who received "excessive heparin" (defined by a greatly prolonged whole blood clotting time), bled, whereas bleeding occurred in only 16% of the patients who did not receive "excessive heparin" (relative risk 3.5).

In summary, although none of the studies were designed to compare the effects on bleeding of either different doses of heparin or different levels of heparin response, there is a suggestion that bleeding is more likely to occur when an in vitro test of coagulation is prolonged excessively, but by no means is this evidence definitive.

<u>Relationship between Risk of Bleeding and Method of Administering Heparin</u>

The evidence for a relationship between the risk of bleeding and the method of administering heparin comes from six randomized trial in which heparin was either administered by continuous intravenous infusion or intermittent intravenous injection[53–58] and five randomized trials in which heparin was either administered by continuous infusion or twice daily subcutaneous injection[7,8,59–61] (Table 2). A meta–analysis which compared continuous with intermittent intravenous heparin showed an average incidence of major bleeding of 6.8% in the continuous infusion group and 14.2% in the intermittent heparin group, an odds ratio of 0.42 ($P = 0.01$) (Table 3). Patients randomized to receive heparin by intermittent injection received a higher dose of heparin than those allocated to the continuous intravenous group. Therefore, the difference in bleeding between methods could be explained by differences in dose. A meta–analysis of the studies comparing continuous intravenous heparin with subcutaneous heparin revealed an incidence of bleeding of 5.2% and 4.1% respectively, an odds ratio of 1.1 (Table 3).

TABLE 2

RISK OF BLEEDING ASSOCIATED WITH THERAPEUTIC HEPARIN

Study	Method of Administration	Patient No.	Major** Hemorrhage No. (%)	24 Hour Heparin Dose (Units)
Salzman (1975)	Continuous vs. Intermittent	69 72	1 (1.0) * 6 (8.3)	24,480 31,740
Glazier (1976)	Continuous vs. Intermittent	20 21	0 * 7 (33.0)	25,488 32,808
Mant (1977)	Continuous vs. Intermittent	40 36	6 (15.0) 4 (11.0)	33,074 29,861
Wilson (1979)	Continuous vs. Intermittent	40 40	2 (5.0 4 (10.0)	27,695 37,015
Fagher (1981)	Continuous vs. Intermittent	15 13	3 (20.0) 4 (30.8)	37,371 41,380
Wilson (1981)	Continuous vs. Intermittent	36 29	3 (8.0) 5 (17.0)	28,440 43,570
Bentley (1980)	Subcutaneous vs. Continuous	50 50	2 (4.0) 4 (8.0)	36,998 36,814
Andersson (1982)	Subcutaneous vs. Continuous	72 69	2 (2.7) 2 (2.8)	35,000 31,500
Hull (1986)	Subcutaneous vs. Continuous	57 58	2 (3.5) 2 (3.4)	32,317 29,674
Doyle (1987)	Subcutaneous vs. Continuous	51 52	4 (7.8) 2 (3.8)	29,180 29,260
Pini (1990)	Subcutaneous vs. Continuous	138 133	5 (3.6) 9 (6.8)	33,800 31,700

* $p < 0.05$ ** 5 fatal bleeds in all studies combined

Relationship between the Risk of Bleeding and Patient Risk Factors

There is good evidence that comorbid conditions, particularly recent surgery or trauma, are very important risk factors for heparin induced bleeding[62–64]. This association was demonstrated in the recent study by Hull and associates in patients with proximal vein thrombosis[62]. Patients without clinical risk factors for bleeding were treated with a starting dose of 40,000 units of heparin by continuous infusion while those with well recognized risk factors for bleeding (recent surgery, trauma) received a starting dose of 30,000 units. Bleeding occurred in one of 88 patients (1%) who received 40,000 units and 12 of 111 patients (11%) who received 30,000 units.

TABLE 3

META–ANALYSIS OF BLEEDING INCIDENCE IN STUDIES COMPARING DIFFERENT METHODS OF HEPARIN ADMINISTRATION*

No. of Studies	Method of Administration	No. of Patients	% Major Bleeding	Odds Ratio
6	Continuous I.V. Intermittent I.V.	220 221	6.8 14.2	0.42
5	Continuous I.V. Subcutaneous	362 368	5.2 4.1	1.1

* From studies presented in Table 2

The concomittant use of aspirin was identified as a risk factor in early retrospective studies[63]. The association of aspirin ingestion with heparin induced bleeding was confirmed by Goldman and associates in their study in patients undergoing aortocoronary by–pass surgery[64]. In this study, the pre–operative use of aspirin caused excessive operative bleeding in patients who received high doses of heparin as part of the routine for by–pass procedures. Other studies have reported that older patients had a higher risk of bleeding[27,65].

CLINICAL USE OF HEPARIN

Heparin is effective in the prevention and treatment of venous thrombosis and pulmonary embolism, in the prevention of mural thrombosis after myocardial infarction, in the treatment of patients with unstable angina and acute myocardial infarction, and in the prevention of coronary artery rethrombosis after thrombolysis.

As noted above, the anticoagulant response to heparin varies widely between patients with thromboembolic disease and the clinical efficacy of heparin is optimized if the anticoagulant effect is maintained in a therapeutic range[7,66]. For these reasons, heparin treatment is usually monitored to maintain the activated partial thromboplastin time test (APTT) equivalent to a heparin level of 0.2 to 0.4 U/ml by protamine titration or an anti–Factor Xa level of 0.35 to 0.7 U/ml. For many APTT reagents this is equivalent to a

TABLE 4

RELATIONSHIP BETWEEN FAILURE TO REACH LOWER LIMIT OF THERAPEUTIC RANGE AND THROMBOEMBOLIC EVENTS FROM SUBGROUP ANALYSIS OF PROSPECTIVE STUDIES

Study	Type of Patients	Outcome	N	Relative Risk
Hull et al. (7)	DVT	Recurrent VTE	115	15.0
Basu et al. (27)	DVT	Recurrent VTE	157	10.7
Turpie et al. (28)	AMI	LVMT	112	22.2[1]
Kaplan et al. (29)	AMI	Recurrent MI/AP	75	6.0[2]
Camilleri et al. (30)	AMI	Recurrent MI/AP	70	13.3

DVT = Deep Vein Thrombosis
AMI = Acute Myocardial Infarction
LVMT = Left Ventricular Mural Thrombosis
AP = Angina Pectoris

1 Estimated by assuming a normal distribution of the reported heparin levels.

2 Kaplan used a PTT measurement and reported the relative risk associated with PTT values less than 50 seconds compared with PTT values of more than 100 seconds.

ratio (test APTT/lab control APTT) of 1.5 to 2.5; referred to as the therapeutic range. The recommended therapeutic range[7,27] is supported by evidence from animal studies[67] and by subgroup analysis of prospective cohort studies of the treatment of deep vein thrombosis, of the prevention of mural thrombosis after myocardial infarction and in the prevention of recurrent ischemia following coronary thrombolysis[29,30] (Table 4). The heparin regimens which have been shown to be effective in trials of venous and arterial thrombosis are summarized in Table 5.

TABLE 5
CLINICAL USE OF HEPARIN

Condition	Recommendations
VENOUS THROMBOEMBOLISM	
a) Prophylaxis of DVT & PE	5,000 U S/C 12 hourly or 8 hourly or adjusted low dose heparin*
b) Treatment of DVT	5,000 U I/V bolus followed by 30,000–35,000 U/24 hours by I/V infusion or 35,000–40,000 U/24 hours S/C adjusted to maintain APTT at 1.5–2.5 times control*+
CORONARY HEART DISEASE	
a) Unstable Angina and Acute Myocardial Infarction (Including Post Thrombolytic Therapy)	5,000 U I/V bolus followed by 24,000 U/24 hours I/V infusion adjusted to maintain APTT at 1.5–2.5 times control*+

* 3,500 U heparin S/C 8 hourly adjusted to an APTT in upper normal range.

*+ Equivalent to heparin level of 0.2–0.4 U/ml by protamine titration or an anti-factor Xa level of 0.35–0.7 U/ml.

REFERENCES

1. Preissner KT, Muller–Berghaus G, Neutralization and binding of heparin by S–protein/vitronectin in the inhibition of factor Xa by antithrombin III, J Biol Chem 262:12247 (1987).
2. Lijnen HR, Hoylaerts M, Collen D, Heparin binding properties of human histidine–rich glycoprotein. Mechanism and role in the neutralization of heparin in plasma, J Biol Chem 258:3803 (1983).
3. Boneu B, Caranobe C, Cadroy Y, et al., Pharmacokinetic studies of standard unfractionated heparin, and low molecular weight heparins in the rabbit, Sem Thromb Hemost 14:18 (1988).
4. De Swart CAM, Nijmeyer B, Roelofs JMM, Sixma JJ, Kinetics of intravenously administered heparin in normal humans, Blood 60:1251 (1982).
5. Olsson P, Lagergren H, Ek S, The elimination from plasma of intravenous heparin. An experimental study on dogs and humans, Acta Med Scand 173:619 (1963).
6. Bjornsson TO, Wolfram BS, Kitchell BB, Heparin kinetics determined by three assay methods, Clin Pharmacol Ther 31:104 (1982).
7. Hull RD, Raskob GE, Hirsh J, et al., Continuous intravenous heparin compared with intermittent subcutaneous heparin in the initial treatment of proximal–vein thrombosis, N Engl J Med 315:1109 (1986).
8. Pini M, Pattacini C, Quintavalla R, et al., Subcutaneous vs intravenous heparin in the treatment of deep venous thrombosis – A randomized clinical trial, Thromb Haemost 64:222 (1990).

9. Young E, Petrowski P, Hirsh J, Glycosaminoglycans displace anticoagulantly–active heparin from plasma protein binding sites. XIIIth Congress of the International Society on Thrombosis and Haemostasis. Amsterdam 1991. Abst.#844, Thromb Haemost 65(6):933 (1991).
10. Hirsh J, van Aken WG, Gallus AS, Dollery CT, Cade JF, Yung WG, Heparin kinetics in venous thrombosis and pulmonary embolism, Circulation 53:691 (1976).
11. Chui HM, van Aken WG, Hirsh J, Regoeczi E, Horner AA, Increased heparin clearance in experimental pulmonary embolism, J Lab Clin Med 90:204 (1977).
12. Habbab MA, Haft JI, Heparin resistance induced by intravenous nitroglycerin, Circulation 74(Suppl II):321 (1986).
13. Pizzulli L, Nitsch J, Luderitz B, Inhibition of the heparin effect by nitroglycerin, Deutsche Med Wochenschrift 133:1837 (1988).
14. Bode V, Welzel D, Franz G, Polensky U, Absence of drug interaction between heparin and nitroglycerin: Randomized placebo–controlled crossover study, Arch Intern Med 150:2117 (1990).
15. Marciniak E, Factor X_a inactivation by antithrombin III: evidence for biological stabilization of factor X_a by factor V–phospholipid complex, Br J Hematol 24:391 (1973).
16. Walker FJ, Esmon CT, The effects of phospholipid and factor V_a on the inhibition of factor X_a by antithrombin III, Biochem Biophys Res Comm 90:641 (1979).
17. Holt JC, Niewiarowski S, Biochemistry of α–granule proteins, Sem Hematol 22:151 (1985).
18. Weitz JI, Hudoba M, Massel D, Maraganore J, Hirsh J, Clot–bound thrombin is protected from inhibition by heparin–antithrombin III but is susceptible to inactivation by antithrombin III independent inhibitors, J Clin Invest 86:385 (1990).
19. Hogg PJ, Jackson CM, Fibrin monomer protects thrombin from inactivation by heparin–antithrombin III: Implications for heparin efficacy, Proc Nat Acad Sci 86:3619 (1989).
20. Heras M, Chesebro JH, Penny WJ, et al, Effects of thrombin inhibition on the development of acute platelet–thrombus deposition during angioplasty in pigs. Heparin versus recombinant hirudin, a specific thrombin inhibitor, Circulation 79:657 (1989).
21. Agnelli G, Pascucci C, Cosmi B, Nenci GG, The comparative effects of recombinant hirudin (CGP 39393) and standard heparin on thrombus growth in rabbits, Thromb Haemost 63:204 (1990).
22. Hirsh J, Ofosu FA, Levine MN, The development of low molecular weight heparins for clinical use, in: Thrombosis and Haemostasis, Verstraete M, Vermylen J, Lijnen R, Amout J eds., Leuven University Press, Brussels (1987).
23. Eisenberg PR, Sherman L, Rich M, et al, Importance of continued activation of thrombin reflected by fibrinopeptide A to the efficacy of thrombolysis, J Am Coll Cardiol 7:1255 (1986).
24. Owen J, Friedman KD, Grossman BA, Wilkins C, Berke AD, Powers ER, Thrombolytic therapy with tissue plasminogen activator or streptokinase induces transient thrombin activity, Blood 72:616 (1988).
25. Rapold JH, Kuemmerli H, Weiss M, et al, Monitoring of fibrin generation during thrombolytic therapy of acute myocardial infarction with recombinant

tissue-type plasminogen activator, Circulation 79:980 (1989).
26. Bar-Shavit R, Eldor A, Vlodavsky I, Binding of thrombin to subendothelial extracellular matrix. Protection and expression of functional properties, J Clin Invest 84:1096 (1989).
27. Basu D, Gallus A, Hirsh J, Cade JF, A prospective study of the value of monitoring heparin treatment with the activated partial thromboplastin time, N Engl J Med 287:324 (1972).
28. Turpie AGG, Robinson JG, Doyle DJ, et al, Comparison of high-dose with low-dose subcutaneous heparin to prevent left ventricular mural thrombosis in patients with acute transmural anterior myocardial infarction, N Engl J Med 320:352 (1989).
29. Kaplan K, Davison R, Parker M, et al, Role of heparin after intravenous thrombolytic therapy for acute myocardial infarction, Am J Cardiol 59:241 (1987).
30. Camilleri JF, Bonnet JL, Bouvier JL, et al, Thrombolyse intraveineuse dans l'infarctus du myocarde. Influence de la qualite de l'anticoagulation sur le taux de recidives precoces d'angor ou d'infarctus, Arch Mal Coeur 81:1037 (1988).
31. Wessler S, Morris LE, Studies in intravascular coagulation. IV. The effect of heparin and dicoumarol on serum-induced venous thrombosis, Circulation 12:563 (1956).
32. Soloway HB, Cornett BM, Grayson JW, Comparison of various activated partial thromboplastin reagents in the laboratory control of heparin therapy, Am J Clin Pathol 59:587 (1973).
33. Spector I, Corn M, Control of heparin therapy with activated partial thromboplastin times, JAMA 201:75 (1967).
34. Struver GP, Bittner DL, The partial thromboplastin time (cephalin time) in anticoagulant therapy, Am J Clin Pathol 38:473 (1962).
35. Zucker S, Cathey MH, Control of heparin therapy. Sensitivity of the activated partial thromboplastin time for monitoring the antithrombotic effects of heparin, J Lab Clin Med 73:320 (1969).
36. Shapiro GA, Huntzinger SW, Wilson JE, Variations among commercial activated partial thromboplastin time reagents in response to heparin, Am J Clin Pathol 67:477 (1977).
37. Shojania A, Tetreault J, Turnbull G, The variations between heparin sensitivity of different lots of activated partial thromboplastin time reagent produced by the same manufacturer, Am J Clin Pathol 89:29 (1988).
38. Barrowcliffe TW, Gray E, Studies of phospholipid reagents used in coagulation II: Factors influencing their sensitivity to heparin, Thromb Haemost 46(2):634 (1981).
39. Stevenson KJ, Daston AC, Curry A, Thomson JM, Poller L, The reliability of activated partial thromboplastin time methods and the relationship to lipid composition and ultrastructure, Thromb Haemost 55(2):250 (1986).
40. van den Besselaar AMHP, Meeuwisse-Braun J, Jansen-Gruter R, Bertina RM, Monitoring heparin therapy by the activated partial thromboplastin time - The effect of preanalytical conditions, Thromb Haemost 57:226 (1987).
41. Triplett DA, Harms CS, Koepke JA, The effect of heparin on the activated partial thromboplastin time, Am J Clin Pathol 70:556 (1978).
42. van den Besselaar AMHP, Meeuwise-Braun J, Bertina RM, Monitoring heparin therapy: Relationships between the activated partial thromboplastin time and heparin assays based on ex-vivo heparin samples, Thromb Haemost 63:16 (1990).
43. Cruickshank MK, Levine MN, Hirsh J, Roberts RS, Siguenza M, A standard heparin nomogram for the management of heparin therapy, Arch Intern Med 151:333 (1991).

44. Rosenberg RD, Lam L, Correlation between structure and function of heparin, Proc Natl Acad Sci USA 76(2):1218 (1979).
45. Rosenberg RD, The heparin–antithrombin system: a natural anticoagulant mechanism, in: Hemostasis and Thrombosis: Basic Principles and Clinical Practice, Second Edition, Colman RW, Hirsh J, Marder VJ, Salzman EW, eds., J.B. Lippincott Co., Philadelphia, (1987).
46. Lindahl U, Backstrom G, Hook M, Thunberg L, Fransson L–A, Linker A, Structure of the antithrombin-binding site of heparin, Proc Natl Acad Sci USA 76(4):3198 (1979).
47. Fernandez F, Nguyen P, van Ryn J, et al, Hemorrhagic doses of heparin and other glycosaminioglycans induce a platelet defect, Thromb Res 43:491 (1986).
48. Blajchman MA, Young E, Ofosu FA, Effects of unfractionated heparin, dermatan sulfate and low molecular weight heparin on vessel wall permeability in rabbits, Ann NY Acad Sci 556:245 (1989).
49. Levine MN, Hirsh J, Kelton JG, Heparin–induced bleeding. in: Heparin, Chemical and Biological Properties, Clinical Applications, Lane DA, Lindahl U, eds., Edward Arnold, London/England (1989).
50. Green D, Lee MY, Ito VY, et al, Fixed vs adjusted–dose heparin in the prophylaxis of thromboembolism in spinal cord injury, JAMA 260:1255 (1988).
51. Urokinase–Pulmonary Embolism Trial, Morbidity and mortality, Circulation 158:66 (1973).
52. Norman CS, Provan JL, Control and complications of intermittent heparin therapy, Surg Gynecol Obstet 145:338 (1977).
53. Wilson JR, Lampman J, Heparin therapy: a randomized prospective study, Am Heart J 97:155 (1979).
54. Salzman EW, Deykin D, Shapiro RM, Rosenberg R, Management of heparin therapy, N Eng J Med 292:1046 (1975).
55. Glazier RL, Corwell EB, Randomized prospective trial of continuous vs intermittent heparin therapy, JAMA 236:1365 (1976).
56. Mant MJ, O'Brien BD, Thong KL, et al, Haemorrhagic complications of heparin therapy, Lancet I:1133 (1977).
57. Fagher B, Lundh B, Heparin treatment of deep vein thrombosis, Acta Med Scand 210:357 (1981).
58. Wilson JE, Bynum LJ, Parkey RW, Heparin therapy in venous thromboembolism, Am J Med 70:808 (1981).
59. Bentley PG, Kakkar VV, Scully MF, et al, An objective study of alternative methods of heparin administration, Thromb Res 18(1–2):177 (1980).
60. Andersson G, Fagrell B, Holmgren K, et al, Subcutaneous administration of heparin: A randomized comparison with intravenous administration of hep rin to patients with deep vein thrombosis, Thromb Res 27:631 (1982).
61. Doyle DJ, Turpie AGG, Hirsh J, et al, Adjusted subcutaneous heparin or continuous intravenous heparin in patients with acute deep vein thrombosis: A randomized trial, Ann Intern Med 107(4):441 (1987).
62. Hull RD, Raskob GE, Rosenbloom D, et al, Heparin for 5 days versus for 10 days in the initial treatment of proximal venous thrombosis, N Engl J Med 322:1260 (1990).
63. Yett HS, Skillman JJ, Salzman EW, The hazards of heparin plus aspirin, N Engl J Med 298:1092 (1978).
64. Sethi GK, Copeland JG, Goldman S, et al, Implications of preoperative administration of aspirin in patients undergoing coronary artery bypass grafting, JACC 15(1):15 (1990).

65. Jick H, Slone D, Borda IT, Chapiro S, Efficacy and toxicity of heparin in relation to age and sex, N Engl J Med 279:284 (1968).
66. Levine MN, Hirsh J, Hemorrhagic complications of anticoagulant therapy, Sem Thromb Hemost 12:39 (1986).
67. Chiu HM, Hirsh J, Yung WL, Rogoeczi E, Gent M, Relationship between the anticoagulant and antithrombotic effects of heparin in experimental venous thrombosis, Blood 49:171 (1977).

REGULATION OF PROTEASE NEXIN-1 ACTIVITY BY HEPARIN AND HEPARAN SULFATE

Dennis D. Cunningham, Steven L. Wagner, and David H. Farrell

Department of Microbiology and Molecular Genetics, College of Medicine
University of California
Irvine, CA 92717

INTRODUCTION

Protease nexin-1 (PN-1) is a 43 kDa protease inhibitor that is synthesized and released by a variety of cultured cells including fibroblasts, smooth muscle cells, astrocytes and to a lesser extent, neurons (1-4). It is not present at significant concentrations in plasma, although it is found in platelets and is released upon platelet activation (5,6). PN-1 inhibits certain serine proteases by forming a complex with their catalytic site serine residue. The complexes then bind back to the cells, via a receptor for the PN-1 moiety of the complex, and are rapidly internalized and degraded. This provides a mechanism for inhibiting and clearing these proteases from the extracellular environment (7).

Several studies have contributed to our understanding of which proteases are likely to be physiolgical targets of PN-1. Examination of the inhibition of various proteases by PN-1 showed that it is a rapid inhibitor of thrombin, urokinase, plasmin and plasma kallikrein; the association rate constants for inhibition of these proteases by PN-1 are greater than 1×10^5 M^{-1} s^{-1} (8,9). It is noteworthy that PN-1 binds to heparin and that heparin increases its rate constant for inhibition of thrombin by about 200-fold (1,8). PN-1 also binds to the extracellular matrix (ECM) of fibroblasts; this interaction accelerates its inhibition of thrombin (10,11). Surprisingly, the interaction of PN-1 with the ECM also regulates its target protease specificity since ECM-bound PN-1 does not inhibit urokinase or plasmin (12). Together, these results suggest that thrombin is a likely physiological target of PN-1 in the extracellular environment.

Studies on the biological activities of PN-1 are in concert with this suggestion, since the activities of PN-1 that have been observed depend on its ability to inhibit thrombin. Early studies showed that PN-1 could modulate the mitogenic activity of thrombin on fibroblasts (13). More recent

Heparin and Related Polysaccharides
Edited by D.A. Lane *et al.*, Plenum Press, New York, 1992

studies showed that PN-1 is identical to a previously identified glial-derived neurite promoting factor or glial-derived nexin which had been shown by Monard and his colleagues to stimulate neurite outgrowth in cultured neuroblastoma cells (14,15). The first indication that the neurite outgrowth activity of PN-1/glial-derived nexin depends on its ability to inhibit thrombin came from experiments which showed that thrombin blocks this activity (16). Subsequently it was demonstrated that picomolar concentrations of thrombin bring about retraction of neurites (17). The neurite outgrowth activity of PN-1 is due to blocking the retraction of neurites by thrombin (18). Indeed, the neurite outgrowth can be brought about by adding other thrombin inhibitors or simply by removing thrombin from the cultures by rinsing it away (17,18). PN-1 and thrombin can similarly regulate the outgrowth of processes on cultured astrocytes; here also, the effect of PN-1 depends on its ability to inhibit thrombin (19).

In view of findings that PN-1 binds to heparin and that heparin acclerates the inactivation of thrombin by PN-1, it has been important to study the effects of glycosaminoglycans on PN-1. These findings have also prompted experiments on the regulation of PN-1 activity and target protease specificity by the ECM and the identification of ECM molecules responsible for these effects. These studies are summarized below and suggest the importance of future detailed studies in this area.

RESULTS AND DISCUSSION

Acceleration of PN-1-mediated thrombin inhibition

Studies on the localization of PN-1 in cultured human fibroblasts showed that it is present not only in the culture medium but also on the cell surface and ECM (10). The PN-1 present on the ECM is tightly bound since it is not removed by rinsing with 1.0 M NaCl. Thus, it should be localized to the ECM under physiological conditions. In view of the previous finding that heparin accelerates the inactivation of thrombin by PN-1 (1,8), and the presence of heparan sulfate in the ECM and cell surface, it seemed important to examine the possibility that the localization of PN-1 to the cell surface and ECM could regulate its inhibition of proteases.

The first experiments on the regulation of PN-1 by the ECM and cell surface were conducted with cultured human fibroblasts. In these studies, PN-1 was bound to either fixed cells or to membranes or ECM prepared from unfixed cells (11). Incubation of PN-1 with each of these preparations accelerated its inactivation of thrombin. Interestingly, incubation of antithrombin III with these preparations did not accelerate its inactivation of thrombin. In view of the studies briefly summarized above which showed that PN-1 and thrombin can reciprocally regulate the processes on neuroblastoma cells and astrocytes and the finding that PN-1 is abundant in human brain, we subsequently examined the ability of cultured glioblastoma and neuroblastoma cells to accelerate the inactivation of thrombin by PN-1 (4). As shown in Figure 1, incubation of PN-1 with either glioblastoma cells (top panel) or neuroblastoma cells (bottom panel) significantly increases its inactivation of thrombin judged by the formation of PN-1-thrombin complexes that are stable in SDS. It should be emphasized that the fixation of the cells with paraformaldehyde completely inactivates any PN-1 bound to them. Thus, the difference in activity observed in the presence of cells (plus symbols) or absence of cells (open or closed circles) is due to activation of added purified PN-1 and not to active PN-1 contributed by the cells.

In view of the ability of certain glycosaminoglycans to regulate the activity of protease inhibitors such as antithrombin III and heparin cofactor II, and the ability of heparin to accelerate the inactivation of thrombin by PN-1,

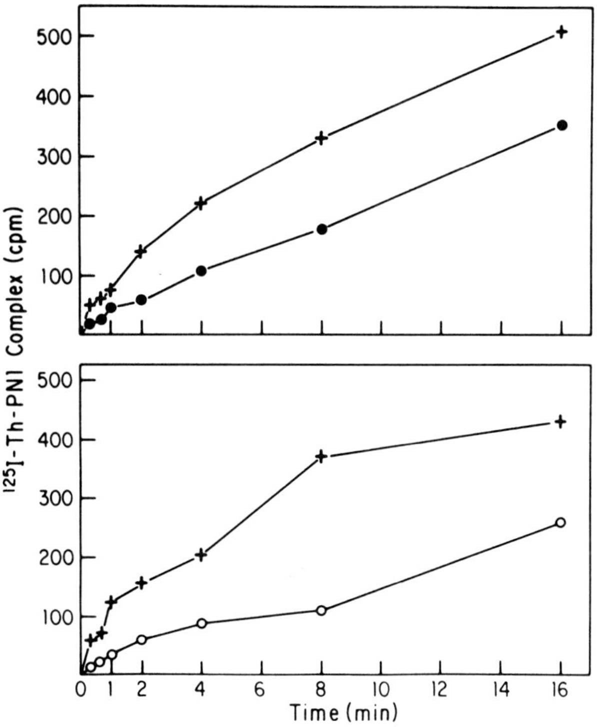

Fig. 1. *Paraformaldehyde-fixed human glioblastoma and neuroblastoma cells accelerate the inhibition of thrombin by PN-1.* Confluent cultures of glioblastoma cells (top) and neuroblastoma cells (bottom) were incubated in serum-free medium for 48 h, rinsed with PBS, and then fixed with 2% paraformaldehyde for 15 min at room temperature. The cells were then rinsed extensively with PBS. PN-1 (13.7 nM) was incubated with (+) or without (○ or ●) fixed cells in 0.5 ml of fresh serum-free medium for various times. ^{125}I-Thrombin (1.37 nM) was then added to initiate the reaction. The reactions were quenched by adding an equal volume of electrophoresis sample dilution buffer at the indicated times. Aliquots of the samples were analyzed by SDS-PAGE, and the radioactivity in bands corresponding to PN-1-^{125}I-Thrombin complexes was quantified using a γ-counter. Reprinted with permission from reference 4.

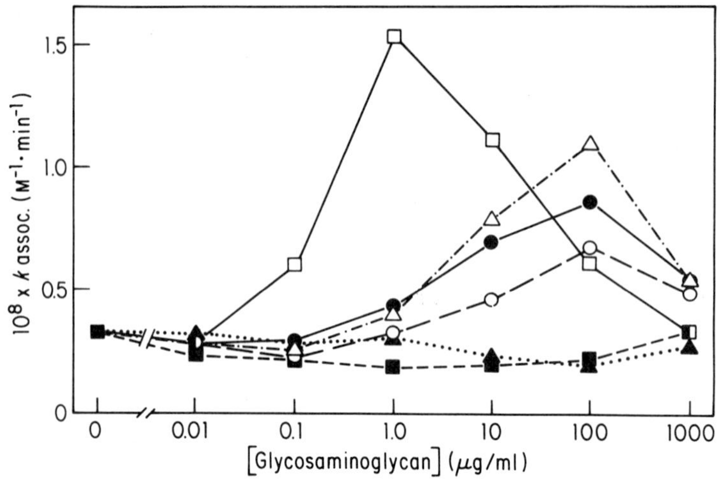

Fig. 2. *Effect of purified glycosaminoglycans on the second-order rate constant for PN-1-mediated thrombin inhibition.* PN-1 (30 nM) was incubated with 4.4 nM-thrombin in 200 µl of Tris/saline/0.1% poly(ethylene glycol) for 20 s at 37°C in the presence of the following glycosaminoglycans at the indicated concentrations: chondroitin sulfate A (○), chondroitin sulfate C (Δ), heparan sulfate from bovine liver (□), heparan sulfate from human aorta (●), hyaluronic acid (▲) and keratan sulfate (■). The glycosaminoglycans were precipitated by adding 2 mg of Polybrene/ml, and the remaining thrombin activity was quantified by measuring the amount of Bz-Arg-[^{3}H]OEt hydrolysis (0.66 µCi/ml) over a 5 min period. The reaction was quenched with 2mM-phenylmethanesulphonyl fluoride and the released [^{3}H]ethanol was extracted and counted for radioactivity. Reprinted with permission from reference 20.

we examined the effects of various glycosaminoglycans on the second-order rate constant for PN-1-mediated inhibition of thrombin (20). As shown in Figure 2, several glycosaminoglycans significantly increased this rate constant. A large effect was observed for heparan sulfate prepared from bovine liver (open squares). Chondroitin sulfate C (open triangles) and heparan sulfate prepared from human aorta also increased the second order rate constant for PN-1 mediated thrombin inhibition, although much higher concentrations of these glycosaminoglycans were required to see significant effects. In order to ensure that the accelerative activities of these glycosaminoglycans were not due to other contaminating activities, specific glycosidases were used to digest the glycosaminoglycans before use in the assays. In each case, the glycosidase removed virtually all of the activity of its corresponding glycosaminoglycan (20).

Although the data in Figure 2 show that the saccharide subunits of glycosaminoglycans are critical for determining their ability to accelerate the inactivation of thrombin by PN-1, it was interesting to evaluate also the effect of degree of sulfation (20). The importance of sulfation within a class of glycosaminoglycans is demonstrated by the relationship between the activities of the heparin/heparan sulfates and their degree of sulfation. Heparin was most active, followed by heparan sulfate from bovine liver which was more active than heparan sulfate from human aorta; their sulfer to hexosamine ratios are respectively 2.33, 0.97 and 0.66. This series shows a marked correlation between sulfation of the glycosaminoglycan and its ability to accelerate the inactivation of thrombin by PN-1 (20).

The results summarized above suggested that heparan sulfate and to a lesser degree, chondroitin sulfate C might be responsible for the ability of intact cells and preparations of ECM and membranes to accelerate the inactivation of thrombin by PN-1. This was tested by incubating membrane preparations with highly purified heparitinase and chondroitinase ABC. Then, ability of the treated membranes was compared with the ability of untreated membranes to accelerate thrombin inhibition by PN-1 (20). Heparitinase treatment of the membranes decreased their accelerative effect on PN-1-mediated thrombin inhibition by about 80 percent. In similar experiments, chondroitinase ABC treatment of the membranes decreased their effect on the reaction between PN-1 and thrombin by about 20 percent. When the two glycosidases were mixed together, essentially all of the accelerative activity was removed. (In control experiments, heparitinase and chondroitinase ABC did not cause a decrease in the basal rate of PN-1-thrombin complex formation in the absence of membranes). Thus, the effects of heparan sulfate and chondroitin sulfate are additive, and this suggests that heparan sulfate accounts for about 80 percent of the accelerative activity and that chondroitin sulfate accounts for the remaining 20 percent (20).

Regulation of PN-1 target protease specificity

The next step in this analysis was to determine if the ECM, cell surface or glycosaminoglycans might also regulate the inhibition of urokinase or plasmin by PN-1. Previous studies employing highly purified proteases and PN-1 in solution showed that thrombin, urokinase and plasmin are rapidly inactivated by PN-1 with second order association rate constants of 6.0×10^5 $M^{-1}s^{-1}$, 1.5×10^5 $M^{-1}s^{-1}$ and 1.3×10^5 M^{-1} s^{-1} respectively (8). When PN-1 was bound to the ECM, however, it no longer formed complexes with urokinase or plasmin (12). This is illustrated in the autoradiogram of Figure 3. In each panel, lane a shows the labeled protease employed. Lane b shows the PN-1-protease complexes that were formed after incubating each labeled protease with with PN-1. In the remaining lanes, ECM prepared from cultured human fibroblasts was incubated alone (lane 0) or with increasing

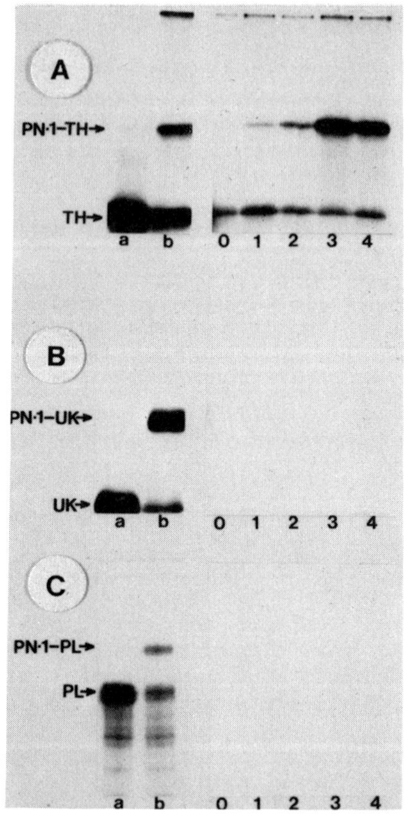

Fig. 3. *Effect of binding purified PN-1 to ECM on its reactivity with ^{125}I-proteases.* ECM was prepared from confluent fibroblasts as described in reference 12. For *lanes 0* to *4*, the dishes containing ECM were incubated for 60 min at 37°C with either no PN-1 (*lane 0*) or PN-1 at 250 ng/ml (*lane 1*), 500 ng/ml (*lane 2*), 750 ng/ml (*lane 3*) or 1 µg/ml (*lane 4*). After removal of the PN-1 solution, the dishes were rinsed and then incubated for 30 min with 150 ng/ml ^{125}I-thrombin (*panel A*), 500 ng/ml ^{125}I-urokinase (*panel B*), or 1.0/µg/ml ^{125}I-plasmin (*panel C*). After removal of the ^{125}I-proteases, the ECM preparations were rinsed and solubilized in 250 µl of Laemmli sample buffer. The solubilized proteins were visualized by autoradiography following SDS-PAGE *Lane a*, ^{125}I-protease; *lane b*, ^{125}I-protease plus purified PN-1. The *arrows* denote positions of the ^{125}I-protease and the corresponding PN-1-^{125}I-protease complexes. Reprinted with permission from reference 12.

concentrations of PN-1 (lanes 1 to 4). After a brief rinse, the appropriate labeled protease was added. The ECM was then solubilized in SDS and the proteins were analyzed by SDS-PAGE. As shown in the autoradiogram of Figure 3, the PN-1 that was bound to the ECM formed complexes with thrombin, but not with urokinase or plasmin. Control experiments showed that the absence of PN-1-urokinase or PN-1-plasmin complexes in panels B and C was not due to formation and then release of the complexes into the incubation medium. Similar studies were conducted with intact fibroblasts; these experiments showed that PN-1 that is bound to the cell surface is also

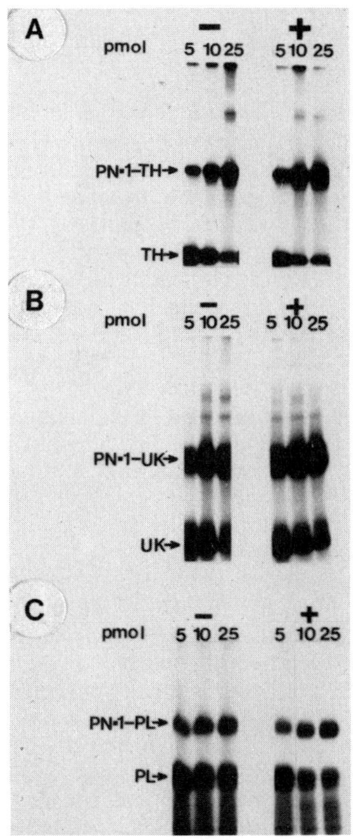

Fig. 4. *Effect of heparin on the reactivity of PN-1 with ^{125}I-proteinases.* The indicated quantities of PN-1 (picomoles) were incubated for 2 min at 37°C in 25μl of 0.01% bovine serum albumin in the presence (+) or absence (-) of a 1500 molar excess of heparin. To each solution was added 1 pmol of either ^{125}I-thrombin (*panel A*), ^{125}I-urokinase (*panel B*), or ^{125}I-plasmin (*panel C*). The samples were incubated for 30 min at 37°C; the reactions were stopped by adding an equal volume of Laemmli gel sample dilution buffer. The labeled proteins were visualized by autoradiography following SDS-PAGE. The *arrows* denote positions of the ^{125}I-protease and the corresponding PN-1-^{125}I-proteinase complexes. Reprinted with permission from reference 12.

blocked in its ability to form complexes with urokinase or plasmin. Together, these experiments showed that even though PN-1 in solution inactivates thrombin, urokinase and plasmin at about the same rates, it will not inactivate urokinase or plasmin when it is bound to the ECM or cell surface (12).

Since the rate of inactivation of thrombin by PN-1 can be regulated by heparin and heparan sulfate, it was of interest to determine if these molecules might inhibit the ability of PN-1 to form complexes with urokinase or plasmin. The results in Figure 4 show this is not the case. When PN-1 was incubated with heparin, it slightly increased the formation of complexes between PN-1 and urokinase or plasmin. Thus, ECM molecules other than heparan sulfate appear to be responsible for the interactions with PN-1 that block its ability to inhibit urokinase or plasmin (12).

FUTURE STUDIES

The experiments described above have provided some broad outlines of the regulation of PN-1 activity and target protease specificity by the cell surface and the ECM. Much remains to be done in terms of defining the molecular interactions that mediate this regulation. In fact, little is known about the ECM components that are responsible for its binding of PN-1. PN-1 binds to heparin, and studies on the elution of PN-1 from heparin-Sepharose have shown that it elutes at about 0.5 M NaCl (1,21). From these results, it seems likely that heparan sulfate participates in the binding of PN-1 to the ECM. However, PN-1 binds very tightly to the ECM. Studies with ECM prepared from human fibroblasts showed that about 2.0 M NaCl is required to elute PN-1 (10). This tight binding probably involves several ECM components including heparan sulfate.

The studies summarized above showed that the ECM acceleration of thrombin inactivation by PN-1 is mostly due to heparan sulfate and to a lesser degree to chondroitin sulfate. These experiments also showed that heparan sulfate and chondroitin sulfate accelerated inhibition of thrombin by PN-1. (20) In future studies, it will be interesting to define the particular sulfated saccharides species in these preparations that are responsible for the acceleration. Much remains to be done in terms of defining the components of the ECM and cell surface that block the ability of PN-1 to inhibit urokinase or plasmin. The limited studies carried out so far indicate that it is probably not due to heparan sulfate alone since heperin did not block this activity of PN-1 (12). Of course, heparan sulfate might participate with other components of the ECM/cell surface to bring about this regulation of PN-1 target protease specificity, but this remains to be defined. It is important to elucidate the cellular components that mediate this regulation, for it will provide additional insights into the mechanisms by which cells participate in the control of regulatory proteases in their extracellular environment. Some of these proteases are involved in regulating the proliferation, differentiation and movements of cells. Thus, understanding their regulation is fundamental to unraveling the control of these complex cellular activities.

ACKNOWLEDGMENTS

We thank Alice L. Lau for excellent technical assistance. This work was supported by NIH Research Grant GM 31609 and American Cancer Society Grant BC-602.

REFERENCES

1. J.B. Baker, D. A. Low, R. L. Simmer, and D. D. Cunningham, Protease nexin: a cellular component that links thrombin and plasminogen activator and mediates their binding to cells. **Cell** 21:37 (1980).
2. W E. Laug, R. Aebersoll, A. Jong, W. Rideout, B. L. Bergman, and J. B. Baker, Isolation of multiple types of plasminogen activator inhibitors from vascular smooth muscle cells. **Thrombosis & Hemostasis,** 61:517 (1989).
3. D.E. Rosenblatt, C. W. Cotman, M. Nieto-Sampedro, J. W. Rowe, and D. J. Knauer, Identification of a protease inhibitor produced by astrocytes that is structurally and functionally homologous to human protease nexin 1. **Brain Res.** 415:40 (1987).
4. S.L. Wagner, A. L. Lau, A. Nguyen, J. Mimuro, D. J. Loskutoff, P. J. Isackson, and D. D. Cunningham, Inhibitors of urokinase and thrombin in cultured neural cells. **J. Neurochem.** 56:234 (1991).
5. R.S. Gronke, B. L. Bergman, and J. B. Baker, Thrombin interaction with platelets. Influence of a platelet protease nexin. **J. Biol. Chem.** 262:3030 (1987).
6. R.S. Gronke, D. J. Knauer, S. Veeraraghavan, and J. B. Baker, A form of protease nexin-1 is expressed on the platelet surface during platelet activation. **Blood,** 73:472 (1989).
7. D.A. Low, J. B. Baker, W. C. Koonce, and D. D. Cunningham, Released protease nexin regulates cellular binding, internalization and degradation of serine proteases. **Proc. Natl. Acad. Sci. USA** 78:2340 (1981).
8. R.W. Scott, B. L. Bergman, A. Bajpai, R. T. Hersh, H. Rodriguez, B. N. Jones, C. Barreda, S. Watts, and J. B. Baker, Protease nexin: properties and a modified purification procedure. **J. Biol. Chem.** 260:7029 (1985).
9. W.E. Van Nostrand, L. D. McKay, J. B. Baker, and D. D. Cunningham, Functional and structural similarities between protease nexin-1 and C1 inhibitor. **J. Biol. Chem.** 263:3979 (1988).
10. D.H. Farrell, S. L. Wagner, R. H. Yuan, and D. D. Cunningham, Localization of protease nexin-1 on the fibroblast extracellular matrix. **J. Cell. Physiol.** 134:179 (1988).
11. D.H. Farrell and D. D. Cunningham, Human fibroblasts accelerate the inhibition of thrombin by protease nexin. **Proc. Natl. Acad. Sci. USA** 83:6858 (1986).
12. S.L. Wagner, A. L. Lau, and D. D. Cunningham, Binding of protease nexin-1 to the fibroblast surface alters its target proteinase specificity. **J. Biol. Chem.** 264:611 (1989).

13. D.A. Low, R. W. Scott, J. B. Baker, and D. D. Cunningham, Cells regulate their mitogenic response to thrombin through release of protease nexin. **Nature** 298:476 (1982).
14. S. Gloor, K. Odink, J. Guenther, N. Hanspeter, and D. Monard, A glia-derived neurite promoting factor with protease inhibitory activity belongs to the protease nexins. **Cell** 47:687 (1986).
15. M. McGrogan, J. Kennedy, M. Li, C. Hsu, R. Scott, C. Simonsen, and J. Baker, Molecular cloning and expression of two forms of human protease nexin 1. **Bio/Technology** 6:172 (1988).
16. D. Monard, E. Niday, A. Limat, and F. Solomonson, Inhibition of protease activity can lead to neurite extension in neuroblastoma cells. **Prog. Brain Res.** 56:359 (1983).
17. D. Gurwitz and D. D. Cunningham, Thrombin modulates and reverses neuroblastoma neurite outgrowth. **Proc. Natl. Acad. Sci. USA** 85:3440 (1988).
18. D. Gurwitz and D. D. Cunningham, Neurite outgrowth activity of protease nexin-1 on neuroblastoma cells requires thrombin inhibition. **J. Cell. Physiol.** 142:155 (1990).
19. K. Cavanaugh, D. Gurwitz, D. D. Cunningham, and R. A. Bradshaw, Reciprocal modulation of astrocyte stellation by thrombin and protease nexin-1. **J. Neurochem.** 54:1735 (1990).
20. D.H. Farrell and D. D. Cunningham, Glycosaminoglycans on fibroblasts accelerate thrombin inhibition by protease nexin-1. **Biochem. J.** 245:543 (1987).
21. W.E. Van Nostrand, S. L. Wagner, and D. D. Cunningham, Purification of a form of protease nexin 1 that binds heparin with a low affinity. **Biochemistry**, 27:2176 (1988).

NEW APPROACHES FOR DEFINING SEQUENCE SPECIFIC SYNTHESIS OF HEPARAN SULFATE CHAINS

R.D. Rosenberg*§ and A.I. de Agostini*

*Department of Biology
Massachusetts Institute of Technology
Cambridge, Massachusetts 02139

§Department of Medicine
Beth Israel Hospital and Harvard Medical School
Boston, Massachusetts 02215

ABSTRACT

Mammalian cells synthesize heparan sulfate proteoglycans (HSPG) which consist of core proteins with covalently linked glycosaminoglycans (GAGs) of 50-150 disaccharide units. The GAGs exhibit great structural diversity which arise from differing arrangements of alternate disaccharide units. It has been hypothesized that HSPG may be involved in regulating the most basic aspects of cell biologic systems such as adhesion, proliferation and differentiation. However, considerable doubt exists about the specific nature of the above interactions because of a failure to isolate GAGs of unique monosaccharide sequence with appropriate biologic activities.

We have demonstrated that mouse LTA cells synthesize cell surface heparan sulfate proteoglycans with regions of defined monosaccharide sequence that specifically interact with antithrombin ($HSPG^{act}$). However, it remains unclear how $HSPG^{act}$ can be generated by a biosynthetic pathway with no simple template for directing the ordered assembly of monosaccharide units. To examine this issue, we treated LTA cells with ethylmethane sulfonate and then identified mutants that exhibit decreased antithrombin binding to heparan sulfate chains but possess no gross defects in glycosaminoglycan biosynthesis. After screening 40,000 colonies, we isolated 7 stable mutants which synthesize 8-27% of the wild type $HSPG^{act}$ but produce normal amounts of other HSPG. These mutants are recessive in nature, and fall into at least two different complementation groups. The delineation of the molecular basis of these defects should greatly improve our understanding of how cells synthesize HSPG with regions of defined monosaccharide sequence.

INTRODUCTION

Heparin preparations which are a natural product of mast cells contain about 30% of GAGs which bind tightly to AT and dramatically accelerate complex formation[1]. These molecular species can only be isolated by affinity chromatography with the protease inhibitor. The GAGs that interact with AT exhibit a unique sequence of sulfated and nonsulfated uronic acid and glucosamine residues which are required to

Heparin and Related Polysaccharides
Edited by D.A. Lane *et al.*, Plenum Press, New York, 1992

complex with the protease inhibitor and induce a conformational change in the protein which accelerates blood coagulation enzyme neutralization[2-9].

The highly specific interaction outlined above serves as the basis of a natural anticoagulant mechanism of the blood vessel wall. The endothelial cells synthesize HSPG of which only 1%-10% exhibit heparan sulfate chains with appropriate monosaccharide sequences to bind to AT and accelerate neutralization of blood coagulation enzymes ($HSPG^{act}$)[10-12]. The delineation of a specific biochemical function for an HSPG that possesses regions of defined monosaccharide sequence suggests that a similar situation may exist in many other biologic systems. However, it remains unclear how $HSPG^{act}$ can be generated by a biosynthetic pathway with no simple template for directing the ordered assembly of different disaccharide units[13]. One approach to this problem would be to isolate mutants which are defective in the synthesis of $HSPG^{act}$ but possess no other gross defect in the overall production of these components. This class of mutants might eventually allow the pinpointing of genes which regulate the monosaccharide sequence of GAGs. In this communication, we describe a cell line and screening technique which permits the isolation of mutants of this type. Our overall approach is based upon the studies of Esko and coworkers who have demonstrated that the GAG biosynthetic pathway is dissectable by genetic methods[14].

RESULTS

The use of somatic cell genetic techniques to investigate the synthesis of anticoagulantly active HSPG ($HSPG^{act}$) required a cell line which grows rapidly from clonal density, is easily manipulated to generate stable mutants, and produces large amounts of cell surface $HSPG^{act}$. Several lines of evidence suggested that LTA cells which had been previously utilized in somatic cell genetic analyses[15-17] synthesized cell surface $HSPG^{act}$.

Firstly, LTA cells produce heparan sulfate chains which interact specifically with AT and possess a protease inhibitor binding site sequence similar to that of anticoagulantly active heparin/heparan sulfate. To demonstrate that this was the case, LTA monolayers were metabolically labelled with $Na_2{}^{35}SO_4$, heparan sulfate chains were isolated and then affinity fractionated with AT and Concanavalin A. The anticoagulantly active and anticoagulantly inactive fractions were obtained by affinity chromatography according to a minor modification of the technique of Jordan *et al*[18]. To this end, labelled GAGs were admixed with a 100 fold molar excess of AT, Concanavalin A-Sepharose (1ml of gel/mg AT) was added, the mixture was incubated for 1h, and the anticoagulantly active heparan sulfate-AT/AT lectin matrix complex was collected by low speed contrifugation. The bound polysaccharide was eluted from the Concanavalin A-Sepharose with 1 M NaCl in 10 mM Tris-HCl, pH 7.5 which allows the protease inhibitor to remain associated with the lectin, and then obtained in the supernatant fraction by low speed contrifugation. The entire process was repeated three times with the unadsorbed heparan sulfate in order to remove small amounts of anticoagulantly active GAGs. The two types of heparan sulfates were extensively dialyzed against 10mM ammonium bicarbonate, and evaporated to dryness on a speed vac concentrator.

In two separate experiments, an average of 5.5% of the heparan sulfate chains bound tightly to immobilized AT, and could be eluted with a high salt wash. The remaining heparan sulfate chains with minimal affinity for the protease inhibitor did not complex with immobilized AT and were found in the supernatant fraction. The nonspecific binding of radiolabeled heparan sulfate to the affinity matrix was less than 0.5%.

To establish the structural differences between the high affinity and depleted heparan sulfate chains, we quantitatively cleaved both populations of GAGs to disaccharides and then examined these species by ion exchange HPLC[12,19,20-21]. The average results of two separate experiments are summarized in Table 1. It is readily apparent that the high affinity heparan sulfate chains are enriched with regard to GlcA-AMN-3-O-SO_3 and GlcA-AMN-3,6-O-SO_3 and that the depleted GAG chains possess minimal amounts of these disaccharides. This is to be expected since the two disaccharides are markers for the primary AT binding domain of heparin and heparan sulfate[2-9].

Secondly, the LTA cell surface exhibits significant amounts of $HSPG^{act}$ which can bind tightly to AT. To demonstrate the presence of these components, cell monolayers were incubated with varying concentrations of ^{125}I-AT, and the levels of bound protease inhibitor were determined. Fig. 1 shows a typical binding curve after subtraction of nonspecific binding. Inspection of the isotherm suggests that AT interacts with about 480,000 binding sites per cell with a K_{diss} of about 50 nM which is similar to that previously obtained with cloned microvascular and macrovascular endothelial cells[10-12]. However, the binding curve does not reach a true plateau which makes rigorous calculation of binding parameters questionable. To show that the protease inhibitor is bound specifically to cell surface $HSPG^{act}$, cell monolayers were preincubated for 1h at 37° with purified *Flavobacterium* heparitinase (0.5U/ml) which virtually eliminates the subsequent binding of ^{125}I-AT (data not shown).

The above results show that LTA cells synthesize $HSPG^{act}$ which is present on the surface of these cells. These observations suggested that this proteoglycan could be detected *in situ* in LTA colonies immobilized on polyester cloth. In preliminary experiments, we demonstrated that LTA cells are able to bind ^{125}I-AT as judged by autoradiography in proportion to the protein content of the colony as quantitated by Coomassie staining intensity or GAG content of the colony as determined by incorporation of $^{35}SO_4$. To generate mutants with decreased ability to synthesize $HSPG^{act}$, exponentially growing LTA cells were treated with 400 ug/ml of ethylmethane sulfonate (EMS) in

TABLE 1. DISACCHARIDE ANALYSES OF HEPARAN SULFATE FROM LTA CELL

Disaccharide	Percentage	
	Affinity Fractionated	Depleted
GlcA → AMN 2-O-SO_3	1.5±0.36	1.4±0.21
GlcA → AMN 6-O-SO_3	13.4±2.05	12.6±0.87
IdA → AMN 6-O-SO_3	15.6±1.75	21.9±0.12
IdA → AMN 2-O-SO_3	25.1±2.69	31.5±0.92
GlcA → AMN 3-O-SO_3	7.0±1.04	0.2±0.29
IdA → AMN 2-O-SO_3 6-O-SO_3	32.7±0.59	32.5±1.10
GlcA → AMN 3,6-O-$(SO_3)_2$	4.1±0.6	ND

GlcA, glucuronic acid; IdA, iduronic acid; AMN, anhydromannitol

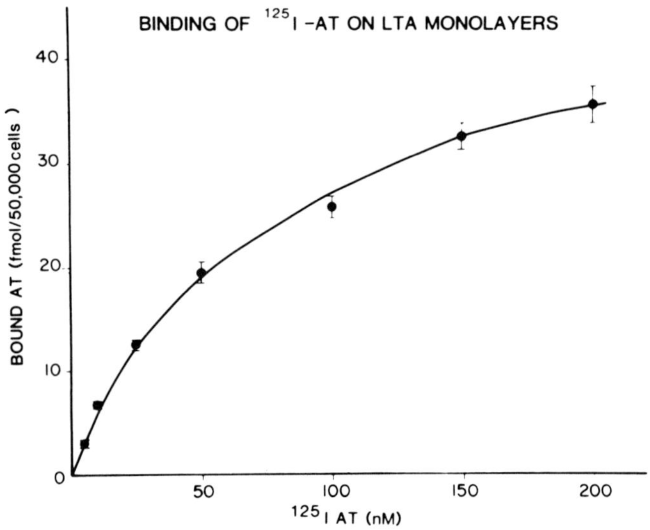

Figure 1. Binding of ^{125}I-AT to LTA monolayers. The extent of binding of ^{125}I-AT to LTA cells was determined on confluent cell monolayers preincubated for 1 hr at 37°C with media containing 1% Nutridoma SP (Boehringer Mannheim, Indianapolis, In.). The cell monolayers were then incubated for 1 hr at 4°C with various concentrations of ^{125}I-AT diluted in PBS containing 1% Nutridoma SP and 50 µg/ml BSA with or without a hundred fold molar excess of unlabeled protease inhibitor. The cell monolayers were then washed five times with PBS containing 100 µg/ml BSA and bound protease inhibitor was quantitated by removing cells with a cotton swab with subsequent counting of ^{125}I. Specific binding is calculated as the difference between binding in the presence and absence of a 100 fold molar excess of unlabeled protease inhibitor. Nonspecific binding averaged about 10% of specific binding. Cell numbers were estimated by trypsinizing control wells and then counting cells on a Coulter counter. All determinations were performed in triplicate. The mean values and error bars represent the results of one typical experiment with associated standard errors. Similar results were obtained in three other independent experiments.

supplemented DMEM [Dulbecco's minimal essential medium supplemented with 10% (vol/vol) fetal calf serum (GIBCO), 100 units of penicillin per ml, and 100 µg of streptomycin per ml] for 16h at 37°. The cells were washed three times with the above media and then grown to confluence. This method of mutagenesis was selected after determining the concentration of EMS and time of incubation with the alkylating agent which increased the frequency of the ouabain resistance phenotype by 2 to 3 orders of magnitude[14].

Mutagenized LTA cells were then examined by a minor modification of the replica plating technique of Raetz *et al*[22]. To this end, cells were seeded at a clonal density of 500 cells/100mm plastic tissue culture dish in supplemented DMEM with 2.5 ug/ml fungizone added, allowed to grow for about five days until the colonies reached the 5-10 cell stage, and then overlayed with two 17 um cutoff polyester filters, as well as one 1 um cutoff filter. The filters were held in contact with the plastic dish by a layer of glass beads (4mm in diameter). The cells were grown for an additional two weeks until colonies of about 1-3 mm were formed on the upper 17 um filter. The filters were removed, placed for two days in separate culture dishes, and screened for colony position, AT binding, and GAG synthesis as outlined below. The tissue culture dishes (master plates) were overlayed with a sterile Whatman #42 filter paper, as well as beads and were maintained at 37° until used for isolation of selected colonies.

The mutagen-treated colonies yielded an occasional variant that failed to complex with the protease inhibitor. Putative mutants identified in this way were harvested, and then passed through a second round of screening to enrich their numbers as well as to confirm their altered phenotypes (Fig. 2; column 1). The above colonies were also screened for their ability to incorporate $^{35}SO_4$ since our goal was to isolate mutants which exhibited restricted defects in the synthesis of $HSPG^{act}$ but possessed no gross abnormalities in GAG biosynthesis (Fig. 2; column 2). As shown for mutant VI-51, replica filters obtained during the second round of screening exhibited interspersed wild type and mutant colonies with regard to AT binding normalized to Coomassie staining but both types of colonies possess the same intensity of $^{35}SO_4$ incorporation normalized to Coomassie staining. All other isolated mutants showed the same characteristics with regard to AT binding and $^{35}SO_4$ incorporation. The mutants were finally cloned by limiting dilution and exhibited a homogenous population of colonies with regard to minimal AT binding as evaluated on replica filters. The screening of 40,000 colonies yielded 7 mutant lines each derived from five separate mutagenized cell populations. These mutants exhibited stable phenotypes for a minimum of 6 months in culture, and possessed no obvious alterations with regard to gross morphology, trypsin sensitivity, or adhesion.

We wished to determine the extent to which the various mutants are able to synthesize cell surface $HSPG^{act}$. To this end, the binding of different concentrations of radiolabeled protease inhibitor to mutant cell monolayers was quantitated. The data revealed that the mutants exhibited a residual ability to bind AT which ranged from 8%-27% of the wild type cells when protease inhibitor concentrations of 10 nM were utilized (Fig. 3). Similar results were apparent at AT concentrations of 10 nM, 50 nM and 100 nM, respectively (data not shown). Additional studies were carried out which demonstrated that the extent of $HSPG^{act}$ shedding into the media is identical to the amount of cell surface $HSPG^{act}$ present on the various mutant cell lines (data not shown). The heparan sulfate chains were also isolated from several of the mutants as described in Materials and Methods and affinity fractionation with AT revealed that less than 0.5%-1.0% of the labelled GAGs bound to the protease inhibitor. Therefore, the observed reduction in AT binding by the mutants is caused by a restricted defect in the synthesis of $HSPG^{act}$ and not by a general reduction of cell surface HSPG.

We then attempted to ascertain whether the various mutants exhibited alterations in overall GAG biosynthesis. The previous observation of identical incorporation of $^{35}SO_4$ by mutant and wild type LTA cell colonies as judged by replica plating suggests that the two types of cells possess a qualitatively similar capacity to produce GAGs. To detect biosynthetic abnormalities in a more definitive fashion, we incubated the various cell lines with $Na_2{}^{35}SO_4$ under conditions which label GAGs to constant specific activity, quantitatively released the glycoconjugates with trypsin, and then determined the amounts of chondroitin sulfate and heparan sulfate harvested relative to cell protein. The extent of $^{35}SO_4$ labelled material released from wild type LTA cells averaged 1.2 x 10^6 cpm/ug of cell protein (n=3) whereas the same parameter in the 7 mutant cell lines ranged from 0.9 x 10^6 cpm/µg of cell protein to 1.6 x 106 cpm/µg of cell protein. The labeled GAGs were then characterized by enzymatic degradation with *Flavobacterium*

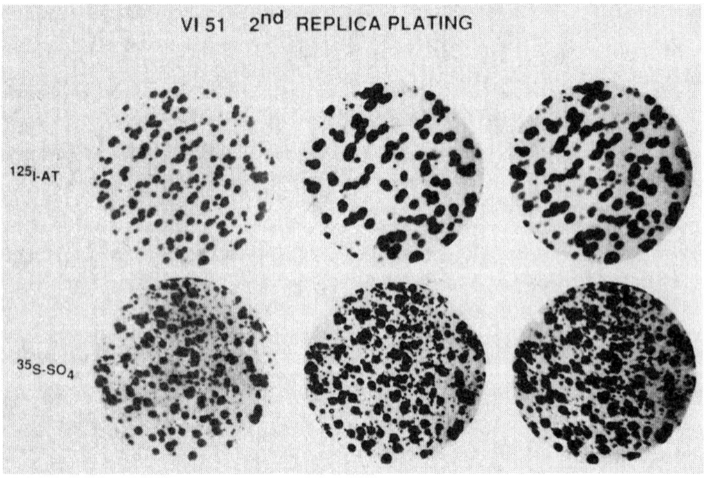

Figure 2. The second round of replica filter analyses of mutant VI-51 as judged by AT binding and synthesis of GAGs. Column 1 depicts Coomassie staining of duplicate filters for the second round of screening of mutant VI-51. Column 2 depicts autoradiographs of ^{125}I and ^{35}S counts on the same replica filters. Column 3 is the superimposition of the autoradiographs and Coomassie stained filters. The duplicate replica filters for the second round of screening of mutant VI-51 are depicted with respect to AT binding (top filter) and GAG biosynthesis (bottom filter). The filter assay for AT binding was carried out with cells grown on replica filters as outlined in Materials and Methods. The technique employed for AT binding is identical to that described in Figure 1 except that the concentration of ^{125}I-AT were set at 0.2 nM, filters were constantly agitated, and BSA was omitted in the the final wash. Upon completion of the above procedure, cell colonies were fixed, stained with Coomassie brilliant blue (2 gm/L) dissolved in MeOH/CH_3COOH/H_2O (4.5/1/4.5), destained in the same solvent mixture, and air dried. The filter assay for GAG synthesis was conducted as outlined for the labelling of cell surface glycoconjugates except that filters were incubated with 50 μCi/ml of $Na_2{}^{35}SO_4$ for 24 h, washed 5 times with ice cold 10% TCA, air dried, and stained with Coomassie brilliant blue as outlined above. The stained filters were exposed for autoradiography using Kodak X-OMAT AR films (Eastman Kodak, Rochester, NY) with one Cronex Xtra Life intensifier screen (Dupont, Wilmington, Del.) at -70°C for 18 to 24 h, and then developed using a Kodak RP X-OMAT Processor.

heparitinase and/or chondroitinase ABC and subsequent quantitation of the digested products by HPLC gel filtration. The above analyses were carried out for wild type LTA cells which demonstrated that 70% ± 4% of the labeled material was degraded to tetrasaccharides/ disaccharides with *Flavobacterium* heparitinase (n=4) whereas 32% ± 4% of the labeled material was cleaved to the same size fragments with chondroitinase ABC (n=4). Similar studies conducted with the 7 mutant cell lines showed that 63%-80% of the labelled material is degraded to tetrasaccharides/disaccharides with *Flavobacterium* heparitinase whereas 20%-37% of the labelled material is cleaved to the same size fragments with chondroitinase ABC. It is also of interest to note that heparan sulfate and chondroitin sulfate chains of the wild type and mutant cells exhibited the same approximate molecular sizes as judged by HPLC gel filtration. Thus, it would appear that the 7 mutant cell lines with defective binding of ^{125}I-AT possess the ability to synthesize normal amounts and sizes of heparan sulfate and chondroitin sulfate chains.

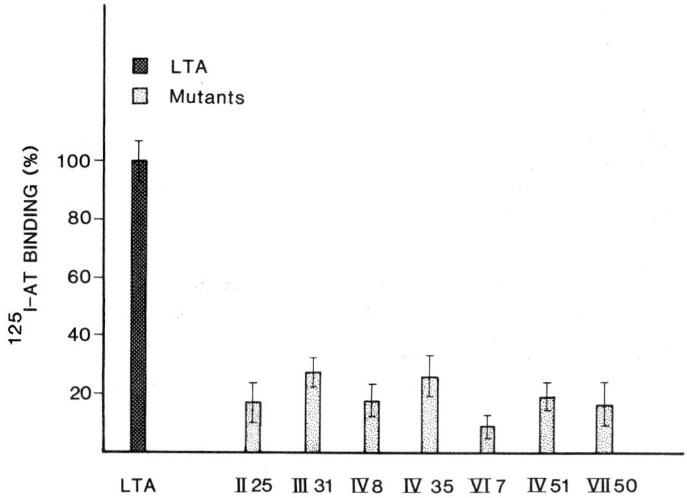

Figure 3. The binding of AT to wild type and mutant LTA cell monolayers (fmol/50,000 cells). The binding of ^{125}I-AT to wild type and mutant LTA cell monolayers were determined as outlined in Figure 1 except that the protease inhibitor concentration was 10 nM. The mean values and error bars represent the results of 3-5 separate experiments with associated standard errors. Mutant cell lines were cultured continuously for up to 6 months prior to assay.

We also initiated investigations to determine the numbers of different mutations present within our collection of mutants. To this end, we carried out complementation analyses on two of our mutants as well as the wild type cells. Pairs of mutants and wild type cells were fused by treating mixed cell monolayers with polyethylene glycol, the resulting cell populations containing some hybrids were plated at high density on filters, and the frequency of hybrids able to bind ^{125}I-AT were evaluated by autoradiography. Figure 4 depicts the results obtained after fusing wild type LTA cells with themselves (positive control), mutant VI-7 and mutant VII-50 with themselves (negative controls), and mutant VI-7 with mutant VII-50. It is readily apparent that self fusion of VI-7 generates hybrids which can not bind AT, and that self fusion of VII-50 produces a similar picture except for rare blotchy areas of residual uptake of protease inhibitor. It is clearly evident that fusion of VI-7 with VII-50 as compared to the self fusion of the same mutants generates a large number of small colonies which are able to bind AT as well as the self fusion of wild type cells. Examination of the same filters by Coomassie staining showed no significant differences between the various cell populations except for a dramatically increased cell density in regions that correspond to the rare areas of residual uptake of protease inhibitor by the self-fusion of VII-50 (data not shown). Thus, the above data suggests that the mutations that lead to VI-7 and VII-50 are recessive in nature and belong to two different complementation groups.

DISCUSSION

The synthesis of HSPG^{act} requires the production of a core protein, assembly of a linkage region of four monosaccharide units coupled to specific serine residues, generation of a precursor polysaccharide chain of alternating N-acetyl glucosamine and glucuronic acid residues, and subsequent remodeling of this repetitive structure by variable epimerization of glucuronic acid residues, variable sulfation of the glucuronic and iduronic acid residues, variable N-deacetylation and N-resulfation of glucosamine residues in conjunction with variable

6-O- and 3-O- ester sulfation[13]. The end result of this complex process is the generation of a region of unique carbohydrate structure which at the very least contains the monosaccharide sequence nonsulfated uronic acid--N-acetyl(N-sulfate)glucosamine 6-O-sulfate--glucuronic acid--N-sulfate glucosamine 3-O-sulfate,(6-O-sulfate)--iduronic acid 2-O-sulfate--N-sulfate glucosamine-6-O-sulfate[2-9]. The use of natural and synthetic fragments of heparin, in conjunction with fast reaction kinetics and equilibrium dialysis, demonstrate that the 6-O- sulfate on residue 2 and the 3-O-sulfate on residue 4 act in a thermodynamically linked fashion to bind to AT and trigger the conformational change of the protease inhibitor which is essential for rapid neutralization of some coagulation enzymes whereas sulfate groups at residues 5 and 6 dramatically augment interaction with the protease inhibitor[8-9]. Carboxy groups on uronic acid residues within this sequence may also be of biologic importance. Finally, it is also apparent that regions outside the above primary AT binding domain outlined above are essential for accelerating the inhibition of other blood coagulation enzymes[23-24].

The biosynthetic enzymes which carry out the post translational modifications of the oligosaccharide chains are not available in

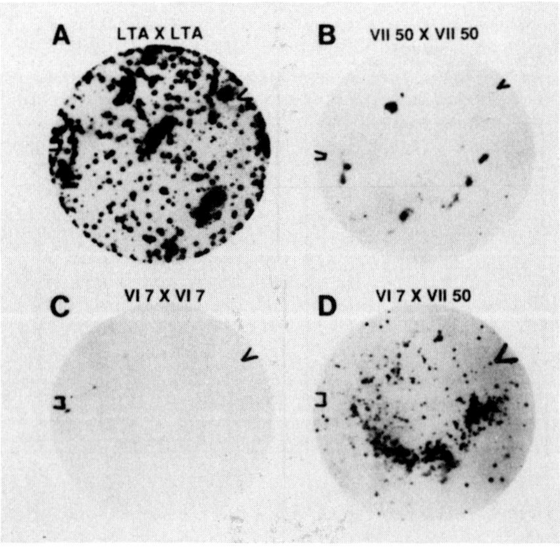

Figure 4. Complementation analysis of mutants VI-7 and VII-50. The various wild type LTA or mutant cell pairs were seeded on tissue culture dishes at a density of 10^5 cells per cm^2. After 24 h, the cells were washed two times with 1 ml of DMEM with no added serum, fused by incubating for 1 min with 50% polyethylene (PEG 1450, Kodak, Rochester, NY), and washed three additional times as described above. The cells were fed with DMEM supplemented with 10% FBS/antibiotics, and then allowed to recover for 24 h. The total cell populations were trypsinized, 10^4 cells from each fusion were directly seeded on polyester 17-μm mesh filters floating in supplemented DEME, and the cells were allowed to grow for 12 days to generate about 500 individual colonies per filter. The various filters were then assayed for AT binding and Coomassie stained as described in Figure 2. The autoradiographs of the filters for the various fused cell populations are provided. (A) fusion of LTA cells with LTA cells; (B) fusion of mutant VII-50 with mutant VII-50; (C) fusion of mutant VI-7 with mutant VI-7; (D) fusion of mutant VI-7 with mutant VII-50.

purified form although much recent progress has been made in this area[25] and little is known about how these enzymes are coordinately regulated within the Golgi complex. It has been postulated that the synthesis of heparan sulfate chains with regions of unique monosaccharide sequence might be achieved by specific interactions between core protein regions and multimolecular arrays of biosynthetic enzymes/modulating proteins within the Golgi complex or alternatively by the random action of early biosynthetic enzymes with restricted substrate preference of later biosynthetic enzymes for certain subsets of the randomly generated monosaccharide sequences. Resolution of this issue is important to our understanding of how cells might synthesize $HSPG^{act}$ with regions of unique monosaccharide sequence, and whether GAGs are likely to modulate complex biologic systems via specific interactions with target molecules.

We have attempted to establish a somatic cell genetic system for investigating the synthesis of $HSPG^{act}$ which may be helpful in resolving the above question. To this end, we demonstrated that LTA cells synthesize cell surface $HSPG^{act}$; isolated stable mutants of this cell type which are defective in the synthesis of the unique monosaccharide sequence required to bind AT but have no gross alterations in the overall pathways of GAG biosynthesis; and showed that the above mutations are due to alterations in at least two different genes. Our overall approach to this problem is based upon the pioneering studies of Esko and coworkers who mutagenized CHO cells that synthesized GAG chains with no ability to bind to AT, and identified mutants in the GAG biosynthetic pathway by replica plating methods involving $^{35}SO_4$ incorporation[14,26-28]. These investigators were able to isolate at high frequency a variety of mutants with reduced ability to transport SO_4 ion, synthesize the linkage region required for coupling of heparan sulfate or chondroitin sulfate chains to core proteins, or N-sulfate glucosamine residues. The isolation of the above mutants at high frequency clearly indicate that the GAG biosynthetic pathway may be amenable to dissection by genetic approaches. It is of interest to note that our investigations also document a relatively high frequency of mutation with regard to the biosynthetic pathways needed to synthesize the AT binding site but none of our mutations is caused by alterations in SO_4 transport, or a decreased ability to generate the linkage region as described above. This is most likely due to the use of different screening approaches to identify our mutants.

At the present time, we have no detailed knowledge of the molecular basis of our mutants except that none exhibits a defect in the GAG chain elongation mechanism. It is possible that alterations have been directly or indirectly induced in core protein(s), N- or O-sulfotransferase enzymes, or potential modulatory proteins involved in establishing monosaccharide sequence. The ongoing investigations of the detailed structure of the heparan sulfate chains of the mutants and direct enzyme assays of the various steps in the biosynthetic pathway of the mutants should pinpoint the abnormalities. The correlation of these changes with the loss of AT binding sites should greatly improve our understanding of how $HSPG^{act}$ with regions of defined monosaccharide sequence is generated.

REFERENCES

1. Lam, L., Silbert, J. E. & Rosenberg, R. D. (1976) Biochem. Biophys. Res. Commun. 69, 570-577.

2. Rosenberg, R. D. & Lam, L. (1979) Proc. Nat. Acad. Sci. USA 76, 1218-1222.

3. Rosenberg, R. D., Armand, G. & Lam, L. (1978) Proc. Nat. Acad. Sci. USA 75, 3065-3069.

4. Lindahl, U., Backstrom, G., Thunberg, L. & Leder, I. G. (1980) Proc. Nat. Acad. Sci. USA 77, 6551-6555.

5. Lindahl, U., Backstrom, G. & Thunberg, L. (1983) J. Biol. Chem. 258,9826-9830.

6. Choay, J., Petitou, M., Lormeau, J. C., Sinay, P., Casu, B. & Gatti, G. (1983) Biochem. Biophys. Res. Commun. 116, 492-499.

7. Atha, D. H., Stephens, A. W., Rimon, A. & Rosenberg, R. D. (1984) Biochem. 23, 5801-5812.

8. Atha, D. H., Lormeau, J. C., Petitou, M., Rosenberg, R. D. & Choay, J. (1985) Biochemistry 24, 6723-6729.

9. Atha, D. H., Lormeau, J. C., Petitou, M., Rosenberg, R. D. & Choay, J. (1987) Biochemistry 26, 6454-6461.

10. Marcum, J. A., McKenny, J. B. & Rosenberg, R. D. (1984) J. Clin. Invest. 74, 341-350.

11. Marcum, J. A. & Rosenberg, R. D. (1985) Biochem. Biophys. Res. Commun. 126, 365-372.

12. Marcum, J. A., Atha, D. H., Fritze, L. M. S., Nawroth, P., Stern, D. & Rosenberg, R. D. (1986) J. Biol. Chem. 261, 7507-7517.

13. Lindahl, U. & Kjellén, L. (1987) *In* The Biology of the Extracellular Matrix; Proteoglycans (Wight T.N. and Mecham R. Eds.) pg. 59-104, Academic Press N.Y.

14. Bame, K. J., Reddy, R. V. & Esko, J.D. (1991) J. Biol. Chem. 266, 12461-12468.

15. Kit, S., Dubbs, D. R., Piekarski, L. J. & Hsu, T. C. (1963) Exp. Cell Res. 31, 297-312.

16. Gum, J. R. & Raetz, C. R. H. (1983) Proc. Natl. Acad. Sci. USA 80, 3918-3922.

17. Gum, J. R. & Raetz, C. R. H. (1985) Mol. Cell Biol. 5, 1184-1187.

18. Jordan, R. E., Favreau, L. V., Braswell, E. H. & Rosenberg, R. D. (1982) J. Biol. Chem. 257, 400-406.

19. Shively, J. E. & Conrad, H. E. (1970) Biochem. 9, 33-43.

20. Dmitriev, B. A., Knirel, Y. A. & Kochetkov, N. K. (1985) Carbohyd. Res. 40, 365-372.

21. Bienkowski, M. J. & Conrad, H. E. (1985) J. Biol. Chem. 260, 356-365.

22. Raetz, C. R. H., Wermuth, M., Mcintyre, T. M., Esko, J. D. & Wing,D. C. (1982) Proc. Natl. Acad. Sci. USA 79, 3223-3227.

23. Oosta, G. M., Gardner, W. T., Beeler, D. L. & Rosenberg, R. D. (1981) Proc. Natl. Acad. Sci. USA 78, 829-833.

24. Stone, A. L., Beeler, D. L., Oosta, G. M. & Rosenberg, R. D. (1982) Proc. Natl. Acad. Sci. USA 79, 7190-7194.

25. Kjellén, L. et al. This symposium.

26. Bame, K. J. & Esko, J. D. (1989) J. Biol. Chem. 264, 8059-8065.

27. Esko, J. D. (1986) Methods Enzymol. 129, 237-253.

28. Esko, J. D., Elgavish, A., Prasthofer, T., Taylor, W. H. & Weinke, J. L. (1986) J. Biol. Chem. 261, 15725-15733.

MODULATION OF NEOVASCULARIZATION AND METASTASIS BY SPECIES OF HEPARIN

I. Vlodavsky*, R. Ishai-Michaeli*, M. Mohsen*, R. Bar-Shavit*, R. Catane*, H.-P. T. Ekre+ and C.M. Svahn +

*Department of Oncology, Hadassah University Hospital, Jerusalem, Israel and +R&D Cardiovascular, Kabi Pharmacia, Stockholm, Sweden

We have characterized the importance of size, sulfation and anticoagulant activity of heparin in: i) release of basic fibroblast growth factor (bFGF) from the subendothelial extracellular matrix (ECM), and ii) inhibition of heparanase (endo-ß-D-glucuronidase) and tumor metastasis. Oligosaccharides derived from depolymerized heparin and containing as little as 8-10 sugar units were, on a molar basis, equivalent to whole heparin in their ability to release bFGF from ECM. Low sulfate oligosaccharides were less effective than medium and high sulfate oligosaccharides. N-desulfation or N-acetylation resulted in an almost complete inhibition of bFGF release. Heparin fractions with high and low affinity to antithrombin III exhibited the same high bFGF releasing activity. Similar requirements were observed for release of ^{125}I-bFGF sequestered by the luminal surface of blood vessels. Heparanase expressed by intact cells (i.e. platelets, mast cells, neutrophils, lymphoma cells) released active bFGF from ECM and basement membranes. Heparanase activity also correlated with the ability of blood-borne tumor cells to extravasate and metastasize. Unlike release of bFGF, efficient inhibition of both heparanase and tumor metastasis was best achieved by heparin species containing 16 sugar units or more and was only slightly affected by substitution of N-sulfates with acetyl groups. These results indicate that various non-anticoagulant heparin species of different size, sulfation and substituted groups can be designed to elicit specific effects such as release of bFGF and inhibition of heparanase, resulting in induction of neovascularization and inhibition of tumor metastasis, respectively.

Heparin and Related Polysaccharides
Edited by D.A. Lane *et al.*, Plenum Press, New York, 1992

Introduction

Extravasation of blood-borne cells, whether normal or malignant, is initiated by specific adhesive interactions between circulating cells and the vascular endothelium. These interactions are mediated by an array of a diverse family of cell surface adhesion molecules expressed by endothelial cells (EC), cells of the immune system and tumor cells. Metastatic tumor cells often attach at or near the intercellular junctions between adjacent EC followed by rupture of the junctions, retraction of the EC borders and migration through the breach in the endothelium toward the exposed underlying basal lamina (1). Once enveloped between EC and the basal lamina, the invading cells must degrade the subendothelial glycoproteins and proteoglycans in order to migrate out of the vascular compartment. It appears that normal and malignant blood-borne cells may utilize the same enzymatic machinery in order to rupture the EC junctions and degrade various constituents of the subendothelial ECM. Our studies on the extravasation of normal and malignant blood-borne cells focus on the ability of cells to degrade heparan sulfate proteoglycans (HSPG) in a basement membrane-like extracellular matrix (ECM) produced by cultured corneal and vascular endothelial cells (2,3). This ECM closely resembles the subendothelium in vivo in its morphological appearance and molecular composition. It contains primarily collagens (mostly type III and IV, with smaller amounts of types I and V), proteoglycans (mostly heparan sulfate- and dermatan sulfate-proteoglycans, with smaller amounts of chondroitin sulfate proteoglycans), laminin, fibronectin and elastin (4). Because the ECM is secreted in a polar fashion, exclusively underneath the endothelial cell monolayer, and because it is firmly attached to the entire area of the tissue culture dish, the endothelial cell layer can be removed while leaving the underlying ECM intact and free of nuclei, cytoskeletal elements and cellular debris (4). HSPG have been isolated from a variety of basement membranes and cell surfaces of normal and malignant cells (5,6). The ability of HSPG to interact with ECM macromolecules such as collagen, laminin and fibronectin and with different attachment sites on plasma membranes suggests a key role for this proteoglycan in the self-assembly and insolubility of ECM components as well as in cell adhesion and locomotion. HSPG are prominent components of blood vessels. In large vessels they are concentrated mostly in the intima and inner media whereas in capillaries they are found mainly in the subendothelial basement membrane where they support proliferating and migrating endothelial cells and stabilize the structure of the capillary wall. Cleavage of heparan sulfate (HS) may therefore result in disassembly of the subendothelial ECM and hence may play a decisive role in extravasation of blood-borne cells. The ability of cells to degrade HS in the ECM was studied by allowing cells to interact with a metabolically sulfate labeled ECM, followed by gel filtration (Sepharose 6B) analysis of degradation products released into the culture medium (2,7). While intact HSPG is eluted next to the void volume of the column ($Kav<0.2$, $Mr \sim 0.5 \times 10^6$), labeled degradation fragments of HS side chains are eluted more toward the V_t of the column (0.5<Kav<0.8, $Mr=4\text{-}7\times10^3$) (2). Expression of a HS degrading endoglucuronidase (heparanase) was found to correlate with the metastatic potential of various tumor cells (2,7) and with the ability of activated cells of the immune system to leave the circulation and elicit both inflammatory and autoimmune responses (8).

Involvement of Heparanase in Cell Invasion and Metastasis

A common and most important route for the dissemination of neoplastic cells within the body involves invasion and penetration of tumor cells into blood vessels and/or lymphatics. Circulating tumor cells arrested in the capillary beds of different organs must invade the endothelial cell (EC) lining and degrade its underlying basement membrane in order to escape into the extravascular tissue(s), where they establish metastasis (9). Heparanase activity was found to correlate with metastatic potentials of mouse lymphoma (2), fibrosarcoma and melanoma (7). Moreover, elevated levels of heparanase were detected in sera from metastatic tumor bearing animals and melanoma patients (7) and in tumor biopsies of cancer patients (10). Immunohistochemical staining of frozen tissue sections revealed that heparanase is localized preferentially in metastatic murine and human melanomas (11). Heparanase mediated degradation of HS is inhibited by heparin, both when exerted by intact cells or soluble heparanase (7,12). We investigated the heparanase inhibitory effect of various non-anticoagulant species of heparin that might be of potential use in preventing extravasation of blood-borne cells. Inhibition of heparanase depended on the size and degree of sulfation of the heparin molecule, the position of sulfate groups and the occupancy of the N-position of the hexosamines. Inhibition of heparanase was best achieved by heparin species containing 16 sugar units or more and having sulfate groups at both the N and O positions. Low sulfate oligosaccharides were less effective heparanase inhibitors than medium and high sulfate fractions of the same size. While O-desulfation abolished the heparanase inhibiting effect of heparin, O-sulfated N-acetylated heparin retained a high inhibitory activity, provided that the N-substituted molecules had a molecular size of about 4000 daltons or more. A synthetic pentasaccharide, representing the binding site to antithrombin III, was devoid of inhibitory activity (3,12). A further indication that the heparanase inhibitory and anticoagulant activities of heparin are unrelated was obtained by using heparin fractions with high and low affinity for antithrombin III. These heparins differed about 200 folds in their anticoagulant activity, but had a similar high heparanase inhibitory activity.

Treatment of tumor cells and animals with heparanase inhibitors markedly reduced the incidence of lung metastases induced by B16 melanoma, Lewis lung carcinoma and mammary adenocarcinoma cells (3,13,14). A single injection (I.V. or S.C.) of native or modified heparins, decreased the number of melanoma lung metastases to about 5% of control, but there was no effect to totally desulfated heparin. Heparin fractions with high and low affinity to antithrombin III exhibited a comparable high anti-metastatic activity, indicating that the heparanase inhibiting activity of heparin rather than its anticoagulant activity plays a role in the anti-metastatic properties of the polysaccharide (3). Most efficient inhibition of tumor cell metastasis was obtained when the melanoma cells and heparin were injected at the same time. Around 75% inhibition of metastasis was achieved when heparin was given 6 h before or 2 h after the tumor cells, suggesting that the polysaccharide interferes with the passage of tumor cells across the capillary wall. Similar results were reported by Irimura et al. (13) and Parish et al (14).

Heparanase Activity Expressed by Normal and Malignant Cells Releases Active bFGF from ECM

Fibroblast growth factors are a family of structurally related polypeptides characterized by high affinity to heparin (15). They are highly mitogenic for vascular EC and are among the most potent inducers of neovascularization and mesenchyme formation (15,16). Basic fibroblast growth factor (bFGF) has been extracted from the subendothelial ECM produced *in vitro* (17) and from basement membranes of the cornea (18), suggesting that ECM may serve as a reservoir for bFGF. Immunohistochemical staining revealed the localization of bFGF in basement membranes of diverse tissues (19) and blood vessels (20). Despite the ubiquitous presence of bFGF in normal tissues, EC proliferation in these tissues is usually very low, suggesting that bFGF is somehow sequestered from its site of action. Studies on the interaction of bFGF with ECM revealed that bFGF binds to HSPG in the ECM and can be released by heparin-like molecules, HS degrading enzymes (21,22), or plasmin (23). These results suggest that the ECM HSPG provide a natural storage depot for bFGF and possibly other growth promoting factors (3,24). Displacement of bFGF from its storage within basement membranes and ECM may provide a novel mechanism for induction of neovascularization in normal and pathological situations (3,24). To investigate whether heparanase, expressed by various normal and malignant cells, is involved in release of bFGF from ECM, we first identified molecules which inhibit the enzyme but do not release the ECM-bound bFGF. Using these inhibitors (i.e. carrageenan lambda, N-acetylated heparin), we have demonstrated that heparanase activity expressed by platelets, neutrophils, and lymphoma cells is involved in release of active bFGF from ECM and basement membranes of bovine corneas (22). Regardless of the source of heparanase and of whether release of bFGF was brought about by a pure enzyme, intact cells, or cell lysates, inhibition of bFGF-release correlated with inhibition of heparanase activity, measured by release from ECM of sulfate labeled HS degradation products (22) . We suggest that heparanase activity expressed by tumor cells may not only function in cell migration and invasion, but at the same time may also elicit an indirect neovascular response by means of releasing the ECM-resident FGF. Likewise, platelets, mast cells and activated cells of the immune system (i.e.

macrophages, neutrophils, T lymphocytes) that are often attracted by tumor cells may indirectly stimulate tumor angiogenesis by means of their heparanase activity. These cells may also elicit an angiogenic response in the process of inflammation and wound healing.

Several studies indicate that μg quantities of heparin and HS inhibit the mitogenic activity of bFGF, but at the same time stabilize and protect the molecule from inactivation (25,26). It is therefore proposed that bFGF is stored in ECM in a highly stable but relatively inactive form, as also indicated by the highly stable ECM-resident growth-promoting activity, as compared to that of bFGF in a fluid phase. Release from ECM of bFGF as a complex with HS fragment is likely to yield a form of bFGF that is more stable than free bFGF and yet capable of binding to high affinity plasma membrane receptors (23). A recent study indicated that cell surface and/or soluble heparin and HS are required for binding of bFGF to high affinity cell surface receptors. While bFGF failed to bind to HS-deficient CHO mutant cells, binding to high affinity receptor sites was restored by the addition of ng quantities of heparin and HS (27). It was therefore suggested that binding of heparin or HS imposes on the bFGF molecule the conformation necessary for optimal interaction with its high affinity cell surface receptor (27). We propose that restriction of EC growth factors in ECM prevents their systemic action on the vascular endothelium, thus maintaining a very low rate of EC turnover and vessel growth. On the other hand, release of bFGF from storage in ECM may elicit a localized EC proliferation and neovascularization in processes such as wound healing, inflammation and tumor development (3,24).

Structural Requirements for Release of ECM-bound bFGF by Heparin

Heparin exhibits a high degree of heterogeneity due to variations in the size of the polysaccharide chains and in the degree and distribution of sulfate groups. In the following experiments we investigated structural requirements for release of ECM- and cell surface-bound bFGF by heparin and heparin-like molecules. For this purpose subendothelial ECM and cultured vascular EC were incubated with ^{125}I-bFGF, washed free of unbound bFGF and exposed to various chemically modified species of heparin and to homogenously sized oligosaccharides derived from depolymerized heparin.

Oligosaccharides derived from depolymerized heparin. Maximal release of ^{125}I-bFGF (i.e. 50-60% of the ECM-bound bFGF) was achieved by 5 μg/mL of the octasaccharide. Exposure of ECM to higher oligosaccharides containing up to 16 sugar units yielded results which were, on a weight basis, similar to those obtained with the octasaccharide (28). Size dependent differences in bFGF releasing activity were more pronounced when the various oligosaccharides were compared on a molar basis. As demonstrated in figure 1, half maximal release of ECM-bound ^{125}I-bFGF was obtained at 0.1 μM of the octasaccharide, as compared to 2 μM of the tetrasaccharide. Oligosaccharides containing 10, 12 and 14 sugar units yielded a half maximal release at about 0.06 μM (Figure 1).

We compared the effect of oligosaccharides derived from nitrous acid degraded heparin and having varying degrees of sulfation. While medium and high sulfate fractions of a given oligosaccharide (ratio of sulfur: carbon ~ 0.18 and ~ 0.21, respectively) (29) showed virtually the same bFGF releasing activity, low sulfate oligosaccharides (sulfur : carbon = 0.11) (29) were less efficient releasers of bFGF, especially when compared at low concentrations (0.1-1 μg/mL). This effect was more evident with oligosaccharides containing 4-8 sugar units as compared to 12 sugar units or more (28).

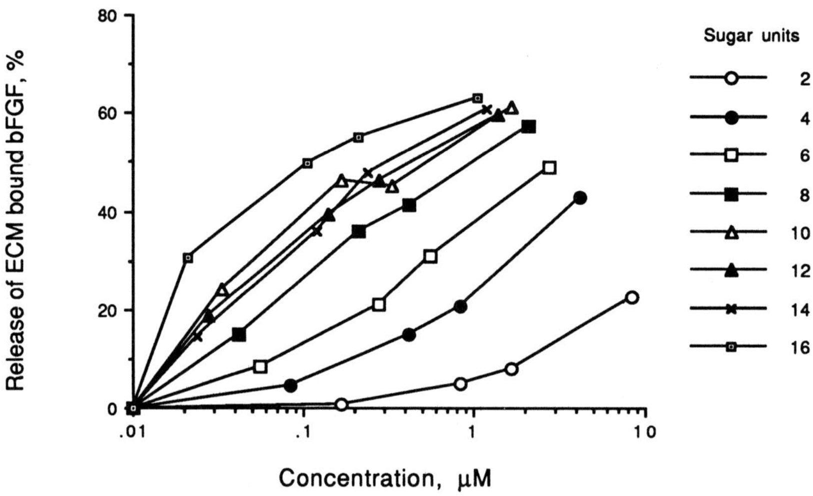

Figure 1. Release of ECM-bound bFGF by heparin derived oligosaccharides of varying sizes. ECM-coated wells of 4 well plates were incubated (24°C, 3 h) with ^{125}I-bFGF (2x10^4 cpm/well). The ECM was washed 4 times and incubated (24°C, 3 h) with various concentrations of homogenously sized oligosaccharides prepared by nitrous acid depolymerization. (The number of sugar units is presented next to each symbol). Released radioactivity is expressed as percent of total ECM-bound ^{125}I-bFGF.

<u>Chemically modified species of heparin.</u> In order to analyze the involvement of O-sulfate and N-sulfate residues of heparin in release of ECM-bound bFGF, both heparin and low Mr heparin were either totally desulfated, N-desulfated, or partially N-desulfated. The N-sulfate groups of these heparins and low Mr heparins were substituted with acetyl or hexanoyl groups. Both totally desulfated and N-desulfated heparin failed to release the ECM-bound bFGF. In contrast, N-resulfation of totally desulfated heparin restored its bFGF releasing activity to a large extent. Total N-acetylation of heparin or low Mr heparin

resulted in an almost complete inhibition of their bFGF releasing activity . Over sulfation of these N-acetylated molecules resulted in partial restoration of their ability to release bFGF from ECM. These results indicate that N-sulfate groups of heparin are critical for efficient release of ECM-bound bFGF (28). The ability of a given compound to release bFGF from ECM was dependent primarily on the position of the sulfate group rather than on the total level of sulfation (i.e. % sulfur). Thus, N-resulfated, O-desulfated low Mr heparin (% sulfur= 5.3) exhibited a much higher bFGF releasing activity than either N-desulfated-, or N-acetylated- heparin containing 9.7% and 8.7% sulfur, respectively. O-sulfate groups of heparin contributed to its bFGF releasing activity, since oversulfation of N-acetylated low Mr heparin (% sulfur= 13.9) resulted in partial restoration of its bFGF releasing activity (28).

In order to correlate the anticoagulant activity of heparin to its bFGF releasing activity, heparin was separated on an antithrombin-Sepharose column into a nonbinding fraction of virtually no anti-factor Xa (FXa) activity and a binding fraction of high anti-FXa activity (30). Heparin fractions with high and low affinity to antithrombin III exhibited a similar high bFGF releasing activity, despite a nearly 200 fold difference in their anti-FXa activity (28).

Different structural requirements for release of ECM-bound bFGF and for inhibition of heparanase activity by heparin. Different structural properties of heparin were required for release of ECM-bound bFGF and for inhibition of heparanase activity. For example, substitution of the N-sulfates of heparin with acetyl groups had little or no effect on its ability to inhibit heparanase, as opposed to a greatly reduced ability of the substituted heparin molecules to release ECM-bound bFGF. On the other hand, heparin derived oligosaccharides containing 8-10 sugar units exhibited a high bFGF releasing activity, but failed to inhibit the heparanase enzyme. Likewise, release of bFGF from ECM, but little or no inhibition of heparanase activity were brought about by O-desulfated, N-resulfated heparin (28).

Our results indicate that different effects of heparin are mediated by different sugar sequences and that specific heparin-like molecules can be designed to elicit or inhibit a specific effect. For example, N-substituted species of heparin, rather than native heparin, could be applied to inhibit tumor metastasis, since their efficient inhibition of heparanase activity was not associated with a significant release of active bFGF from cells and ECM. These compounds are therefore expected to inhibit metastases formation by certain tumor cells, correlated with their inhibition of heparanase activity (3,7,13,14), with little or no potential induction of tumor angiogenesis in response to bFGF release.

Our present and previous studies on the sequestration and release of bFGF suggest that bFGF may acquire an immobilized storage form that is stable but relatively inactive, and a soluble form that is labile but highly active. Release of HS-bound bFGF by heparin derived oligosaccharides containing 8-10 sugar units, as well as by heparin- and HS-degrading enzymes may yield an intermediate type of molecule that is relatively stable (25,26) and may readily diffuse through the stroma (31), as compared to free bFGF. This released form is also capable of binding to high affinity plasma membrane receptors (23), resulting in proliferative and differentiation responses in endothelial cells and other mesoderm derived cells. Sequestration and release of FGF-like factors may thus provide a novel mechanism for regulation of capillary blood vessel growth. Under normal conditions it may prevent them from acting on the vascular endothelium, while perturbation of the ECM and/or exposure to heparin-like molecules may elicit localized EC proliferation and neovascularization.

Other ECM-Immobilized Growth Factors, Enzymes and Active Molecules

The list of growth factors that bind to heparin and HS is growing all the time. Some of the recently identified members of this group are not related to the FGF family. These include the hepatocyte growth factor, members of the EGF family such as amphiregulin and the recently identified heparin binding EGF-like growth factor (HB-EGF) (32) as well as neutrite promoting factors such as pleiotropin and the heparin binding neurotrophic factor. Preliminary results indicate that some of these growth factors are associated with the ECM (33).

HSPG may function in localizing granulocyte-macrophage colony stimulating factor (GM-CSF) and interleukin 3 (IL-3) to the stromal cell matrix where interaction with stem cells occurs. These colony-stimulating factors, once bound and localized, can be presented in an active form to hamatopoietic cells (34). Studies on the interaction of interferon-γ with a reconstituted basement membrane, revealed a high affinity binding of this multifunctional cytokine to HSPG (35). Osteogenin, an ECM-associated bone-inductive protein, was isolated from demineralized bone matrix by heparin affinity chromatography. Other heparin-binding growth factors that are tightly associated with bone matrix have been identified, suggesting that the bone matrix is a repository for chemotactic, mitogenic and differentiation inducing factors involved in endochondral bone formation (36). Transforming growth factor β (TGF-β) binds to the core protein of a cell-surface proteoglycan called betaglycan (37). TGF-β also binds to an ECM proteoglycan, decorin, through its core protein. Binding of decorin to TGF-β is reversible and can also neutralize the activity of the growth factor (33).

We have identified active tissue-type plasminogen activator (t-PA) and urokinase-type PA (u-PA) in the subendothelial ECM (3,24) and demonstrated that heparanase and the ECM-resident PA participate synergistically in sequential degradation of HSPG in the ECM (24). Moreover, exposure of ECM to heparanase resulted in release of PA activity, suggesting sequestration of PA by the ECM-HS. In addition to molecules that are secreted by the ECM producing cells, the ECM may serve as a storage depot for active molecules produced by other cell types. We have recently demonstrated that thrombin, a serine protease which exerts growth-promoting and chemotactic activities, binds with high affinity to dermatan sulfate in the subendothelial ECM (38). While ECM-bound thrombin retained its biological activities (e.g. esterolytic activity, platelet activation and stimulation of vascular smooth-muscle cell proliferation), it was protected from inactivation by its physiological circulating inhibitor, antithrombin III. Likewise, we have demonstrated that lipoprotein lipase (LPL) binds to the subendothelial ECM and retains its catalytic activity for a longer period than the soluble enzyme (39). This enzyme is situated at the luminal surface of capillary endothelium where it hydrolyzes triacylglycerols present in the core of plasma chylomicrons and very low density lipoproteins. We suggest that the subendothelial ECM sequesters and stabilizes LPL secreted by non-endothelial cells, providing a solid phase reservoir of LPL on its way from its site of synthesis to its site of action (39).

Cellular activities depend on multifactorial events involving soluble-phase molecules such as cytokines and growth factors, and solid phase extracellular matrix macromolecules. It is now evident that the ECM provides a storage depot for growth factors, enzymes and plasma proteins and that some of the effects of ECM can be attributed to the combined action of structural components and ECM-immobilized molecules that are thereby protected and stabilized. This may allow a more localized, regulated and persistent mode of action, as compared to the same molecules in a fluid phase. Proteoglycans have been found to function as the principal binders of growth factors and cytokines and more than 20 known heparin-binding proteins contain a consensus structural motif which participate in this binding. Because proteoglycans are abundant and ubiquitous components of ECM and cell surfaces, they are likely to be able to capture most of those growth factors and cytokines that have affinity for the GAG or protein part of proteoglycans. It may be that growth factors and some enzymes were meant to act on their target cells over short ranges only and that their immobilization in basement membranes, ECMs and cell surfaces through proteoglycan binding accomplishes that goal. The result is a tight spatial control of growth factor and enzyme activities which is likely to be of major significance in the formation and maintenance of tissue architecture (33).

Acknowledgements

This work was supported by Public Health Service Grant CA-30289 by the National Cancer Institute, DHHS, and by grants from the USA-Israel Binational Science Foundation and the German-Israel Foundation for Scientific Research and Development.

References

1. Nicolson, G.L. Organ specificity of tumor metastasis: Role of preferential adhesion, invasion and growth of malignant cells at specific secondary sites. Cancer Met. Rev. 7:143-188 (1988).
2. Vlodavsky, I., Fuks, Z., Bar-Ner, M., Ariav, Y., Schirrmacher, V. Lymphoma cell mediated degradation of sulfated proteoglycans in the subendothelial extracellular matrix: Relationship to tumor cell metastasis. Cancer Res. 43:2704-2711 (1983).

3. Vlodavsky, I., Korner, G., Ishai-Michaeli, R., Bashkin, P., Bar-Shavit, R., Fuks, Z. Extracellular matrix-resident growth factors and enzymes: possible involvement in tumor metastasis and angiogenesis. Cancer Met. Rev. 9:203-226 (1990).
4. Vlodavsky, I., Liu, G.M., Gospodarowicz, D. Morphological appearance, growth behavior and migratory activity of human tumor cells maintained on extracellular matrix vs plastic. Cell 19: 607-616 (1980).
5. Hassell, J.R., Kimura, J.H. , Hascall, V.C. Proteoglycan core protein families. Ann. Rev. Biochem., 55: 539-567 (1986).
6. Hook, M., Kjellen, L., Johansson, S., Robinson, J. Cell surface glycosaminoglycans. Ann. Rev. Biochem. 53:847-869 (1984).
7. Nakajima, M., Irimura, T., Nicolson, G.L. Heparanase and tumor metastasis. J. Cell Biochem. 36:157-167 (1988).
8. Lider, O., Baharav, E., Mekori, Y., Miller, T., Naparstek, Y., Vlodavsky, I., Cohen, I.R. Suppression of experimental autoimmune diseases and prolongation of allograft survival by treatment of animals with heparinoid inhibitors of T lymphocyte heparanase. J. Clin. Invest. 83:752-756 (1989).
9. Liotta, L.A., Rao, C.N., Barsky, S.H. Tumor invasion and extracellular matrix. Lab. Invest. 49:636-649 (1983).
10. Vlodavsky, I., Ishai-Michaeli, R., Bar-Ner, M., Fridman, R., Horowitz, A.T., Fuks, Z., Biran, S. Involvement of heparanase in tumor metastasis and angiogenesis. Is. J. Med. 24:464-470 (1988).
11. Jin, L., Nakajima, M., Nicolson, G.L. Immunochemical localization of heparanase in mouse and human melanomas. Int. J. Cancer 45:1088-1095 (1990).
12. Bar-Ner, M., Eldor, A., Wasserman, L., Matzner, Y., Vlodavsky, I. Inhibition of heparanase mediated degradation of extracellular matrix heparan sulfate by modified and non-anticoagulant heparin species. Blood 70:551-557 (1987).
13. Irimura,T., Nakajima, M., Nicolson, G.L. Chemically modified heparins as inhibitors of heparan sulfate specific endo-b-glucuronidase (heparanase) of metastatic melanoma cells. Biochemistry 25:5322-5328 (1986).
14. Parish, C.R., Coombe, D.R., Jakobsen, K.B., Underwood, P.A. Evidence that sulphated polysaccharides inhibit tumor metastasis by blocking tumor cell-derived heparanase. Int. J. Cancer 40:511-517 (1987).
15. Burgess, W.H., Maciag, T. The heparin-binding (fibroblast) growth factor family of proteins. Annu. Rev. Biochem. 58:575-606 (1989).
16. Folkman, J., Klagsbrun, M. Angiogenic factors. Science 235:442-447 (1987).
17. Vlodavsky, I., Folkman, J., Sullivan, R., Fridman, R., Ishai-Michaeli, R., Sasse, J., Klagsbrun, M. Endothelial-cell-derived basic fibroblast growth factor: synthesis and deposition into subendothelial extracellular matrix. Proc. Nat. Acad. Sci. USA, 84:2292-2296 (1987).
18. Folkman, J., Klagsbrun, M., Sasse, J., Wadzinski, M., Ingber, D., Vlodavsky, I. A heparin-binding angiogenic protein - basic fibroblast growth factor - is stored within basement membrane. Am. J. Pathol. 130:393-400 (1988).
19. Gonzalez, A.-M., Buscaglia, M., Ong, M., Baird, A. Distribution of basic fibroblast growth factor in the 18-day rat fetus: localization in the basement membranes of diverse tissues. J. Cell Biol. 110:753-765 (1990).
20. Cardon-Cardo, C., Vlodavsky, I., Haimovitz-Friedman, A., Hicklin, D. Fuks, Z. Expression of basic fibroblast growth factor in normal human tissues. Lab. Invest. 63:832-840 (1990).
21. Bashkin, P., Klagsbrun, M., Doctrow, S., Svahn, C.-M., Folkman, J., Vlodavsky, I. Basic fibroblast growth factor binds to subendothelial extracellular matrix and is released by heparanase and heparin-like molecules. Biochemistry 28:1737-1743 (1989).
22. Ishai-Michaeli, R., Eldor, A., Vlodavsky, I. Heparanase activity expressed by platelets, neutrophils and lymphoma cells releases active fibroblast growth factor from extracellular matrix. Cell Reg. 1:833-842 (1990).
23. Saksela, O., Rifkin, D.B. Release of basic fibroblast growth factor-heparan sulfate complexes from endothelial cells by plasminogen activator-mediated proteolytic activity. J. Cell Biol. 110:767-775 (1990).

24. Vlodavsky, I., Fuks, Z., Ishai-Michaeli, R., Bashkin, P., Levi, E., Korner, G., Bar-Shavit, R., Klagsbrun M. Extracellular matrix-resident basic fibroblast growth factor: Implication for the control of angiogenesis. J. Cell Biochem. 45: 167-176 (1991).
25. Gospodarowicz, D., Cheng, J. Heparin protects basic and acidic FGF from inactivation. J. Cell. Physiol. 128:475-484 (1986).
26. Saksela, O., Moscatelli, D., Sommer, A., Rifkin, D.B. Endothelial cell-derived heparan sulfate binds basic fibroblast growth factor and protects it from proteolytic degradation. J. Cell Biol. 107:743-751 (1988).
27. Yayon, A., Klagsbrun, M., Esko, J.D., Leder, P., Ornitz, D. Cell surface, heparin-like molecules are required for binding of basic fibroblast growth factor to its high affinity receptor. Cell 64: 841-848 (1991).
28. Ishai-Michaeli, R., Svahn, C-M., Chajek-Shaul, T., Korner, G., Ekre, H.-P.T., and Vlodavsky, I. Importance of size and sulfation of heparin in release of basic fibroblast growth factor from the vascular endothelium and extracellular matrix. Biochemistry. In press (1991).
29. Sudhalter, J., Folkman, J., Svahn, C-M., Bergendal, K., D'Amore, P.A. Importance of size, sulfation and anticoagulant activity in the potentiation of acidic fibroblast growth factor by heparin. J. Biol. Chem. 264:6892-6897 (1989).
30. Ekre, H.-P.T., Fjellner, B., and Hagermark, O. Inhibition of complement dependent experimental inflammation in human skin by different heparin fractions. Int. J. Immunopharmac. 8:277-286 (1986).
31. Flaumenhaft, R., Moscatelli, D., and Rifkin, B. Heparin and heparan sulfate increase the radius of diffusion and action of basic fibroblast growth factor. J. Cell Biol. 111:1651-1659 (1990).
32. Higashiyama, S., Abraham, J.A., Miller, J., Fiddes, J.C., and Klagsbrun, M. A heparin-binding growth factor secreted by macrophage-like cells that is related to EGF. Science 251, 936-939 (1991).
33. Ruoslahti, E., and Yamaguchi, Y. Proteoglycans as modulators of growth factor activities. Cell 64, 867-869 (1991).
34. Roberts, R., Gallagher, J., Spooncer, S., Allen, T.D., Bloomfield, F., Dexter, T.M. Heparan sulphate bound growth factors: a mechanism for stromal cell mediated haemopoiesis. Nature 332, 376-378 (1988).
35. Lortat-Jacob, H., Kleinman, H.K., Grimaud, J-A. High affinity binding of interferon to a basement membrane complex (Matrigel). J. Clin. Invest. 87, 878-883 (1991).
36. Hauschka, P.V., Chen, T.L., Mavrakos, A.E. Polypeptide growth factor in bone matrix. Ciba Found. Symp. 136, 207-225 (1988).
37. Massague, J. The transforming growth factor-β family. Annu. Rev. Cell Biol. 6, 597-641 (1990).
38. Bar-Shavit, R., Eldor, A., Vlodavsky, I. Binding of thrombin to subendothelial extracellular matrix: Protection and expression of functional properties. J. Clin. Invest. 84, 1096-1104 (1989).
39. Chajek-Shaul, T., Friedman, G., Bengtsson-Olivercrona, A., Vlodavsky, I., Bar-Shavit, R. Interaction of lipoprotein lipase with subendothelial extracellular matrix. Biochim Biophys Acta 1042, 168-175 (1990).

ANTI-INFLAMMATORY EFFECTS OF HEPARIN AND ITS DERIVATIVES INHIBITION OF COMPLEMENT AND OF LYMPHOCYTE MIGRATION

Hans-Peter Ekre*[1], Yaakov Naparstek[2], Ofer Lider[3], Pia Hydén[1], Östen Hägermark[4], Torbjörn Nilsson[5], Israel Vlodavsky[6], and Irun Cohen[3]

[1]R&D Immunobiology, Kabi Pharmacia Biopharma, Stockholm, Sweden, [2]Clinical Immunology Unit Department of Medicine and [6]Department of Oncology, Hadassah Hospital, Jerusalem, Israel, [3]Department of Cell Biology, The Weizmann Institute of Science, Rehovot, Israel, [4]Department of Dermatology, Karolinska Hospital, Solna, Sweden, and [5]Department of Clinical Chemistry, Umeå University, Sweden

INTRODUCTION

Inflammation is initiated through immune recognition by T lymphocytes or antibodies, or by non-immune tissue injury. The effector functions thus triggered, activation and extravasation of leucocytes and humoral cascade systems, lead to the reactions we collectively call inflammation. The physiological functions of this powerful response are to combat microbial invasion, to eliminate damaged tissue and eventually to repair damaged tissue; the self-limiting control mechanisms normally lead to health. *Pathological inflammation*, acute or chronic, is often the result of inadequate control within the immune system leading to allergies (hyper-reactivity) or autoimmune diseases. Inflammation also contributes to morbidity in persistant infections (e.g. leprosy, tuberculosis), persistant tissue injury (overstrain, atherosclerosis) or acute tissue injury (trauma, acute myocardial infarction (AMI), shock/adult respiratory distress syndrome (ARDS)). Graft rejection represents a special case of unwanted inflammation.

The fact that heparin is stored in mast cell granules, which are released by several inflammatory effector mechanisms, suggests a role for endogenous heparin in inflammation. Amelioration of pain and swelling shortly after injection of heparin to patients with DVT is one

---

*Address correspondence to H.-P. Ekre at new affiliation: Pharmacology 1, Astra Draco AB, P.O.B. 34, S-221 00 Lund, Sweden

Heparin and Related Polysaccharides
Edited by D.A. Lane *et al.*, Plenum Press, New York, 1992

example in favor of an in vivo anti-inflammatory effect of heparin. Experimental studies have demonstrated several mechanisms through which heparin may inhibit inflammatory reactions. Examples are interference with granulocyte activities (Laghi Parsini et al 1984, Matzner et al 1984, Ley et al 1991), reduced cytotoxic potential of eosinophil mediators (Frigas et al 1980, Motojima et al 1989) and NK-cells (Yamamoto et al 1985), inhibition of complement activation (Ecker & Gross 1929), and interference with lymphocyte recirculation and extravasation/delayed type hypersensivity, DTH, (Bradfield & Born 1969, Cohen et al 1967). Hence, the role for heparin in inflammation appears to be that of a modulator.

The main reason why pharmaceutical heparin has not been clinically evaluated for control of pathological inflammation is the risk for bleeding. Progress in the understanding of structure function relationship regarding its anticoagulant effects (i.e. affinity for antithrombin III, AT (Andersson et al, 1976, Höök et al, 1976, Lam et al, 1976, Lindahl et al, 1980)) and of its anti-inflammatory mechanisms have resulted in increased interest in heparin as a basis for new therapeutic principles in the treatment of inflammatory diseases. We have focused our studies in this field on inhibition of complement activation and of lymphocyte migration, with special reference to inhibition of heparanase or its secretion from T lymphocytes, by heparin derivatives. Here, we review the findings (published and about to be published) from these studies including results from experimental models in man.

MODULATION OF COMPLEMENT ACTIVATION

The complement cascade, schematically depicted in Fig.1, can be activated by antibodies bound to their antigen, the classical pathway, and by surface structures on certain microbes, the alternative pathway. Both pathways can also be activated by damaged tissue. If activation occurs on a cell-surface this will be damaged by the pores formed by the membrane attack complex, MAC, and the cell eventually dies. This complex may also form in solution; the terminal complement complex, TCC. Several intermediates generated in the cascade amplifies inflammation. Especially powerful and fast acting mediators are the anaphylatoxins C3a and C5a that promote extravasation and attract, activate and aggregate granulocytes. Also the C5b67 has chemotactic activity. The biology of complement is reviewed in Immunology Today, September 1991, and the role of complement in disease was recently reviewed by Morgan (1990) .

Heparin's ability to inhibit complement was early described (Ecker & Gross 1929) and several points of action have been identified, as depicted in Fig 1 (see references in Ekre 1985). The rate limiting steps are suggested to be enhancement of the C1INH and C1s interaction in the classical pathway (Caughman et al, 1982) and prevention of formation of the amplification convertase of the alternative pathway (Maillet et al, 1983). The ability of heparin to inhibit the early steps in the cascade and thus to prevent formation of anaphylatoxins makes it attractive as a complement inhibitor. Consequently our finding that heparin with low affinity for AT, LA-heparin, was as potent as standard and high AT-affinity heparin, HA-heparin, in complement inhibition (Ekre, 1979) encouraged us to further investigate the structure function relationships and in vivo effects of heparin fractions.

Methods

Heparin fractions were prepared from pharmaceutical grade porcine heparin by affinity chromatography on antithrombin-Sepharose (Andersson et al 1979, Ekre, 1985) and by gelfiltration on Sephacryl S-200 (Holmer et al 1981, Ekre, 1985) into fractions differing in AT-affinity (HA- and LA-heparin) or in average MW. Low MW heparin fragments were prepared

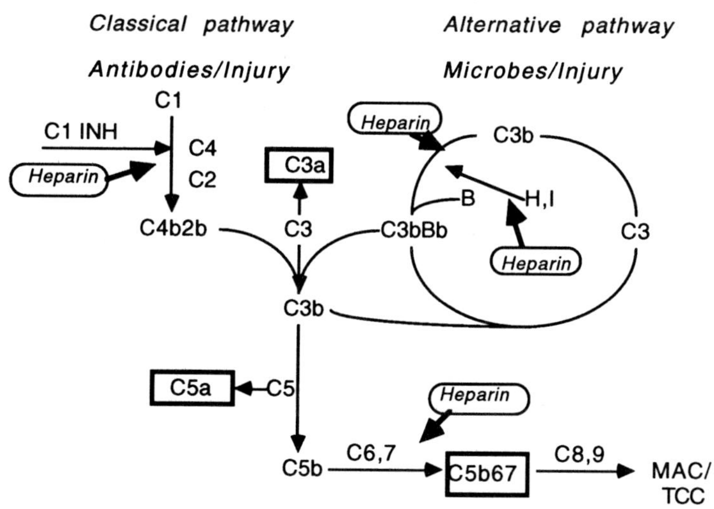

Figure 1. Schematic drawing of the complement cascade. Intermediates amplifying inflammation are boxed. Sites of heparin inhibition are depicted.

by nitrous acid depolymerization of porcine or bovine heparin (Thunberg et al, 1980) and further separated by ion-exchange chromatography and gelfiltration on Sephadex G-50SF (Mattsson et al 1989). Various chemical derivatives were prepared and provided by Drs M. Weber and C.-M. Svahn.

Inhibition of complement activation by heparin derivatives in vitro was studied by i) measuring the inhibition of C3 cleavage, using crossed immuno-electrophoresis, after activation of normal human serum with aggregated IgG or zymosan (classical and alternative pathway respectively) (Ekre, 1985), ii) inhibition of haemolytic activity of serum complement using antibody coated sheep erythrocytes as targets (Ekre, 1985), and iii) by determining the acceleration of C1s inhibition by C1INH using a chromogenic substrate, S-2314, for detection (Nilsson & Wiman, 1983).

To study the modulating effect of heparin fractions in vivo we developed a model (Ekre at al, 1986a) based on Ishisaka's findings (Ishizaka et al, 1961) that heat-aggregated human IgG, HAGG, can substitute for preformed immune complexes to induce a rapid arthus-type reaction. Intradermal injection of 0.1 mg HAGG in 10 µl into the the lateral aspect of the upper arm of healthy volunteers induced a reaction characterized by erythema, wheal and itch, which was fully developed after 15 min. Simultaneous injection of heparin inhibited the reactions in a dose-dependent manner as did the histamine antagonist mepyramine. To determine if complement was the target for heparin we investigated its effect on i.d. injection of trypsin (a bypass generator of anaphylatoxins), histamine and the histamine liberator compound 48/80. Mepyramine inhibited the reactions induced by these mediators, but heparin did not, strongly suggesting that heparin excerts its effect on HAGG provocation by inhibition of complement activation in this model (for details see Ekre et al, 1986a).

Results

Heparin fractions prepared from pharmaceutical heparin, differing in AT-affinity or average MW (range 4800-17000), were equally potent inhibitors of human complement on a weight basis. This independence of anticoagulant activity and size was the same for inhibition of C3 activation (both pathways) and inhibition of haemolysis. A somewhat surprising observation was that in experiments employing guinea pig complement inhibition of haemolysis showed a MW-dependence and LA-heparin had a decreased potency, whereas all fractions were equipotent inhibitors of classical pathway guinea pig C3 activation. This was not further investigated, but possible species differences should be kept in mind when complement inhibition by heparin and its derivatives is studied in vitro and in vivo.

To see if the fractions were equally potent also in vivo we used an experimental model in human skin; inhibition of the inflammation induced by i.d. injection of 0.1 mg HAGG, as described above. After determination of suitable doses, experiments were designed to compare LA- to HA-heparin and low-MW to high-MW fractions each at two doses; 1 and 10 µg. In a third experiment all four fractionswere compared at 10 µg. All fractions inhibited the reactions in a dose-dependent manner and they were all found to be equally effective; partial inhibition with 1 µg and almost complete inhibition with 10 µg (Ekre et al, 1986a). Fig. 2 shows the effects on erythema by LA-heparin (LAH) and HA-heparin (HAH).

LMW heparin fragments, LMWH, have entered clinical practice and are likely to replace heparin as the antithrombotic drug of choice. Their improved bio-availability, duration and also safety argue in favor of their evaluation also for modulation of inflammation. Therefore we also compared LMWH (Fragmin®) to heparin for the abilty to inhibit activation of human

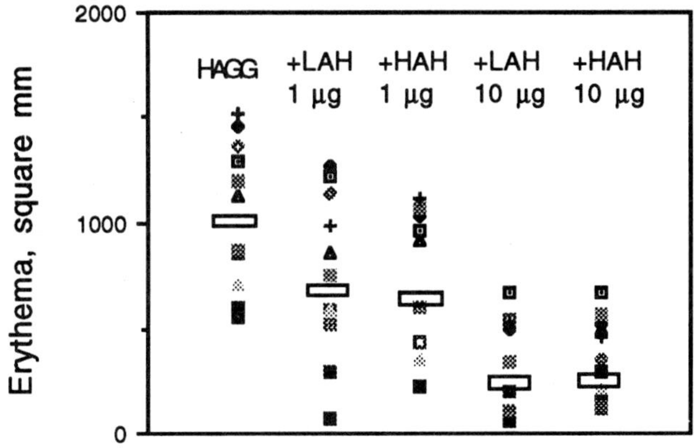

Figure 2. Dose-dependent inhibition of HAGG-induced skin inflammation by LA-heparin (LAH) and HA-heparin (HAH) as indicated. The horisontal bars show the mean values from the 11 human volunteers.

complement in vitro and in human skin. Not unexpectedly, they were found to be equally effective (Ekre et al, 1986b, Ekre et al in preparation).

The structural requirements on heparin for inhibition of the classical pathway in vitro were studied. In addition to inhibition of haemolysis and of C3 activation we also determined the rate enhancing effect of heparin derivatives on the C1INH-C1s interaction, previously shown to be independent of AT-affinity (Nilsson & Wiman, 1983). By using heparin fragments of different average MW, prepared by stepwise elution from DEAE-Sepharose, the minimal size for a potency similar to heparin was found to be ~3100, whereas the potency offragments of lower MW was reduced to 50% or less (Nilsson et al, 1988). This was verified with homogenous oligosaccharide preparations showing that a chain length of 10 to 14 monosaccharide units, depending on charge, is necessary. Similar results were obtained with the three methods arguing in favor of the C1INH-C1s interaction as the rate limiting step (Ekre et al, in preparation). The potency of chemically modified fragments of average MW in the range 3000-4000 showed that the most critical groups for activity are the O-sulfates (Ekre et al, in preparation).

MODULATION OF T LYMPHOCYTE MIGRATION

T lymphocytes are the directors and initial effector cells (through the release of cytokines) of the delayed type hypersensitivity (DTH) type of inflammatory reactions, characterized by mononuclear cell infiltration. Such reactions dominate the scene in contact allergies and autoimmune diseases like rheumatoid arthritis and multiple sclerosis. The mechanisms involved in the migration of cells between the vascular and the extravascular compartments are not fully understood. However, this represents a field of intense research exemplified by the finding that cells able to extravasate produce a heparan sulfate (HS) degrading endoglycosidase, heparanase. The observation that this enzyme enables them to cross the subendothelial extracellular matrix, ECM, a major barrier between vessels and tissues in which HS is an important structural component, was first made in studies of tumor cells with metastasising potential (Vlodavsky et al, 1983, Nakajima et al, 1983). It was subsequently found that activated but not resting T lymphocytes produced heparanase and that antigen recognition in the context of ECM triggered the release of this enzyme suggesting an important mechanism for homing and for penetration of ECM (Naparstek et al, 1984), Fig. 3. An endoglycosidase that can degrade the HS of ECM has previously been found in platelets (Wasteson et al, 1977), and more recent studies show that heparanase is produced also by other cells of the haematopoietic lineage (Savion et al 1984, Matzner et al, 1985, Bashkin et al, 1990).

Heparin was found to inhibit heparanase degradation of HS , and this effect was independent of its anticoagulant activity (Irimura et al, 1986, Bar-Ner et al, 1987). Injection of heparin is known to affect the recirculation of lymphocytes as evidenced by lymphocytosis (Bradfield & Born, 1969) and has been shown to inhibit DTH in experimental animals (Cohen et al, 1967). Inhibition of heparanase may contribute to this effect. For these reasons it was of interest to test the ability of heparin to modulate T lymphocyte dependent inflammation in animal models of autoimmune diseases, graft rejection, and contact allergy (DTH).

Methods

For experimental details on animal models see (Lider et al, 1989, 1990). Adjuvant arthritis, AA, was induced in Lewis rats by intradermal injection at the tail base of 0.1 ml complete Freund's adjuvant containing 10 mg/ml of killed *Mycobacterium tuberculosis*. Adoptive experimental autoimmune encephalomyelitis, EAE, was induced in naive Lewis rats by i.v. injection of 1.5 x 10^7 lymph node cells collected from syngeneic rats immunized with myelin basic protein, BP, in adjuvant. Skin allografting from C57BL/6 mice to BALB/c recipient mice was performed as described (Lider et al, 1990). Contact hypersensivity , DTH, was induced in BALB/c mice by painting the shaved abdomen with oxazolone in acetone/olive oil and evaluated as ear swelling after challenge (ear painting) with the same substance.

Heparin treatment was by s.c. daily injections in various doses. The N-desulfated, N-acetylated heparin used in AA, EAE and graft rejection was prepared by Institut Choay.

Effect of heparin on DTH in humans was evaluated by comparing the skin response, erythema and induration, to intradermally injected recall antigens (mumps and tetanus toxoid) before and during daily s.c. injections of heparin (Naparstek et al, in preparation).

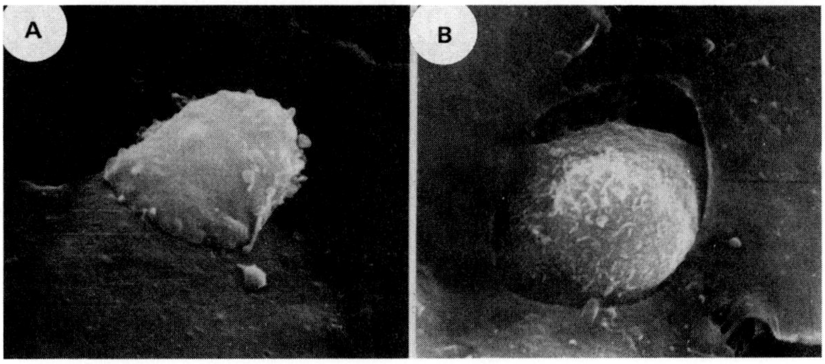

Figure 3. Activated T lymphocytes invade cultured vascular endothelial cells. Anti-myelin basic protein specific T cells were separated by Ficoll density centrifugation and incubated (2 x 10^6 cells per 35-mm dish for 6 hr) on top of intact endothelial cell monolayers. The cultures were then washed and processed for scanning electron microscopy. About 30% of the cells adhering to endothelial cells were seen at various stages of penetration through the vascular endothelium (x6000).

Heparanase activity was determined by incubation of samples in metabolically ^{35}S-labelled ECM tissue culture dishes and subsequent measurement of heparan sulfate degradation using gel filtration on Sepharose 6B monitored by determining the position of radioactive peaks in the eluate (Vlodavsky et al, 1983, Naparstek et al, 1984).

Results

In the studies performed by Lider et al (1989) initial experiments were designed to prevent adjuvant arthritis in rats, a model resembling rheumatoid arthritis. Symptoms appear at about day 12after immunization (see above) and heparin was given by daily subcutaneous injections from day 8 to 14. A wide dose range was tested and a clear effect on arthritis was observed at 20 µg per rat (100 µg/kg). This proved to be optimal and already at half or double that dose the effect was less pronounced, a finding that was quite surprising. Twenty µg daily was also effective on established AA (treatment from day 21 and until day 50) with prompt amelioration of arthritis and no recurrence of disease after cessation of heparin. In adoptively transferred EAE (an animal model of MS), where signs of paralysis appear 4 days after inoculation of BP-stimulated lymphocytes, 20 µg heparin given daily from day -1 and during the 10 days disease course, reduced the severity of disease. In mice the effect of heparin on rejection of skin allografts was studied. Also here a dose window was observed, the optimal dose beeing 5 µg per mouse daily, which prolonged the survival of the graft from 15 to 24 days.

Considering the low doses it appeared unlikely that the anticoagulant activity of heparin would be involved and this was also shown in experiments employing N-desulfated, N-acetylated heparin, devoid of anticoagulant activity, that gave similar effects on AA, EAE and allograft rejection as standard heparin at the expected doses (Lider et al, 1989).

Taken together these results indicate that heparin interferes with the migration or with the effector functions of lymphocytes rather than induction of immunity. This was further adressed in DTH experiments in mice (Lider et al, 1990). Ear swelling after challenge of sensitized mice could be inhibited with 5 μg heparin daily. In adoptive transfer experiments, where lymphocytes from sensitized animals were injected into naive recipients, DTH appeared after challenge. Heparin inhibited the reaction if donor cells were injected i.v., but not if the cells were injected into the site of antigen challenge (Lider et al, 1990). Thus the effect of heparin at

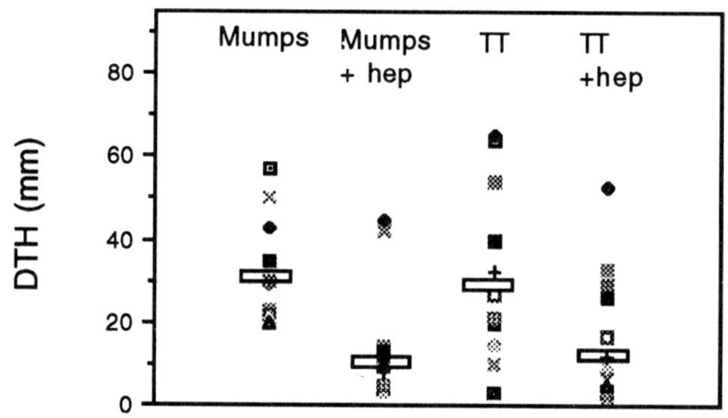

Figure 4. Effect of low-dose heparin (3 mg daily) on DTH reactions to the recall antigens mumps and tetanus toxoid (TT) as indicated. The horisontal bars show the mean values from the 12 human volunteers.

this low dose is modulatation of T lymphocyte migration since the induction and effector functions of the cellular immune response remain unaffected. Inhibition of heparanase activity seemed a less probable mechanism at this dose level, but instead inhibition of heparanase secretion from T lymphocytes by heparin may be involved since spleen cells from naive mice make heparanase in response to the mitogen Con A in vitro whereas cells from heparin treated mice do not (Lider et al, 1990).

The effect of low dose heparin was tested on DTH skin reactions to recall antigens in humans (Naparstek et al, in preparation). A tentative dose of 3 mg (500 IU) daily, given s.c., was found effective in a pilot experiment, but 45 mg (7500 IU) was ineffective. DTH, induration and erythema, to mumps and tetanus toxoid antigens were determined in 12 healthy volunteers and three weeks later daily s.c. injections of 3 mg heparin were given for 10 days during which the DTH tests were repeated. A marked decrease in the responses was recorded, as shown in Fig. 4.

DISCUSSION

The key findings in our investigations, comprising experimental studies in humans, are that heparin's inhibition of complement activation is independent of its anticoagulant activity and that T lymphocyte mediated inflammatory reactions can be modulated by heparin at a low(subanticoagulant), narrow dose range.

Several groups have studied complement inhibition by heparin fractions in vitro and most (Ekre, 1979, 1985, Kazatchkine et al, 1981, Linhardt et al, 1988) but not all (Bianchini et al, 1982) report that this effect is independent of its anticoagulant activity. Our results on complement dependent skin inflammation, equal potency of LA- and HA-heparin, are strong arguments in favor of an inhibitory mechanism that does not involve antithrombin and that does not require an intact antithrombin binding sequence in heparin. The effects of very LMW fragments and chemical derivatives thereof that we have found in classical pathway inhibition show a very good agreement with what others have found in studies of alternative pathway inhibition (Linhardt et al, 1988, Maillet et al, 1988). Similar structural requirements for control of both pathways are compatible with a feed-back inhibitor function of natural heparin, which is released by anaphylatoxins. It is also an advantage in the context of pharmacologic control of complement activation where a potential drug should modulate both pathways.

There is limited experience of complement inhibitors in treatment of inflammation because there are no such drugs available. Plausible applications of a complement inhibitor would be in acute inflammation or in acute phases of chronic inflammatory diseases. Recently there has been increased interest in modulation of complement in AMI and cerebral stroke, and recombinant soluble CR1 (complement receptor type 1), which has a control function that prevents anaphylatoxin formation, has been shown to reduce infarct size by 44% in a rat model of reperfusion injury of ischemic myocardium (Weisman et al, 1990). Heparin, in a similar model, reduced infarct size by 67% (Wright et al, 1988) and has also been shown to be a good adjunct to fibrinolytic drugs in AMI (Melandri et al, 1990), an effect that was suggested to involve antithrombotic as well as anti-inflammatory properties. LMWH and LA-heparin seem as attractive candidates to further explore the potential of complement inhibition in pathological inflammation; they are active complement inhibitors in humans, as described here, and should be relatively predictable regarding safety, although their effective systemic dose remains to be determined.

The treatments of experimental T lymphocyte mediated autoimmune diseases with heparin described here, were initiated based on the findings that activated T cells make and use heparanase for migration through ECM (Naparstek et al, 1984) and that heparanase can be inhibited by heparin (Bar-Ner et al, 1987). It was thus a surprise that the active dose range of heparin in these models proved to be very low and narrow, in contrast to inhibition ofheparanase in vitro and inhibition of metastasis in mice, where heparin shows an expected dose response with total inhibition at high doses (see Vlodavsky et al, this volume). Similarly, recent studies from others have shown that EAE in rats can be reduced in severity by high doses of heparin given continously with osmotic minipumps, an effect that was ascribed to inhibition of heparanase (Willenborg & Parish, 1988). In contrast, the low dose effect observed by us appeared to be attributable to prevention of heparanase production or secretion rather than inhibition of the enzyme itself (Lider et al, 1990). A possible explanation is that heparin gives a signal by crosslinking receptors on lymphocytes, a mechanism that can be anticipated to occur at a narrow concentration range and that could be a natural feed-back mechanism. Further studies to define the mechanism(s) involved are in progress.

There is a need for efficacious and safe therapy in T lymphocyte mediated autoimmune diseases, such as rheumatoid arthritis, multiple sclerosis, type 1 diabetes, and also to prevent graft rejection. Pharmacological intervention of lymphocyte migration may be an approach to accomplish this, provided that the defence against infections is not compromised. Antagonists to adhesion or homing molecules is one possibilty. Another, as demonstrated here, may be to interfere with the lymphocyte invasive machinery by heparin or a heparin derivative. The protocol that proved effective, very low dose heparin once daily, appears attractive and safe. Even though this regimen worked well in an experimental study in humans and has recently been successfully applied in confirmatory animal studies by others (Lagodzinski et al, 1990, Sai et al, 1991) it may be difficult to apply as a therapeutic routine due to the narrow dose window in combination with the heterogeneity of heparin (which also interacts with multiple components in the circulation) and the individual variation in a patient population. Our efforts are now also directed towards defining the heparin molecules and structures responsible for the modulation of lymphocyte migration to overcome this problem.

In conclusion: We are beginning to understand the anti-inflammatory principles and mechanisms of heparin and it seems appropriate to regard natural heparin as a physiological modulator of inflammation. Heparin can also be regarded as a basis for drug development (as fractions, derivatives or templates for new molecules) where modulation of complement activation and of T lymphocyte migration represent new therapeutic principles for inflammatory diseases where there is a great demand for improved therapy.

ACKNOWLEDGEMENTS

Parts of this work were supported by the Gablinger fund for clinical Immunology, a grant from the Israeli Ministry of Health, a Public Health Service Grant CA-30289 by the National Cancer Institute, DHHS, and by Kabi Pharmacia Cardiovascular. Y.N. holds the Leiferman chair in Immunology and I.C. is the incumbent of the Mauerberger Chair in Immunology.

REFERENCES

Andersson, L.-O., Barrowcliffe, T.W., Holmer, E., Johnsson, E.A., and Sims, G.E.C., 1976, Anticoagulant properties of heparin fractionated by affinity chromatography on matrix-

bound antithrombin III and by gel filtration, Thromb. Res., 9, 575.
Bar-Ner, M., Eldor, A., Wasserman, L., Matzner, Y., Cohen, I.R., Fuks, Z., and Vlodavsky I., 1987, Inhibition of heparanase-mediated degradation of extracellular matrix heparan sulfate by non-anticoagulant heparin species, Blood, 70, 551.
Bashkin, P., Razin, E., Eldor, A., and Vlodavsky, I., 1990, Degranulating mast cells secrete an endoglycosidase that degrades heparan sulfate in subendothelial extracellular matrix, Blood, 75, 2204
Bianchini, P., Osima, B., Parma, B., Nader, H.B., and Dietrich, C.P., 1982, Pharmacological activities of heparins obtained from different tissues: Enrichment of heparin fractions with high lipoprotein lipase, antihemolytic and anticoagulant activities by molecular sieving and antithrombin III affinity chromatography, J. Pharmacol.Exp. Ther., 220,406.
Bradfield, J.W.B., and Born, G.V.R., 1969, Inhibition of lymphocyte recirculation by heparin, Nature (Lond.), 222, 1183.
Caughman, G.B., Boackle, R.J., and Vesely, J., 1982, A postulated mechanism for heparin's potentiaition of C1 inhibitor function, Mol. Immun., 19, 287.
Cohen, S., Benacerraf, B., McCluskey, R.T., and Ovary, Z., 1967, Effect of anticoagulants on delayed hypersensitivity reactions, J. Immunol., 98,351.
Ecker, E.E., and Gross, P., 1929, Anticomplementary power of heparin, J. infect. Dis., 44, 250.
Ekre, H.-P.T., 1979, Anticomplementary activity of heparin fractions with high and low anticoagulant activity and with different molecular weights, Scand. J. Immun., 10, 364.
Ekre, H.-P.T., 1985, Inhibition of human and guinea pig complement by heparin fractions differing in affinity for antithrombin III or in average molecular weight, Int. J. Immunopharmac., 7, 271.
Ekre, H.-P.T., Fjellner, B., and Hägermark, Ö., 1986a, Inhibition of complement dependent experimental inflammation in human skin by different heparin fractions, Int. J. Immunopharmac., 8, 277.
Ekre, H.-P.T., Hydén, P., Fjellner, B., and Hägermark, Ö., 1986b, Effects of a low molecular weight heparin fragment on activation of human complement in vitro and in vivo. Complement 3, 4.
Frigas, E., Loegering, D.A., and Gleich, G.J., Cytotoxic effects of the guinea pig eosinophil major basic protein on tracheal epithelium, Lab. Invest., 42, 35.
Holmer, E., Kurachi, K., and Söderström, G., 1981, The molecular weight dependence of the rate-enhancing effect of heparin on the inhibition of thrombin, Factor Xa, Factor IXa, Factor XIIa and kallikrein by antithrombin, Biochem. J., 193, 395.
Höök, M., Björk, I., Hopwood, J., and Lindahl, U., 1976, Anticoagulant activity of heparin: separation of high-activity and low-activity heparin-species by affinity chromatography on immobilized antithrombin, FEBS Lett., 66, 90.
Irimura, T., Nakajima, M., and Nicolson, G.L., 1986, Chemically modified heparins as inhibitors of heparan sulfate specific endo-ß-glucuronidase (heparanase) of metastatic melanoma cells, Biochemistry, 25, 5322.
Ishizaka, K., Ishizaka, T., and Sugahara, T., 1961, Biological activity of aggregated gammaglobulin. III. Production of Arthus-like reactions, J. Immunol., 86, 220.
Kazatchkine, M.D., Fearon, D.T., Metcalfe, D.D., Rosenberg, R.D., and Austen, K.F., 1981, Structural determinants of the capacity of heparin to inhibit the formation of the human amplification C3 convertase, J. Clin. Invest., 67, 223.
Laghi Parsini, F., Pasqui, A.L., Ceccatelli, L., Capecchi, P.L., Orrico, A., and Di Perri, T., 1984, Heparin inhibition of polymorphonuclear leucocyte activation in vitro. A possible pharmacological approach to granulocyte-mediated vascular damage, Thromb. Res., 35, 527.
Lagodzinski, Z., Gorski. A., and Wasik, M., 1990, Immunosuppressive action of low-dose heparin. Effect on skin allograft survival, Transplantation, 50, 714.
Lam, L.H., Silbert, J.E., and Rosenberg, R.D., 1976, The separation of active and inactive forms of heparin, Biochem. Biophys. Res. Commun., 69, 570-577.
Ley, K., Cerrito, M., and Arfors, K.-E., 1991, Sulfated polysaccharides inhibit leucocyte rolling in rabbit mesentery venules, Am. J. Physiol., 260, H1667.
Lider, O., Baharav, E., Mekori, Y., Miller, T., Naparstek, Y., Vlodavsky, I., and Cohen, I.R., 1989, Suppression of experimental autoimmune diseases and prolongation of allograft survival by treatment of animals with low doses of heparins, J. Clin. Invest., 83, 752.
Lider, O., Mekori, Y., Miller, T., Bar-Tana, R., Vlodavsky, I., Baharav, E., Cohen, I.R., and Naparstek, Y., 1990, Inhibition of T lymphocyte heparanase by heparin prevents T cell

migration and T cell-mediated immunity, Eur. J. Immunol., 20, 493.
Lindahl, U., Bäckström, G., Höök, M., Thunberg., Fransson, L.-Å., and Linker, A., 1980, Structure of antithrombin binding site in heparin, Proc. Natl. Acad. Sci. USA 76, 3198.
Linhardt, R.J., Rice, K.G., Kim, Y.S., Engelken, J.D., and Weiler, J.M., Homogeneous, structurally defined heparin oligosaccharides with low anticoagulant activity inhibit the generation of the amplification pathway C3 convertase in vitro, J. Biol. Chem., 263, 13090.
Maillet, F., Kazatchkine, M.D., Glotz, D., Fischer, E., and Rowe, M., 1983, Heparin prevents formation of the human C3 amplification convertase by inhibiting the binding site for B on C3b, Molec. Immunol., 20, 1401.
Maillet, F., Petitou, M., Choay, J., and Kazatchkine, M., 1988, Structure-function relationship in the inhibitory effect of heparin on complement activation: independency of the anti-coagulant and anti-complementary sites on the heparin molecule, Molec. Immunol., 25,917.
Mattsson, C., Palm, M., Söderberg, K., and Holmer, E., 1989, Antithrombotic effects of heparin oligosaccharides, Ann. N.Y. Acad. Sci., 556, 323.
Matzner, Y., Bar-Ner, M., Yahalom, J., Ishai-Micheli, R., Fuks, Z., and Vlodavsky, I., 1985, Degradation of heparan-sulfate in the subendothelial extracellular matrix by a readily released heparanase from human neutrophils: Possible role in invasion through basement membranes, J. Clin. Invest., 76, 1306.
Matzner, Y., Marx, G., Drexler, R., and Eldor, A., 1984, The inhibitory effect of heparin and related glycosaminoglycans on neutrophil chemotaxis, Thromb. Haemostas., 52, 134.
Melandri, G., Branzi, A., Semprini, F., Cervi, V., Galiè, N., and Magnani, B., 1990, Enhanced thrombolytic efficacy and reduction of infarct size by simultaneous infusion of streptokinase and heparin, Br. Heart J., 64, 118.
Morgan, B.P., 1990, "Complement. Clinical aspects and relevance to disease", Academic Press, London.
Motojima, S., Frigas, E., Loegering, D.A., and Gleich, G.J., 1989, Toxicity of eosinophil cationic proteins for guinea pig tracheal epithelium in vitro, Am. Rev. Respir. Dis., 139, 801.
Nakajima, M., Irimura, T., Di Ferrante, D., Di Ferrante, N., and Nicolson, G.L., 1983, Heparan sulfate degradation: Relation to tumor invasion and metastatic properties of mouse B 16 melanoma sublines, Science, 220, 611.
Naparstek, Y., Cohen, I.R., Fuks, Z., and Vlodavsky, I., 1984, Activated T lymphocytes produce a matrix-degrading heparan sulfate endoglycosidase, Nature (Lond.), 310, 241.
Nilsson, T., Siösteen, Y., and Ekre, H.-P., 1988, Low molecular weight heparin fractions with good anticomplement but poor anticoagulant activity. Complement 5, 210.
Nilsson, T., and Wiman, B., 1983, Kinetics of the reaction between human C1-esterase inhibitor and C1r or C1s, Eur. J. Biochem., 129, 663.
Sai, P., Pogu, S., and Ouary, M., 1991, Heparin attenuates low-dose streptozotocin-induced immune diabetes in mice and inhibits the beta-cell binding of T-splenocytes in vitro, Diabetologica, 34, 212.
Savion, N., Fuks, Z., and Vlodavsky, I., 1984, T lymphocyte and macrophage interaction with cultured vascular endothelial cells: Attachment, invasion and subsequent degradation of the subendothelial extracellular matrix, J. Cell. Physiol., 118, 169.
Thunberg, L., Bäckström, G., Grundberg, H., Riesenfeld, J., and Lindahl, U., 1980, The molecular size of the antithrombin-binding sequence in heparin. FEBS Lett., 117, 203.
Vlodavsky, I., Fuks, Z., Bar.Ner, M., Ariav, Y., and Schirrmacher, V., 1983, Lymphoma cell-mediated degradation of sulfated proteoglycans in the subendothelial extracellular matrix: Relationship to tumor cell metastasis, Cancer Res., 43, 2704.
Wasteson, Å, Glimelius, B., Busch, C., Westermark, B., Heldin, C.-H. and Norling, B., 1977, Effect of a platelet endoglycosidase on cell surface associated heparan sulphate of human cultured endothelial and glial cells, Thromb. Res., 11, 309
Weisman, H., Bartow, T., Leppo, M., Marsh Jr, H.C., Carson, G.R., Concino, M.F., Boyle, M.P., Roux, K.H., Weisfeldt, M.L., and Fearon, D.T., 1990, Soluble human complement receptor type 1: In vivo inhibitor of complement suppressing post ischemic myocardial inflammation and necrosis, Science, 249, 146.
Willenborg, D.O., and Parish, C.R., 1988, Inhibition of allergic encephalomyelitis in rats by treatment with sulfated polysaccharides, J. Immunol., 140, 3401.
Wright, G.J., Kerr, J.C., Valeri, C.R., and Hobson II, R.W., 1988, Heparin decreases ischemia-reperfusion injury in isolated canine gracilis model, Arch. Surg., 123, 470.
Yamamoto, H., Fuyama, S., Arai, S., and Sendo, F., Inhibition of mouse natural killer cytotoxicity by heparin, Cell. Immunol., 96, 409.

HEPARAN SULFATE GLYCOSAMINOGLYCANS AS PRIMARY CELL SURFACE RECEPTORS FOR HERPES SIMPLEX VIRUS

Patricia G. Spear, Mei-Tsu Shieh, Betsy C. Herold, Darrell WuDunn and Thomas I. Koshy

Microbiology-Immunology Department
Northwestern University Medical School
Chicago, IL 60611, USA

INTRODUCTION

The purposes of this communication are (i) to summarize the evidence that heparan sulfate moieties of proteoglycans serve as cell surface receptors for herpes simplex virus (HSV) and (ii) to describe the heparin-binding viral glycoproteins that mediate the binding of virus to cells. These topics are discussed in the context of the molecular interactions required for entry of HSV into cells and, also, the pathology of HSV infections.

HERPES SIMPLEX VIRUSES TYPES 1 AND 2: DISEASES AND BIOLOGY

The most common form of disease caused by HSV in humans is manifested as mucocutaneous lesions, which occur usually in or near the mouth (cold sores or fever blisters) or eyes (keratitis) or on genital tissues. Because the virus that causes primary lesions establishes a latent infection in the ganglia of sensory nerves and can be reactivated by appropriate stimuli, periodic recurrence of herpetic lesions is common and is one of the most troublesome aspects of infections with HSV. Although it happens less frequently, HSV can also cause life-threatening disease affecting vital organs, including encephalitis in apparently normal adults and disseminated disease in infants and immunocompromised individuals.

There are two distinct forms of HSV, designated types 1 and 2 (HSV-1 and HSV-2). The viruses isolated from cases of adult encephalitis, keratitis and facial lesions are usually HSV-1 whereas those isolated from cases of neonatal disease and genital lesions are usually HSV-2. This apparent preference of the two HSV types for different parts of the human anatomy could be due principally to the usual source of inoculum for infection at each site. Transmission of virus from one individual to another occurs most readily when there is close contact with mucosal surfaces from which virus is being shed. In addition, HSV-1 and HSV-2 may differ intrinsically in their ability to establish infection at different sites, perhaps due in part to preferential use of different cell surface receptors that are not equally represented at the different sites. Before any information was available as to the nature of HSV receptors, evidence was presented that the cell surface receptors for HSV-1 and HSV-2 are not identical (Vahlne *et al.,* 1979; Vahlne *et al.,* 1980; Addison *et al.,* 1984).

The genomes of HSV-1 and HSV-2 are closely related in that each gene of HSV-1 has an homologous counterpart in HSV-2 and the genes are arranged on the DNA in

Heparin and Related Polysaccharides
Edited by D.A. Lane *et al.*, Plenum Press, New York, 1992

the same colinear arrangement (McGeoch, 1989). Recombination readily occurs between HSV-1 and HSV-2 to yield viable progeny, implying funtional as well as sequence homology of related genes. Nevertheless, divergence between homologous genes of the two types has been global and significant, such that overall identity or close similarity of DNA sequences is about at the level of 50%. Different HSV strains of the same type also exhibit genetic diversity, although to a much lesser degree. One of the questions to be raised in this communication is whether genetic variations in the viral proteins that mediate the binding of virus to cells may influence receptor recognition and tissue tropism.

Although HSV infects principally humans outside the laboratory, animals of many species can be infected experimentally. Most vertebrate cells that can be propagated in cell culture, especially adherent cells, are also susceptible to HSV infection. The extremely broad host range of this virus implies that receptors for the virus must be highly conserved in evolution and widely distributed on many cell types.

The virus particles (or virions) of HSV-1 and HSV-2 have the typical morphology of herpesviruses. The virions consist, from the inside out, of a core containing the DNA genome, an icosahedral protein shell that encloses and protects the core, a poorly defined layer of protein located between the outer surface of the protein shell and the inner surface of the limiting envelope and, finally, the outer limiting envelope itself. This envelope is a lipid bilayer containing approximately ten viral membrane glycoproteins.

Events in the replication of HSV by a cell may be summarized as follows. Specific viral membrane glycoproteins mediate the binding of HSV to the surface of a cell. Entry of HSV into the cell occurs by fusion of the viral envelope with the cell plasma membrane. Endocytosis of HSV may occur but appears to play little or no role as a pathway of entry that can lead to infection of the cell. After fusion of the virus with the cell and disassembly of the nucleocapsid (the icosahedral protein shell containing the viral genome), the viral DNA is transported to the cell nucleus where it serves as template for viral gene transcription and for genome replication. Progeny nucleocapsids are assembled in the cell nucleus and acquire an envelope by budding through virus-modified patches of the inner nuclear membrane. The progeny virions are then transported out of the cell through a secretory pathway that appears to include the Golgi apparatus.

Relevant literature on these topics has been reviewed by Corey and Spear (1988) and Roizman and Sears (1990).

CELL SURFACE RECEPTORS FOR HSV

The hypothesis that cell surface heparan sulfate serves as receptor for HSV has its origins in results reported about thirty years ago. It was shown in these early studies that heparin and other sulfated polysaccharides could inhibit HSV infection (Nahmias and Kibrick, 1964; Takemoto and Fabisch, 1964; Vaheri, 1964). Recent investigation revealed that heparin inhibits the binding of HSV (both HSV-1 and HSV-2) to cells and that virions can bind to heparin affinity columns (WuDunn and Spear, 1989). In addition, it was shown that treatment of cells with heparitinase or heparinase, but not chondroitin lyase, destroys receptors for HSV (WuDunn and Spear, 1989). Binding of HSV-1 or HSV-2 to cells treated with the former enzymes was significantly impaired and the cells were resistant to HSV infection. Control experiments showed that the treated cells remained fully susceptible to other viruses.

Cell mutants defective in various aspects of glycosaminoglycan synthesis have been used to explore further the role of heparan sulfate in HSV infection. Mutants isolated and characterized by Esko and colleagues were derived from the Chinese hamster ovary (CHO) cell line and are described in Table 1. Studies with these cell mutants (Shieh *et al.,* 1991) revealed the following (Table 1): (i) Whereas various strains of HSV-1 and HSV-2 could bind to wild-type CHO cells and could infect the

Table 1. HSV binding to, and infection of, wild-type CHO cells and mutants

Cell line	Biochemical deficiency	GAGs produced: Heparan sulfate	GAGs produced: Chondroitin sulfate	References on cell lines	HSV binding and infection (% of control)[a]
Wild-type K1	None	Yes	Yes		(100)
Mutants:					
pgsA-745	Xylosyltransferase	No	No	Esko et al. (1985)	<1
pgsB-761	Galactosyltransferase	No	No	Esko et al. (1987)	<1
pgsD-677	N-acetylglucosaminyl & glucuronosyltransferases	No	Yes[b]	Esko et al. (1988) Lidholt et al. (1991)	<1
pgsE-606	N-sulfotransferase	Yes[c]	Yes	Bame & Esko (1989) Bame et al. (1991)	30-40

[a]Compared with the wild-type K1 cells, the mutant cells were impaired as indicated, both for the binding of radiolabeled HSV and for susceptibility to HSV infection, the latter as judged by quantitating the expression of an immediate-early viral protein (Shieh *et al.*, 1991).
[b]Chondroitin sulfate accumulates to levels three times higher than in wild type cells.
[c]The heparan sulfate produced is undersulfated by a factor of two to three.

cells with varying degrees of efficiency, all the virus strains bound very poorly if at all to, and failed to infect, the CHO mutants that are defective for heparan sulfate synthesis. (ii) The CHO mutant that is defective for heparan sulfate synthesis, but overproduces chondroitin sulfate, was no more susceptible to HSV infection than the cell mutants that produce neither heparan sulfate nor chondroitin sulfate. (iii) The CHO mutant that produces undersulfated heparan sulfate had fewer HSV receptors than did wild-type cells and was intermediate in susceptibility to HSV infection.

All these results taken together show that the binding of HSV to cells and infection of the cells requires the presence of heparan sulfate on the cell surface. Other glycosaminoglycans cannot substitute for heparan sulfate and the level of sulfation of the heparan sulfate influences the amount of virus that can bind to the cells.

Does this mean that heparan sulfate is the receptor for HSV? For other kinds of cell surface molecules, the evidence summarized above would be considered sufficient to draw this conclusion. For heparan sulfate, however, certain properties of the molecule complicate interpretation of the results. First, heparan sulfate, like other glycosaminoglycans, is highly charged and binds with varying affinities and degrees of specificity to many proteins. The fact that HSV requires heparan sulfate on the cell surface, and cannot make do with chondroitin sulfate, implies that non-specific ionic interactions alone cannot be responsible for the binding of virus to cells. An additional indication of specificity is the finding that one viral glycoprotein can be assigned primary responsibility for the binding of virus to cells and that this glycoprotein has affinity for heparin (Herold *et al.,* 1991).

Second, heparin and heparan sulfate are best known as modulators of protein-protein interactions, as is amply demonstrated by other contributions in this volume and elsewhere. For example, it was recently shown that high-affinity binding of basic fibroblast growth factor (bFGF) to cells requires the dual interaction of bFGF with cell surface heparan sulfate (or exogenous heparin) and a protein receptor for bFGF. bFGF failed to bind to an appropriate protein receptor on cells deficient in heparan sulfate or on cells deficient in the sulfation of heparan sulfate, unless exogenous heparin was added (Yayon *et al.,* 1991; Rapraeger *et al.,* 1991). We investigated the possibility that the role of cell surface heparan sulfate in HSV infection might be to facilitate binding of the virus to some other receptor, by testing whether exogenous heparin could substitute for cell surface heparan sulfate. We found, however, that there was no enhancement in the binding of HSV to heparan sulfate-deficient cells, or in infection of the cells, over a wide range of heparin concentrations (Shieh *et al.,* 1991). Therefore, it seems that HSV requires cell-bound heparan sulfate, presumably in the form of membrane proteoglycan.

The simplest interpretation of the findings summarized here is that heparan sulfate serves as receptor for the initial binding of HSV to cells. This binding of virus to heparan sulfate could serve to initiate a cascade of interactions between components of the viral envelope and the cell surface. Because the viral envelope contains as many as ten viral glycoproteins, of which at least half play some role in virion binding and penetration, the possibility exists for several kinds of virion-cell surface interactions to occur between the initial attachment of the virus to the cell and fusion of the virion envelope with the plasma membrane. If any of the viral glycoproteins engages in interactions with receptors other than heparan sulfate, however, these interactions must usually occur secondary to the binding of virus to heparan sulfate and/or must not be of high enough affinity to enable efficient binding of virus to cells in the absence of heparan sulfate. It should be noted here that the protein receptor for bFGF plays no role in the binding of HSV to cells or in infection of cells (Shieh and Spear, 1991; Mirda *et al.,* 1991) despite previous reports suggesting this possibility (Kaner *et al.,* 1990; Baird *et al.,* 1990).

The cell surface heparan sulfate moieties required for HSV infection are presumably constituents of proteoglycans. It remains to be determined whether any heparan sulfate proteoglycan can serve as receptor or whether a particular proteoglycan is required. Specific proteoglycans may be preferred as HSV receptors due to characteristic modifications of their heparan sulfate chains. Also, HSV may bind to diverse heparan sulfate proteoglycans but be activated for penetration of the cell only after interaction with particular proteoglycans, due perhaps to interactions of the virus with molecular determinants of both heparan sulfate and the protein core or to requirement for proteoglycans that are physically linked to other cell components involved in viral penetration.

DO HSV VIRIONS CONTAIN A HEPARINASE OR HEPARITINASE?

Some viruses that use cell surface polysaccharides as receptors also encode enzymes that can hydrolyze the viral receptors. For example, the envelope of influenza virus contains two glycoproteins, designated hemagglutinin and neuraminidase. The hemagglutinin mediates the binding of virus to cells by binding to sialic acid on cell surface glycoproteins and glycolipids. The neuraminidase can cleave off the sialic acid moieties recognized by the hemagglutinin. The biological role of the neuraminidase is not fully understood. It is believed to play an important role in permitting the release of progeny virus from infected cells. It has also been suggested that the neuraminidase may be important during the initial stages of infection, to permit virus bound to non-productive receptors to be released for re-binding to other receptors. Influenza neuraminidase is the topic of a recent review (Air and Laver, 1989).

This feature of other viruses suggested the possibility that HSV might have an envelope-associated enzyme capable of cleaving the heparan sulfate chains to which virus binds. To test this possibility, both heparin and heparan sulfate were incubated with purified virions or with commercially available heparinase or heparitinase. Whereas degradation of both substrates by the appropriate enzyme was observed, no degradation of either substrate was detectable in the samples incubated with purified virions (Fig. 1). We have calculated that, if there were only one molecule of enzyme per virion with specific activity comparable to that of heparinase, enzymatic activity should have been detectable at the highest concentration of virions tested.

Although the methods used may not have permitted detection of a hydrolase that generates only large fragments of heparan sulfate, it appears that there is no heparinase-like or heparitinase-like enzyme associated with HSV virions.

Alternatively, the enzyme is latent and requires special conditions (for example, binding to cells) for activation. It should be noted that absence of this kind of enzymatic activity is not inconsistent with the biology of HSV. Progeny virions tend to remain tightly associated with the surfaces of the cells that produced them, suggesting that the virus has no specific mechanism for release.

THE HSV GLYCOPROTEINS REQUIRED FOR INFECTIVITY

Four of the ten known membrane glycoproteins encoded by HSV have been shown to participate in the binding of virus to cells or in penetration of virus into cells. This is not to imply that the other six glycoproteins are non-participants in these processes. Rather, insufficient information is available to permit judgments as to their roles in infectivity.

The HSV glycoprotein designated gC plays a principal role in the binding of virus to cells. This conclusion is based on findings that gC has affinity for heparin and that mutant virions devoid of gC are significantly impaired in ability to bind to cells (Fig. 2) (Herold *et al.,* 1991). The reduced specific infectivity of gC-negative virions (less than 1/10th that of wild-type virus) can be explained by the reduced binding of virions to cells and also by the less efficient penetration of the virus into cells. Interestingly, gC is not absolutely required for HSV infectivity. Mutant virions devoid of gC can bind to cells and initiate infection, albeit inefficiently. What are the requirements for infection of cells by gC-negative virions? We have found that the presence of cell surface heparan sulfate *is* required and that exogenous heparin can inhibit binding and infection, just as for wild-type virus (Herold *et al.,* 1991). Because a second HSV glycoprotein (designated gB) also binds to heparin, we suspect that gB mediates the binding of virus to cells when gC is absent.

The other three glycoproteins known to influence HSV infectivity (gB, gD and gH) appear to be essential for viral penetration into cells. Mutations in any of these

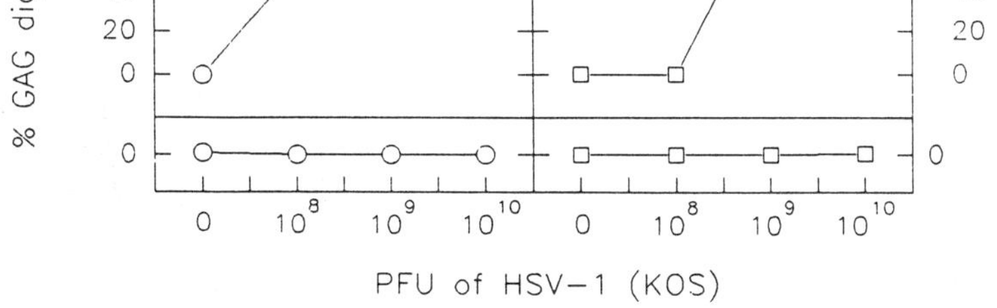

Fig. 1. Absence of heparinase-like or heparitinase-like activities on HSV-1 virions. Reaction mixtures (1 ml in 2.5 mM calcium acetate) containing heparin (circles) or heparan sulfate (squares) at 10 ng/ml and purified HSV-1(KOS) virions (lower panels) or heparinase or heparitinase (upper panels) at the concentrations indicated (units per ml) were incubated at 37oC for 1 hr. Samples of 0.1 ml were removed and added to 0.9 ml of Azure A (10 ug/ml). The metachromatic shift of the dye in response to heparin or heparan sulfate (Jaques et al., 1974) was monitored at 620 nm to determine the concentrations of undigested GAG. The response of the Azure A to both GAGs was linear over the range of 0 to 1 ng. Similar results were obtained using an alternative buffer (PBS plus 1 mM CaCl2). This assay was adapted from one described by Galliher et al. (1981). Unpublished results obtained by T. Koshy.

glycoproteins can reduce infectivity by several orders of magnitude (Desai *et al.,* 1988; Ligas and Johnson, 1988; Little *et al.,* 1981; Sarmiento *et al.,* 1979). Mutant virions devoid of gB or gD have been shown to bind to cells normally but to be defective in penetration (Cai *et al.,* 1988; Ligas and Johnson, 1988). Moreover, monoclonal antibodies specific for any of these three glycoproteins can block the penetration of virus into cells without impairing the binding of virus to cells (Fuller and Spear, 1987; Highlander *et al.,* 1987; Highlander *et al.,* 1988; Fuller *et al.,* 1989). Recent studies have shown that isolated truncated forms of gD can bind to cells, suggesting that gD interacts with its own, yet to be identified, cell surface receptor (Johnson *et al.,* 1990). The fact that gD-negative virions bind to cells with normal efficiency, however, indicates that gD probably does not play any role in the initial attachment of virus to cells. The fact that gB-negative virions also exhibit no obvious impairment in binding to cells (Fig. 2) (Cai *et al.,* 1988; Herold *et al.,* 1991) indicates that the heparin-binding activity of gB is not important for virus binding when gC is present.

The following sections describe more fully the available information about gB and gC, the two viral envelope glycoproteins known to have affinity for heparin (Herold *et al.,* 1991).

Glycoprotein B

Glycoprotein B is present in infected cells and virions as an oligomer, probably a homodimer or homotrimer of gB (Sarmiento and Spear, 1979; Claesson-Welsh and Spear, 1986). This oligomer is very stable, even in the presence of ionic detergent, but can be dissociated by heat. The virion-associated oligomer can be visualized by electron microscopy as a rod-like spike projecting about 14 nm from the surface of the envelope (Stannard *et al.,* 1987).

Some of the features of the gB gene product that can be deduced from nucleotide sequence analysis are summarized in Fig. 3. The N-terminal hydrophobic segment has characteristics of a cleavable signal peptide and was shown experimentally to function as such (Claesson-Welsh and Spear, 1987). Other hydrophobic segments are present

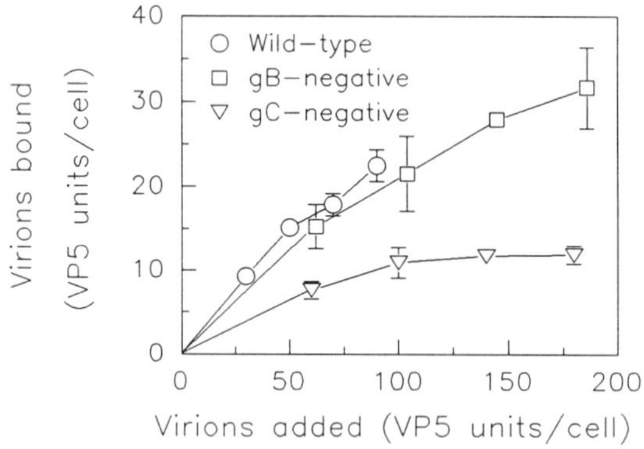

Fig. 2. Binding of wild-type and mutant HSV-1 virions to human HEp-2 cells. The wild-type virus was strain KOS and the mutants were K082 (gB-negative) and $gC^{-}3$ (gC-negative), both derived from strain KOS. Various concentrations of purified [^{35}S]methionine-labeled virions were added to the cells, which had been plated in 96-well round-bottom plates. Binding was for 2 hr at 4°C. After unbound virus was washed away, the cells were solubilized and cell-bound radioactivity was quantitated. The relative concentrations of input virions were determined by densitometry of a silver-stained SDS-PAGE gel and are expressed as VP5 units. VP5 is the highly conserved major protein of the capsid and is present in a constant number of copies per virion. The input and bound virus was determined from the known ratios of radioactivity and VP5 units for each virus preparation. Each point is the average of triplicate determinations, and the error bars represent standard deviation. Modified from Fig. 4 presented by Herold *et al.* (1991).

toward the C-terminus. It has been suggested that gB spans the membrane three times, with the three most C-terminal hydrophobic domains serving as the membrane-spanning domains (Pellett *et al.,* 1985). This is consistent with the results of experiments done to determine the orientation of the termini of the gB translation product (Claesson-Welsh and Spear, 1987). Glycoprotein B is modified by the addition of both N-linked and O-linked glycans (Johnson and Spear, 1983). The potential sites for the addition of N-linked glycans are shown in Fig. 3. Also noted in Fig. 3 is a region that has high local concentrations of basic amino acids, particularly lysine, and proline. This region, which is located at the N-terminus of the mature protein, is a good candidate for a heparin-binding domain.

Glycoprotein C

Glycoprotein C exists in infected cells and virions probably as a monomeric protein (Sarmiento and Spear, 1979). This glycoprotein cannot readily be visualized in virions by standard techniques of electron microscopy but its presence in virions can be demonstrated by the use of monoclonal anti-gC antibodies coupled to colloidal gold. Antigenic determinants of gC extend as far as 15 to 20 nm from the virion envelope surface, suggesting that this glycoprotein is present as a very slender extended molecule (Stannard *et al.,* 1987).

Fig. 4 summarizes some of the known features of gC. Its N-terminus has properties of a cleavable signal sequence. The actual N-terminus of the mature protein has not yet been determined, however. The hydrophobic segment near the C-terminus is a membrane-spanning region. The larger N-terminal ectodomain is modified by the addition of N-linked glycans and a considerable number of O-linked glycans as well

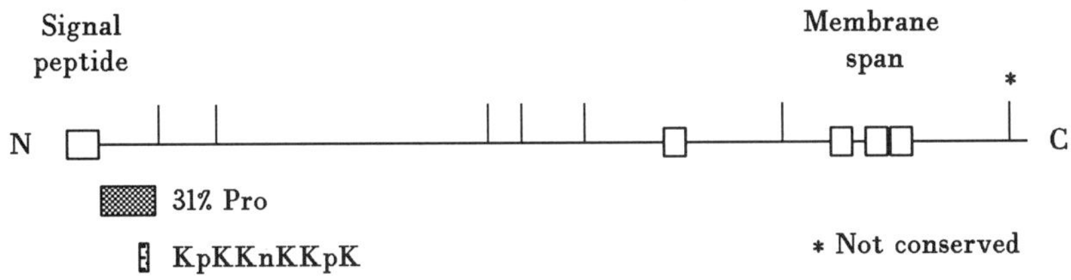

Fig. 3. Structural features of HSV-1 gB. Hydrophobic regions that could be membrane-spanning or membrane-associated are indicated by the open boxes. One of these hydrophobic regions is a signal peptide that is cleaved off during processing of the translation product. The three hydrophobic regions near the C-terminus have been proposed to enable this protein to span the membrane three times. The vertical lines indicate the positions of potential sites for the addition of N-linked glycans (the site near the C-terminus is not conserved in HSV-2, is on the cytoplasmic side of the membrane and therefore is not likely to be modified by the addition of carbohydrate). The lysine-rich and proline-rich region that might interact with heparin is indicated. Based on several published nucleotide sequences and interpretations (Bzik *et al.,* 1984; Bzik *et al.,* 1986; Pellett *et al.,* 1985; McGeoch *et al.,* 1988).

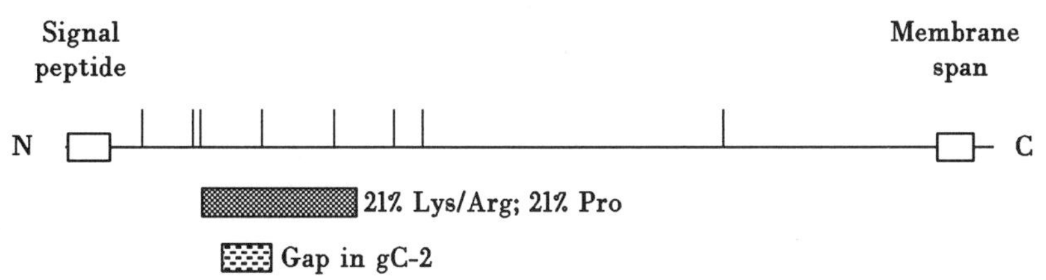

Fig. 4. Structural features of HSV-1 gC. The features indicated are as described in Fig. 3. This glycoprotein spans the membrane only once and has a very short C-terminal cytoplasmic domain. The region rich in basic amino acids and proline is indicated. The HSV-2 form of gC (gC-2) is shorter than gC-1, due to an apparent deletion of part of the domain that is rich in basic amino acids. Based on several published nucleotide sequences and interpretations (Swain *et al.,* 1985; Dowbenko and Lasky, 1984; Frink *et al.,* 1983; Draper *et al.,* 1984; McGeoch *et al.,* 1988).

(Johnson and Spear, 1983; Olofsson *et al.,* 1981; Olofsson *et al.,* 1983; Dall'Olio *et al.,* 1985). Similar to gB, gC has a region near the N-terminus that is rich in basic amino acids and proline and is a good candidate for a heparin-binding domain. The putative heparin-binding domains of gB and gC are not similar or related in actual amino acid sequence.

COMPARISONS OF HSV-1, HSV-2 AND OTHER HERPESVIRUSES

HSV-1 and HSV-2

HSV-1 and HSV-2 are biologically quite similar. From the symptoms of HSV disease in an individual patient, it is not possible to tell whether the causative virus is HSV-1 or HSV-2. Nevertheless, the two HSV types have diverged significantly, with respect to gene sequences, perhaps due in part to occupying different niches in human tissues. It is interesting to consider whether HSV-1 and HSV-2 might differ in specificity for cell surface receptors.

It seems clear that heparan sulfate moieties of cell surface proteoglycans serve as initial receptors for both HSV-1 and HSV-2. Yet there is evidence to suggest that cells can have different numbers and ratios of receptors for the two different types of HSV (Vahlne *et al.,* 1980). At least two hypotheses can be proposed to reconcile these apparently contradictory findings. First, the possibility exists that HSV-1 and HSV-2 forms of the viral heparin-binding glycoproteins recognize distinct and different structural features of heparan sulfate (i.e. different receptors present on heparan sulfate chains). This implies that either gB or gC or both bind to heparan sulfate with some degree of specificity and that the HSV-1 and HSV-2 forms of gB and/or gC differ in this specificity. Second, the possibility exists that the amount of virus bound to a cell depends not only on the amount and properties of cell surface heparan sulfate but also on the number and properties of other secondary cell surface receptors. HSV-1 and HSV-2 might differ in their specificities for interaction with these putative secondary receptors. These two hypotheses are not mutually exclusive.

If the first hypothesis is correct, then the heparin-binding domains of gB and/or gC should differ in sequence for HSV-1 and HSV-2. Although the heparin-binding domains have not yet been identified, the lysine-rich regions near the N-termini of both gB and gC do exhibit considerably more divergence between HSV-1 and HSV-2 than do other regions of these proteins. The amino acid sequences of gB for strains belonging to the same type are 98-99% identical. The same is true for gC. On the other hand, there is about 85% identity of aligned sequences for the HSV-1 and HSV-2 forms of gB and about 64% identity of aligned sequences for the HSV-1 and HSV-2 forms of gC. For the N-terminal 100 amino acids of the mature glycoproteins, however, the sequence identities are only 60% and 30%, respectively. The actual amino acid sequences at the N-termini and the alignments are shown in Fig. 5. Note that alignment of the HSV-1 and HSV-2 gC sequences requires introduction of a large gap into the HSV-2 sequence right in the middle of the lysine-rich region (Figs. 4 and 5). Also, even though the four sequences available for HSV-1 forms of gB are highly conserved, almost all the sequence variation seen is at the N-terminus, as indicated in Fig. 5. These sequence comparisons are provocative and will guide efforts, at least initially, to explore the specificity and affinity of interactions between gB/gC and heparan sulfate, with attention focused on the sequence requirements of the glycoproteins as well as the heparan sulfate.

HSV and other herpesviruses

Members of the large herpesvirus family have been divided into three sub-families (alpha, beta, and gamma) on the basis of biological properties. HSV is the prototype of the alpha sub-family.

Results obtained with two other members of the alpha sub-family (pseudorabies virus [PRV] of pigs and bovine herpesvirus type 1 [BHV-1]) reveal striking similarity

```
                                                pl pa   t    K  n
gB-1 (17)    apsspgtp---gvaaatqaanggpatpappapgapptgdpKpKKnRKpKp - 77
             ::..: .:   :  ::: :::::::.  :: :    :   : :  . ::
gB-2 (333)   apaapaapRasggvaatvanggpasRpppvpspattKaRKRKtKKppKRp - 72

             p
             t
gB-1 (17)    pKppRpagdnatvaaghatlRehlRdiKaentdanfyvcppptgapvvqf -127
             :    :   ::::::::::: :::.:: ::.::.::::::::::::::::
gB-2 (333)   peatpppdanatvaaghatlRahlReiKvenadaqfyvcppptgapvvqf -122

---------------------------------------------------------------------

gC-1 (17)    mapgRvglavvlwsllwlgagvsggsetastgptitagavknaseaptsg - 50
             :: ::::::: :: :::.:  :        :: : ::: :   ::: :  :
gC-2 (333)   malgRvglavglwgllwvgvvvvlan--aspgRtitvgpRgnasnaapsa - 48

gC-1 (17)    spgsaaspevtptstpnpnnvtqnKttptepaspptttpKptstpKsppts -100
             ::  :..:  :::  :.:   :  :.. . :: ::
gC-2 (333)   spRnasapRttptp-pqpRKatKsKastaKpappp--------------- - 82

gC-1 (17)    tpdpKpKnnttpaKsgRptKppgpvwcdRRdplaRygsRvqiRcRfRnst -150
                          :.: :     :: : : ::::::::::::::: :::
gC-2 (333)   -------------KtgppKtssepvRcnRhdplaRygsRvqiRcRfpnst -119

gC-1 (17)    RmefRlqiwRysmgpsppiapapdleevltnitappggllvydsapnltd -200
             : : :::::::.      :  :: ::::. :..::::: :::::::: ::
gC-2 (333)   RtesRlqiwRyatatdaeigtapsleevmvnvsappggqlvydsapnRtd -169
```

Fig. 5. Amino acid sequences near the N-termini of gB and gC. Numbering is from the first amino acid of the translation product including the signal sequence. For gB the first amino acid shown is the beginning of the mature protein after cleavage of the signal sequence. Alignments of the HSV-1(strain 17) and HSV-2(strain 333) forms of each protein (e.g., gB-1 and gB-2) were done using the PALIGN program in the PCGENE suite. The double dots indicate sequence identity and the single dots, sequence similarity; the dashes represent gaps introduced to maximize alignment. Analysis of these sequences and others in Version 18 of the Swiss Protein Database revealed that, in the regions shown here, gB-2 sequences for two strains were identical as were gC-1 sequences and gC-2 sequences for two strains each. gB-1 sequences for four strains revealed the sequence variations indicated by the substitutions shown above the line of strain 17 sequence. The basic amino acids are shown in underlined capital letters.

with HSV in requirements for binding to cells. For both PRV and BHV-1, (i) heparin inhibits the binding of virus to cells and infection; (ii) treatment of cells with heparitinase or heparinase destroys receptors for virus and renders the cells resistant to infection; and (iii) a viral glycoprotein homologous to HSV gC has heparin-binding activity and mediates the attachment of virus to cells (Schreurs *et al.,* 1988; Mettenleiter, 1989; Zuckermann *et al.,* 1989; Mettenleiter *et al.,* 1990; Sawitzky *et al.,* 1990; Zsak *et al.,* 1991; Okazaki *et al.,* 1991). In the case of BHV-1, the viral glycoprotein homologous to HSV gB also has heparin-binding activity (Okazaki *et al.,* 1991). Similar results were reported for PRV gII (gB) by one group (Sawitzky *et al.,* 1990) but not by another (Mettenleiter *et al.,* 1990).

Although the heparin-binding glycoproteins specified by HSV, PRV and BHV-1 have diverged considerably in amino acid sequence, common features of the proteins are discernable. For example, HSV gC and the homologous glycoproteins of PRV and BHV-1 all exhibit a very hydrophilic region, with a run of basic amino acids, near their N-termini. These herpesvirus glycoproteins provide an evolutionarily related, but divergent, set of proteins that should be very useful for studies of the specificity and physiological significance of affinity for heparin and related glycosaminoglycans.

SUMMARY

Our current incomplete picture of the earliest events in HSV infection may be summarized as follows. The initial interaction of virus with cells is the binding of virion gC to heparan sulfate moieties of cell surface proteoglycans. Stable binding of virus to cells may require the interaction of other virion glycoproteins with other cell surface receptors as well (including the interaction of gB with heparan sulfate). Penetration of virus into the cell is mediated by fusion of the virion envelope with the cell plasma membrane. Events leading up to this fusion require the action of at least three viral glycoproteins (gB, gD and gH), one or more of which may interact with specific cell surface components. It seems likely that binding of gB to cell surface heparan sulfate may occur and may be important in the activation of some event required for virus penetration.

Heparan sulfate is present not only as a constituent of cell surface proteoglycans but also as a component of the extracellular matrix and basement membranes in organized tissues. In addition, body fluids contain both heparin and heparin-binding proteins, either of which can prevent the binding of HSV to cells (WuDunn and Spear, 1989). As a consequence, the spread of HSV infection is probably influenced, not only by immune responses to the virus, but also by the probability that virus will be entrapped or inhibited from binding to cells by extracellular forms of heparin or heparan sulfate.

ACKNOWLEDGMENTS

We thank J.D. Esko for providing the CHO cells and mutants and for advice. Studies described here that were done in the laboratory of P.G. Spear were supported by grants from the American Cancer Society and the National Cancer Institute (CA 21776). T.I. Koshy is supported by fellowship F32 AI08526 from the National Institutes of Health.

REFERENCES

Addison, C., Rixon, F. J., Palfreyman, J. W., O'Hara, M., and Preston, V. G. (1984). Characterisation of a herpes simplex virus type 1 mutant which has a temperature-sensitive defect in penetration of cells and assembly of capsids. *Virology* **138**, 246-259.

Air, G. M. and Laver, W. G. (1989). The neuraminidase of influenza virus. *Proteins* **6**, 341-356.

Baird, A., Florkiewicz, R. Z., Maher, P. A., Kaner, R. J., and Hajjar, D. P. (1990). Mediation of virion penetration into vascular cells by association of basic fibroblast growth factor with herpes simplex virus type 1. *Nature* **348**, 344-346.

Bame, K. J., Lidholt, K., Lindahl, U., and Esko, J. D. (1991). Biosynthesis of heparan sulfate: Coordination of polymer-modification reactions in a chinese hamster ovary cell mutant defective in N-sulfotransferase. *J. Biol. Chem.* **266**, 10287-10293.

Bame, K. J. and Esko, J. D. (1989). Undersulfated heparan sulfate in a Chinese hamster ovary cell mutant defective in heparan sulfate N-sulfotransferase. *J. Biol. Chem.* **264**, 8059-8065.

Bzik, D. J., Fox, B. A., DeLuca, N. A., and Person, S. (1984). Nucleotide sequence specifying the glycoprotein gene, gB, of herpes simplex virus type 1. *Virology* **133**, 301-314.

Bzik, D. J., Debroy, C., Fox, B. A., Pederson, N. E., and Person, S. (1986). The nucleotide sequence of the gB glycoprotein gene of HSV-2 and comparison with the corresponding gene of HSV-1. *Virology* **155**, 322-333.

Cai, W., Gu, B., and Person, S. (1988). Role of glycoprotein B of herpes simplex virus type 1 in viral entry and cell fusion. *J. Virol.* **62**, 2596-2604.

Claesson-Welsh, L. and Spear, P. G. (1986). Oligomerization of herpes simplex virus glycoprotein B. *J. Virol.* **60**, 803-806.

Claesson-Welsh, L. and Spear, P. G. (1987). Amino-terminal sequence, synthesis, and membrane insertion of glycoprotein B of herpes simplex virus type 1. *J. Virol.* **61**, 1-7.

Corey, L. and Spear, P. G. (1988). Infections with herpes simplex viruses. *N. Engl. J. Med.* **314**, 686-689;-749-757.

Dall'Olio, F., Malagolini, N., Speziali, V., Campadelli-Fiume, G., and Serafini-Cessi, F. (1985). Sialylated oligosaccharides O-glycosidically linked to glycoprotein C from herpes simplex virus type 1. *J. Virol.* **56**, 127-134.

Desai, P. J., Schaffer, P. A., and Minson, A. C. (1988). Excretion of non-infectious virus particles lacking glycoprotein H by a temperature-sensitive mutant of herpes simplex virus type 1: evidence that gH is essential for virion infectivity. *J. Gen. Virol.* **69**, 1147-1156.

Dowbenko, D. J. and Lasky, L. A. (1984). Extensive homology between the herpes simplex virus type 2 glycoprotein F gene and the herpes simplex virus type 1 glycoprotein C gene. *J. Virol.* **52**, 154-163.

Draper, K. G., Costa, R. H., Lee, G. T. -Y., Spear, P. G., and Wagner, E. K. (1984). Molecular basis of the glycoprotein C-negative phenotype of herpes simplex virus type 1 macroplaque strain. *J. Virol.* **51**, 578-585.

Esko, J. D., Stewart, T. E., and Taylor, W. H. (1985). Animal cell mutants defective in glycosaminoglycan biosynthesis. *Proc. Natl. Acad. Sci. U. S. A.* **82**, 3197-3201.

Esko, J. D., Weinke, J. L., Taylor, W. H., Ekborg, G., Roden, L., Anantharamaiah, G., and Gawish, A. (1987). Inhibition of chondroitin and heparan sulfate biosynthesis in Chinese hamster ovary cell mutants defective in galactosyltransferase I. *J. Biol. Chem.* **262**, 12189-12195.

Esko, J. D., Rostand, K. S., and Weinke, J. L. (1988). Tumor formation dependent on proteoglycan biosynthesis. *Science* **241**, 1092-1096.

Frink, R. J., Eisenberg, R. J., Cohen, G. H., and Wagner, E. K. (1983). Detailed analysis of the portion of the herpes simplex virus type 1 genome encoding glycoprotein C. *J. Virol.* **45**, 643-647.

Fuller, A. O., Santos, R. E., and Spear, P. G. (1989). Neutralizing antibodies specific for glycoprotein H of herpes simplex virus permit viral attachment to cells but prevent penetration. *J. Virol.* **63**, 3435-3443.

Fuller, A. O. and Spear, P. G. (1987). Anti-glycoprotein D antibodies that permit adsorption but block infection by herpes simplex virus 1 prevent virion-cell fusion at the cell surface. *Proc. Natl. Acad. Sci. U. S. A.* **84**, 5454-5458.

Galliher, P. M., Cooney, C. L., Langer, R., and Linhardt, R. J. (1981). Heparinase production by Flavobacterium heparinum. *Appl. Environ. Microbiol.* **41**, 360-365.

Herold, B. C., WuDunn, D., Soltys, N., and Spear, P. G. (1991). Glycoprotein C of herpes simplex virus type 1 plays a principal role in the adsorption of virus to cells and in infectivity. *J. Virol.* **65**, 1090-1098.

Higgins, D. G. and Sharp, P. M. (1988). CLUSTAL: a package for performing multiple sequence alignment on a microcomputer. *Gene* **73**, 237-244.

Higgins, D. G. and Sharp, P. M. (1989). Fast and sensitive multiple sequence alignments on a microcomputer. *CABIOS* **5**, 151-153.

Highlander, S. L., Sutherland, S. L., Gage, P. J., Johnson, D. C., Levine, M., and Glorioso, J. C. (1987). Neutralizing monoclonal antibodies specific for herpes simplex virus glycoprotein D inhibit virus penetration. *J. Virol.* **61**, 3356-3364.

Highlander, S. L., Cai, W., Person, S., Levine, M., and Glorioso, J. C. (1988). Monoclonal antibodies define a domain on herpes simplex virus glycoprotein B involved in virus penetration. *J. Virol.* **62**, 1881-1888.

Jaques, L. B., Bruce-Mitford, M., and Ricker, A. G. (1974). The metachromatic activity of heparin. *Can. Rev. Biol.* **6**, 740-754.

Johnson, D. C., Burke, R. L., and Gregory, T. (1990). Soluble forms of herpes simplex virus glycoprotein D bind to a limited number of cell surface receptors and inhibit virus entry into cells. *J. Virol.* **64**, 2569-2576.

Johnson, D. C. and Spear, P. G. (1983). O-linked oligosaccharides are acquired by herpes simplex virus glycoproteins in the Golgi apparatus. *Cell* **32**, 987-997.

Kaner, R. J., Baird, A., Mansukhani, A., Basilico, C., Summers, B. D., Florkiewicz, R. Z., and Hajjar, D. P. (1990). Fibroblast growth factor receptor is a portal of cellular entry for herpes simplex virus type 1. *Science* **248**, 1410-1413.

Lidholt, K., Weinke, J. L., Kiser, C. S., Lugemwa, F. N., Bame, K. J., Lindahl, U., and Esko, J. D. (1991). Chinese hamster ovary cell mutants defective in heparan sulfate synthesis. *Submitted*

Ligas, M. W. and Johnson, D. C. (1988). A herpes simplex virus mutant in which glycoprotein D sequences are replaced by ß-galactosidase sequences binds to but is unable to penetrate into cells. *J. Virol.* **62**, 1486-1494.

Little, S. P., Joffre, J. T., Courtney, R. J., and Schaffer, P. A. (1981). A virion-associated glycoprotein essential for infectivity of herpes simplex virus type 1. *Virology* **115**, 149-160.

McGeoch, D. J., Dalrymple, M. A., Davison, A. J., Dolan, A., Frame, M. C., McNab, D., Perry, L. J., Scott, J. E., and Taylor, P. (1988). The complete DNA sequence of the long unique region in the genome of herpes simplex virus type 1. *J. Gen. Virol.* **69**, 1531-1574.

McGeoch, D. J.(1989). The genomes of the human herpesviruses: contents,relationships, and evolution. *Annu. Rev. Microbiol.* **43**, 235-265.

Mettenleiter, T. C.(1989). Glycoprotein gIII deletion mutants of pseudorabies virus are impaired in virus entry. *Virology* **171**, 623-625.

Mettenleiter, T. C., Zsak, L., Zuckermann, F., Sugg, N., Kern, H., and Ben-Porat, T. (1990). Interaction of glycoprotein gIII with a cellular heparin-like substance mediates adsorption of pseudorabies virus. *J. Virol.* **64**, 278-286.

Mirda, D. P., Navarro, D., Paz, P., Lee, P. L., Pereira, L., and Williams, L. T. (1991). The fibroblast growth factor receptor is not required for herpes simplex virus type 1 infection. *Submitted*

Myers, E. W. and Miller, W. (1988). Optimal alignments in linear space. *CABIOS* **4**, 11-17.

Nahmias, A. J. and Kibrick, S. (1964). Inhibitory effect of heparin on herpes simplex virus. *J. Bacteriol.* **87**, 1060-1066.

Okazaki, K., Matsuzaki, T., Sugahara, Y., Okada, J., Hasebe, M., Iwamura, Y., Ohnishi, M., Kanno, T., Shimizu, M., and Honda, E. (1991). BHV-1 adsorption is mediated by the interaction of glycoprotein gIII with heparinlike moiety on the cell surface. *Virology* **181**, 666-670.

Olofsson, S., Blomberg, J., and Lycke, E. (1981). O-glycosidic carbohydrate-peptide linkages of herpes simplex virus glycoproteins. *Arch. Virol.* **70**, 321-329.

Olofsson, S., Sjoblom, I., Lundstrom, M., Jeansson, S., and Lycke, E. (1983). Glycoprotein C of herpes simplex virus type 1: characterization of O-linked oligosaccharides. *J. Gen. Virol.* **64**, 2735-2747.

Pellett, P. E., Kousoulas, K. G., Pereira, L., and Roizman, B. (1985). Anatomy of the herpes simplex virus 1 strain F glycoprotein B gene: primary sequence and predicted protein structure of the wild type and of monoclonal antibody-resistant mutants. *J. Virol.* **53**, 243-253.

Rapraeger, A., Kubota, Y., and Olwin, B. B. (1991). Requirement of heparan sulfate for bFGF-mediated fibroblast growth and myoblast differentiation. *Science* **252**, 1705-1708.

Roizman, B. and Sears, A. (1990). *Virology* (Fields, B. N., Ed.) 2nd edition. Raven Press,Ltd., New York. 1795-1841.

Sarmiento, M., Haffey, M., and Spear, P. G. (1979). Membrane proteins specified by herpes simplex viruses. III. Role of glycoprotein VP7 (B_2) in virion infectivity. *J. Virol.* **29**, 1149-1158.

Sarmiento, M. and Spear, P. G. (1979). Membrane proteins specified by herpes simplex viruses. IV. Conformation of the virion glycoprotein designated VP7(B_2). *J. Virol.* **29**, 1159-1167.

Sawitzky, D., Hampl, H., and Habermehl, K. -O. (1990). Comparison of heparin-sensitive attachment of pseudorabies virus (PRV) and herpes simplex virus type 1 and identification of heparin-binding PRV glycoproteins. *J. Gen. Virol.* **71**, 1221-1225.

Schreurs, C., Mettenleiter, T. C., Zuckermann, F., Sugg, N., and Ben-Porat, T. (1988). Glycoprotein gIII of pseudorabies virus is multifunctional. *J. Virol.* **62**, 2251-2257.

Shieh, M. -T., WuDunn, D., Montgomery, R. I., Esko, J. D., and Spear, P. G. (1991). Cell surface receptors for herpes simplex virus are heparan sulfate proteoglycans. *Submitted*

Shieh, M. -T. and Spear, P. G. (1991). Fibroblast growth factor receptor: Does it have a role in the binding of herpes simplex virus? *Science* **253**, 208-210.

Stannard, L. M., Fuller, A. O., and Spear, P. G. (1987). Herpes simplex virus glycoproteins associated with different morphological entities projecting from the virion envelope. *J. Gen. Virol.* **68**, 715-725.

Swain, M. A., Peet, R. W., and Galloway, D. A. (1985). Characterization of the gene encoding herpes simplex virus type 2 glycoprotein C and comparison with the type 1 counterpart. *J. Virol.* **53**, 561-569.

Takemoto, K. K. and Fabisch, P. (1964). Inhibition of herpes simplex virus by natural and synthetic acid polysaccharides. *Proc. Soc. Exp. Biol. Med.* **116**, 140-144.

Vaheri, A.(1964). Heparin and related polyionic substances as virus inhibitors. *Acta Path. Microbiol. Scand. Suppl.* **171**, 7-97.

Vahlne, A., Svennerholm, B., and Lycke, E. (1979). Evidence for herpes simplex virus type-selective receptors on cellular plasma membranes. *J. Gen. Virol.* **44**, 217-225.

Vahlne, A., Svennerholm, B., Sandberg, M., Hamberger, A., and Lycke, E. (1980). Differences in attachment between herpes simplex type 1 and type 2 viruses to neurons and glial cells. *Infect. Immun.* **28**, 675-680.

WuDunn, D. and Spear, P. G. (1989). Initial interaction of herpes simplex virus with cells is binding to heparan sulfate. *J. Virol.* **63**, 52-58.

Yayon, A., Klagsbrun, M., Esko, J. D., Leder, P., and Ornitz, D. M. (1991). Cell surface, heparin-like molecules are required for binding of basic fibroblast growth factor to its high affinity receptor. *Cell* **64**, 841-848.

Zsak, L., Sugg, N., Ben-Porat, T., Robbins, A. K., Whealy, M. E., and Enquist, L. W. (1991). The gIII glycoprotein of pseudorabies virus is involved in two distinct steps of virus attachment. *J. Virol.* **65**, 4317-4324.

Zuckermann, F., Zsak, L., Reilly, L., Sugg, N., and Ben-Porat, T. (1989). Early interactions of pseudorabies virus with host cells: functions of glycoprotein gIII. *J. Virol.* **63**, 3323-3329.

CONTROL OF ANGIOGENESIS BY HEPARIN AND OTHER SULFATED POLYSACCHARIDES

Judah Folkman and Yuen Shing

Harvard Medical School and Children's Hospital
Boston
Massachusetts, U.S.A.

Introduction

Angiogenesis, the growth of new capillary blood vessels, plays an important role in normal development and in physiologic functions such as occur in the female reproductive system (Koos and LeMaire, 1983). Angiogenesis is also essential to the repair of wounds, peptic ulcers and myocardial infarctions (Hunt et. al, 1984). In these physiologic and repair conditions, angiogenesis is regulated and switched on and off at predictable times. In a variety of disease processes however, angiogenesis is unabated and unregulated. Tumor growth and metastasis are angiogenesis-dependent (Folkman, 1990; Folkman 1991). Continuous induction of neovascularization is necessary for progressive tumor growth and increasing neovascularization of certain tumors correlates with increasing metastatic potential (Weidner et. al, 1991). Many non-neoplastic diseases are also dominated by unregulated angiogenesis, for example, diabetic retinopathy and hemangioma (Folkman, 1987).

Role of Heparin in Endothelial Cell Growth and Angiogenesis

During the elucidation of the biologic and biochemical properties of angiogenesis over the past two decades, it is surprising how often heparin, heparan sulfate and their related polysaccharides are involved in the angiogenic process (Folkman and Ingber, 1989; Folkman, et al, 1989). At first, these associations were phenomenological, but now mechanistic roles have been uncovered for heparin and related molecules in the regulation of neovascularization.

The earliest indication that heparin was involved in the regulation of angiogenesis was the finding that mast cells accumulate at the site of tumor angiogenesis before capillary ingrowth (Kessler et. al, 1976). Conditioned medium from mast cells was solely responsible for this activity (Azizkhan et. al, 1980). Protamine and platelet factor 4, both of which bind and inactivate heparin, inhibited angiogenesis in the chick embryo and the rabbit cornea (Taylor et. al, 1982). Tumor growth in animals was inhibited by protamine administered systemically and by platelet factor 4 when administered intralesionally (Maione et. al, 1990, 1991). Heparin converted certain steroids such as cortisone or hydrocortisone to angiogenesis inhibitors (Foldman, et. al, 1983) and was found to potentiate the low level of antiangiogenic activity of other steroids such as tetrahydrocortisol (Crum et. al, 1985). Heparin administered orally was also effective as a potentiator of angiostatic steroids, and was degraded into small heparin fragments which appeared in plasma. These fragments are the size of octasaccharides or smaller (Larsen et. al, 1986). While anti-thrombin activity was not enhanced, anti-factor Xa activity was. Furthermore, an arylsulfatase inhibitor that appears to augment endogenous heparin-like molecules by slowing their desulfation, could be substituted for heparin in the potentiation of steroids as angiogenesis inhibitors (Chen et. al, 1988). On the basis of this finding, a family of "angiostatic" steroids was discovered which inhibit angiogenesis in the absence of glucocorticoid or mineralocorticoid activity. Angiostatic steroids are now being studied in many laboratories (Sakamoto et. al, 1986; Sakamoto et. al, 1987; Wilks et. al, 1991; Maragoudakis et. al, 1989).

Heparin also played a key role in the purification of the first known angiogenic factor. The first complete purification of an angiogenic factor was based on heparin-affinity chromatography (Shing et. al, 1984; Klagsbrun et. al, 1985). An extract of chondrosarcoma yielded an 18kD protein after a 500,000-fold purification to homogeneity was achieved by two cycles through a heparin-Sepharose column. This result capped many years of frustrating attempts to purify a tumor-derived angiogenic factor. When the protein was subsequently sequenced, it was found to be basic fibroblast growth factor (bFGF) (Esch et. al, 1985). Acidic FGF (aFGF) was also purified by heparin affinity (Maciag et. al, 1984; Thomas et. al, 1984). Because aFGF also binds to heparin, the FGFs were among the first molecules to be classified as heparin-binding growth factors (HBGF) (Lobb, et. al, 1986). Subsequently, six other angiogenic factors have been identified and completely sequenced (Folkman and Klagsbrun, 1987). Nevertheless, bFGF is one of the most potent of these angiogenic molecules.

It soon became apparent that the heparin-affinity of the fibroblast growth factors was more than just a useful technique for purification. The affinity of the FGFs for heparin regulates the function of this growth factor in vivo. For example, heparin potentiates the mitogenic effect of acidic FGF on vascular endothelial cells *in vitro* (Thornton et. al, 1983), protects fibroblast growth factors from inactivation (Gospodarowicz and Cheng, 1986; Rosengurt et. al, 1988; Saksela et. al, 1988) and increases the binding of aFGF to its receptor (Schreiber et. al, 1985). Heparin binds to endothelial surfaces (Barzu et. al, 1985; Gajdusek, 1986), and releases cell-associated aFGF. Heparan sulfate produced by vascular endothelial cells contains heparin sequences (Nader et. al, 1987). Endothelial cells also metabolize heparin. They internalize exogenous heparin and release its low molecular weight degradation products (Vannuchi et. al, 1986). Specific sequences of heparan sulfate have been found in the nucleus (Fedarko and Conrad, 1986) and bFGF can enter the nucleolus where it stimulates the transcription of ribosomal genes in endothelial cells undergoing the transition from G_0 to G_1 growth state (Bouche et. al, 1987). Based on these findings, it was postulated that a "heparin shuttle" could chaperone FGF into the nucleus (Folkman and Ingber, 1989). The heparin affinity of the fibroblast growth factors also appears to be the basis for their storage in extracellular matrix (Vlodavsky et. al, 1987; Folkman et. al, 1988; Vlodavsky et. al, 1990; Courty, 1985), where they are bound to heparan sulfate and can be released by heparin or its substituent fragments, as well as by heparinase, but not by chondroitin sulfate or chondroitinase. Of interest is that when bFGF is injected intravenously, it disappears rapidly from the circulation with a half-life of minutes. However, when intravenous heparin is injected up to 8 hours later, the iodinated bFGF reappears in the circulation (Thompson, et. al, 1990).

In addition to conventional high-affinity cell surface receptors for FGF, many cells also have low-affinity receptors which are thought to be heparin-like molecules (Moscatelli, et. al, 1987). Syndecan, a cell surface proteoglycan consisting of chains of heparan sulfate, heparin and chondroitin sulfate linked to a 31 kD membrane protein, may be a candidate for these low-affinity receptors (Bernfield and Sanderson, 1990; Moscatelli, et. al, 1988). Recent evidence suggests that binding of fibroblast growth factors to the heparin-like low-affinity cell surface receptors is a prerequisite of binding to the high affinity receptor (Yayon et. al, 1991).

Heparin and its fragments potentiate angiogenesis initiated by other angiogenic mechanisms (Folkman et. al, 1983; Norrby, 1991). This potentiation may be mediated by the ability of heparin to mobilize FGF from extracellular matrix. The potentiation of

endothelial cell proliferation has also been demonstrated with small fragments of heparin (Sudhalter, et al, 1989). Tumor angiogenesis occurs at a slower rate in mast cell deficient mice (Dethlefsen, et al, 1990). However, one report suggests that under certain conditions, heparin by itself may inhibit angiogenesis (Jakobson et. al, 1989).

Other Sulfated Polysaccharides Which Regulate Angiogenesis

Certain sulfated polysaccharides can substitute for heparin as a regulator of angiogenesis. Because antiangiogenesis by heparin and steroids was critically dependent upon an optimum ratio of these two compounds, Paul Weisz hypothesized that heparin could form a hydrophobic cage around the steroid, sulfates being on the outside of the complex. This heparin-steroid complex could then gain facilitated entry into vascular endothelial cells. This idea led to the finding that certain sulfated cyclodextrins when administered with angiostatic steroids inhibited angiogenesis up to 1000 times more effectively than heparin (Folkman et. al, 1989). Beta-cyclodextrin tetradecasulfate was the most effective of the sulfated cyclodextrins tested. Non-sulfated cyclodextrin was without effect. Complexes of beta-cyclodextrin tetradecasulfate and hydrocortisone potently inhibit endothelial growth and migration *in vitro* (Pereles, et. al, 1989) and inhibit corneal neovascularization in the rabbit eye (Li et. al, 1991). Beta-cyclodextrin tetradecasulfate diffuses efficiently across the cornea into the aqueous fluid of the anterior chamber and this may facilitate the delivery of angiostatic steroids complexed with it to new vessels in the cornea (Rupnick et. al, 1990).

Beta-cyclodextrin tetradecasulfate has also been used to purify fibroblast growth factor because of the high affinity of this polysaccharide for the peptide (Shing et. al, 1990). Non-sulfated cyclodextrins are not effective substitutes for heparin. Certain peptidoglycans can also potentiate angiostatic steroids without heparin, and may inhibit angiogenesis alone (Inoue et. al, 1988; Tanaka et. al, 1991). Suramin (Wilks et. al, 1991) and pentosan polysulfate also inhibit angiogenesis, although they are more toxic than other non-heparin compounds.

Sucrose Octasulfate and its Aluminum Salt

The field of angiogenesis research has now evolved to the extent that laboratory findings are moving toward clinical trials, some of which are already underway. For example in the diagnostic area, neovascularization has been quantitated in biopsies of breast cancer and sensitively predicts metastatic risk (Weidner et. al, 1991). Stimulation of angiogenesis by application of topical bFGF to chronic wounds accelerates healing (Robson, 1991 personal communication). Inhibition of angiogenesis by alpha-interferon has proved life-saving in the treatment of severe hemangiomas in babies (White et. al, 1989; Orchard et. al, 1989; Folkman, 1989; Ezekowitz, Mulliken, and Folkman, 1991 submitted for publication).

Other applications of angiogenesis research are being prepared for future clinical trial. One of these is an attempt to accelerate the healing of experimental duodenal ulcers. Peptic ulcers in the gastrointestinal tract have many similarities to chronic wounds. The ulcer base contains exposed collagen covered by an inflammatory exudate which includes macrophages. Neovascularization is usually sparse and healing is slow.

We recently showed that oral administration of an acid stable bFGF-mutein (Seno et. al., 1988), induced a 10-fold increase of angiogenesis in the bed of duodenal ulcers in rats (Szabo, 1974) and significantly increased the rate of healing of these ulcers compared to untreated controls or to cimetidine, the most widely used H2 blocker (Szabo, et. al, 1989; Folkman, et. al., 1990; Foldman, et al, 1991). bFGF was not absorbed into the circulation. Gastric acid and pepsin were not decreased by this treatment. Because the rationale for almost all successful anti-ulcer therapy has been the reduction of gastric acid, it was surprising to us that duodenal ulcers could be rapidly healed in the face of high acid levels.

This apparent paradox led us to examine sucralfate (Carafate^{R}, the aluminum salt of sucrose octasulfate), a widely used anti-ulcer drug that also does not reduce gastric acid and for which the precise mechanism of action is unknown, Figure 1.

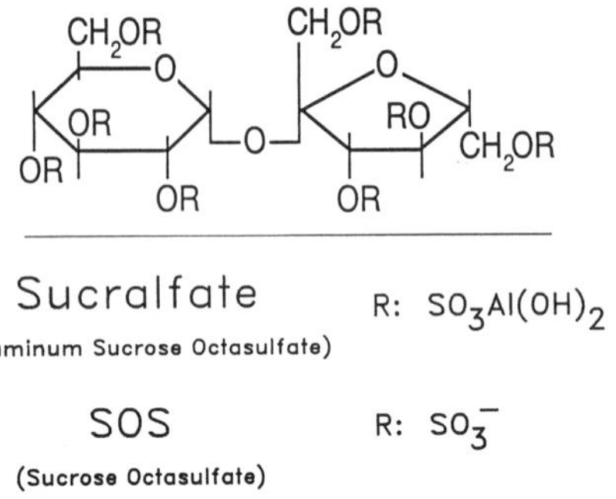

Figure 1. Structure of sucralfate. Sucralfate (aluminum sucrose octasulfate) is poorly soluble compared with its soluble counterpart, potassium sucrose octasulfate (SOS).

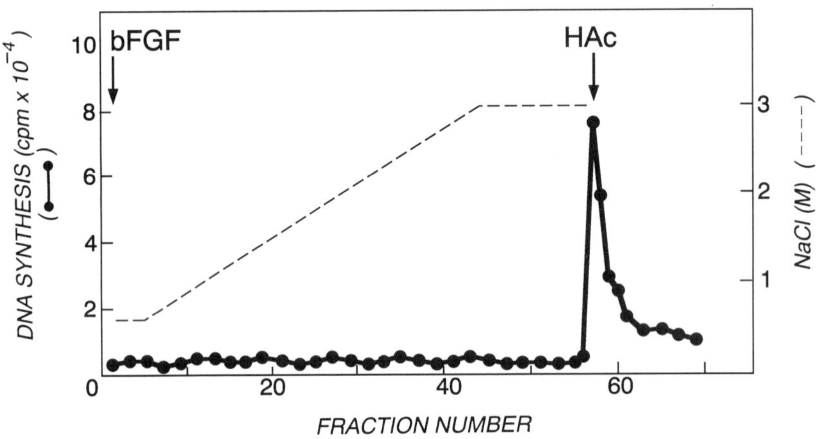

Figure 2. Elution of bFGF from a sucralfate-Sepharose column by acetic acid. A sucralfate-Sepharose column was prepared by mixing 0.1g of sucralfate and 0.8 ml (bed volume) of Sepharose CL-6B (Pharmacia) in Tris buffer with 0.6 M NaCl; 2ug of bFGF in 0.2ml of saline containing 0.1mg of BSA was loaded onto the sulcralfate column. The column was rinsed with 10ml Tris buffer and 0.6 M NaCl and eluted consecutively at a flow rate of 15ml/hr: (1) A gradient of NaCl (80ml) from 0.6 M to 3.0 M, (2) 20ml of 3 NaCl, and (3) 30ml of 1.5 M acetic acid (pH 2.2). Fractions of 2ml were collected. Aliquots from each fraction were diluted, neutralized with Tris buffer, and analyzed for growth factor activity. The bFGF could not be eluted from a sucralfate-Sepharose column even with 3.0 M NaCl. However, biologically active bFGF was eluted by 1.5 M acetic acid.

The drug coats the ulcer bed. Because this disaccharide is highly sulfated, we tested its affinity for basic fibroblast growth factor and found it to be significantly greater than the affinity of heparin for bFGF. In fact, while 1.5 to 1.8 M NaCl efficiently elutes bFGF from heparin-Sepharose, even 3 M NaCl cannot elute bFGF from a column of sucralfate-Sepharose, Figure 2.

Acetic acid (1.5 M, pH 2.2) was required to elute bFGF from sucralfate-Sepharose. Despite these acidic conditions, the eluted bFGF was biologically active. Furthermore, sucrose octasulfate (the soluble form of sucralfate), protected bFGF from inactivation by acetic acid for up to 1 hour, Figure 3.

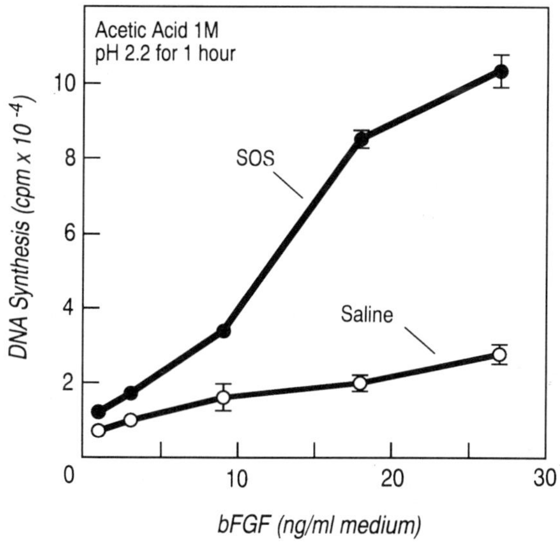

Figure 3. Stabilization of bFGF by potassium sucrose octasulfate under acidic conditions. The bFGF (0.25 ug/60 uL) was incubated at pH 2.2 with 1 M acetic acid in the presence of (•) and absence of (0) 2mg potassium sucrose octasulfate at 37 C for 1 hour. At the end of incubation, bFGF samples were diluted and analyzed for their ability to stimulate DNA synthesis on 3T3 fibroblasts. (from Folkman et. al, 1991 with permission of the Publisher).

These findings suggested that the anti-ulcer activity of sucralfate could possibly depend upon its ability to protect endogenous bFGF in the stomach from degradation and inactivation (Folkman, et. al, 1991). This hypothesis proposed that naturally occurring endogenous bFGF is produced by the gastric and duodenal mucosa, and in the ulcer bed possibly by macrophages. Endogenous bFGF may be necessary for ulcer healing, however it is highly susceptible to continuous inactivation by gastric acid. Because there was no previous report of the presence of "endogenous" bFGF in gastrointestinal mucosa, we extracted rat gastric and duodenal mucosa with sucrose octasulfate. Glandular mucosa contained biologically active bFGF as proved by heparin-Sepharose chromatography, Western blot analysis, mitogenic activity for fibroblasts and endothelial cells, and immunohistochemical localization (Folkman et. al, 1991). Human duodenum also contained bFGF. Implants of sucralfate in the rabbit cornea and in the chick embryo extracted bFGF from extracellular matrix and stimulated angiogenesis. When sucralfate was allowed to complex to bFGF before implantation into the cornea, intense and prolonged angiogenesis was induced. Furthermore, when cimetidine was fed to rats with a gastric pH of 3.4 - 4.3, stomach pH rose to 6.4 - 7.2 and this was followed by a significant elevation of bFGF levels in the ulcer bed (Folkman, et al, 1991). Taken together, these findings are consistent with the hypothesis that endogenous bFGF while present in the mucosa and in the ulcer bed, undergoes continual degradation by gastric acid. Conventional anti-ulcer therapies, because they neutralize gastric acid (e.g., antacids), reduce its production (e.g., H2 blockers), or protect bFGF from acid degradation (e.g., sucralfate), may all operate through a common mechanism of increasing the availability of "endogenous" bFGF for ulcer healing. In this conceptual framework, conventional anti-ulcer therapies may be FGF-dependent. Oral administration of bFGF made acid stable by site-specific mutagenesis, or acid-resistant by mixing with sucralfate, could be consikdered a form of replacement therapy for the treatment of duodenal ulceration.

This new concept of the mechanism of duodenal ulcer formation, arose directly from the investigation of the affinity of an angiogenic peptide for a sulfated polysaccharide. Several questions for future research are generated by this new concept. For example, we have recently found that bFGF is present in gastric mucous. (i) Does gastric mucous contain sulfated polysaccharides which may normally protect bFGF from degradation? (ii) Do ulcers arise in some patients because these putative protective polysaccharides are diminished? (iii) Could certain diets be ulcerogenic because they contain excessive concentrations of highly soluble sulfated polysaccharides which could extract bFGF from the mucosa? (iv) Can colonic ulceration be healed by intracolonic bFGF? (Satoh et. al, 1990).

Other Heparin-Binding Growth Factors and Growth Inhibitors

It is now clear that other growth factors have an affinity for heparin. Vascular endothelial growth factor (VEGF) (Ferrara et. al, 1989), is an angiogenic endothelial mitogen which was recently purified by heparin-affinity chromatography. Thrombospondin also binds to heparin. It is an angiogenesis inhibitor which inhibits endothelial cell proliferation and appears to be under the control of a tumor suppressor gene in normal cells (Buock et. al, 1990). It is down-regulated when hamster cells are transformed to neoplastic cells by carcinogenesis.

Summary

Heparin and its related polysaccharides are revealed to have important new functions as regulators of blood vessel growth and regression. This regulatory activity may be explained in part by at least five mechanisms in which heparin and heparan sulfate interact with peptide growth factors: (1) Heparin and heparan sulfate have a high affinity for angiogenic growth factors such as the fibroblast growth factors and VEGF, as well as for angiogenic inhibitors such as thrombospondin and platelet factor 4. (2) Heparin and its related polysaccharides stabilize bFGF and other growth factors. (3) FGFs and thrombospondin are stored in the extracellular matrix bound to heparan sulfate; fragments of heparin or heparan sulfate may act as natural chaperones to shuttle bFGF or other growth factors to different cellular compartments. (5) Heparin-like low-affinity receptors on the surface of endothelial cells (and other cells), prepare FGFs for binding to their specific high affinity receptors; and (6) Heparin and its related polysaccharides potentiate angiostatic steroids.

It is likely that future investigations will uncover even more fundamental regulatory roles for heparin as well as for other polysaccharides in the normal function of growth factors, especially in the complex process of angiogenesis.

REFERENCES

Azizkhan, R., Azizkhan, J., Zetter, B., and Folkman, J. (1980), Mast cell heparin stimulates migration of capillary endothelial cells in vitro, Journal of Experimental Medicine, 152:931-944.

Barzu, T., Molho, P., Tobelem, G., Petitou, M., and Caen, J. (1985), Binding and endocytosis of heparin by human endothelial cells in culture, Biochem. Biophys. Acta, 845:193-203.

Bernfield, M., and Sanderson, R.D. (1990), Syndecan, a developmentally regulated cell surface proteoglycan that binds extracellular matrix and growth factors, Phil. Trans. R. Soc. London, 327:171-196.

Bouche, N., Gas, N., Prats, H., Baldin, V., Tauber, J., Teissie, J., and Armalric, F. (1987), Basic fibroblast growth factor enters the nucleolus and stimulates the transcription of ribosomal genes in ABAE cells undergoing Go-G1 transcription, Proc. Natl. Acad. Sci. USA, 84:6770-6774.

Bouck, N. (1990), Tumor angiogenesis: The role of oncogenes and tumor suppressor genes, Cancer Cells, 2:179-185.

Chen, N.T., Corey, E.J., and Folkman, J. (1988), Potentiation of angiostatic steroids by a synthetic inhibitor of arylsulfatase, Lab. Invest. 59:453-459.

Courty, J., Loret, C., Moenner, M., Chevallier, B., Lagente, O., Courtois, Y., and Barritault, D. (1985), Bovine retina contains three growth factor activities with different affinity to heparin: eye-derived growth factor I, II, III, Biochemie, 67:265-269.

Crum, R., Szabo, S., and Folkman, J. (1985), A new class of steroids inhibits angiogenesis in the presence of heparin or a heparin fragment, Science, 230:1375-1378.

Dethlefsen, S.M., Matsuura, N., and Zetter, B.R. (1990), Tumor growth and angiogenesis in wild type and mast cell deficient mice, FASEB Journal, 4: A623 (Abstract 2070).

Esch, F., Baird, A., Ling, N., Ueno, N., Hill, F., Denoroy, L., Klepper, R., Gospodarowicz, D., Bohlen, P., and Guillemin, R. (1985), Primary structure of bovine pituitary basic fibroblast growth factor (FGF) and comparison with the amino-terminal sequence of bovine acidic FGF, Proc. Natl. Acad. Sci. USA, 82:6507-6511.

Ezekowitz, R.A., Mulliken, J.B., and Foldman, J. (1991), Interferon alpha 2A therapy for alarming hemangiomas in infancy. Submitted for publication.

Fedarko, N., and Conrad, E. (1986), A unique heparan sulfate in the nuclei of hepatocytes: Structural changes with the growth state of the cells, J. Cell Biol., 102:587-599.

Ferrara, N., and Henzel, W.J. (1989), Pituitary follicular cells secrete a novel heparin-binding growth factor specific for vascular endothelial cells, Biochemical and Biophysical Research Communications, 161: 851-855.

Folkman, J., Taylor, S., and Spillberg, C. (1983), The role of heparin in angiogenesis, Development of the Vascular System, 132-149.

Folkman, J., and Klagsbrun, M. (1987), Angiogenic factors, Science, 235:442-447.

Folkman, J. and Ingber, D.E. (1987), Angiostatic steroids: Method of discovery and mechanism of action, Annals of Surgery, 206:374-384.

Folkman, J. (1987), Angiogenesis, X1th Congress of Thrombosis and Haemostasis, 583-596.

Folkman, J., Klagsbrun, M., Sasse, J., Wadzinski, M., Ingber, D., and Vlodavsky, I. (1988), Heparin-binding angiogenic protein - basic fibroblast growth factor - is stored within basement membrane, Am. J. Pathol., 130:393-400.

Folkman, J., and Ingber, D.E. (1989), Angiogenesis: regulatory role of heparin and related molecules. In: Lane DA., Lindahl U, eds., Heparin, London: Edward Arnold, 317-333.

Folkman, J., and Ingber, D.E. (1989), Angiostatic steroids, Anti-inflammatory Steroid Action: Basic and Clinical Aspects, 330-350.

Folkman, J., Szabo, S., and Shing, Y. (1990), Sucralfate affinity for fibroblast growth factor, Journal of Cell Biology, 111(5):part 2, 223a.

Folkman, J. (1990), What is the evidence that tumors are angiogenesis dependent?, Journal Natl. Canc. Inst., 82: 4-6.

Folkman, J. (1991), Tumor angiogenesis, Cancer Medicine, 3rd Edition, Holland JF, Frei E, Bast RC, Kufe DW, Morton DI, Weichselbaum RR, eds. In press.

Folkman, J., Szabo, S., Stovroff, M., McNeil, P., Li, W., and Shing, Y. (1991), Duodenal Ulcer: Discovery of a new mechanism and development of angiogenic therapy which accelerates healing, Annals of Surgery, 214:414-427.

Gadjdusek, C. (1986), Release of endothelial cell-derived growth factor (ECDGF) by heparin, J. Cell Physiology, 121:13-21.
Glowacki, J., and Mulliken, J. (1982), Mast cells in hemangiomas and vascular malformation, Pediatrics, 70:48-51.
Gospodarowicz, D., and Cheng, J. (1986), Heparin protects basic and acidic FGF inactivation, J. Cell Physiology, 128:475-484.
Hunt, T.K., Knighton, D.R., Thakral, KK., Goodson III, W.H., Andrews, W.S. (1984), Studies on inflammation and would healing: Angiogenesis and collagen synthesis stimulated in vivo by resident and activated wound macrophages, Surgery, 96: 48-54.
Inoue, K., Korenaga, H., Tanaka, N., Sakamoto, N., and Kadoya, S. (1988), The sulfated polysaccharide-peptidoglycan complex potently inhibits embryonic angiogenesis and tumor growth in the presence of cortisone acetate, Carbohydrate Research, 181:135-142.
Jacobsson, A.M., Hahnenberger, R., and Magnusson, A. (1989), A simple method for shell-less cultivation of chick embryos, Pharmacology and Toxicology, 64: 193-195.
Kessler, D., Langer, R., Pless, N., and Folkman, J. (1976), Mast cells and tumor angiogenesis, Int. J. Cancer, 18:703-709.
Klagsbrun, M., and Shing, Y. (1985), Heparin affinity of anionic and cationic capillary endothelial cell growth factors: Analysis of hypothalamus-derived growth factors and fibroblast growth factors, Proc. Natl. Acad. Sci. USA., 82:805-809.
Koos, R.D., and LeMaire, W.J. (1983), Factors that may regulate the growth and regression of blood vessels in the ovary.
Larsen, A.K., Lund, D.P., Langer, R., and Folkman, J. (1986), Oral heparin results in the appearance of heparin fragments in the plasma of rats, Proc. Natl. Acad. Sci. USA, 83:2964-2968.
Li, W.W., Casey, R., Gonzales, E.M., and Folkman, J. (1991), Angiostatic steroids potentiated by sulfated cyclodextrins inhibit corneal neovascularization, Investigate Ophthalmology & Visual Science, 32: 2898-2905.
Lobb, R.R., Sasse, J., Sullivan, R., Shing, Y., D'Amore, P., Jacobs, J. and Klagsbrun, M. (1986), Purification and characterization of heparin-binding endothelial cell growth factors, J. Biol. Chem, 261:1924-1928.
Maciag, T., Mehlman, R., Friesel, R. and Schreiber, A.B. (1984), Heparin binds endothelial cell growth factor, the principal cell mitogen in bovine brain., Science, 225:932-935.
Maione, T.E., Gray, G.S., Petro, J., Hunt, A.J., Donner, A.L., Bauer, S.I., Carson, H.F., and Sharpe, R.J. (1990), Inhibition of angiogenesis by recombinant human platelet factor-4 and related peptides, Science, 247:77-79.
Maione, T.E., Gray, G.S., Hunt, A.J., and Sharpe, R.J. (1991), Inhibition of tumor growth in mice by an analogue of platelet factor 4 that lacks affinity for heparin and retains potent angiostatic activity, Cancer Research, 51: 2077-2083.
Maragoudakis, M.E., Sarmonika, M., and Panoutsacopoulou, M. (1989), Antiangiogenic action of heparin plus cortisone is associated with decreased collagenous protein synthesis in the CAM system, J. Pharmacol. Exp. Ther., 251:679-682.
Moscatelli, D., Silverstein, J., Manejias, R., and Rifkin, D. (1987), M_r 25,000 heparin binding protein from guinea pig brain is a high molecular weight form of basic fibroblast growth factor, Proc. Natl. Acad. Sci. USA., 84: 5778-5782.
Moscatelli, D., and Rifkin, D.B. (1988), Membrane and matrix localization of proteinases: A common theme in tumor cell invasion and angiogenesis, Biochim. Biophys. Acta, 948: 67-85.
Nader, H., Dietrich, C., Buonassisi, V., and Colburn, P. (1987), Heparin sequences in the heparan sulfate chains of an endothelial cell proteoglycan, Proc. Natl. Acad. Sci. USA, 84:3565-3569.
Norrby, K., Jakobsson, A., Sorbo, J. (1990), On mast-cell-mediated angiogenesis in the rat mesenteric window assay, Agents and Actions, 30: 231-233.

Orchard, P.J., Smith III, C.M., Woods, W.G., Day, D.L., Dehner, L.P., and Shapiro, R. (1989), Treatment of haemangioendotheliomas with alpha interferon, Lancet, 2(8662)565-567.
Pereles, T., Ingber, D.E., and Folkman, J. (1989), Inhibition of capillary endothelial cell outgrowth: The role of complex formation between angiostatic steroid and beta-cyclodextrin tetradecasulfate, J. Cell Biol. 109(Part II):311a.
Rosengurt, T.K., Johnson, W.V., Friesel, R., Clark, R., and Maciag, T. (1988), Heparin protects heparin-binding growth factor-1 from proteolytic inactivation in vitro, Biochem. Biophys. Res. Commun. 152:432-440.
Rupnik, M., Casey, R., Gonzalez, E., Li, W., and Folkman, J. (1990), Corneal penetration of B-cyclodextrin tetradecasulfate, Invest. Ophthal. & Vis. Sci., 31: 559.
Sakamoto, N., Tanaka, N., Tohgo, A., and Ogawa, H. (1986), Heparin plus cortisone acetate inhibit tumor growth by blocking endothelial cell proliferation, Cancer Journal, 1:55-58.
Sakamoto, N., and Tanaka, N.G. (1987), Effect of angiostatic steroid with or without glucocorticoid activity on metastasis, Invasion and Metastasis, 7:208-216.
Satoh, H., Takami, K., Kato, K., Folkman, J., and Szabo, S. (1990), Effect of bFGF and its mutein on healing of colonic ulcers induced by N-ethylmaleimide in rats, Gastroenterol., 98(5):A203.
Saksela, O., Moscatelli, D., Sommer, A., and Rifkin, D.B. (1988), Endothelial cell-derived heparan sulfate binds basic fibroblast growth factor and protects it from proteolytic degradation, J. Cell Biol., 107:743-751.
Schreiber, A., Kenney, J., Kowalski, W., Friesel, R., Mehlman, T., and Maciag, T. (1985), Interaction of endothelial cell growth factor with heparin: Characterization by receptor and antibody recognition, Proc. Natl. Acad. Sci. USA., 82:6138-6142.
Shing, Y., Folkman, J., Sullivan, R., Butterfield, C., Murray, J., and Klagsbrun, M. (1984), Heparin-affinity: purification of a tumor-derived capillary endothelial cell growth factor, Science, 223:1926-1299.
Shing, Y., Folkman, J., Weisz, P.B., Joullie, MM., and Ewing, W.R. (1990), Affinity of fibroblast growth factors for beta-cyclodextrin tetradecasulfate, Anal. Biochem., 185:108-111.
Sudhalter, J., Folkman, J., Svahn, C.M., Bergendal, K., and D'Amore, P. (1989), Importance of size, sulfation, and anticoagulant activity in the potentiation of acidic fibroblast growth factor by heparin, J. Biol. Chem., 264:6892-6897.
Szabo, S. (1974), Animal Model: cysteamine-induced chronic duodenal ulcer in the rat, Amer. J. Pathol., 93:273-276.
Szabo, S., Vattay, P., Morales, R.E., Johnson, B., Kato, K., and Folkman, J. (1989), Orally administered bFGF mutein: Effect on healing of chronic duodenal ulcers in rats, Digestive Diseases and Science, 34: 1323.
Tanaka, N.G., Sakamoto, N., Inoue, K., Korenaga, H., Kadoya, S., Ogawa, H., and Osada, Y. (1989), Antitumor effects of an antiangiogenic polysaccharide from an arthrobacter species with or without a steroid, Cancer Research, 49:6727-6730.
Tanaka, N.G., Sakamoto, N, Korenaga, H., Inoue, K., Ogawa, H., and Osaka, Y. (1991), The combination of a bacterial polysaccharide and tamoxifen inhibits angiogenesis and tumour growth, Int. J. Radiat. Biol., 60: 79-83.
Taylor, S., and Folkman J. (1982), Protamine is an inhibitor of angiogenesis, Nature, 97:307-312.
Thomas, K.A., Rios-Candelore, M., and Fitzpatrick S. (1984), Purification and characterisation of acidic fibroblast growth factor, Proc. Natl. Acad. Sci. USA., 81:357-361.
Thompson, R.W., Whales, G.F., Saunders, K.B., Hores, T., and D'Amore, P. (1990), Heparin-mediated release of fibroblast growth factor-like activity into the circulation of rabbits, Growth Factors, 3:221-229.
Thornton, S., Mueller, S., and Levine, E. (1983), Human endothelial cells: Use of heparin in cloning and long-term serial cultivation, Science, 222:623-625.
Vannucchi, S., Pasquali, F., Chiarugi, V., and Ruggiero, M. (1986), Internalization and metabolism of endogenous heparin by cultured endothelial cells, Biochem. Biophys. Res. Commun., 140: 294-301.

Vlodavsky, I., Folkman, J., Sullivan, R., Fridman, R., Ishai-Michaeli, R., Sasse, J., and Klagsbrun, M. (1987), Endothelial cell-derived basic fibroblast growth factor: synthesis and deposition into sub-endothelial extracellular matrix, Proc. Natl. Acad. Sci. USA., 84:2292-2296.

Vlodavsky, I., Korner, G., Ishai-Michaeli, R., Baskin, P., Bar-Shavit, R., and Fuks, Z. (1990), Extracellular matrix-resident growth factors and enzymes: possible involvement in tumor metastasis and angiogenesis, Cancer Metastasis Reviews. 93: 203-226.

Weidner, N., Semple, J.P., Welch, W.R., and Folkman, J. (1991), Tumor angiogenesis and metastasis - correlation in invasive breast carcinoma, New Engl. J. Med., 324: 1-8.

White, C.W., Sondheimer, H.M., Crouch, E.C., Wilson, H., Fan, L.L (1989), Therapy of pulmonary hemangiomatosis with recombinant interferon alfa-2a, New Engl. J. Med., 320: 1197-1200.

Wilks, J.W., Scott, P.S., Vrba, L.K., and Cocuzza, J.M. (1991), Inhibition of angiogenesis with combination treatments of angiostatic steroids and suramin, Internat. Jour. Rad. Biol., 60:73-77.

Yayon, A., Klagsbrun, M., Esko, J., Leder, P., and Ornitz, D. (1991), Cell surface, heparin-like molecules are required for binding basic fibroblast growth factor to its high affinity receptor, Cell, 64:841-848.

INDEX